Ecological Sustenance and Management

Ecological Sustenance and Management

Editor: Jeffery Clarke

Callisto Reference

www.callistoreference.com

Callisto Reference,
118-35 Queens Blvd., Suite 400,
Forest Hills, NY 11375, USA

Visit us on the World Wide Web at:
www.callistoreference.com

ISBN: 978-1-63239-828-4 (Hardback)

Printed in China.

Cataloging-in-publication Data

Ecological sustenance and management / edited by Jeffery Clarke.
p. cm.
Includes bibliographical references and index.
ISBN 978-1-63239-828-4
1. Ecology. 2. Animal ecology. 3. Habitat (Ecology). 4. Sexual behavior in animals. I. Clarke, Jeffery.
QH541 .E26 2017
577--dc23

Table of Contents

Preface

Sustainable development is defined as a process of sustaining the environment by meeting the needs of the present generation without compromising the ability of the future to meet their own needs. In the current scenario, ecological sustenance and management is of utmost importance because of the rapid depletion of resources. This book will shed light on the various techniques for ecological sustenance and management. The text is compiled in such a manner, that it will provide an in-depth knowledge about the theory and practice of ecological-management through processes which involve strategies and techniques like adaptive management, natural resource management, and strategic management, etc. With state-of-the-art inputs by acclaimed experts of this field, this book targets students and professionals. As this field is emerging at a fast pace, this book will help the readers to better understand the concepts and applications of ecological sustenance.

This book has been an outcome of determined endeavour from a group of educationists in the field. The primary objective was to involve a broad spectrum of professionals from diverse cultural background involved in the field for developing new researches. The book not only targets students but also scholars pursuing higher research for further enhancement of the theoretical and practical applications of the subject.

It was an honour to edit such a profound book and also a challenging task to compile and examine all the relevant data for accuracy and originality. I wish to acknowledge the efforts of the contributors for submitting such brilliant and diverse chapters in the field and for endlessly working for the completion of the book. Last, but not the least; I thank my family for being a constant source of support in all my research endeavours.

Editor

1

Taking a Comparative Approach: Analysing Personality as a Multivariate Behavioural Response across Species

Alecia J. Carter[1¤], **William E. Feeney**[2,3*]

1 Fenner School of Environment and Society, Australian National University, Canberra, Australian Capital Territory, Australia, 2 School of Marine and Tropical Biology, James Cook University, Townsville, Queensland, Australia, 3 Research School of Biology, Australian National University, Canberra, ACT, 0200, Australia

Abstract

Animal personality, repeatable behaviour through time and across contexts, is ecologically and evolutionarily important as it can account for the exhibition of sub-optimal behaviours. Interspecific comparisons have been suggested as important for understanding the evolution of animal personality; however, these are seldom accomplished due, in part, to the lack of statistical tools for quantifying differences and similarities in behaviour between groups of individuals. We used nine species of closely-related coral reef fishes to investigate the usefulness of ecological community analyses for the analysis of between-species behavioural differences and behavioural heterogeneity. We first documented behavioural carryover across species by observing the fishes' behaviour and measuring their response to a threatening stimulus to quantify boldness. Bold fish spent more time away from the reef and fed more than shy fish. We then used ecological community analysis tools (canonical variate analysis, multi-response permutation procedure, and permutational analysis of multivariate dispersion) and identified four 'clusters' of behaviourally similar fishes, and found that the species differ in the behavioural variation expressed; some species are more behaviourally heterogeneous than others. We found that ecological community analysis tools are easily and fruitfully applied to comparative studies of personality and encourage their use by future studies.

Editor: Claudia Mettke-Hofmann, Liverpool John Moores University, United Kingdom

Funding: This research was funded by the Australian Research Council Centre of Excellence for Coral Reef Studies (Grant Number: CE0561432). The funders had no role in study design, data collection and analysis, decision to publish, or preparation of the manuscript.

Competing Interests: The authors have declared that no competing interests exist.

* E-mail: william.feeney@anu.edu.au

¤ Current address: Department of Zoology, University of Cambridge, United Kingdom

These authors contributed equally to this work.

Introduction

Selection should theoretically favour individuals that adapt their behaviour to their current ecological and social circumstances, however this is not always observed in nature [1]. Animal personality theory can explain these observations and suggests behaviour is more 'constrained' to certain behavioural types [1,2]. While the definition of animal personality can vary between studies [2,3], it is agreed that animal personality refers to low within-individual variation in behaviour relative to between-individual variation in behaviour; behaviour has to be repeatable through time to be considered a personality trait [2,4,5,6,7]. By being able to explain the existence and exhibition of both optimal and suboptimal behaviours, animal personality has been recognised as an ecologically and evolutionarily important theory [1,2,6,8].

A fundamental question within behavioural ecology concerns the evolutionary basis for the existence of animal personalities [2]. Recent conceptual work has suggested that intraspecific population and/or interspecific comparisons could be used to investigate these types of questions [9,10,11]. For example, intraspecific population comparisons have identified predation as a driving force behind the maintenance of correlated behavioural traits [12,13]. However, studies of interspecific differences in personality traits have produced mixed results. For example, Mettke-Hofmann et al. [11] successfully related differences in exploration and neophobia in 61 species of parrots (Psittacidae) to variation in diet and ecology. Webster et al. [10], however, found that three- and nine-spined sticklebacks (*Gasterosteus aculeatus* and *Pungitius pungitius*, respectively) differed significantly in the time spent in open water, but there was no difference in activity or latency to attack prey (boldness) between the species. Further, Mettke-Hofmann et al. [9] showed that resident and migratory warbler species (*Sylvia melanocephala momus* and *S. borin*, respectively) exhibited differences in the repeatability of and correlations between exploration and boldness. Thus, while the importance of comparative studies to animal personality research is recognised [2], few studies have tested these ideas, and few traits have been investigated.

Comparative approaches towards animal personality face two primary problems: 1) finding a behavioural assay that is comparable between species [2] and 2) identifying similarities and differences between groups of individuals, especially for multivariate responses such as behaviour. Finding a test that is comparable between species can be somewhat overcome by focusing on groups of closely related species. However, techniques to identify similarities and differences in behaviour are less established in behavioural ecology. We suggest that descriptive and statistical techniques used in multivariate ecological community analysis could be useful when studying interspecies differences in behaviour. For example, canonical variate analysis (CVA) is a data reduction technique similar to principal components analyses (PCA) that further uses a categorical grouping factor to separate groups [14]. Previously, it has been used to distinguish species based on morphological measurements [14,15,16]. Further, community composition analyses allow grouping of, in this case, species into similar and dissimilar 'clusters' by identifying compositional dissimilarity and heterogeneity. By identifying behaviourally similar and distinct species, and behaviourally homogeneous and heterogeneous species, we suggest that these statistical tools may provide a base to investigate the factors driving between-species differences in behaviour.

Here we apply ecological community analysis tools to analyse multivariate behaviour in nine species of predator-naïve recruit coral reef fishes (Pomacentridae). We chose this group of species (see Methods) because they are closely related to one another, often occur in sympatry but comprise different ecological niches, and are locally abundant. Further, life-history bottlenecks, such as that found following settlement in coral reef fishes, are ideal systems in which to investigate the evolutionary basis of animal personality as selection (primarily predation) is very high [17,18,19]. Accordingly, recent interest has been placed on the importance of personality for survival in these systems, and traits that have received attention (primarily boldness) have been shown to be repeatable through time [20,21,22,23,24]. In this study we first investigate the ecological relevance of boldness behaviour across species. Second, we use descriptive techniques to identify whether there are differences between species' behaviour, and finally we identify differences in behavioural dissimilarity and heterogeneity between species using community analysis tools 'borrowed' from the ecological literature.

Methods

Ethics statement

All research was conducted in accordance with the James Cook University Animal Ethics guidelines (Permit Number: A1520).

Study area and species

We studied nine species of pre-settlement juvenile damselfishes (Pomacentridae: *Chrysiptera rollandi* ($n = 12$), *Dischistodus perspicillatus* ($n = 10$), *Neopomacentrus azysron* ($n = 10$), *Pomacentrus amboinensis* ($n = 10$), *P. chrysurus* ($n = 12$), *P. coelestis* ($n = 15$), *P. moluccensis* ($n = 13$), *P. nagasakiensis* ($n = 10$), *P. wardi* ($n = 12$)) on the shallow reef flats surrounding Lizard Island (14°40′S, 145°27′E), Northern Great Barrier Reef, Australia in November and December 2009. Like most species of coral reef fishes, these species have a dispersive/pelagic larval stage followed by recruitment to the reefs for a more benthic juvenile/adult stage [25]. We caught all fishes during the recruitment process but prior to settlement to the reef using light traps [20,21,22,26,27]. This collecting technique minimizes phenotypic bias of the experimental population, as at recruitment the population is drastically augmented, most notably by extremely high rates of predation which can result in over 60% mortality of a cohort within 48 h of settlement [17,19,28]. Further, mixing of populations during the pelagic stage [29] and the high number of individual recruits helps ensure independence of replicates. Following capture, the fishes were transported to the research station at Lizard Island where they were sorted by species using a hand-net [20,21,22] and kept overnight in flow-through aquaria to recover prior to experimentation.

Experimental design

Nine patch reefs were created out of coral fragments and coral rubble on the shallow (<5 m) sand banks surrounding Lizard Island. Patch reefs were constructed approximately 10 m apart and approximately 10 m from the reef edge [20,21,22]. Water temperatures during this time varied between 26.2°C to 29.5°C.

Behavioural assays

Experimental fish were randomly selected from the previous day's catch [20,21,22]. Experimental fish were transported to the field and one fish was placed on each patch reef and allowed 30 min to acclimate before behavioural observations. Following acclimation, each individual fish was observed for a 3 min period by one observer (WEF) using self contained breathing apparatus (SCUBA) to obtain behavioural assays and a commonly used boldness measure [20,21,22]. Observed behaviours during assays comprised: total distance moved during the 3 min observation period, percentage of time spent 0, 2, 5 and 10 cm from the patch reef, proportion of time spent feeding (actively striking), lurking (not feeding) and observing (actively looking around environment while slowly rotating), furthest distance travelled from the patch reef, and proportion of time spent at the bottom third, middle third and top third (including above) of the patch reef. We measured the furthest distance travelled from the reef and proportion of time at different heights on the reef as the predatory assemblage differs at different parts of the reef [30,31], which may have implications for fish behaviour. In order to assess whether the recorded behaviours, which were used to assess species' differences in behaviour, were ecologically relevant for animal personality studies, we further recorded strike rate (a measure of feeding success) and individual boldness (using a standard boldness assay) [21].

We defined boldness as the response to risk [13]. Boldness was measured at the end of the 3 min observation periods using a startle response: the observer extended a pencil toward the fish until the fish responded (see below) to its approach [21]. Boldness was recorded on a four-point scale from 0–3 in increments of 1. A boldness measure of zero was defined as the fish hiding in the coral and seldom emerging following the approach of the pencil; 1 was retreating in response to the pencil and not emerging for more than 5 seconds; 2 was retreating to the coral upon approach of the pencil but quickly emerging (<5 sec) and resuming normal behaviours (e.g. feeding); and 3 was readily venturing away from the patch reef, dodging the pencil without returning to the patch and continuing normal behaviours.

Statistical methods

Describing a behavioural syndrome. To verify that the recorded behaviours were important in personality studies (i.e. that they correlated with an experimental measure of boldness), we used a 'traditional' approach (principal component analysis, PCA) to reduce the 12 recorded behaviours to fewer variables to investigate the effect of boldness on these traits. PCAs are commonly used in studies of animal personality [12,32] as a data reduction technique, however we note that this method may potentially violate the assumption of independent samples due to replication within species (if there were indeed between-species differences in behaviour). We use it here to demonstrate the relationship between behaviour, boldness and foraging success, assuming independence of observations. Maximum distance travelled and distance moved in 3 min were log transformed to satisfy assumptions of normality and variables recorded as a proportion were ranked in order to break the bounds of the proportions. All data were then mean-centred and standardised for analysis in the PCA. We used Horn's parallel analysis [33] to determine the number of components to be retained for further analysis.

We next used linear mixed effects (LME) models in order to investigate the relationship between PC1 (boldness-activity behaviour), PC2 (time budgets), PC3 (time spent observing at 0 cm from the reef) and PC4 (time spent in the middle of the reef) and individual boldness measured using a traditional experimental procedure (startle response) and feeding success (strike rate). Species was included as a random effect in all the models. Strike rate was log-transformed in order to satisfy assumptions of normality. We performed three LME models to investigate the relationship between a) boldness score as a response and PCs 1–4 as predictors, b) strike rate as a response and PCs 1–4 as predictors and c) boldness score as a response and strike rate as a predictor. We sequentially dropped non-significant ($t<2$) terms for a) and b) above until the minimal model remained. Residuals of the models were checked for normality. We further investigated these relationships within each species by running models a–c above using linear models (models a, c) or a generalized linear model with a log link (model b). However, we investigated only the effects of PCs 1 and 2, and did not included PCs 3 and 4 (as they had no effect on boldness or strike rate, see Results below), to avoid overparameterisation.

Applying ecological community analysis tools. In order to assess the application of ecological community tools to comparative personality studies, we first investigated the recorded behaviours using multivariate descriptive techniques. We used canonical variates analysis (CVA) to group species according to behaviour [15]. CVA first analyses a matrix of variable values in much the same way as PCA, however CVA maximises the variance explained between groups while PCA maximises the variation among individuals (for a comparison of the results of these methods, see Supplementary Fig. S1). It then calculates the group's means, in this case species' centroids (the geometrical mean), and plots the first two components as a bivariate plot. Like PCAs, CVAs are descriptive techniques that can be used for data reduction, and are interpreted in a similar fashion. However, unlike PCAs, variables are more or less represented by each CV on a scale of 0 (poorly represented) to 1 (well represented) rather than 'loading' on components. Well-represented variables suggest differences between groups in those variables. The 'directions' of each variable (whether ascending or decending, represented as axes on the plot) can be determined by the biplot: the labels are associated with the positive end of the variable. Axes that are closer together are more correlated.

Next, we calculated behavioural dissimilarity between species as a measure of behavioural distinctiveness (how different are species behaviourally from one another?) and behavioural heterogeneity (how variable are species behaviourally from one another?). We calculated behavioural distinctiveness using multi-response permutation procedure (MRPP) [34]. MRPP is a non-parametric multivariate statistical technique used to calculate whether there is a difference between *a prioi* defined groups of entities, in our case, species. It is a quantitative statistic, robust to unequal variance, non-normally distributed data and unbalanced designs through use of permutation between groups. Groups that are clustered in multidimensional space have lower average distances to their group centroid than their inter-group centroids; they are dissimilar to the other species. The average distance, weighted by group size, is referred to as a lower case delta, δ. Thus, if species' points are distinct from other species, the intraspecies average distances will be small compared to interspecies average distances. To test this, the data are permuted in MRPP; in each permutation, group 'membership' is randomised and the resulting δ values (δ_{exp}) are compared to the observed δ (δ_{obs}) to calculate a *p*-value. Finally, within-group agreement, *A*, is calculated as 1-(δ_{obs})/(δ_{exp}) as a measure of how well the individuals 'fit' within their groups, in this case, species. That is, within-group agreement would be 1 if all individuals within each species behaved identically to one another, and 0 if all individuals behaved according to chance. Community ecology values of *A* are commonly $A\approx0.1$, while values >0.3 are high [35]. We used Bray-Curtis distances on the untransformed (count) data in the package "vegan" with 1000 permutations [36]. Finally, we tested for differences in behavioural heterogeneity using permutational analysis of multivariate dispersion in the program PERMDISP2 [37,38] with 999 permutations. Similar to MRPP, PERMDISP2 calculates within-group average distances to the group centroid and compares these values to other group's average distances. Thus, species with higher average distances to the group centroid relative to other species have higher behavioural heterogeneity than species with small average distances.

Results

Describing a behavioural syndrome

Fishes' average scores for each of the behaviours are listed in Table 1. We used Horn's parallel analysis to determine principal components (PCs) 1–4 be retained for further analysis. PCs 1–4 explained 39.1, 16.7, 12.4 and 10.2% of the variation, respectively (78.3% of the total variation). Proportion of time spent in the bottom third of the reef loaded positively on PC1, while proportion of time spent within 5 and 10 cm of the reef, in the top third of the reef and total and maximum distances moved loaded negatively on PC1 (Table 2). This suggests that lower values of PC1 indicate individuals that spent more time in 'riskier' areas (far from and high on the patch reef) and more time moving, and higher values indicate individuals that spent time in 'safer' areas (close to the bottom of the patch reef). PC1 is thus suggestive of a boldness-activity personality trait. PC2 explained conflicts in activity budgets, with low values indicating individuals that spent more time feeding and high values indicating individuals that spent more time lurking. Time spent at 0 cm from the reef and time spent observing loaded positively on PC3 while time spent at 2 cm from the reef and time spent lurking loaded negatively. This partly reflects the definitions of the activity budget behaviours and is again representative of a conflict in these budgets. Finally, PC4 explained time spent in the middle third of the patch reef, with high values indicative of individuals spending more time in the middle.

We investigated the ecological relevance of an experimental boldness assay on naturally occurring behaviour. The minimal models investigating the effects of PCs 1–4 on the experimental measure of boldness (the startle response) and strike rate included PCs 1 and 2 as explanatory variables in both models (Table 3). Across species, fish that spent more time active, feeding and in riskier areas of the patch reef had higher boldness scores (startle responses) and higher strike rates. Secondly, bolder individuals gained a potential fitness benefit through increased feeding rates (Table 3), although this is unsurprising given the above correlations. Finally, we investigated these relationships at the species level (see Supplementary Table S1), but found variable correlations between the recorded behaviours between species.

Applying ecological community analysis tools

We used CVA to group species according to the behaviour of the individuals. CVA revealed three distinct groups of species that

Table 1. Summary statistics (mean ± s.e.) for the behaviours collected during focal observations.

				Proportion of time spent (from reef)				Proportion of time spent at (of reef)			Proportion of time spent				
Species	*N*	Maximum distance travelled from reef	Total distance moved in 3 min	0 cm	2 cm	5 cm	10 cm	Bottom third	Middle third	Top third	Feeding	Lurking	Observing	Strike rate	Boldness
C. rollandi	12	5.33±1.97	12.91±7.82	1.67±3.26	51.67±36.39	42.50±34.08	4.17±11.65	60.42±45.15	26.25±29.70	13.33±21.98	84.58±11.95	12.50±10.55	0.83±2.89	107.00±23.36	2.58±0.51
D. perspicillatus	10	5.50±2.27	12.40±8.80	14.50±28.13	56.50±36.75	28.00±34.17	1.00±3.16	72.00±40.77	27.00±39.17	1.00±3.16	78.50±21.35	21.50±21.35	1.00±3.16	53.60±12.10	1.90±0.88
N. azysron	10	32.70±13.44	73.00±37.51	0.00±0.00	3.00±9.49	20.50±23.15	76.50±29.82	0.00±0.00	11.50±20.00	88.50±20.00	72.00±13.98	28.00±13.98	0.00±0.00	67.40±9.48	2.60±0.70
P. amboinensis	10	6.80±2.35	22.70±12.00	13.00±3.16	61.00±18.38	24.50±24.56	1.50±3.37	25.50±29.86	63.00±26.58	11.50±11.32	66.00±17.13	30.00±11.55	4.00±12.65	69.40±20.44	2.00±0.15
P. chrysurus	12	5.50±4.44	11.92±5.80	58.33±36.57	33.33±29.95	7.50±15.45	0.83±1.95	94.58±8.91	5.41±8.91	0.00±0.00	50.83±17.30	11.67±8.35	37.50±19.13	43.42±14.57	0.33±0.65
P. coelestis	15	15.80±18.34	52.33±46.50	8.67±15.52	49.33±30.52	28.67±20.04	13.33±23.50	32.67±42.34	43.33±36.04	24.00±32.41	84.33±18.60	14.33±15.68	1.33±5.16	125.07±38.26	2.60±0.63
P. moluccensis	13	8.00±4.98	23.85±17.76	15.38±3.16	54.23±28.27	27.31±26.66	3.08±5.22	6.54±22.12	49.23±37.52	44.23±38.51	70.00±25.17	20.77±19.35	9.23±18.01	53.85±22.21	2.38±0.65
P. nagasakiensis	10	6.70±2.50	27.50±17.99	8.00±8.23	73.50±10.01	17.00±11.35	1.50±3.38	10.00±13.94	58.50±30.37	58.50±30.37	73.00±18.89	16.00±9.37	5.00±7.45	84.50±10.55	2.50±0.71
P. wardi	12	9.33±12.23	36.50±44.80	9.17±2.24	73.33±34.07	7.50±11.18	10.00±23.35	85.42±26.92	10.42±19.82	4.17±7.93	60.00±28.60	38.33±26.57	1.67±5.77	64.41±31.42	1.67±0.89

differed in their behaviours on canonical variates 1 and 2 (Fig. 1) and thus warranted further investigation. Behaviours that were well represented by CV1 included the maximum distance travelled, the proportion of time spent 10 cm from the reef, in the top third of the reef and, in the opposite direction, in the bottom third of the reef (Table 3, Fig. 1). Time spent observing was well represented by CV2 (Table 3, Fig. 1). That is, species of fish differed most in the 'risky' parts of the reef they used (CV1) and the behavioural state they were in (CV2).

We next found that species behaviour differed significantly between groups (MRPP: $A = 0.26$, observed $\delta = 0.35$, expected $\delta = 0.48$, $p<0.01$). MRPP found four group clusterings, with *N. azysron's* behaviours showing the highest dissimilarity to all other species, and *P. chrysurus* also showing high dissimilarity from the remaining species (Fig. 2). The remaining species could be further broken into two behaviourally dissimilar clusters (cluster 1: *D. perspicillatus*, *C. rollandi* and *P. wardi*, cluster 2: *P. nagasakiensis*, *P. moluccensis*, *P. coelestis* and *P. amboinensis*), however, this dissimilarity was less marked than the dissimilarity of *N. azysron* and *P. chrysurus*. Finally, we investigated behavioural heterogeneity within species using PERMDISP2. We found significant differences in behavioural heterogeneity between species ($F_{9,\ 104} = 5.9$, $p<0.01$). *P. nagasakiensis* had the lowest behavioural heterogeneity (all individuals within the species behaved relatively similarly) and *P. moluccensis* had the highest (all individuals within the species behaved relatively dissimilarly; Table 4).

Discussion

The importance of interspecific comparative analyses for understanding the evolutionary basis of animal personality is recognised [9,10,11], however, few systems can accommodate these questions and adequate statistical tools for their investigation are lacking. At settlement, a plethora of coral reef fish species simultaneously recruit to the reef, providing an ideal system for interspecific comparative studies. Here, we investigated the usefulness of ecological community analysis tools (CVA, MRPP and PERMDISP2) for investigating interspecific variation in animal personality. First, we confirmed a behavioural syndrome across species, suggesting that animal personality is ecologically relevant in these species. We found that bolder individuals spent more time in risky areas of the patch reef, move more and had higher strike rates than shy individuals. We then used canonical variate analysis (CVA) to visually investigate species' differences in behaviour, and finally, we calculated behavioural dissimilarity between species as a measure of behavioural distinctiveness (how different are species behaviourally from one another?) and behavioural heterogeneity (how variable are species behaviourally from one another?) using MRPP and PERMDISP2, respectively. Using these techniques, we found significant variation between species in both the average behaviour species express and the variation in the behaviour they express.

These statistical techniques, although developed and widely used for the analysis of ecological communities, are quantitative, informative and easy to apply to analyses of animal behaviour. Previous studies that have compared between-species differences in behaviours have been 'restricted' to comparing one behaviour at a time, which are usually experimentally measured personality traits (such as exploration) [9,11] or a 'simplified' behavioural score (such as time spent in open vs. vegetated water) [10]. Ecological community analysis tools can compare multiple behaviours and multiple species simultaneously and quantitatively, and we therefore encourage their use by future studies.

Table 2. Behaviours measured during focal watches that were entered into a principal component analysis (PCA) and canonical variate analysis (CVA) and their loadings on PCs 1–4 and representation by CVs 1 and 2, respectively.

Behaviour	PC1	PC2	PC3	PC4	CV1	CV2
Maximum distance travelled	**−0.37**	0.14	0.00	0.00	**0.73**	0.09
Distance moved in 3 min	**−0.40**	0.15	0.00	−0.18	0.49	0.18
Proportion of time spent within 0 cm of the reef	0.25	0.21	**0.42**	0.24	0.28	0.53
Proportion of time spent within 2 cm of the reef	0.23	−0.18	**−0.53**	0.00	0.59	0.08
Proportion of time spent within 5 cm of the reef	−0.30	−0.27	0.00	**0.31**	0.03	0.14
Proportion of time spent within 10 cm of the reef	**−0.37**	0.21	0.15	−0.28	**0.87**	0.03
Proportion of time spent in the bottom third of the reef	**0.34**	−0.13	0.15	−0.25	**0.68**	0.01
Proportion of time spent in the middle third of the reef	−0.18	0.00	0.00	**0.78**	0.01	0.01
Proportion of time spent in the top third of the reef	**−0.36**	0.20	−0.11	0.00	**0.71**	0.03
Time spent feeding	−0.17	**−0.60**	0.12	−0.15	0.01	0.30
Time spent lurking	0.00	**0.51**	**−0.48**	0.00	0.01	0.32
Time spent observing	0.22	0.30	**0.47**	0.18	0.05	**0.89**
Variation explained (PCA)	39%	17%	12%	10%		
Adjusted eigenvalue (PCA)	4.10	1.58	1.19	1.03		

Bold type indicates behaviours that load on each principal component and behaviours that are well represented by the canonical variates. Note that the canonical variates give no indication of the direction of the relationships. The percentage of variation explained and the adjusted eigenvalues for each principal component are listed.

We found that the behaviours of *N. azysron* and *P. chrysurus* were significantly different to the behaviours of the seven other assayed species. Both expressed relatively low behavioural heterogeneity. *P. chrysurus* expressed, on average, the shyest responses to the boldness assay and spent relatively more time observing than the other species, while *N. azysron* spent relatively more time in the top third of the patch reef and at 10 cm from the patch reef than the other species (Fig. 1). These behaviours appear primarily responsible for the differentiation of the 'out group' species from the other clusters of species, which behaved more similarly to each other (Fig. 2). Although it was not our goal in this study to identify the reasons these fishes may behave differently to other fish species, determining the factors driving these behavioural tendencies may have implications for understanding the evolution of animal personality.

Table 3. The estimates of the variables in the minimal models (linear mixed effects models) investigating the relationships between the startle response and natural behaviour, strike rate and natural behaviour, and the startle response and strike rate.

Response	Predictor	β ± s.e.	N	df	*t*
Startle response	Intercept	2.07±0.16	109	92	12.51
	PC1 (Boldness-Activity)	−0.18±0.04			−5.00
	PC2 (Feeding/Lurking)	−0.11±0.05			−2.18
Strike rate	Intercept	66.68±4.06	109	92	3.04
	PC1 (Boldness-Activity)	−1.08±1.01			−5.89
	PC2 (Feeding/Lurking)	−1.13±1.02			−6.78
Startle response	Intercept	1.48±0.29	109	92	5.07
	Feeding rate	0.01±0.003			2.90

The observed differences in behaviour between *N. azysron* and *P. chrysurus* and the other species may be a result of their life history strategy [11], their external environment [39,40] or a combination of both. For example, these two species appear the most distinct of the nine investigated species with respect to their aggregative tendencies at settlement. *N. azysron* tends to settle and live in schools [41], while newly settled *P. chrysurus* tend to avoid areas with adults (with whom they often have agonistic interactions) and be more solitary [42]; all other species measured tend to inhabit a wider variety of social environments [41,42,43]. Further, these two species may vary in their habitat choices. *N. azysron* tends to inhabit areas above the coral, feeding more in the water column [41], while *P. chrysurus* tends to live more amongst coral rubble [42,44]; the other species vary predominantly between live corals, coral rubble and sand [41,42,43]. However, when considered together in the context of predation, which at this life history stage is extremely high and a strong selective force [17,18,19,28,45], it may be beneficial for the more solitary *P. chrysurus* to be shy and cautious of active predators which are abundant in these habitats (e.g. ambush predators such as the dottyback, *Pseudochromis fuscus*) [30] compared to the much bolder, schooling *N. azysron*, which sits on top of the reef and may benefit from a dilution effect against fast striking, long distance predators (e.g. the lizardfish, *Synodus englemani*) [31]. However, while this is an alluring and simple explanation, more targeted studies are required to disentangle the drivers of behavioural differences between these species.

We also found differences in average behavioural heterogeneity between the 9 species (Table 4). Previous work in other systems has found that selective forces, such as predation, can affect behavioural heterogeneity. For example, Edgell [46] showed that snails, *Litorina obtusata*, that had been exposed to predation by crabs, *Carcinus maenus*, for longer (100 *versus* 60 years) showed higher canalization in antipredator response, that is, lower behavioural flexibility, than snails exposed for shorter time periods [47,48]. As predation is a strong selective force in coral reef systems at settlement [17,19], it is tempting to suggest that this, perhaps coupled with variation of microhabitat choice (and

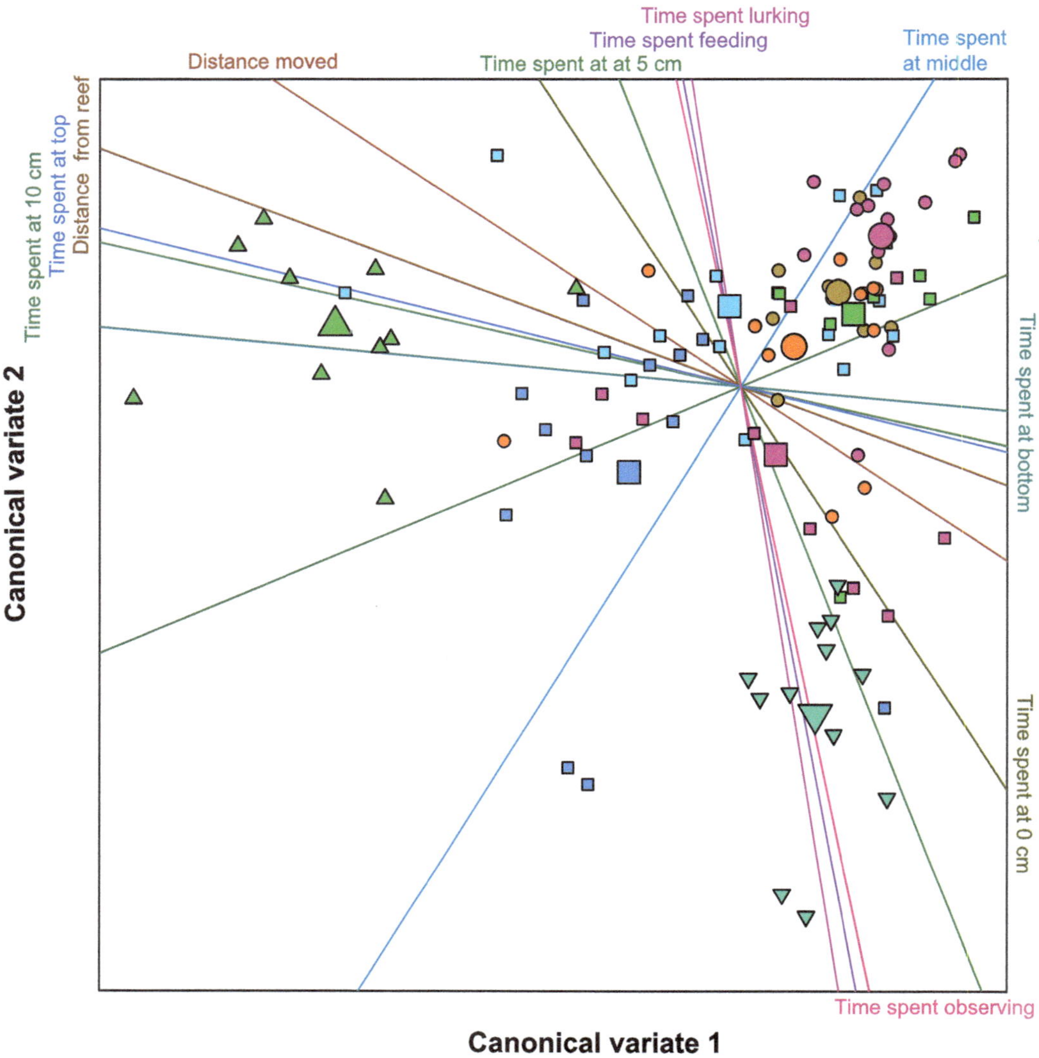

Figure 1. Position of individual fish and group centroids of fishes in multivariate behavioural space. Individual fish plotted on canonical variates 1 and 2. Coloured lines on the plot represent behavioural variables with labels attached to the positive end of each variable. Each smaller point represents an individual fish, with larger points indicating the centroids for each species. Each symbol represents fish that were classified *a posteriori* to belong to a group of similarly behaving fish (see Fig. 2); each colour represents a different fish species: *Chrysiptera rollandi* (orange circle), *Dischistodus perspicillatus* (brown circle), *Neopomacentrus azysron* (green upward pointing triangle), *Pomacentrus amboinensis* (light green square), *P. chrysurus* (aqua downward pointing triangle), *P. coelestis* (light blue square), *P. moluccensis* (blue square), *P. nagasakiensis* (purple square), *P. wardi* (pink circle).

therefore variation in commonly encountered predators) [30,31] by these species [41,42,43,44] may have resulted in the observed differences in behavioural heterogeneity between species. However, causal explanations of behavioural heterogeneity must be treated with caution, especially in samples taken from the wild. While canalising selection can decrease behavioural heterogeneity [46], other explanations, unrelated to selection *per se*, may equally generate heterogeneous/homogeneous behaviour in a species. Differences in the age, size or other latent variables of the sampled fish in our study may have artificially inflated the behavioural variation in some species. Further, differences in plastic responses to local environmental variation [49] may increase the heterogeneity of observed behaviours. Despite this, behavioural heterogeneity may be important in understanding species' differences in behaviour and personality and could be a useful tool in future studies.

Although we found evidence for a between-species behavioural syndrome, we found variable correlations between behavioural variables within species (see Supplementary Table S1). Interestingly, the two species with the lowest behavioural heterogeneity (Table 4) showed no evidence for a behavioural syndrome while three of the four species with highest behavioural heterogeneity (Table 4) showed the greatest evidence of a behavioural syndrome. Although the sample sizes within each species are relatively low, this finding raises two important points. First, if behavioural correlations or personality occur because of underlying neurological or hormonal 'constraints' or adaptive combinations of behaviours [50,51,52,53,54], then investigating where a 'lack of personality' occurs may be instructive in understanding the evolutionary requirements for these mechanisms [55]. Second, our results highlight the need to consider 'situational strength' during behavioural assays. Situational strength refers to how much an individual's behaviour is influenced by the test situation [56].

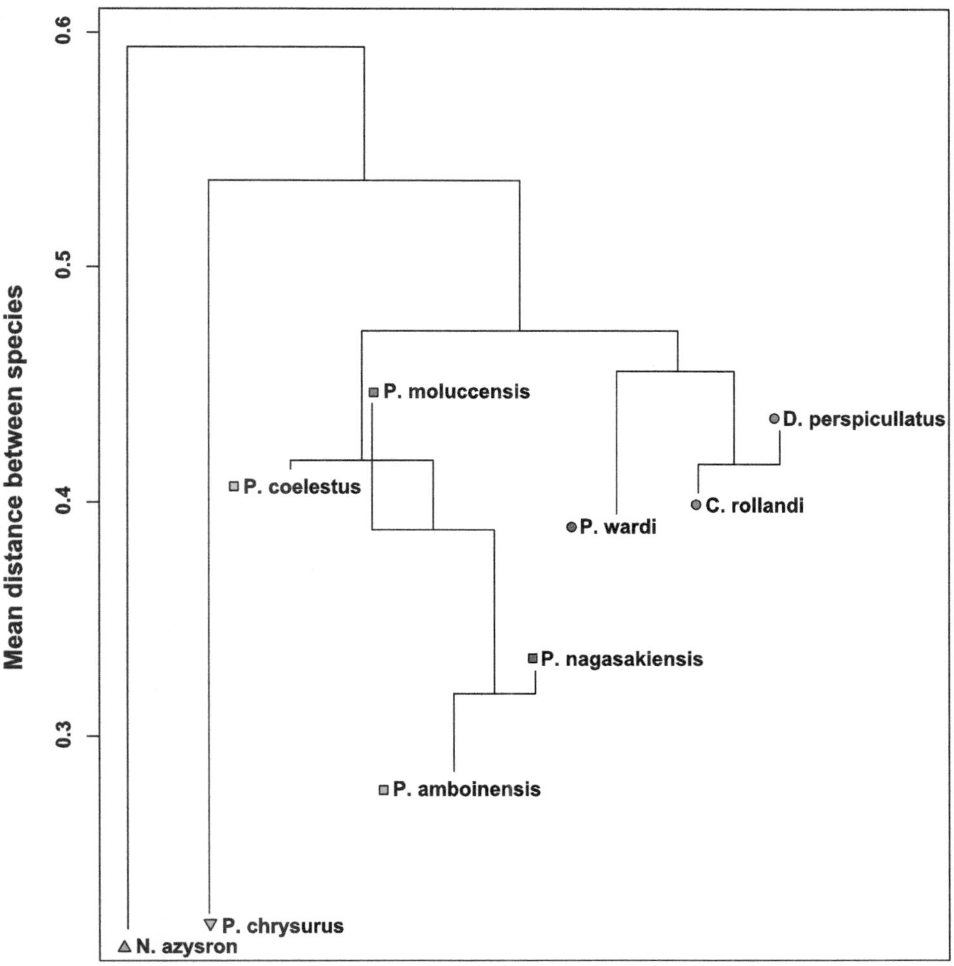

Figure 2. A dendrogram of species of fishes that behave similarly. Mean distances between species in multidimensional space plotted as a dendrogram to group similarly behaving species (small between-group distances). Termini indicate within-cluster dissimilarity for the species; horizontal lines indicate mean between-cluster dissimilarity. Termini that finish above the mean between-cluster line indicate species that are more heterogeneous than the combined cluster. Distance here refers to the positions of groups (group averages) within multidimensional behavioural space.

Table 4. Species' average dispersion (behavioural heterogeneity).

Species	Average distance from centroid	s.e.
C. rollandi	29.76	2.35
D. perspiculatus	29.64	1.96
N. azysron	22.36	2.02
P. amboinensis	17.94	3.90
P. chrysurus	15.50	1.93
P. coelestus	27.61	3.18
P. moluccensis	30.64	2.40
P. nagasakiensis	14.74	1.81
P. wardi	28.15	3.56

Dispersion is indicated by the average distance of each individual to each individual's species' centroid. Large values indicate high behavioural heterogeneity; low values indicate low behavioural heterogeneity.

Strong situations may leave little variation between individual's behaviour whereas weaker situations may allow more inter-individual differences to show [57,58]. In our study it is possible that species 'experienced' different situational strengths during the behavioural assays, which may result in differences in behavioural heterogeneity and thus correlations between behaviours. This is an important consideration for future comparative studies of animal personality.

Finally, comparisons of average behavioural characters, such as personality, across a variety of species are fundamental for investigations of how anthropogenic changes will affect marine communities. Coral reef systems are particularly vulnerable to anthropogenic changes such as climate change, and there is interest on how these processes affect fish behaviour and species' interactions [44,59,60,61,62]. These studies generally investigate the effects of different climate regimes on single species [59,61,62], however multiple-species comparisons are essential for investigating how these processes will affect communities [44] and drive the evolution of phenotypes, such as personality. Techniques such as those highlighted in this study may provide a useful 'first-step' for investigating how these processes affect communities, rather than the effects on single species, and affect the evolution of behavioural phenotypes.

In summary, intraspecific population and interspecific comparative analyses have the potential to help explain the nature and evolutionary significance of animal personalities. We suggest that using techniques such as ecological community analysis can help identify where similarities and differences in behaviour lie between species, after which more targeted studies can be performed to help gain and understanding of the drivers of animal personality differentiation between species.

Supporting Information

Figure S1 Biplots of A) principal components and B) canonical variates analysis of the behavioural data. Individual fish plotted on A) principal components 1 and 2 and B canonical variates 1 and 2. CVA first analyses a matrix of variable values in much the same way as principal components analysis (PCA), however CVA maximises the variance explained between groups while PCA maximises the variation among individuals. The result is that species are more dispersed in the PCA (A), while species are more closely grouped with one another in the CVA (B). Further details can be found in: McCune, B., J.B. Grace, and D.L. Urban. 2002. ***Analysis of Ecological Communities.*** MjM Software Design. Each coloured line represents a behavioural variable; the labels are attached to the positive termini of each variable. Each symbol represents fish that were classified *a posteriori* to belong to a group of similarly behaving fish (see Fig. 2, main text); each colour represents a different fish species: *Chrysiptera rollandi* (orange circle), *Dischistodus perspicillatus* (brown circle), *Neopomacentrus azysron* (green upward pointing triangle), *Pomacentrus amboinensis* (light green square), *P. chrysurus* (aqua downward pointing triangle), *P. coelestis* (light blue square), *P. moluccensis* (blue square), *P. nagasakiensis* (purple square), *P. wardi* (pink circle). Large symbols in B indicate species centroids.

Acknowledgments

We are grateful to Mark McCormick and Susannah Leahy for assistance and advice in the field; Mark McCormick, Mark Meekan, Oona Lonnstedt, Susannah Leahy, Brynn Devine, Talisa Kath, Crystal Neligh Harte and Danielle Dixson for assisting in the collection of fish used in this study; and the staff at the Lizard Island Research Station for logistic support. We also thank Karen Iken and Philip Barton for explaining community research with such patience and Naomi Langmore and Karen Iken for reading early drafts of the manuscript.

Author Contributions

Conceived and designed the experiments: AJC WEF. Performed the experiments: WEF. Analyzed the data: AJC. Contributed reagents/materials/analysis tools: AJC WEF. Wrote the paper: AJC WEF.

References

1. Wilson DS, Clark AB, Coleman K, Dearstyne T (1994) Shyness and boldness in humans and other animals. Trends in Ecology & Evolution 9: 442–446.
2. Réale D, Reader SM, Sol D, McDougall PT, Dingemanse NJ (2007) Integrating animal temperament within ecology and evolution. Biological Reviews 82: 291–318.
3. Sih A, Bell A, Johnson JC (2004) Behavioral syndromes: an ecological and evolutionary overview. Trends in Ecology & Evolution 19: 372–378.
4. Dingemanse NJ, Kazem AJN, Réale D, Wright J (2010) Behavioural reaction norms: animal personality meets individual plasticity. Trends in Ecology & Evolution 25: 81–89.
5. Stamps JA (2007) Growth-mortality tradeoffs and 'personality traits' in animals. Ecology Letters 10: 355–363.
6. Biro PA, Stamps JA (2008) Are animal personality traits linked to life-history productivity? Trends in Ecology & Evolution 23: 361–368.
7. Carter AJ, Heinsohn R, Goldizen AW, Biro PA (2012) Boldness, trappability and sampling bias in wild lizards. Animal Behaviour 83: 1051–1058.
8. Réale D, Dingemanse NJ, Kazem AJN, Wright J (2010) Evolutionary and ecological approaches to the study of personality. Philosophical Transactions of the Royal Society B: Biological Sciences 365: 3937–3946.
9. Mettke-Hofmann C, Ebert C, Schmidt T, Steiger S, Stieb S (2005) Personality traits in resident and migratory warbler species. Behaviour 142: 1357–1375.
10. Webster MM, Ward AJW, Hart PJB (2009) Individual boldness affects interspecific interactions in sticklebacks. Behavioral Ecology and Sociobiology 63: 511–520.
11. Mettke-Hofmann C, Winkler H, Leisler B (2002) The significance of ecological factors for exploration and neophobia in parrots. Ethology 108: 249–272.
12. Dingemanse NJ, Wright J, Kazem AJN, Thomas DK, Hickling R, et al. (2007) Behavioural syndromes differ predictably between 12 populations of three-spined stickleback. Journal of Animal Ecology 76: 1128–1138.
13. Bell AM, Stamps JA (2004) Development of behavioural differences between individuals and populations of sticklebacks, *Gasterosteus aculeatus*. Animal Behaviour 68: 1339–1348.
14. Campbell NA, Atchley WR (1981) The geometry of canonical variate analysis. Systematic Zoology 30: 268–280.
15. la Grange A, le Roux N, Gardner-Lubbe S (2009) BiplotGUI: interactive biplots in R. Journal of Statistical Software 30: 1–30.
16. Holt C, Hirst D, Sutherland A, Macdonald F (1995) Discrimination of species in the genus Listeria by fourier-transform infrared-spectroscopy and canonical variate analysis. Applied and Environmental Microbiology 61: 377–378.
17. Almany G, Webster M (2006) The predation gauntlet: early post-settlement mortality in reef fishes. Coral Reefs 25: 19–22.
18. Holmes T, McCormick MI (2010) How selective are reef fish predators for prey size? Marine Ecology Progress Series 399.
19. Doherty PJ, Dufour V, Galzin R, Hixon MA, Meekan MG, et al. (2004) High mortality during settlement is a population bottleneck for a tropical surgeonfish. Ecology 85: 2422–2428.
20. Fuiman L, Meekan M, McCormick M (2010) Maladaptive behavior reinforces a recruitment bottleneck in newly settled fishes. Oecologia 164: 99–108.
21. McCormick MI, Meekan MG (2010) The importance of attitude: the influence of behaviour on survival at an ontogenetic boundary. Marine Ecology Progress Series 407: 173–185.
22. Meekan MG, von Kuerthy C, McCormick MI, Radford B (2010) Behavioural mediation of the costs and benefits of fast growth in a marine fish. Animal Behaviour 79: 803–809.
23. Biro PA, Beckmann C, Stamps JA (2010) Small within-day increases in temperature affects boldness and alters personality in coral reef fish. Proceedings of the Royal Society of London, Series B: Biological Sciences 277: 71–77.
24. Eriksson A, Booth D, Biro P (2010) 'Personality' in two species of temperate damselfish. Marine Ecology Progress Series 420: 273–276.
25. McCormick MI, Makey LJ (1997) Post-settlement transition in coral reef fishes: overlooked complexity in niche shifts. Marine Ecology Progress Series 153: 247–257.
26. Meekan MM, Wilson SW, Halford AH, Retzel AR (2001) A comparison of catches of fishes and invertebrates by two light trap designs, in tropical NW Australia. Marine Biology 139: 373–381.
27. Doherty PJ (1987) Light-traps: selective but useful devices for quantifying the distributions and abundances of larval fishes. Bulletin of Marine Science 41: 423–431.
28. McCormick MI, Hoey AS (2004) Larval growth history determines juvenile growth and survival in a tropical marine fish. Oikos 106: 225–242.
29. Jones G, Almany G, Russ G, Sale P, Steneck R, et al. (2009) Larval retention and connectivity among populations of corals and reef fishes: history, advances and challenges. Coral Reefs 28: 307–325.
30. Feeney WE, Lönnstedt O, Bosiger Y, Martin J, Jones G, et al. (2012) High rate of prey consumption in a small predatory fish on coral reefs. Coral Reefs: Online first 11th March 2012.
31. Sweatman HPA (1984) A field study of the predatory behavior and feeding rate of a piscivorous coral reef fish, the lizardfish *Synodus englemani*. Copeia 1984: 187–194.
32. Carter A, Marshall H, Heinsohn R, Cowlishaw G (2012) Evaluating animal personalities: do observer assessments and experimental tests measure the same thing? Behavioral Ecology and Sociobiology 66: 153–160.
33. Horn JL (1965) A rationale and test for the number of factors in factor analysis. Psychometrika 30: 179–185.

34. Zimmerman GM, Goetz H, Mielke PW (1985) Use of an improved statistical-method for group comparisons to study effects of prairie fire. Ecology 66: 606–611.
35. McCune B, Grace J, Urban D (2002) Analysis of ecological communities. MjM Software Design.
36. Oksanen J, Blanchet FG, Kindt R, Legendre P, Minchin PR, et al. (2011) vegan: Community Ecology Package.
37. Anderson MJ (2004) PERMDISP: a FORTRAN computer program for permutational analysis of multivariate dispersions (for any two-factor ANOVA design) using permutation tests. University of Aukland, New Zealand: Department of Statistics.
38. Anderson MJ, Ellingsen KE, McArdle BH (2006) Multivariate dispersion as a measure of beta diversity. Ecology Letters 9: 683–693.
39. Tolon V, Dray S, Loison A, Zeileis A, Fischer C, et al. (2009) Responding to spatial and temporal variations in predation risk: space use of a game species in a changing landscape of fear. Canadian Journal of Zoology 87: 1129–1137.
40. Vail AL, McCormick MI (2011) Metamorphosing reef fishes avoid predator scent when choosing a home. Biology Letters 7: 921–924.
41. Leis JM, Carson-Ewart BM, Webley J (2002) Settlement behaviour of coral-reef fish larvae at subsurface artificial-reef moorings. Marine and Freshwater Research 53: 319–327.
42. Öhman MC, Munday PL, Jones GP, Caley MJ (1998) Settlement strategies and distribution patterns of coral-reef fishes. Journal of Experimental Marine Biology and Ecology 225: 219–238.
43. Leis JM, Carson-Ewart BM (2002) In situ settlement behaviour of damselfish (Pomacentridae) larvae. Journal of Fish Biology 61: 325–346.
44. Ferrari MCO, Dixson DL, Munday PL, McCormick MI, Meekan MG, et al. (2011) Intrageneric variation in antipredator responses of coral reef fishes affected by ocean acidification: implications for climate change projections on marine communities. Global Change Biology 17: 2980–2986.
45. Leis JM, McCormick MI (2002) The biology, behavior, and ecology of the pelagic, larval stage of coral reef fishes. In: Sale PF, editor. Coral reef fishes - dynamics and diversity in a complex ecosystem. London: Academic Press. pp. 171–199.
46. Edgell Timothy C, Lynch Brian R, Trussell Geoffrey C, Palmer A R (2009) Experimental Evidence for the Rapid Evolution of Behavioral Canalization in Natural Populations. The American Naturalist 174: 434–440.
47. Bell AM, Sih A (2007) Exposure to predation generates personality in threespined sticklebacks (*Gasterosteus aculeatus*). Ecology Letters 10: 828–834.
48. Dingemanse NJ, Van der Plas F, Wright J, Réale D, Schrama M, et al. (2009) Individual experience and evolutionary history of predation affect expression of heritable variation in fish personality and morphology. Proceedings of the Royal Society B-Biological Sciences 276: 1285–1293.
49. Nussey DH, Wilson AJ, Brommer JE (2007) The evolutionary ecology of individual phenotypic plasticity in wild populations. Journal of Evolutionary Biology 20: 831–844.
50. Careau V, Thomas D, Humphries MM, Réale D (2008) Energy metabolism and animal personality. Oikos 117: 641–653.
51. Koolhaas JM, de Boer SF, Coppens CM, Buwalda B (2010) Neuroendocrinology of coping styles: towards understanding the biology of individual variation. Frontiers in Neuroendocrinology 31: 307–321.
52. Stamps J, Groothuis TGG (2010) The development of animal personality: relevance, concepts and perspectives. Biological Reviews 85: 301–325.
53. Lessells CM (2008) Neuroendocrine control of life histories: what do we need to know to understand the evolution of phenotypic plasticity? Philosophical Transactions of the Royal Society B-Biological Sciences 363: 1589–1598.
54. Herczeg G, Garamszegi LZ (2012) Individual deviation from behavioural correlations: a simple approach to study the evolution of behavioural syndromes. Behavioral Ecology and Sociobiology 66: 161–169.
55. Duckworth RA (2010) Evolution of personality: developmental constraints on behavioral flexibility. Auk 127: 752–758.
56. Uher J (2011) Individual behavioral phenotypes: an integrative meta-theoretical framework. Why "behavioral syndromes" are not analogs of "personality". Developmental Psychobiology 53: 521–548.
57. Mischel W (1973) The interaction of person and situation; Magnusson D, Endler N, editors. New York: John Wiley & Sons.
58. Mischel W (1973) Toward a cognitive social learning reconceptualization of personality. Psychological Review 80: 252–283.
59. Nowicki JP, Miller GM, Munday PL (2012) Interactive effects of elevated temperature and CO2 on foraging behavior of juvenile coral reef fish. Journal of Experimental Marine Biology and Ecology 412: 46–51.
60. Munday PL, Dixson DL, Donelson JM, Jones GP, Pratchett MS, et al. (2009) Ocean acidification impairs olfactory discrimination and homing ability of a marine fish. Proceedings of the National Academy of Sciences 106: 1848–1852.
61. Dixson DL, Munday PL, Jones GP (2010) Ocean acidification disrupts the innate ability of fish to detect predator olfactory cues. Ecology Letters 13: 68–75.
62. Cripps IL, Munday PL, McCormick MI (2011) Ocean acidification affects prey detection by a predatory reef fish. PLoS ONE 6: e22736

2

Phenotypic and Evolutionary Consequences of Social Behaviours: Interactions among Individuals Affect Direct Genetic Effects

Barbora Trubenová*, Reinmar Hager

Computational and Evolutionary Biology, Faculty of Life Sciences, University of Manchester, Manchester, United Kingdom

Abstract

Traditional quantitative genetics assumes that an individual's phenotype is determined by both genetic and environmental factors. For many animals, part of the environment is social and provided by parents and other interacting partners. When expression of genes in social partners affects trait expression in a focal individual, indirect genetic effects occur. In this study, we explore the effects of indirect genetic effects on the magnitude and range of phenotypic values in a focal individual in a multi-member model analyzing three possible classes of interactions between individuals. We show that social interactions may not only cause indirect genetic effects but can also modify direct genetic effects. Furthermore, we demonstrate that both direct and indirect genetic effects substantially alter the range of phenotypic values, particularly when a focal trait can influence its own expression via interactions with traits in other individuals. We derive a function predicting the relative importance of direct versus indirect genetic effects. Our model reveals that both direct and indirect genetic effects can depend to a large extent on both group size and interaction strength, altering group mean phenotype and variance. This may lead to scenarios where between group variation is much higher than within group variation despite similar underlying genetic properties, potentially affecting the level of selection. Our analysis highlights key properties of indirect genetic effects with important consequences for trait evolution, the level of selection and potentially speciation.

Editor: Yury E. Khudyakov, Centers for Disease Control and Prevention, United States of America

Funding: BT is a student of Doctoral Training Centre Integrative Systems Biology at the University of Manchester (http://www.mib.manchester.ac.uk/postgraduate/DTC/), which is partly supported by Biotechnology and Biological Sciences Research Council (50%), Partly by Faculty of Life Sciences UoM (25%)and partly by Faculty of Engineering and Physical Sciences UoM (25%). The funders had no role in study design, data collection and analysis, decision to publish, or preparation of the manuscript.

Competing Interests: The authors have declared that no competing interests exist.

* E-mail: barbora.trubenova@postgrad.manchester.ac.uk

Introduction

Complex social interactions are widespread among animals and are of considerable interest to both behavioural ecologists [1–3] and quantitative geneticists [4–6]. Understanding the role of interactions among individuals for phenotypic variation and fitness is crucial for the study of such diverse areas as behavioural ecology, where IGEs are implicated in the evolution of cooperation and social dominance [2,7–11] and agriculture (inform breeding design, reduction of competition) [12,13]. From an evolutionary perspective, any interaction or behaviour can be regarded as social whenever it influences not only the fitness of the actor, but also the fitness of other individuals [14]. For example, cooperation, altruism, but also aggression and spite are social behaviours, and traits underlying these behaviours are influenced by interactions with conspecifics [15–18].

Standard quantitative genetics theory assumes that the phenotype, or trait values of an individual, are affected by its genes and the environment. Environmental effects are usually considered to have no genetic basis, and therefore be non-heritable. However, every individual living in a social environment is affected by the social behaviours of conspecifics, which is partly given by their genotypes. The environment provided by conspecifics is often the most important component of the environment experienced by individuals, frequently having profound effects on trait expression and fitness [19].

The influence of the genotypes of other individuals on the phenotype of a focal individual is referred to as indirect genetic effect (IGE) [4,8,20–25] or associative effect [26–28], because a social partner's genes influence the trait indirectly - they are expressed in interacting individuals, not in the individual whose phenotype is observed.

The idea that the phenotype of an individual is affected not only by the genes it carries, but also by the genes of other individuals was proposed already by Hamilton [7] in his theory of neighbour modulated fitness. Further development of this concept was strongly linked with research in the field of parental care [4,20,29]. Therefore, IGEs are often illustrated by maternal effects [21,30], where the phenotype of the mother influences expression of traits in her offspring (for example body size) [31,32]. Maternal effects result in IGEs whenever the traits contributing to the environment provided by the mother are heritable. Maternal genetic effects often account for as much as half of the variance in characters expressed early in life [21].

However, social interactions among individuals are ubiquitous and IGEs are not limited to interactions among relatives [33]. Often, unrelated individuals interact and these interactions may strongly influence their respective phenotypes [34]. Behaviours such as aggression, cooperation and courtship are common examples of such interactions [18,27,35,36]. Even the behaviour of primitive organisms, such as bacteria and amoeba, is affected by genes expressed in other individuals [37,38].

There have been very few studies of IGEs within social groups [39]. In *Drosophila melanogaster*, the genotypic composition of social groups (single versus mixed genotypes) was shown to affect expression of clock genes, genes for pheromonal profile on the cuticle and mating frequency [40]. Several recent empirical studies have mapped the quantitative trait loci (QTL) for maternal genetic effects and other IGEs [34,41–44], shedding light on the genetic architecture underlying IGEs.

Unlike effects of the physical environment, social influences have both an environmental and genetic component. Because of this genetic (heritable) component, social influences are subject to selection and can evolve, while the environmental component may also act as a selection factor. By affecting the strength of selective pressure and changing the expected genotype-phenotype relationship, IGEs can change the speed and direction of evolution [4,20,36]. Thus, IGEs and social interactions play an important role in trait evolution.

Models of IGEs are traditionally formulated in two different ways: the trait-based approach developed by Moore [20] focuses on how a phenotype of a focal individual is influenced by specific traits in social partners. The second approach, developed in a series of papers by Griffing [45–48], was first adopted by Muir [26] and later by Bijma and colleagues [12,13,49–51], looks at the total genetic variance and its partitioning into two independent components. The direct variance component arises from an individual's own genotype while the indirect component is caused by interactions among individuals [23]. The second approach is favoured by empiricists because the methods of estimating variance components are relatively straightforward. A detailed comparison of both frameworks was carried out by McGlothlin and Brodie III [23], who demonstrated the compatibility of both approaches.

Here, we provide a detailed analysis of IGEs and their dependence on key parameters: the strength of interaction between individuals, group size and population genotypic characteristics. Unlike Moore [20], we consider interactions of multiple individuals and demonstrate how group size can affect the strength of IGEs, the mean phenotype and its variance. Furthermore, we investigate the effect of interactions on within and between group variation and discuss the possible implication for multilevel selection. In contrast to previous models, we formally distinguish two separate consequences of social interactions: first the indirect genetic effect and, second, the change in the direct genetic effect. Previous trait based models did not distinguish between these two separate effects [20,36], while variance-based models [12,13] did not explore the effect of interactions on direct genetic effects. However, we can show that both direct and IGEs may reach extreme values, either cancelling each other (if the effects are of opposite sign) or increasing effects on phenotypic values (if of the same sign). We further develop our model in a way suitable for agent-based modelling, commonly used to study social evolution [52–55], and use simulations to illustrate our results.

Results

The model

We begin by generalizing Moore et al. 's model [20] for the case with N individuals

$$\mathbf{p}_i = \mathbf{\Gamma}\mathbf{g}_i + \sum_{j\neq i}^{N-1} \mathbf{\Psi}\mathbf{p}_j + \mathbf{e}_i \tag{1}$$

where $\mathbf{g}_i$ (genotype) denotes a column vector of genotypic values of an individual's genes, i, $\mathbf{p}_i$ (phenotype) is a column vector of corresponding trait values (same as z in [20]), and $\mathbf{e}_i$ is a vector of environmental components acting on the i-th individual. As we want to investigate the potential role of social interactions on direct and indirect genetic effects, we will not consider environmental effects other than those caused by social interactions in the following analysis. Therefore, $\mathbf{e_i} = 0$ in our model. However, environmental influences can easily be incorporated into all calculations, by replacing $\mathbf{\Gamma g_i}$ (the genetic component) by $\mathbf{\Gamma g_i + e_i}$ in all following equations (see Methods).

The first term represents direct genetic effects (DGE), with matrix $\mathbf{\Gamma}$ mediating the translation of an individual's own genotype $\mathbf{g}_i$ into its trait values. For the simplest models, where every trait is encoded by a single gene, this matrix is square and diagonal. However, in a more realistic scenario multiple genes affect the same trait or the same gene affects more than one trait (pleiotropy). These effects can be incorporated into the model by populating matrix $\mathbf{\Gamma}$, with size $m \times n$, where m is the number of traits under investigation and n is the number of genes involved.

The second term represents associative effects [13] that may be interpreted as heritable environmental effects provided by social partners of the focal individual [4]. Matrix $\mathbf{\Psi}$ is a square $(m \times m)$ interaction matrix [4,20,36], in which Ψ_{kl} defines the effect of the partner's trait l on the trait k of the focal individual. If Ψ_{kl} equals 0, there is no effect. If it is negative, a higher expression of the partner's trait l lowers the expression of the focal individual's trait k. Positive $\mathbf{\Psi}_{kl}$ means that the expression of trait l enhances the expression of trait k in the focal individual. N denotes the number of individuals in the group, i the focal individual and j all other individuals in the group.

To investigate the dependence of the focal individual's phenotype on the genotypes of its social partners, it is necessary to separate the focal individual's genes from those of its conspecifics and rewrite equation 1 as follows

$$\begin{aligned}\mathbf{p}_i &= (\mathbf{I}+\mathbf{\Psi})^{-1}(\mathbf{I}+\mathbf{\Psi}(\mathbf{I}-N\mathbf{\Psi}+\mathbf{\Psi})^{-1})\mathbf{\Gamma}\mathbf{g}_i \\ &+ (\mathbf{I}+\mathbf{\Psi})^{-1}\mathbf{\Psi}(\mathbf{I}-N\mathbf{\Psi}+\mathbf{\Psi})^{-1}\mathbf{\Gamma}\sum_{j\neq i}^{N-1}\mathbf{g}_j \\ &= \mathbf{\Gamma}'\mathbf{g}_i + \mathbf{\Psi}'\sum_{j\neq i}^{N-1}\mathbf{g}_j\end{aligned} \tag{2}$$

where $\mathbf{I}$ is an identity matrix and

$$\mathbf{\Gamma}' = (\mathbf{I}+\mathbf{\Psi})^{-1}(\mathbf{I}+\mathbf{\Psi}(\mathbf{I}-N\mathbf{\Psi}+\mathbf{\Psi})^{-1})\mathbf{\Gamma} \tag{3}$$

$$\mathbf{\Psi}' = (\mathbf{I}+\mathbf{\Psi})^{-1}\mathbf{\Psi}(\mathbf{I}-N\mathbf{\Psi}+\mathbf{\Psi})^{-1}\mathbf{\Gamma}. \tag{4}$$

The first term of equation 2 ($\mathbf{\Gamma}'\mathbf{g}_i$) focuses only on the focal individual's genes, and therefore represents DGEs [4]. It should be noted that the term $\mathbf{\Gamma}'$ specifying the dependence of an individual's phenotype on its genotype is more complicated than in a case without any interaction (when $\mathbf{\Gamma}' = \mathbf{\Gamma}$). This complication arises from the potential for feedback loops that may occur when a trait affects itself in a social partner [6,20,23] or when a trait affects some other trait in the same individual via interactions with traits expressed in conspecifics. In such a case, the phenotypic value of a particular trait is not only given by genes directly underlying this trait and by genes expressed in conspecifics, but also indirectly by other genes expressed in the focal individual (via the interaction with other individuals). For example, if a trait X affects trait Y in conspecifics, while trait Y affects trait Z, the phenotypic value of Z is given by an individual's own genes for Z, as well as X (see Figure 1 B). This suggests that interactions of individuals do not only change their phenotypes due to IGEs, but they can also change the way the genotype of the focal individual is translated into its own phenotype (direct effect). A similar effect was described as $g \leftrightarrow e$ interdependence: organisms change the environment around them, which in turn changes the expression of their genes, for instance via stress, and are especially common in the case of social interactions [56]. In our case, $g \leftrightarrow e$ interdependence is represented by the coefficient $(\mathbf{I}+\mathbf{\Psi})^{-1}(\mathbf{\Psi}(\mathbf{I}-N\mathbf{\Psi}+\mathbf{\Psi})^{-1})$ that alters gene expression from $\mathbf{\Gamma}$ to $\mathbf{\Gamma}'$.

The second term $\mathbf{\Psi}'\sum_{j\neq i}^{N-1}\mathbf{g}_j$ in equation 2 represents IGEs. IGEs depend only on the genes of interacting individuals and not on the genes of the focal individual. While it is obvious that the IGEs are the product of matrix $\mathbf{\Gamma}$ and the interaction matrix $\mathbf{\Psi}$, it is not obvious why the terms $(\mathbf{I}+\mathbf{\Psi})^{-1}$ and $(\mathbf{I}-N\mathbf{\Psi}+\mathbf{\Psi})^{-1}$ appear. These terms are present due to the possibility of feedback interactions, i.e. when a trait in the focal individual changes a trait in interacting individuals, which, in turn, changes the same or a different trait in the focal individual. Indeed, $\mathbf{\Psi}'$ simplifies to $\mathbf{\Psi}'=\mathbf{\Psi}\mathbf{\Gamma}$ in the absence of feedback interactions. Note, that in order to infer equation 2 we assume that all interactions are symmetric.

In equation 2, $\mathbf{\Gamma}'$ can be interpreted as the relative effect of an individual's (own) genes for its phenotype compared with the effects of genes in conspecifics for the phenotype of the focal individual (given by $\mathbf{\Psi}'$). There is a simple relationship between the relative importance of an individual's own genes and the importance of genes in conspecifics, given as:

$$\mathbf{\Gamma}'=(\mathbf{I}+\mathbf{\Psi})^{-1}\mathbf{\Gamma}+\mathbf{\Psi}'. \quad (5)$$

Three classes of interactions

As stated above, some interactions may contain a feedback loop that occurs when one trait affects the same trait in a social partner or when it alters the expression of a different trait in the focal individual via interactions with social partners [6,20,23,26]. Our analysis shows that interactions involving feedback loops have different properties from those without such feedback (Figure 1 A). Furthermore, interactions with a feedback loop at the individual level, but not the trait level (Figure 1 B), yield different outcomes from interactions when the feedback loop points back to the same trait (Figure 1 C–D).

Interactions without feedback. The first, simplest class involves situations, in which a trait in a focal individual influences some other trait in its social partners, however, these affected traits do not alter the expression of any trait in the focal individual (Figure 1 A). A common example of such a case is the effect of body size on aggressive behaviour, as individuals often mediate their aggressiveness according to the body size of their opponents [57–59]. The first type of interaction between two individuals described in Moore at al. [20] belongs into this class.

It is easily seen whether or not a situation belongs to this class - only unaltered traits may affect other traits, otherwise feedback occurs. Formally, the condition $\mathbf{\Psi}\mathbf{\Psi}=0$ must be fulfilled. For this class of interaction, there is no $g \leftrightarrow e$ interdependence and equation 2 can be simplified to

$$\mathbf{p}_i=\mathbf{\Gamma}\mathbf{g}_i+\mathbf{\Psi}\mathbf{\Gamma}\sum_{j\neq i}^{N-1}\mathbf{g}_j. \quad (6)$$

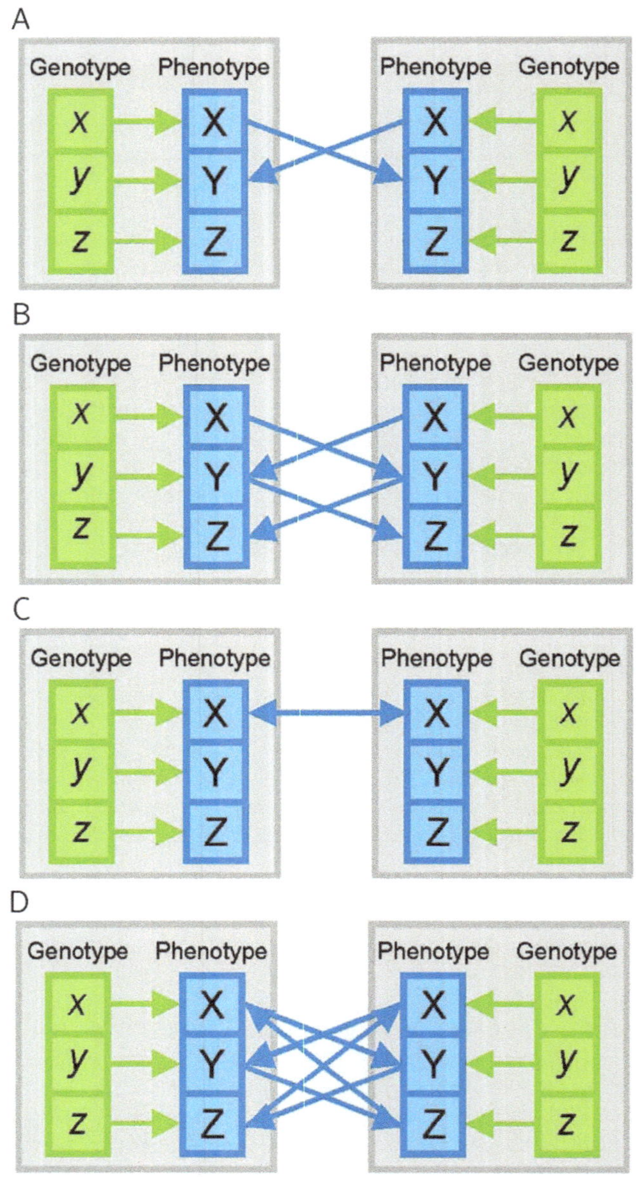

Figure 1. Different classes of interactions. (A) Example of interactions without feedback. (B) Example of interactions with feedback at the individual level. (C, D) Example of interactions with feedback at the trait level: (C) one trait affecting itself. (D) three traits involved in feedback to themselves via multiple interactions. Boxes represent individuals, blue squares represent phenotypic traits, green squares represent genes. Blue arrows represent matrix $\mathbf{\Psi}$, green arrows represent matrix $\mathbf{\Gamma}$.

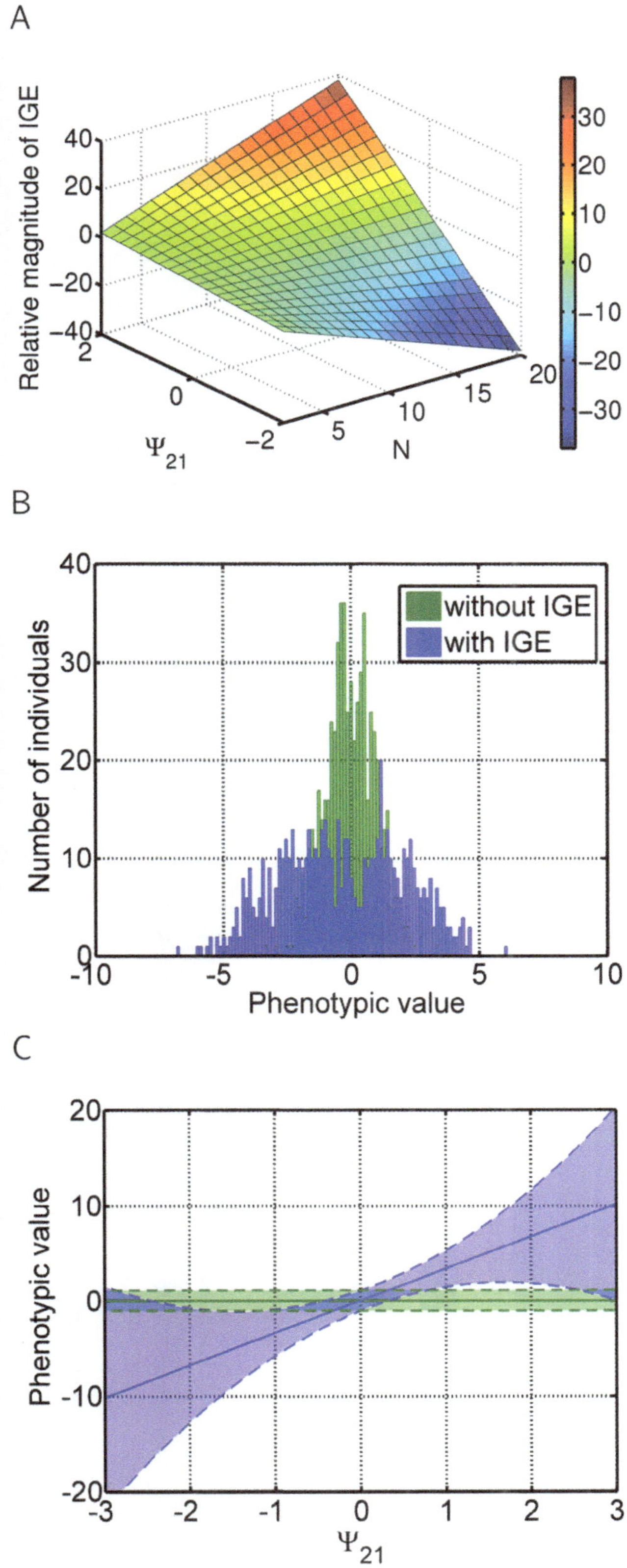

Figure 2. Interactions without feedback. Indirect genetic effects of gene x on trait Y (see Figure 1 A), with trait X affecting trait Y. (A) Relative magnitude of indirect genetic effect, as described in equation 6, with the strength of the interaction $\Psi_{21} = 0.5$. (B) Histogram of the phenotypic values across population. Green - without IGEs, blue - with IGEs. Strength of the interaction set to $\Psi_{21} = 0.5$. (C) Mean phenotypic value (solid line) plus and minus variance (dashed lines). Blue - mean phenotype and its variance if IGEs are considered. Green - no IGEs are considered. Both (B) and (C) are results of the simulations of a group of 30 individuals.

Here, the IGE magnitude depends linearly on the strength of interactions (elements of matrix $\mathbf{\Psi}$). Therefore, the stronger the effect of the other's phenotype on the phenotype of the focal individual, the stronger the indirect genetic effect, as illustrated in Figure 2 A. Direct genetic effects are not altered by these interactions. However, as phenotypes of all individuals are changed by IGEs, there is a change in the distribution of phenotypic values of the affected trait (see Figure 2 B).

The mean phenotype is given by

$$\bar{\mathbf{p}} = (\mathbf{I} + (N-1)\mathbf{\Psi})\mathbf{\Gamma}\bar{\mathbf{g}} \tag{7}$$

and depends linearly on group size N and the strength of interactions, depicted in Figure 2 C. Here, $\mathbf{\Gamma}\bar{\mathbf{g}}$ defines the average phenotype when no interaction occurs, while $\bar{\mathbf{p}}$ is the average phenotype when interactions do occur.

To obtain the relationship between an individual's phenotype and the group mean phenotype, we have to express an individual's genotype as its deviation from the mean genotype

$$\mathbf{g}_i = \bar{\mathbf{g}} + \Delta\mathbf{g}_i \tag{8}$$

and substitute $\mathbf{g}_i$ in equation 6. An individual's phenotype can thus be expressed as

$$\mathbf{p}_i = \underbrace{(\mathbf{I} + (N-1)\mathbf{\Psi})\mathbf{\Gamma}\bar{\mathbf{g}}}_{\bar{\mathbf{p}}} + \underbrace{(\mathbf{I} - \mathbf{\Psi})\mathbf{\Gamma}\Delta\mathbf{g}_i}_{\Delta\mathbf{p}_i}. \tag{9}$$

The first term corresponds to the mean phenotype, while the second term represents an individual's deviation from the mean phenotype. Note that an individual's deviation from the mean phenotype is linearly dependent only on the strength of the interaction and the deviation from mean genotype but not on the number of individuals.

Phenotypic (within group) variance for each trait can be written as

$$\begin{aligned} var\ \mathbf{p} &= \frac{1}{N}\sum_{i}^{N}(\Delta\mathbf{p}_i \circ \Delta\mathbf{p}_i) \\ &= \frac{1}{N}\sum_{i}^{N}((\mathbf{I}-\mathbf{\Psi})\mathbf{\Gamma}\Delta\mathbf{g}_i)\circ((\mathbf{I}-\mathbf{\Psi})\mathbf{\Gamma}\Delta\mathbf{g}_i) \end{aligned} \tag{10}$$

where $var\ \mathbf{p}$ is a vector of variances of all traits (NB. $(\Delta\mathbf{p}_i \circ \Delta\mathbf{p}_i)$ is the Hadamard product of the vectors). Therefore, the variance of a given trait is not only given by the genetic variation of genes directly underlying the trait but also by the variation found in all genes affecting this trait.

Interactions with feedback on the level of individuals. The second class comprises situations, in which a trait in a focal individual (trait X) affects a different trait in its social partners (e.g. trait Y), which in turn influences a trait Z, different from both X and Y in the focal individual. This scenario describes two partial interactions: trait X affecting trait Y and trait Y affecting trait Z. However, no trait affects itself by any set of partial interactions (see Figure 1 B). In other words, a feedback loop does return back to the same individual, but not to the same trait. An example of such a situation may be scent marks that advertise territoriality to competitors, which may change the competitor's behaviour toward the signaller (and hence affect the latter's phenotype) e.g. in rodents [60]. Another possibility is the degree of cooperation among individuals. In such cases, an individual's scent induces cooperative behaviour in more closely related individuals, which in turn may affect body size (trait Z) of the focal individual. Note, that interactions in this class are distinct from those described in Moore et al. [20].

For this class of interaction, equation 2 cannot be simplified any more and IGEs depend linearly on all partial interactions involved. The effect of an individual's own genotype on its phenotype (direct genetic effect) is altered as well (see Figure 3 A). For example, an IGE acting on trait Z depends linearly on both coefficients of interaction of trait X acting on trait Y and trait Y acting on the trait Z (Figure 1 B). However, an IGE on trait Z depends now quadratically on the number of individuals: although linearly dependent on each partial interaction, trait Z is now affected by two subsequent interactions, as depicted in Figure 3 B (as N^{m-1} where m is the number of traits involved).

From equation 2, the mean phenotype $\bar{\mathbf{p}}$ can be expressed as

$$\bar{\mathbf{p}} = (\mathbf{I} - N\mathbf{\Psi} + \mathbf{\Psi})^{-1}\mathbf{\Gamma}\bar{\mathbf{g}}. \qquad (11)$$

To obtain the relationship between an individual and the mean phenotype, we substitute equation 8 in equation 2. An individual's phenotype can then be written as

$$\mathbf{p}_i = \underbrace{(\mathbf{I} - N\mathbf{\Psi} + \mathbf{\Psi})^{-1}\mathbf{\Gamma}\bar{\mathbf{g}}}_{\bar{\mathbf{p}}} + \underbrace{(\mathbf{I} + \mathbf{\Psi})^{-1}\mathbf{\Gamma}\Delta\mathbf{g}_i}_{\Delta\mathbf{p}_i}. \qquad (12)$$

As $det(\mathbf{I}+\mathbf{\Psi})$ and $det(\mathbf{I}-N\mathbf{\Psi}+\mathbf{\Psi})$ are non zero for any value of $\mathbf{\Psi}$ in this class, it is always possible to find inverse matrices of both terms. Therefore, phenotypic values can always be predicted using equation 12 for any combination of parameter values.

The within group phenotypic variance of each trait is given by

$$\begin{aligned} var\ \mathbf{p} &= \frac{1}{N}\sum_{i}^{N}(\Delta\mathbf{p}_i \circ \Delta\mathbf{p}_i) \\ &= \frac{1}{N}\sum_{i}^{N}((\mathbf{I}+\mathbf{\Psi})^{-1}\mathbf{\Gamma}\Delta\mathbf{g}_i)\circ((\mathbf{I}+\mathbf{\Psi}\mathbf{Y})^{-1}\mathbf{\Gamma}\Delta\mathbf{g}_i). \end{aligned} \qquad (13)$$

Interactions with feedback on the trait level. The last class contains situations, in which a particular trait affects its own expression by interacting with other individuals. Both types of interactions described in Moore et al. [20] belong to this class of interactions. An example of such a case is aggression, where aggressive behaviour of one individual induces aggressiveness of its opponent, which in turn may increase aggressive behaviour of the first individual, e.g. in primate groups [61] (Figure 1 C). It is not necessary that only one trait is involved - two distinct traits can reciprocally affect each other, or even more traits can be involved (Figure 1 D). The key point is that the expression of a particular trait feeds back to alter itself.

Such feedback introduces a possibility of increasing trait values to infinity, as depicted in Figure 4. Indeed, unlike in the previous class, the terms $(\mathbf{I}+\mathbf{\Psi})^{-1}$ and $(\mathbf{I}-N\mathbf{\Psi}+\mathbf{\Psi})^{-1}$ can introduce undefined points, when $det(\mathbf{I}+\mathbf{\Psi})=0$ or $det(\mathbf{I}-N\mathbf{\Psi}+\mathbf{\Psi})=0$.

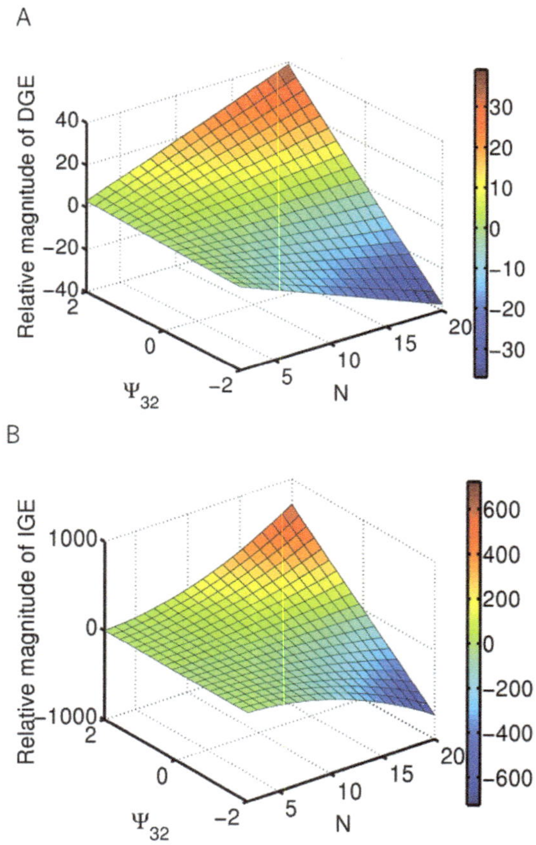

Figure 3. Interactions with feedback at the individual level: direct and indirect genetic effects. Dependence of the direct (A) and indirect (B) genetic effect on the strength of the interaction Ψ_{32} (trait Y acting on trait Z) and group size N. Visualization of DGE and IGE of genes x and y acting on Z, as described in equation 2. Interaction, as depicted in Figure 1 B, with the interaction strength of X acting on Y set to $\Psi_{21}=0.05$.

When elements of matrix $\mathbf{\Psi}$ (specifying the strength of the interactions) are close to these points, DGEs and IGEs reach extreme values. This phenomenon was observed by Moore et al. [20], however, since they dealt only with dyadic interactions, several aspects of this phenomenon could not be explored - for example the dependence of one of the undefined points (and hence extreme phenotypes) on group size. The dependence on group size may be very important for many aspects of behavioural biology such as group size and stability. For instance, if the interaction strength is 0.145, the level of aggressiveness may be acceptable for groups with 5 members, but the arrival of an even mildly aggressive sixth member would increase the overall aggression by more than 5 times.

Note that both direct and indirect genetic effects may reach extreme values under the above conditions as the effects of $g \leftrightarrow e$ interdependence can be very strong. If the direct and indirect genetic effects are of the same sign, they may cause extreme phenotypes. However, if they are of opposite signs and of relatively similar (though extreme) values, they cancel each other and may lead to a phenotype similar to the case of no interaction.

Individual and mean phenotypes, as well as variances are defined as in the case of feedback at the level of the individual (equations 12, 11 and 13, respectively). When considering equation 12, the undefined points are separated - one is

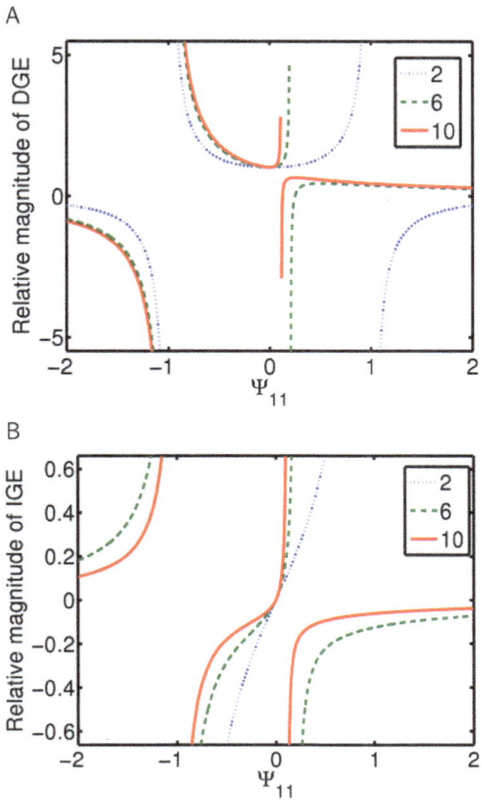

Figure 4. Interactions with feedback at the trait level: direct and indirect genetic effects. Relative magnitude of direct (A) and indirect (B) genetic effects and their dependence on the strength of the interaction, for different numbers of individuals (two, six and ten individuals). Interaction as depicted in Figure 1 C, where Ψ_{11} is the strength of the interaction between trait X of the focal individual and the same trait in its social partner.

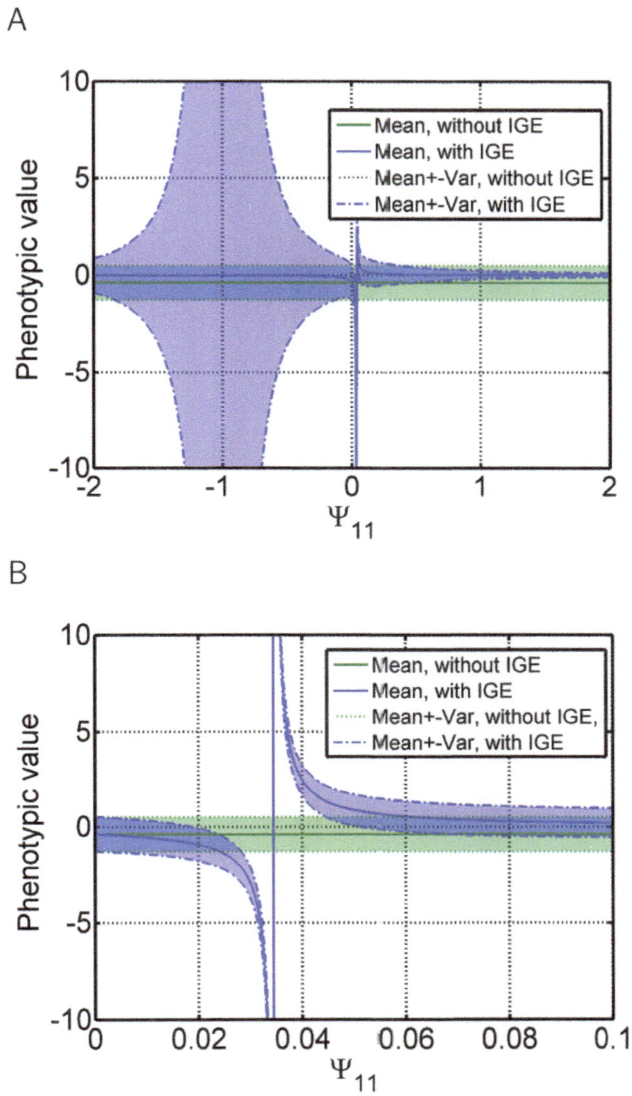

Figure 5. Interactions with feedback at the trait level - mean and variance. Dependence of the mean group phenotypic value and its variance on the interaction strength Ψ_{11}, when one trait affects itself in other individuals (as depicted in Figure 1 C). Mean and variance (A) around $det(\mathbf{I}+\mathbf{\Psi})=0$, (B) around $det(\mathbf{I}-N\mathbf{\Psi}+\mathbf{\Psi})=0$. Blue - mean phenotype and its variance if IGEs are considered. Green - no IGEs are considered. Simulation of 30 interacting individuals.

responsible for extremes of the mean phenotype and one for extreme deviations of individual phenotypes from the mean phenotype. Therefore, an individual can have an extreme phenotype due to two independent factors - either the individual phenotype deviates extremely from a relatively normal mean phenotype, or the mean group phenotype is extreme while the individual's phenotype deviation is relatively small.

The former occurs when values of $\mathbf{\Psi}$ are close to $det(\mathbf{I}+\mathbf{\Psi})=0$. In such a case, all phenotypes are distributed around a normal mean with extremely high variance (Figure 5 A). Thus, differences among individuals within groups are much higher than differences between groups. This term does not depend on the number of individuals in the group, just on the strength of the interactions and genetic deviations.

On the other hand, when $\mathbf{\Psi}$ is close to $det(\mathbf{I}-N\mathbf{\Psi}+\mathbf{\Psi})=0$, mean group phenotypes reach extreme values and even small variances in the mean genotype of the group may cause drastic changes to the mean phenotype. However, the phenotypes of all individuals in each group are distributed around the mean phenotype with relatively small variance (Figure 5 B). Thus, intergroup variation is much higher than intragroup variation. In this scenario, small differences in mean genotypes between groups are translated into large differences between the mean phenotypes of such groups (Figure 6 B). Therefore, if a trait affecting fitness is modified in such a way, individual fitness may strongly depend on the properties of the group, such as genotypic mean and variation among group members and group size. Since all individuals are similar to each other in each of these groups, such low intragroup and large intergroup variance may thus lead to selection on the level of groups and potentially speciation. Note that the range of the interaction strength $\mathbf{\Psi}$, leading to large intergroup variance, is much smaller than the range of values leading to large intragroup variance (Figure 5 A and B). The effect of interaction strength on intra and intergroup variance is shown in Figure 6 A.

The value of the mean phenotype depends on the strength of the interaction, the mean genotype, and the number of interacting individuals. This means that the same type of interaction may have very different effects on phenotypes of individuals belonging to two differently sized groups. This applies, for example, to groups of nine and ten individuals when $\Psi_{11}=0.105$. In this case, the interaction effect on the mean group phenotype in groups of ten

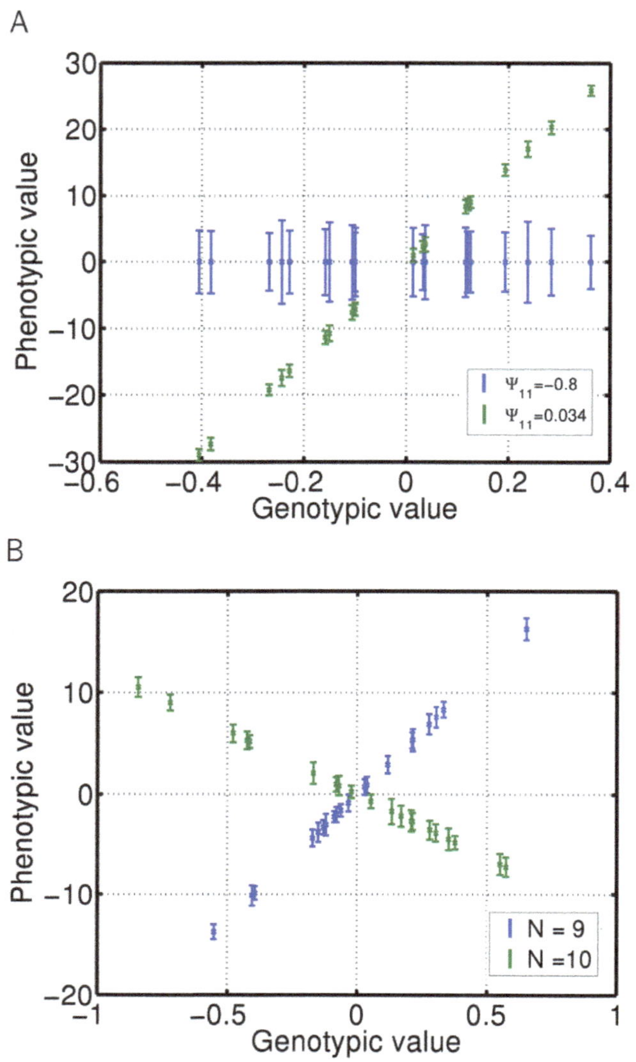

Figure 6. Inter and intragroup phenotypic variance. Mean phenotypic value and variance of each group are calculated and plotted against mean genotypic value of the group. Errorbars represent intragroup variance. Simulation of a scenario when a given trait in a focal individual affects the same trait in other individuals, as shown in Figure 1 (A). Simulation of 20 groups of 30 individuals. Blue - interaction strength $\Psi_{11}=-0.8$, green - interaction strength $\Psi_{11}=0.034$. (B) Simulation of 20 groups of 9 (blue) or 10 (green) individuals; interaction strength $\Psi_{11}=0.105$.

individuals is much stronger than the effect in groups of nine individuals, as illustrated in Figure 6 B. The variance within groups is relatively small and similar for groups with nine and ten members. Note, that the value of interaction strength is smaller than 1/(N-1). This means that the relative influence of all the other individuals is still smaller than individual's own genotype, which is a reasonable assumption. Extreme phenotypic values are not caused by strong interactions.

The effect of group size on IGEs has been investigated mostly focusing on of the dilution effect in larger groups [24,27,36]. Dilution is a decrease of IGEs with group size [24], for instance because individuals interact less often. Without such dilution, individuals in larger groups should experience stronger effects, as more individuals contribute to the IGEs. While this may apply to the first two classes of interaction (where the effects depend linearly on group size) it does not for the third class. When interactions feed back on themselves, the dependence on group size is no longer linear.

Fitness and selection

When a fitness of an individual is affected by a phenotypic trait of conspecifics, social selection is said to occur. However, interacting individuals may affect each other's fitness indirectly as well, via IGEs - affecting those traits of the focal individual that are important for fitness.

If we assume M groups of N interacting individuals, the relative fitness of each individual can be defined as

$$w_{ki}=\frac{\beta'\mathbf{p_{ki}}+C}{\sum_l^M\sum_j^N(\beta'\mathbf{p_{lj}}+C)} \tag{14}$$

where β' is a column vector describing the importance of an individual's own traits for its own fitness and C is a constant. Note that the mean fitness of the population is defined as $1/MN$ as opposed to 1 as often used elsewhere, so the sum of the fitness of all individuals across the population is 1. We use this value for the convenience of the calculations and simulations.

If we normalize values so that $\bar{\bar{\mathbf{p}}}=0$ and use equation 2 for the phenotype of a focal individual, its relative fitness can be written as

$$w_{ki}=\frac{\beta'\mathbf{\Gamma}'\mathbf{g_{ki}}+\beta'\mathbf{\Psi}'\sum_{i\neq j}^{N-1}\mathbf{g_{kj}}+C}{MNC} \tag{15}$$

or

$$w_{ki}=\frac{\beta'(\mathbf{\Gamma}'+(N-1)\mathbf{\Psi}')\Delta\overline{\mathbf{g}_k}+\beta'(\mathbf{\Gamma}'-\mathbf{\Psi}')\Delta\mathbf{g}_{ki}+C}{NMC} \tag{16}$$

where $\Delta\overline{\mathbf{g_k}}$ denotes the difference between the mean genotype of the kth group and mean population genotype, while $\Delta\mathbf{g_{ki}}$ is a difference between individuals own genotype and mean genotype of its group.

Response to selection. To understand the effect of interactions on trait evolution, we have to investigate their effects on the response to selection. The response to selection $\Delta\bar{\bar{\mathbf{p}}}$ is defined as a difference between the mean phenotypic value of offspring and the mean phenotypic value of the parental generation, and can be calculated using Price's equation:

$$\begin{aligned}\Delta\bar{\bar{\mathbf{p}}}^T&=[\beta'(\mathbf{I}+N\mathbf{\Psi}+\mathbf{\Psi})^{-1}\mathbf{\Gamma}\Sigma(\Delta\bar{\mathbf{g}})(\mathbf{I}+N\mathbf{\Psi}+\mathbf{\Psi})^{-1}\mathbf{\Gamma})^T\\&+\beta'(1-\mathbf{\Psi})^{-1}\mathbf{\Gamma}\Sigma(\Delta\mathbf{g})((1-\mathbf{\Psi})^{-1}\mathbf{\Gamma})^T]/C\end{aligned} \tag{17}$$

where $\mathbf{\Sigma}(\Delta\bar{\mathbf{g}})$ is a variance-covariance matrix of mean group phenotypic values of each trait (between group variance-covariance), and $\mathbf{\Sigma}(\Delta\mathbf{g})$ is a variance-covariance matrix of traits within the group (within group variance-covariance).

Equation 17 shows that the response of selection depends not only on the overal genotypic variance of the population, but on the intragroup and intergroup genotypic variances as well. The strength of the indirect genetic effect given by $\mathbf{\Psi}$ is a key factor in determining which of these variances becomes more important for the response to selection. If the strength of interaction is close to the value where $det(\mathbf{I}-N\mathbf{\Psi}+\mathbf{\Psi})=0$, small differences between genotypic distributions of individual groups may translate into substantial differences in the fitness of groups. In such a case, group properties are much more important for individual's fitness, and the selection at the level of groups becomes much stronger than the selection at the level of individuals (Figure 7).

Furthermore, interaction strength $\mathbf{\Psi}$ and the variance between groups and within groups determine whether the response to selection is positive or negative. When within group variance is smaller than between group variance, even traits that are non beneficial to the focal individual, but beneficial to the group, may create a positive response to selection, similar to the effect of social selection. This can be interpreted in terms of Hamilton's rule [62]: term $\Sigma(\Delta\bar{\mathbf{g}})$ describes the relatedness between groups, while $\Sigma(\Delta\mathbf{g})$ is the relatedness between individuals within each group. In this case, smaller variance within groups may be interpreted as higher relatedness in each group, while costs and benefits are given by the strength of interaction $\mathbf{\Psi}$ and the selection gradients β'.

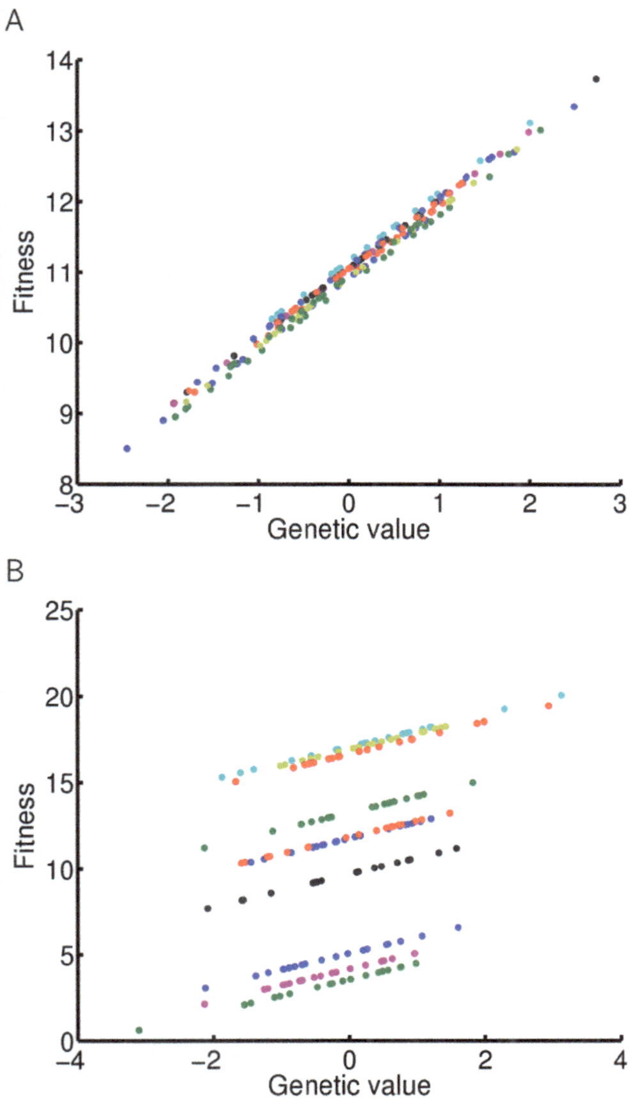

Figure 7. Inter and intragroup variance in fitness. Simulations of 10 groups of 20 interacting individuals; a given trait in a focal individual affects the same trait in other individuals (shown in Figure 1 C). Different colours represent individuals in different groups. (A) The value of $\Psi_{11}=0.02$. Within group variance is high, but between group variance is low. (B) The value of $\Psi_{11}=0.05$. Within group variance is low, but between group variance is high.

Discussion

The key result of our model is a detailed description and analysis of extreme phenotypes caused by feedback when IGEs occur. We show that DGEs can be drastically altered by social interactions. This profoundly influences the relative importance of an individual's genotype for its own fitness when compared to genes of its conspecifics. Furthermore, we can demonstrate that even small differences in the mean genotypes between groups may lead to large phenotypic and fitness differences between groups. This may lead to selection at the level of groups and potentially speciation, especially when considering traits involved in mate choice. Finally, social interaction effects may depend to a large extent on the number of interacting individuals, in a non-linear way.

We advance on previous models by investigating interaction effects among multiple individuals on their phenotypes. Specifically, we can predict the possible range of phenotypic values as well as their dependence on group size and genetic composition of the group. Thus, our predictions can be directly tested in systems that allow manipulation of group size and genetic composition of groups, as well as inter- and intra-group genetic variance (e.g. mixed litters in genetically variable mice [63]). Furthermore, we derive equations that can easily be incorporated into agent based models, e.g. to simulate more complex evolutionary scenarios for which no analytical solution can be derived.

Our analysis has shown that the phenotype of a focal individual is not only affected by IGEs, but importantly also by changing the dependence of an individual's phenotype on its own genotype (direct genetic effect). This phenomenon is akin to $g \leftrightarrow e$ interdependence: organisms change the environment around them, which in turn changes the expression of their genes, for instance via stress [56]. Niche construction is an example where genotypes (i.e. individuals) modify their environment, which in turn affects the selective pressure on these genotypes [64,65]. Clearly, when considering the social environment, $g \leftrightarrow e$ interdependences are especially common [56].

Types of interactions

In our analysis, we distinguish between three classes of interactions that fully describe all possible ways in which IGEs can manifest. In the simplest class of interaction where no feedback occurs, IGEs depend linearly on group size and the strength of interactions. The stronger the interaction or the more individuals are in the group, the stronger the IGEs. However, this assumes that interaction strength does not depend on group size itself, which is not necessarily true. For example, the effect of interactions among particular individuals may be less intense in larger groups [24], where individuals interact less often. This may lead to a decrease in interaction strength ('dilution' - sensu [24]).

The second class comprises of situations, where a feedback loop exists to the same individual, but not to the same trait. An example of such a situation is cooperation based on kin recognition, when the presence of a specific trait increases cooperation (e.g. a green beard [66]), which in turn may affect body size (e.g. through better resource exploitation). Unlike the first class, IGEs will depend non-linearly on the number of interacting individuals N and on the number of traits m involved in the feedback loop (specifically, N^{m-1}).

On the other hand, interactions that feed back to the same trait, resulting in a trait affecting itself, represent our third class of interaction. Moore and colleagues showed that for interactions involving feedback, IGEs may reach extreme values as interaction strength (Ψ) approaches ± 1 [20]. We generalized their findings and showed that both direct and indirect genetic effects are undefined when $det(\mathbf{I}+\boldsymbol{\Psi})=0$ or $det(\mathbf{I}-N\boldsymbol{\Psi}+\boldsymbol{\Psi})=0$. When the strength of interaction $\boldsymbol{\Psi}$ reaches values that satisfy one of these two conditions, $g\leftrightarrow e$ interdependence is very strong and both direct and indirect genetic effects may reach extreme values (extreme phenotypes). When an individual genotype is expressed as the mean group genotype plus the deviation from this mean, we can separate the above two conditions. The separation of these two conditions allows for comparisons of the importance of an individual's own genotype for the expression of its own phenotype relative to properties of its group (e.g. mean genotype). The first condition, ($det(\mathbf{I}+\boldsymbol{\Psi})=0$), refers to an individual's deviation from the group mean phenotype, while the second condition, ($det(\mathbf{I}-N\boldsymbol{\Psi}+\boldsymbol{\Psi})=0$), refers to the mean phenotype of a group. Our results suggest that, in a narrow interval close to $det(\mathbf{I}-N\boldsymbol{\Psi}+\boldsymbol{\Psi})=0$, group properties become more important for an individual's phenotype than its own genes, with potential implications for the level of selection. Moore and colleagues observed that phenotypic trait values are undefined when $det(\mathbf{I}+\boldsymbol{\Psi})=0$ [20]. As they dealt only with dyadic interactions, the dependence of the second undefined point on the number of interacting individuals was not discussed. However, the dependence on group size may be very important for many aspects of behavioural and evolutionary biology, especially group size and group stability.

Evolutionary consequences

From an evolutionary point of view, if $\boldsymbol{\Psi}$ is close to $det(\mathbf{I}+\boldsymbol{\Psi})=0$, an individual's (genotypic) deviation from the mean genotype becomes much more important for its phenotype than its own genotype. Thus, if a trait is related to fitness, individual differences in fitness will be greater within groups than between groups (see Figure 7 A). In such cases, selection at the level of individuals may be very strong. However, if the strength of interaction is close to the second undefined point $det(\mathbf{I}-N\boldsymbol{\Psi}+\boldsymbol{\Psi})=0$, all phenotypes in the interacting group are sensitive to small changes such as the number of individuals or the mean genotype. All individuals in the group will have similar trait values with only small variation, while small intergroup differences in the mean genotype will be enhanced, and may cause large intergroup variation in phenotypic values. Again, if the trait is related to fitness, group properties such as mean genotype or the number of interacting individuals may lead to individuals within a group having higher fitness than individuals in a different group even if they have similar genotypes across groups (Figure 7 B). This may therefore create selection pressure at the group level, similar to the effect of social selection.

When interactions feed back on themselves, the dependence of IGEs on group size is no longer linear. Even lowering the number of interacting individuals while keeping the strength of interaction constant may lead to much stronger effects. If the trait in question (e.g. aggression) impacts on group stability [67,68], the consequences of feedback interactions may affect group stability.

Our study highlights the importance of IGEs for trait evolution and shows the mechanisms by which IGEs manifest. Scenarios where IGEs and group properties become more important for trait evolution than DGEs and the potential for the occurrence of extreme phenotypes further suggests that IGEs may need to be considered for their role in speciation but certainly when quantifying community interactions [69] and their evolutionary consequences.

Methods

We begin by generalizing Moore et al. 's model [20] for the case with N individuals

$$\mathbf{p}_i=\boldsymbol{\Gamma}\mathbf{g}_i+\sum_{j\neq i}^{N-1}\boldsymbol{\Psi}\mathbf{p}_j+\mathbf{e}_i \tag{18}$$

In our model, $\mathbf{g}_i$ (genotype) denotes a column vector of genes of an individual i, $\mathbf{p}_i$ (phenotype) is a column vector of corresponding traits (same as $\mathbf{Z}$ in [20]). Matrix $\boldsymbol{\Gamma}$ mediates the translation of individual's own genes $\mathbf{g}_i$ into its trait values. Matrix $\boldsymbol{\Psi}$ is an interaction matrix, where Ψ_{kl} defines the effect of the partner's trait l on the trait k of the focal individual. For simplification, we do not consider environmental effects other than those caused by social interactions.

We can rewrite equation 18 as

$$\mathbf{p}_i=\boldsymbol{\Gamma}\mathbf{g}_i+\sum_{j}^{N}\boldsymbol{\Psi}\mathbf{p}_j-\boldsymbol{\Psi}\mathbf{p}_i+\mathbf{e}_i \tag{19}$$

Summation over all individuals in a group yields

$$\sum_i^N\mathbf{p}_i=\sum_i^N\boldsymbol{\Gamma}\mathbf{g}_i+\sum_i^N\sum_j^N\boldsymbol{\Psi}\mathbf{p}_j-\sum_i^N\boldsymbol{\Psi}\mathbf{p}_i+\sum_i^N\mathbf{e}_i$$

$$\sum_i^N\mathbf{p}_i=\sum_i^N\boldsymbol{\Gamma}\mathbf{g}_i+\sum_i^N N\boldsymbol{\Psi}\mathbf{p}_i-\sum_i^N\boldsymbol{\Psi}\mathbf{p}_i+\sum_i^N\mathbf{e}_i$$

$$\sum_i^N\mathbf{p}_i-\sum_i^N N\boldsymbol{\Psi}\mathbf{p}_i+\sum_i^N\boldsymbol{\Psi}\mathbf{p}_i=\sum_i^N\boldsymbol{\Gamma}\mathbf{g}_i+\sum_i^N\mathbf{e}_i \tag{20}$$

Now, we can express the sum of all phenotypic values as

$$(\mathbf{I}-N\boldsymbol{\Psi}+\boldsymbol{\Psi})\sum_i^N\mathbf{p}_i=\sum_i^N(\boldsymbol{\Gamma}\mathbf{g}_i+\mathbf{e}_i)$$

$$\sum_i^N\mathbf{p}_i=(\mathbf{I}-N\boldsymbol{\Psi}+\boldsymbol{\Psi})^{-1}\sum_i^N(\boldsymbol{\Gamma}\mathbf{g}_i+\mathbf{e}_i) \tag{21}$$

A substitution of equation 21 into equation 19, we are able to express the phenotype of the focal individual as a function of genotypes of all group members

$$\mathbf{p}_i=\boldsymbol{\Gamma}\mathbf{g}_i+\boldsymbol{\Psi}(\mathbf{I}-N\boldsymbol{\Psi}+\boldsymbol{\Psi})^{-1}\sum_j^N(\boldsymbol{\Gamma}\mathbf{g}_j+\mathbf{e}_j)-\boldsymbol{\Psi}\mathbf{p}_i+\mathbf{e}_i$$

$$\mathbf{p}_i+\boldsymbol{\Psi}\mathbf{p}_i=\boldsymbol{\Gamma}\mathbf{g}_i+\boldsymbol{\Psi}(\mathbf{I}-N\boldsymbol{\Psi}+\boldsymbol{\Psi})^{-1}\sum_j^N(\boldsymbol{\Gamma}\mathbf{g}_j+\mathbf{e}_j)+\mathbf{e}_i$$

$$(\mathbf{I}+\mathbf{\Psi})\mathbf{p}_i=\mathbf{\Gamma g}_i+\mathbf{e}_i+\mathbf{\Psi}(\mathbf{I}-N\mathbf{\Psi}+\mathbf{\Psi})^{-1}\sum_j^N(\mathbf{\Gamma g}_j+\mathbf{e}_j)$$

$$\mathbf{p}_i=(\mathbf{I}+\mathbf{\Psi})^{-1}(\mathbf{\Gamma g}_i+\mathbf{e}_i)+(\mathbf{I}+\mathbf{\Psi})^{-1}\mathbf{\Psi}(\mathbf{I}-N\mathbf{\Psi}+\mathbf{\Psi})^{-1}\sum_j^N(\mathbf{\Gamma g}_j+\mathbf{e}_j) \quad (22)$$

To separate direct genetic effects (effect of individuals own genotype) from IGEs (effect of genotypes of other individuals), we have to subtract the effects of the genes of the focal individual from the second part of the equation 22.

$$\mathbf{p}_i=(\mathbf{I}+\mathbf{\Psi})^{-1}(\mathbf{\Gamma g}_i+\mathbf{e}_i)+(\mathbf{I}+\mathbf{\Psi})^{-1}\mathbf{\Psi}(\mathbf{I}-N\mathbf{\Psi}+\mathbf{\Psi})^{-1}\sum_{j\neq i}^{N-1}(\mathbf{\Gamma g}_j+\mathbf{e}_j)$$
$$+(\mathbf{I}+\mathbf{\Psi})^{-1}\mathbf{\Psi}(\mathbf{I}-N\mathbf{\Psi}+\mathbf{\Psi})^{-1}(\mathbf{\Gamma g}_i+\mathbf{e}_i)$$

$$\mathbf{p}_i=\underbrace{(\mathbf{I}+\mathbf{\Psi})^{-1}(\mathbf{I}+\mathbf{\Psi}(\mathbf{I}-N\mathbf{\Psi}+\mathbf{\Psi})^{-1})\mathbf{\Gamma g}_i}_{DGE}$$
$$\underbrace{+(\mathbf{I}+\mathbf{\Psi})^{-1}\mathbf{\Psi}(\mathbf{I}-N\mathbf{\Psi}+\mathbf{\Psi})^{-1}\sum_{j\neq i}^{N-1}\mathbf{\Gamma g}_j}_{IGE} \quad (23)$$
$$+(\mathbf{I}+\mathbf{\Psi})^{-1}(\mathbf{I}+\mathbf{\Psi}(\mathbf{I}-N\mathbf{\Psi}+\mathbf{\Psi})^{-1})\mathbf{e}_i$$
$$+(\mathbf{I}+\mathbf{\Psi})^{-1}\mathbf{\Psi}(\mathbf{I}-N\mathbf{\Psi}+\mathbf{\Psi})^{-1}\sum_{j\neq i}^{N-1}\mathbf{e}_j$$

Now, the first term of equation 23 depends only on the focal individual's genotype (i.e. DGE) while the second term depends only on the social partners' genotypes (i.e. IGE).

Simulations

We analysed the role of interaction strength and number of individuals for both direct and indirect genetic effects in our model. Unlike previous frameworks, our model was developed with agent based modelling in mind. Equations were derived for individuals, therefore can be directly used in agent based models for calculations of an individual's phenotype, when genotypes are known.

Table 1. Populating of matrix $\mathbf{\Psi}$ for given interactions.

Figure of occurrence	Simulated interaction	Matrix $\mathbf{\Psi}$
Figure 2	As in Figure 1 A	$\begin{pmatrix}0&0&0\\ \Psi_{21}&0&0\\ 0&0&0\end{pmatrix}$
Figure 3	As in Figure 1 B	$\begin{pmatrix}0&0&0\\ \Psi_{21}&0&0\\ 0&\Psi_{32}&0\end{pmatrix}$
Figure 4, 5, 6, 7	As in Figure 1 C	$\begin{pmatrix}\Psi_{11}&0&0\\ 0&0&0\\ 0&0&0\end{pmatrix}$

To illustrate the interaction strength effect on intragroup and intergroup phenotypic variance, we simulated M groups of N individuals (Figure 2 C, D; Figures 5, 6 and 7). All simulations in this study were carried out in Matlab R2010a.

We assume that each individual is haploid and has three genes and three traits. Groups were created by assigning each gene for each individual a random value sampled from a standard normal distribution. Then, the mean genotype of the whole population was calculated and subtracted from the genotype of each individual, thus the population mean genetic value of each gene was set to 0.

Figures 2 A, 3 and 4 were created using equation 2.

Matrix $\mathbf{\Psi}$ (3×3) was populated to describe a given interaction (see Table 1). We calculated phenotypes for all individuals for a given $\mathbf{\Psi}$, as well as mean phenotype of each group and phenotypic variance.

Acknowledgments

We would like to thank Richard Preziosi, Peter Glaus, Rasmus Petersen and Beatrice Gini for helpful comments on a previous draft of this paper.

Author Contributions

Conceived and designed the experiments: BT RH. Performed the experiments: BT RH. Analyzed the data: BT RH. Wrote the paper: BT RH.

References

1. Trivers RL (1974) Parent-offspring conict. Am Zool 14: 249–264.
2. Moore AJ, Haynes KF, Preziosi RF, Moore PJ (2002) The evolution of interacting phenotypes: genetics and evolution of social dominance. Am Nat 160 Suppl: S186–S197.
3. Davies NB, Krebs JR, West SA (2012) An Introduction to Behavioural Ecology. Chichester: Wiley-Blackwell, 4th edition, 416 pp.
4. Wolf JB, Brodie EDI, Cheverud JM, Moore AJ (1998) Evolutionary consequences of indirect genetic effects. Trends Ecol Evol 13: 64–69.
5. Cheverud JM (2003) Evolution in a genetically heritable social environment. PNAS 100: 4357–4359.
6. Kölliker M, Brodie ED, Moore AJ (2005) The coadaptation of parental supply and offspring demand. Am Nat 166: 506–516.
7. Hamilton WD (1964) The genetical evolution of social behaviour. I. J Theor Biol 7: 1–16.
8. Nowak MA (2006) Five rules for the evolution of cooperation. Science 314: 1560–1563.
9. West SA, Griffin AS, Gardner A (2007) Evolutionary explanations for cooperation. Curr Biol 17: R661–R672.
10. Clutton-Brock T (2009) Cooperation between non-kin in animal societies. Nature 462: 51–57.
11. Wilson AJ, Morrissey MB, Adams MJ, Walling Ca, Guinness FE, et al. (2011) Indirect genetics effects and evolutionary constraint: an analysis of social dominance in red deer, Cervus elaphus. J Evol Biol 24: 772–783.
12. Bijma P, Muir WM, Ellen ED, Wolf JB, Van Arendonk JaM (2007) Multilevel selection 2: Estimating the genetic parameters determining inheritance and response to selection. Genetics 175: 289–99.
13. Bijma P, Muir WM, Van Arendonk JAM (2007) Multilevel selection 1: Quantitative genetics of inheritance and response to selection. Genetics 175: 277–288.
14. Wilson DS, Wilson EO (2007) Rethinking the theoretical foundation of sociobiology. The Quarterly Review of Biology 82: 327–348.
15. Gardner A, West SA (2004) Spite and the scale of competition. J Evol Biol 17: 1195–1203.
16. West SA, Griffin AS, Gardner A (2007) Social semantics: altruism, cooperation, mutualism, strong reciprocity and group selection. J Evol Biol 20: 415–432.
17. Wenseleers T, Gardner A, Foster KR (2009) Social evolution theory: a review of methods and approaches. In: Székely T, Moore AJ, Komdeur J, editors, Social Behaviour: Genes, Ecology and Evolution, Cambringe: Cambridge University Press. pp. 132–158.

18. Wilson AJ, Gelin U, Perron MC, Réale D (2009) Indirect genetic effects and the evolution of aggression in a vertebrate system. Proc R Soc Lond [Biol] 276: 533–541.
19. Wolf JB (2003) Genetic architecture and evolutionary constraint when the environment contains genes. PNAS 100: 4655–4660.
20. Moore AJ, Brodie EDI, Wolf JB (1997) Interacting phenotypes and the evolutionary process: I. Direct and indirect genetic effects of social interactions. Evolution 51: 1352–1362.
21. Wolf JB (2000) Indirect genetic effects and gene interactions. In: Wolf JB, Brodie EDI, Wade MJ, editors, Epistasis and the evolutionary process, Oxford: Oxford University Press. pp. 158–176.
22. Agrawal AF, Brodie EDI, Wade MJ (2001) On indirect genetic effects in structured populations. Am Nat 158: 308–323.
23. McGlothlin JW, Brodie EDI (2009) How to measure indirect genetic effects: the congruence of trait-based and variance-partitioning approaches. Evolution 63: 1785–1795.
24. Bijma P (2010) Multilevel selection 4: modeling the relationship of indirect genetic effects and group size. Genetics 186: 1029–1031.
25. Teplitsky C, Mills JA, Yarrall JW, Merilä J (2010) Indirect genetic effects in a sex-limited trait: the case of breeding time in red-billed gulls. J Evol Biol 23: 935–944.
26. Muir WM (2005) Incorporation of competitive effects in forest tree or animal breeding programs. Genetics 170: 1247–1259.
27. Hadfield JD, Wilson AJ (2007) Multilevel selection 3: modeling the effects of interacting individuals as a function of group size. Genetics 177: 667–668.
28. Bijma P (2010) Fisher's fundamental theorem of inclusive fitness and the change in fitness due to natural selection when conspecifics interact. J Evol Biol 23: 194–206.
29. Cheverud JM, Moore AJ (1994) Quantitative genetics and the environment provided by relatives. Chicago: University of Chicago Press. pp. 67–100.
30. Wolf JB, Wade MJ (2009) What are maternal effects (and what are they not)? Proc R Soc Lond [Biol] 364: 1107–1115.
31. Cheverud JM (1984) Evolution by kin selection: a quantitative genetic model illustrated by maternal performance in mice. Evolution 38: 766–777.
32. Mousseau TA, Fox CW (1998) The adaptive significance of maternal effects. Trends Ecol Evol 13: 403–407.
33. Wilkinson GS (1984) Reciprocal food-sharing in the vampire bat. Nature 308: 181–184.
34. Mutic JJ, Wolf JB (2007) Indirect genetic effects from ecological interactions in Arabidopsis thaliana. Mol Ecol 16: 2371–2381.
35. Crespi BJ (2001) The evolution of social behavior in microorganisms. Trends Ecol Evol 16: 178–183.
36. McGlothlin JW, Moore AJ, Wolf JB, Brodie ED (2010) Interacting phenotypes and the evolutionary process. III. Social evolution. Evolution 64: 2558–2574.
37. Buttery NJ, Thompson CRL,Wolf JB (2010) Complex genotype interactions inuence social fitness during the developmental phase of the social amoeba *Dictyostelium discoideum*. J Evol Biol 23: 1664–71.
38. Xavier JaB, Kim W, Foster KR (2010) A molecular mechanism that stabilizes cooperative secretions in *Pseudomonas aeruginosa*. Mol Microbiol 79: 166–179.
39. Keller L (2009) Adaptation and the genetics of social behaviour. Proc R Soc Lond [Biol] 364: 3209–3016.
40. Kent C, Azanchi R, Smith B, Formosa A, Levine JD (2008) Social context inuences chemical communication in *D. melanogaster* males. Curr Biol 18: 1384–1389.
41. Boake CRB, Arnold SJ, Breden F, Meffert LM, G M, et al. (2002) Genetic tools for studying the evolution of behaviour. Am Nat 160: s143–s159.
42. Peripato AC, De Brito RA, Vaughn TT, Pletscher LS, Matioli SR, et al. (2002) Quantitative trait loci for maternal performance for offspring survival in mice. Genetics 162: 1341–1353.
43. Peripato AC, Cheverud JM (2002) Genetic inuences on maternal care. Am Nat 160 Suppl: S173–S185.
44. Hager R, Cheverud JM, Wolf JB (2008) Maternal effects as the cause of parent-of-origin effects that mimic genomic imprinting. Genetics 178: 1755–1762.
45. Griffing B (1967) Selection in reference to biological groups. I. Individual and group selection applied to populations of unordered groups. Aus J Biol Sci 20: 127–139.
46. Griffing B (1969) Selection in reference to biological groups. IV. Application of selection index theory. Aus J Biol Sci 22: 131–142.
47. Griffing B (1976) Selection in Reference to Biological Groups. V. Analysis of Full-Sib Groups. Genetics 82: 703–722.
48. Griffing B (1981) A theory of natural selection incorporating interaction among individuals. II. Use of related groups. J Theor Biol 89: 659–677.
49. Bijma P, Wade MJ (2008) The joint effects of kin, multilevel selection and indirect genetic effects on response to genetic selection. J Evol Biol 21: 1175–1188.
50. Bijma P (2010) Estimating indirect genetic effects: precision of estimates and optimum designs. Genetics 186: 1013–1028.
51. Bijma P, Aanen DK (2010) Assortment, Hamilton's rule and multilevel selection. Proc R Soc Lond [Biol] 277: 673–5; discussion 677–8.
52. Nowak MA, Sigmund K (1992) Tit for tat in heterogeneous populations. Nature 355: 250–253.
53. Nowak MA, Sigmund K (1998) Evolution of indirect reciprocity by image scoring. Nature 393: 573–577.
54. Axelrod R (2000) On six advances in cooperation theory. In: Analyse & Kritik - Special Edition on the Evolution of Cooperation. volume 22, pp. 130–151.
55. Nadell CD, Foster KR, Xavier JaB (2010) Emergence of spatial structure in cell groups and the evolution of cooperation. PLoS Comput Biol 6: e1000716.
56. Sokolowski MB, Levine J (2010) Nature-nuture interactions. In: Székely, T and Komdeur, J and Moore, AJ, editor, Social Behaviour: Genes, Ecology and Evolution, Cambridge: Cambridge University Press. pp. 11–25.
57. Brenner F, Gaetano C, Mauser S, Belowich D (1978) Body weight and social interactions as factors in determining dominance in captive eastern chipmunks. Anim Behav 26: 432–437.
58. Thornhill R (1984) Fighting and assessment in harpobittacus scorpionies. Evolution 38: 204–214.
59. Huntingford F, Turner A (1987) Animal Conict. London: Chapman and Hall, 448 pp.
60. Hurst JL, Beynon RJ (2004) Scent wars: the chemobiology of competitive signalling in mice. BioEssays 26: 1288–1298.
61. Smuts BB (1987) Sexual competition and mate choice. In: Smuts BB, Cheney DL, Seyfarth RM, Wrangham RW, Struhsaker TT, editors, Primate Societies, Chicago: University of Chicago Press, 31. pp. 385–399.
62. Hamilton WD (1964) The genetical evolution of social behaviour. II. J Theor Biol 7: 17–52.
63. Hager R, Johnstone R (2005) Differential growth of own and alien pups in mixed litters of mice: A role for genomic imprinting? Ethol 111: 705–714.
64. Odling-Smee FJ, Laland KN, Feldman MW (2003) Niche construction: the neglected process in evolution, volume 37 of *Monographs in Population Biology*. Princeton, New Jersey: Princeton University Press, 472 pp.
65. Laland KN, Sterelny K, Odling-Smee J, Hoppitt W, Uller T (2011) Cause and Effect in Biology Revisited: Is Mayr's Proximate-Ultimate Dichotomy Still Useful? Science 334: 1512–1516.
66. Dawkins R (1976) The Selfish Gene. Oxford: Oxford University Press, Oxford, 360 pp.
67. Reeve HK (2000) A Transactional Theory of Within-Group Conict. Am Nat 155: 365–382.
68. Tibbetts EA, Reeve HK (2008) Two experimental tests of the relationship between group stability and aggressive conict in Polistes wasps. Die Naturwissenschaften 95: 383–389.
69. Rowntree JK, Shuker DM, Preziosi RF (2011) Forward from the crossroads of ecology and evolution. Proc R Soc Lond [Biol] 366: 1322–1328.

3

Density-Dependent Compensatory Growth in Brown Trout (*Salmo trutta*) in Nature

L. Fredrik Sundström*, Rasmus Kaspersson, Joacim Näslund, Jörgen I. Johnsson

University of Gothenburg, Department of Biological and Environmental Sciences, Göteborg, Sweden

Abstract

Density-dependence is a major ecological mechanism that is known to limit individual growth. To examine if compensatory growth (unusually rapid growth following a period of imposed slow growth) in nature is density-dependent, one-year-old brown trout (*Salmo trutta* L.) were first starved in the laboratory, and then released back into their natural stream, either at natural or at experimentally increased population density. The experimental trout were captured three times over a one-year period. We found no differences in growth, within the first month after release (May-June), between the starved fish and the control group (i.e. no evidence of compensation). During the summer however (July-September), the starved fish grew more than the control group (i.e. compensation), and the starved fish released into the stream at a higher density, grew less than those released at a natural density, both in terms of weight and length (i.e. density-dependent compensation). Over the winter (October-April), there were no effects of either starvation or density on weight and length growth. After the winter, starved fish released at either density had caught up with control fish in body size, but recapture rates (proxy for survival) did not indicate any costs of compensation. Our results suggest that compensatory growth in nature can be density-dependent. Thus, this is the first study to demonstrate the presence of ecological restrictions on the compensatory growth response in free-ranging animals.

Editor: Carlos Garcia de Leaniz, Swansea University, United Kingdom

Funding: LFS was funded by the Swedish Research Council Vetenskapsrådet and as a Marie Curie Outgoing International Fellow under contract MOIF-CT-2005-8141 from the European Community's Sixth Framework Programme. The present work does not necessarily reflect the Community's views and in no way anticipates its future policy in this area. JIJ was funded by the Swedish Research Council for Agriculture Research and Spatial Planning. The funders had no role in study design, data collection and analysis, decision to publish, or preparation of the manuscript.

Competing Interests: The authors have declared that no competing interests exist.

* E-mail: fred.sundstrom@gmail.com

Introduction

Individual growth rates are determined by trade-offs between benefits and immediate and/or delayed costs affecting fitness through effects on behaviour, physiology, morphology, reproduction, and longevity [1–3]. Ecological factors that affect growth rate include food availability and quality, predation risk, competition from conspecifics, photoperiod, and temperature [4–6]. Several of these factors act in a density-dependent way so that, for example, when the number of conspecifics increases, individual growth rate decreases [7–10]. At the same time, many organisms can buffer against environmental variation by adjusting behavior and/or physiology to stabilize their growth trajectory [2,11,12].

One particular response that has been observed in many animal taxa is compensatory growth, which is a period of unusually rapid growth following a period of depressed growth [2,3,13–15]. The increased growth during the compensatory period is commonly achieved through hyperphagia, i.e. an increase in food intake, but it can also arise from a reduced metabolic rate and/or improved conversion efficiency [13,14,16]. Observations of compensatory growth under natural conditions [17–19], suggests that some of the factors apparently restricting normal growth can be circumvented during compensation. However, the extent to which compensatory growth is regulated by the same ecological factors as normal growth is not known. For example, in the laboratory, roach (*Rutilus rutilus*) do not normally grow at temperatures below 12°C, but will do so if the growth is compensatory [20] showing that the behavioral decision to feed and physiological regulation of growth may be different for normal growth compared to compensatory growth.

In the present study we investigated whether compensatory growth is affected by population density, an important ecological factor that has been shown experimentally to reduce growth under normal growth conditions (i.e. non-compensatory) in nature [7]. Hence, we combine the observation that animals are able to compensate at natural density (i.e. the animal can "at will" grow faster than normal) with the observation that growth is density-dependent (i.e. when individuals compete for space and/or food). Because the faster compensatory growth can be observed under normal growth conditions when we know that growth is density-dependent, compensatory growth may be regulated by mechanisms that are different from those acting during normal growth.

To examine whether density-dependence operates on compensatory growth, we carried out a study on a natural population of one-year-old brown trout. First, we deprived trout of food in the laboratory and then returned them to a stream subsection with a natural density (i.e. density at the time of release) or experimentally increased density of conspecifics and monitored their growth, survival, and movement over a year.

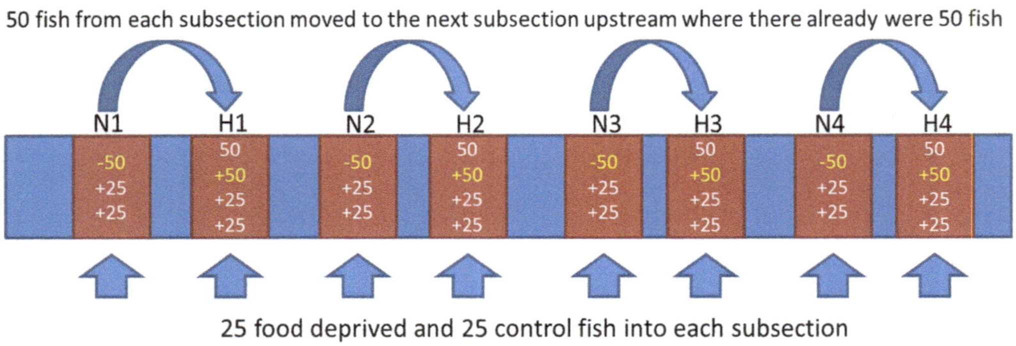

Figure 1. Schematic overview of the experimental stream showing buffer zones (blue) and experimental subsections (red), and the additions of fish into each subsection. Length of each experimental subsection was defined as containing 50 fish. In the natural density subsections (N), these fish (marked in yellow) were transferred to upstream sections (+50), and replaced with 25 treatment and 25 control fish from the laboratory, thus restoring the natural density of 50 individuals. High density subsections (H) had their initial 50 individuals, received 50 from the natural density subsections, and 25 treatment and 25 control fish from the laboratory, resulting in three times higher number than normal.

Methods

Study site and species

The study was conducted in a 700 m experimental section of the small coastal stream Norumsån in south-western Sweden from the spring of 2008 until the spring of 2009. The stream has a width of <1–5 meters and a depth ranging from a few centimetres in shallow habitats at low flow rates to more than a meter in deeper pools. It runs through deciduous forest with much vegetative overhang. The dominant fish species is sea-migratory brown trout (>95% of all individual fish), which spend about two years in the stream before migrating to the sea. Larger trout typically migrate sooner than smaller and size appears to be the main determinant of migration [21,22]. After 1–2 years in the sea, adults return to their natal stream to breed. There is also a varying, but small, proportion of larger resident (non-migratory) trout that mature in the stream [23] and they typically occupy the deeper pools. Predators on juvenile trout are heron (*Ardea cinerea*), eel (*Anguilla anguilla*), larger resident brown trout, and mink (*Neovison vison*). We studied one-year-old (parr) brown trout of which most were expected to migrate after the end of the experiment in the late spring of 2009. On all occasions when fish were captured from the stream we used electro fishing from which the fish usually recover within a minute. This study was performed in accordance with Swedish animal welfare laws and was approved by the Gothenburg Ethical Committee for Animal research (113–2008). Electro-fishing permits were obtained from the County Administrative Board Västra Götaland and all field work was approved by the private land owners.

Laboratory treatment

On 30-Apr-2008, a random sample of 200 brown trout were captured from the lower end of the stream and brought into the laboratory at the Department of Zoology in Gothenburg to become the starved treatment group. The following day, these fish were marked (PIT-tags), weighed and measured, and divided in equal numbers into four 100 liter tanks containing re-circulating freshwater at 10°C, coarse gravel, 5–10 rocks, and plastic plants. These fish weighed 6.9 g±0.2 SE and measured 86.2 mm±0.6, giving a condition factor (calculated as $CF = 10^5 \times W \times L^{-3}$) of 1.05±0.01. Since fish in the starved treatment were presumably a random sample of the population when initially captured, we assumed that fish that remained in the stream, of which some were later to constitute the control treatment, were of the same average weight and length. Fish of the starvation treatment were food deprived for three weeks, because this has previously been shown to induce subsequent compensatory growth without compromising the health of the fish [18]. All fish were in good health upon release and there were no mortalities during this starvation period.

After a period of three weeks, on 18-May-2008, we captured 200 trout (to be control treatment) from the experimental section of the stream and brought them into the laboratory. The following day, these control fish were PIT-tagged, weighed and measured. In addition, starved fish were again weighed and measured. As expected, control fish that had grown in the stream were at this time point significantly larger (8.4 g±0.2 and 92.2 mm±0.7; Student's t-test: $p<0.001$ in both cases) than the starved fish (6.2 g±0.1 SE and 86.5 mm±0.7). Starved fish also had lower condition factor (0.92±0.04 SE) than control fish (1.04±0.04; $p<0.001$).

On 21-May-2008, the 200 control and 200 starved fish in the laboratory were split into eight containers, with 25 control and 25 treatment fish in each, and transported to the stream where fish were released into the center of each of eight subsections (Fig. 1) in the experimental stream.

Stream experiment

On 20-May-2008, one day before returning the fish to the stream, we divided the stream section from which experimental (starved and control) fish had been obtained previously, into eight subsections separated by buffer zones (Fig. 1). The length of each subsection was determined by electro-fishing upstream until 50 individuals had been captured (range of subsections 25–55 m). Lengths of buffer zones (33–51 m) were determined by stream conditions so that they were expected to house about 50 individual trout each. Hence, each subsection contained at least 50 fish before the experiment started. From every other subsection, 50 fish were removed and replaced with the 50 experimental fish (25 starved and 25 control) thereby maintaining a natural density. These four subsections are referred to as natural density subsections (N1 to N4). The high density subsections (H1 to H4) were created by adding the 50 fish removed from sections N1-N4 and the experimental fish (25 starved and 25 control) thereby tripling the number of fish (50 existing, 50 added from downstream subsection, 25 starved, and 25 control fish added). This transfer of fish resulted in an average fish biomass of 1344 g±27 SD and density of 3.36±0.48 individuals per meter in H subsections and 355 g±32 SD and 1.4±0.50 individuals per m in N subsections. These numbers are approximate since not all fish were necessarily captured during this setup phase; fish would be able to emigrate out and into the subsections as well as to and from buffer sections where an unknown number of fish were present, and fish would grow over the experimental year possibly influencing functional density. In fact, during the study, densities in the subsections changed from an initial 2.4× in H subsections relative to N subsections upon release, to 1.25×, 1.15×, and 1.41× for the three recapture points (see below), respectively (including all fish captured in the stream).

Recaptures were made by electro-fishing the entire stream section, including buffer zones and experimental subsections, after 26 days (17/18-Jun-2008; first recapture after a period when fish in the system can grow rapidly [7] and when the potential for compensation would be great), 122 days (29–30 Sep 2008; second recapture after a longer period of slower growth and before adult spawners enter the stream), and 309 days (30-Mar-2008/01-Apr-2009; third recapture following the winter before most individuals initiate their downstream migration to the sea). Since this study was conducted under fully natural conditions, fish could freely move in the stream section and hence between subsections. We did not try to physically prevent movements with nets, because based on previous experience, nets are quickly flushed downstream. A previous experiment in this stream showed that trout normally move very little even at elevated densities [7].

Statistical analysis

First we were interested in detecting whether fish compensated or not, then whether fish at the natural density were able to compensate more than those at the high density. The presence of compensatory growth in the starvation treatment relative to control treatment was assessed by analyzing the natural logarithm (ln) of absolute growth in weight and length (i.e. weight/length at a later recapture minus the weight/length at the preceding recapture, e.g. size at first recapture minus size at release), and changes in condition factor between recaptures using Restricted Maximum Likelihood (REML) in a Linear Mixed Model:

$$RV = D + T + D \times T + CV + S(D)$$

where RV = response variable (i.e. ln-transformed absolute weight change, ln-transformed absolute length change, change in condition factor), T = treatment (starved vs. control), D = density (high vs. natural), CV = covariate (weight at the beginning of a period for weight analyses, and length at the beginning of a period for analyses of length and condition factor), and S(D) = Density nested in random factor release subsection (H1–H4, N1–N4).

To detect a compensatory response, a starved fish would need to increase more in absolute size (weight and length) compared to a control fish of the same initial size after starvation treatment. When the random factor S(D) had p>0.25, data were pooled across subsections to increase power [24]. A high p-value for the random factor would imply that the release subsections do not significantly explain any differences in growth. For the last recapture, too few fish were recaptured from some subsections to include the random factor so this period was analysed on the pooled data only.

If a compensatory growth response was observed (i.e. significant treatment effect), a detailed analysis of the effect of density on compensatory growth was carried out on starved fish only. Given that we were interested only in the density effects on the compensatory response, we used only the starved fish for this analysis. It should be noted that density effects on natural growth has been documented previously [7]. The analysis of density on only starved fish therefore used the same model and selection procedure as described above, but excluded all treatment terms.

Mortality costs of compensation (i.e. recapture rate used as a proxy for survival), were assessed for each recapture time point using the pooled model (random factor removed) as above fitted to a binomial distribution with a logit-link function. To adjust for differences in "survival" prior to the onset of a period, the model-predicted recapture rate at the onset of a period was transformed and used as an offset variable in the analysis of the subsequent period. Hence, recapture rate in September and April was adjusted for the expected number of fish present in June and September, respectively, based on the recapture rate at that time. Because length at the start of the period was used as a covariate, only fish recaptured, which thus had a known length, could be included for assessment of survival to the September and April recapture time points.

Movements were variable and could not be made homogenous nor be made normally distributed. Instead, we grouped data into movements ≤20 m and >20 m as these would roughly represent fish that remained in their release subsection and fish that moved out of their release subsection, respectively. Movement data were then analysed using the same model as for recapture rate above.

By fishing the stream section twice on each recapture occasion we can estimate the probability of capturing a fish that is present [25]. First the population size in the section is estimated as $n = c_1^2/(c_1 - c_2)$ where n is the estimated number of fish in the section, c_1 is the number of fish captured during the first fishing and c_2 is the number of fish captured in the second fishing. Next, the proportion captured (c_1+c_2) is divided with the estimated population size n to give the probability of capturing a fish.For analyses of growth in weight and length, change in condition, and movements, only fish recaptured at subsequent electro-fishing occasions could be used. This resulted in 204, 100, and 47 fish being available for analyses for periods 1, 2, and 3, respectively.

Below, we focus on effects relevant for our hypotheses only, i.e. 1) did fish compensate, and if so 2) did this compensation differ depending on population density? Only the final statistics are provided. Hence, if the random factor subsection had P>0.25, only output from the pooled data is provided. Due to low numbers of fish recaptured in some stream subsections during the last period (Oct-Apr), all analyses on this last period focused on pooled data only. Complete statistical output for interactions, covariates, random factors, and model selection procedures are provided in the online supplementary material (Appendix S1).

Results

Of the 400 fish released in May, 204 were recaptured in June, 141 in September, and 83 the following April. The estimated probability for capturing an experimental fish during the three recaptures was 94.1%, 87.4%, and 99.4%, respectively. These high recapture rates suggest that fish not recaptured where no longer in the stream section that was sampled but had either migrated or died. Of the tagged fish, 85 control (42.5%) and 47 starved (23.5%) fish were never recaptured.

Weight growth

There was no significant difference in weight growth between starved and control fish during the first period (May-June; random factor retained with $P = 0.13$, treatment: $F_{1,\ 15.4} = 0.94$, $P = 0.35$). However, a clear compensatory growth response is observed for the second period (Jul-Sep) when starved fish grew more in weight than did control fish (random factor removed with $P = 0.35$, $F_{1,95} = 11.8$, $P < 0.001$). Analysis of density effects, by including the starved fish only, showed that those fish at a natural density gained on average 17% more in weight than those fish at a higher density ($F_{1,63}$ =4.2, P=0.045), indicating the presence of density-dependent compensatory growth during this period (Fig. 2a). There were no significant differences in weight growth between starved and control fish during the last period (Oct-Apr; $F_{1,42}$ =0.01, P=0.91) which was clearly associated with an increased variation in growth among individuals (Fig. 2a). Final weight of the four groups were similar (overall average 20.5 g±5.9 SD; one-way ANOVA $F_{3,79}$ =0.2, P=0.91) suggesting that a complete catch-up in weight had been achieved by the starved fish at both the high and natural density.

Length growth

There was no significant difference in length growth between starved and control fish during the first period (random factor retained with P=0.08, treatment: $F_{1,\ 13.6}$ =2.0, P=0.18). For the second period (random factor removed with P=0.36), starved fish grew on average more in length than control ($F_{1,95}$ =4.2, P=0.043), indicating the presence of compensatory length growth. When looking at the starved fish only, those at a natural density grew on average 15% more in length than those at a high density ($F_{1,63}$ =3.9, P=0.054), indicating the presence of density-dependent compensatory growth in length similar to that found for weight (Fig. 2b). There were no significant effects of treatment on length growth during the last period ($F_{1,42}$ =0.1, P=0.33). Final length of the four groups were similar (average 125.6 mm±11.9 SD; $F_{3,79}$ =0.2, P=0.87), suggesting that starved fish were eventually able to catch up in length.

Condition factor

For the first period (random factor removed with P=0.35), starved fish increased more in condition than did control fish ($F_{1,199}$ =56.2, P<0.001), but density had no effect on condition in the starved fish ($F_{1,\ 127}$ =0.7, P=0.42; Fig. 3). The effect on condition appears to arise from a relatively faster weight increase and slower length growth in starved fish compared to the control group (Fig. 2a and b). During the second period, starved fish lost less in condition than the fish in the control group (random factor removed with P=0.26; $F_{1,95}$ =7.7, P=0.007) probably because they started this period with a lower condition. Detailed analysis on starved fish revealed no effects of density on condition factor during the second period ($F_{1,63}$ =0.16, P=0.69). There were no significant effects of starvation on change in condition during the last period ($F_{1,42}$ =2.27, P=0.14). Final condition factor was the same among the groups (average 1.01 g±0.08 SD; $F_{3,79}$ =0.3, P=0.83) matching the findings of no differences in weight and length.

Recapture rate

At the first recapture, more starved fish were recaptured than fish from the control group in the experimental section (χ^2 =24.7, P<0.001), and analysis on starved fish only did not show any effects of density on recapture rate (χ^2=0.4, P=0.55; Fig. 4). This difference in recapture rate is likely due to differences in movement rather than survival (see below). At the second recapture, after adjusting for recapture rate during the first period, there were no differences among the fish due to treatment (χ^2 =1.4, P=0.24). Recapture rates after the winter did not differ

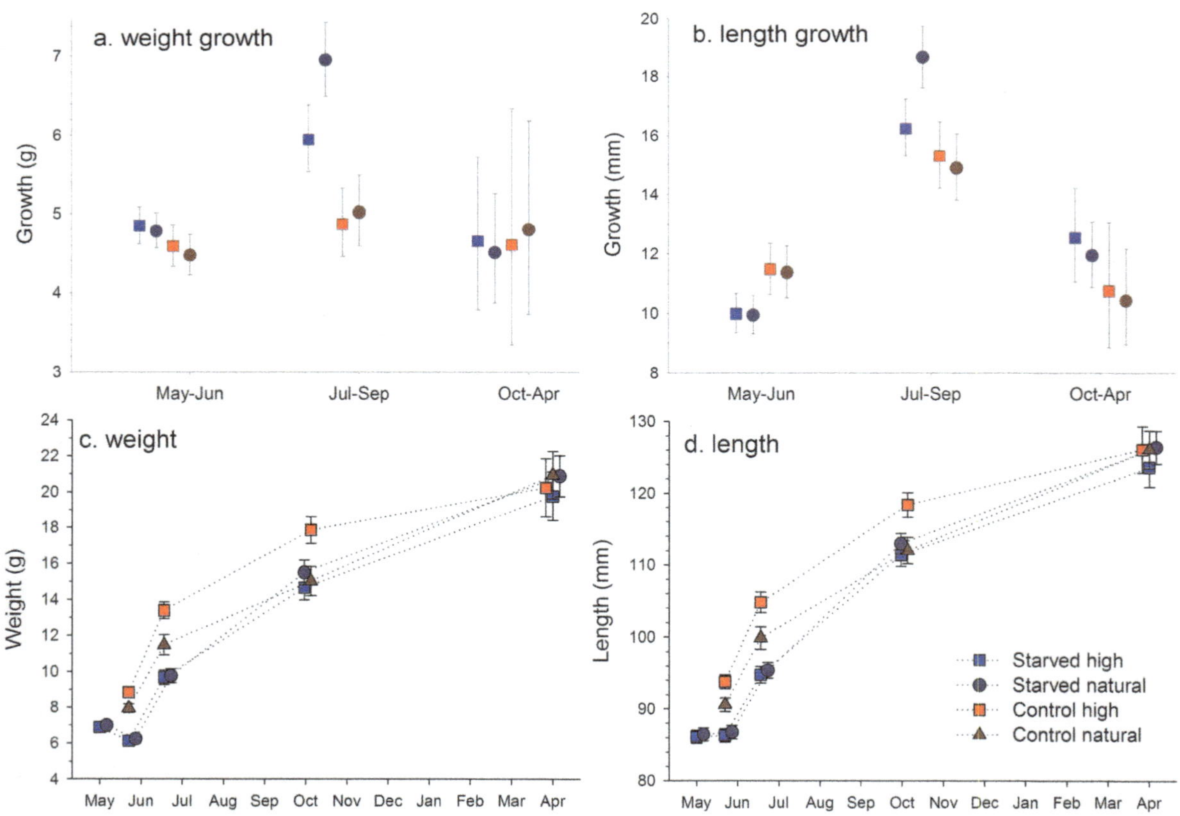

Figure 2. Absolute growth in weight (a) and length (b), and growth trajectories in weight (c) and length (d) in brown trout (*Salmo trutta*) during each of the three periods. Fish were starved (starved) or not (control) in the laboratory followed by release to nature at a natural (natural) or experimentally increased (high) density. Values in (a) and (b) are based on back-transformed estimated marginal means ± SE.

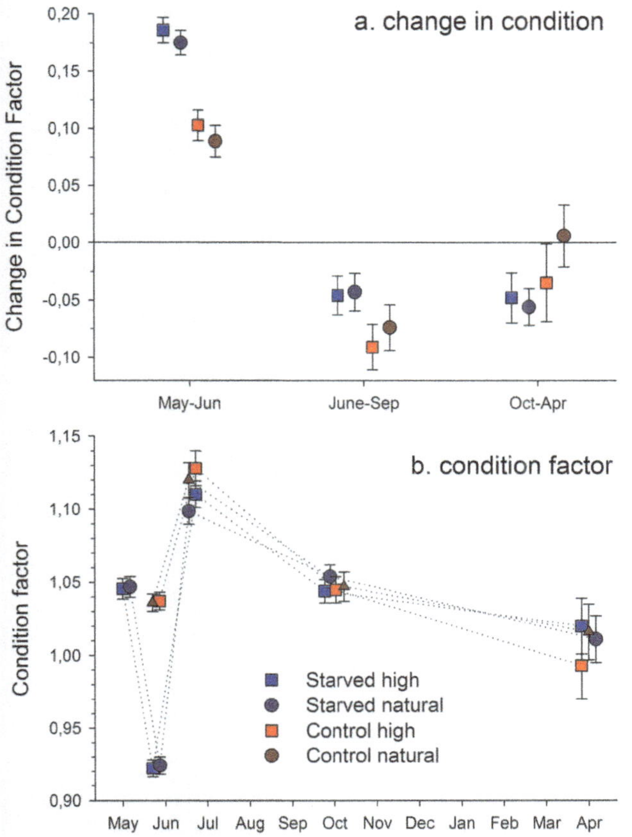

Figure 3. Change in condition factor (a) and condition factor (b) for brown trout (*Salmo trutta*) during each of the three periods. Fish were starved (starved) or not (control) in the laboratory followed by release to nature at a natural (natural) or experimentally increased (high) density. Values in (a) are based on back-transformed estimated marginal means ± SE.

between starved and control fish (χ^2 =2.2, P=0.14) but was higher in fish released into natural density subsection compared to high density subsections (χ^2 =4.0, P=0.046). This finding indicates that long-term costs may be higher due to increased density effects relative to effects of starvation and/or compensatory growth.

Movements

The number of fish in the control group that moved more than 20 m, were higher than the number of starved fish that moved more than 20 m during the first period (χ^2 =9.4, P=0.002; Fig. 5), which may help explain the lower recapture rates in fish from the control group after this period of only one month (Fig. 4). Among the starved fish, individuals at the high density were more likely to move compared to those at the natural density (χ^2 =6.7, P=0.009). During the second period, there was no effect of the starvation treatment (χ^2 =0.73, P=0.39), but overall more fish moved at the high density compared to the low density (χ^2 =6.6, P=0.010). During the last period, all fish from the control group at the high density had moved more than 20 m so this data could not be analyzed using the full model. Analysis of the starved fish did not indicate any effect of density on movement (χ^2 =0.31, P=0.58).

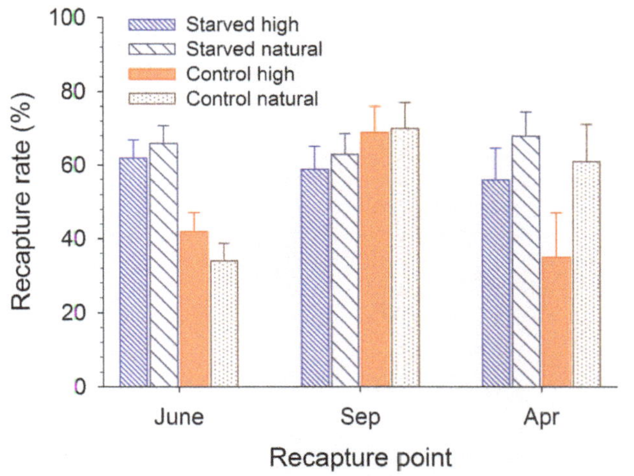

Figure 4. Recapture rate of brown trout (*Salmo trutta*) at each electro-fishing occasion adjusted for assumed survival based on recapture rate during the preceding period. Fish were starved (starved) or not (control) in the laboratory followed by release to nature at a natural (natural) or experimentally increased (high) density. Based on estimated marginal means ± SE.

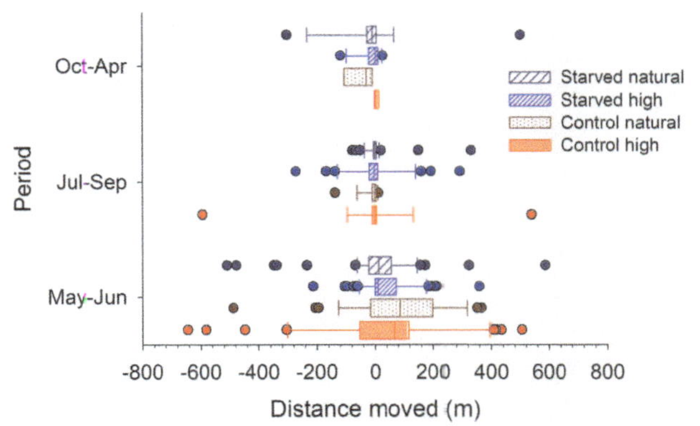

Figure 5. Movements of brown trout (*Salmo trutta*) during each of the three periods. Box-and-whisker plot showing median (vertical line inside box), 25 and 75 percentiles (edge of box), 10 and 90 percentiles (whiskers) and individual data points beyond (filled circle). Fish were starved (starved) or not (control) in the laboratory followed by release to nature at a natural (natural) or experimentally increased (high) density.

Discussion

We found that young brown trout, deprived of food for three weeks in the spring, showed compensatory growth in terms of weight and length over the summer, with a more pronounced response at a natural population density than at an experimentally increased density. This suggests that compensatory growth, just as normal growth in this species [7], is density-dependent. Previous studies, including those in our system, have found that fish first restore weight, when food is typically abundant and potential for growth is high [7,8,18,19]. Indeed, even though we found no effect on weight or length growth during spring we did see a significant increase in condition factor in starved fish relative to fish in the control group, suggesting that starved fish first restored energy reserves (weight) before growing in structure (length), which seems to be the case for most fishes [13]. The somewhat delayed growth response observed in the present work may be due to the unusually low water levels in the spring (noted at the time of first recapture), which we believe also caused extensive movements, particularly in control fish, during this period. Low water levels may also have decreased the feeding ability by, for example, reducing the input of allochthonous energy to the stream (by a reduction in the stream surface area), in addition to reducing the available habitat for the fish, which may increase competition and aggression. The lack of differences in weight, length or condition factor among the four groups after the winter could be due to full compensation by starved fish with no effects of density, although we note that data at this point were highly variable and sample sizes lower than before the winter.

An unexpected finding in this study is the lower recapture rate of control fish relative to starved fish after the first period of only 26 days. It seems unlikely that the control fish, which were larger than the starved fish at this point, would suffer such high mortality during this short period. Rather than experiencing increased mortality, we believe that the control fish moved, to a greater extent than starved fish as suggested by our analysis on movements, to other sections of the stream as water levels dropped. This would also agree with previous work showing that brown trout choose deeper and slower-flowing water with increasing size [26]. Therefore, fish from the control group may have been more likely to emigrate out of the study section during this period in search for deeper pools. However, we also note that natural selection may act against large body-size increasing mortality and/or emigration at low water levels [27,28]. With no differences in recapture rate after the summer, during which most of the compensation occurred, we have no indication that there were direct costs of compensation, such as increased predation often associated with rapid growth [29,30].

In the present study, we did not find any delayed mortality cost of compensatory growth over the winter which is in agreement with a previous field study on brown trout using moderate feed restriction in the laboratory before return to the stream [17]. However, in a more recent study including groups with a feeding restriction comparable to the present study, compensatory growth did result in increased winter mortality [18]. Moreover, in the present study fish released at higher densities had lower apparent winter survival than those released at natural densities, which also suggests that three weeks of starvation per se did not compromise the survival of the fish. In contrast, in a previous field experiment on brown trout parr [7], density affected growth but not survival. These variable effects found in comparable stream model systems suggest that environmental variations can greatly influence the interactive effects of density, compensatory growth, and survival [31].

Movements during the first 26-day period altered the experimental density in each subsection as well as shifted the density experienced by individual experimental fish. Most fish probably experienced the high experimental density for only part of the initial four weeks after release, which suggests that our results are conservative. This is particularly interesting since, despite the limited compensation during this short initial period, we found compensatory growth over the summer. More controlled field studies will be required to evaluate the relative effects of the duration and magnitude of density fluctuations.

In most studies, the compensatory growth response in fish is achieved mainly by hyperphagia and to a lesser extent increased conversion efficiency and reduced metabolism [13]. This provides a link to the effects of density on compensation because a hyperphagic response is likely to increase competition over limited resources and/or increase social interactions resulting in increased energy expenditure [32–34]. As such, it will be more difficult to compensate at a high population density than at a lower density, keeping other factors constant. These effects may be particularly costly for the compensating fish, since they are smaller due to the growth restriction, compared to their non-starved conspecifics, and this may explain why the potential mortality effect (i.e. low recapture rate) of density (assuming that movement did not contribute to reduced recapture at this point) found in the present study has not been observed in normally growing (i.e. non-compensatory) fish [7].

In summary, this study demonstrates that compensatory growth in nature can be density-dependent just as normal routine growth. Thus, restoration of adaptive individual growth trajectories [18] through compensatory growth may be environment-dependent. This adds to the basic understanding of the regulatory mechanisms underlying compensatory growth, but we are far from having a complete picture of the factors eliciting compensatory growth, how it is regulated, and why. A more complete evolutionary and ecological understanding of compensatory growth requires both experimental manipulation and examination under a variety of natural settings.

Acknowledgments

We thank Joyris Paquit, Johan Höjesjö, Peter Waldeck, Olof Tällström, and Daniel Nyqvist for assistance in the field.

Author Contributions

Conceived and designed the experiments: LFS JIJ. Performed the experiments: LFS RK JN. Analyzed the data: LFS. Wrote the paper: LFS RK JN JIJ.

References

1. Auer SK, Arendt JD, Chandramouli R, Reznick DN (2010) Juvenile compensatory growth has negative consequences for reproduction in Trinidadian guppies (*Poecilia reticulata*). Ecology Letters 13: 998–1007.
2. Metcalfe NB, Monaghan P (2001) Compensation for a bad start: grow now, pay later? Trends in Ecology & Evolution 16: 254–260.
3. Dmitriew CM (2011) The evolution of growth trajectories: what limits growth rate? Biological Reviews 86: 97–116.
4. Starck J, Ricklefs R (1998) Avian growth and development: evolution within the altricial-precocial spectrum. New York: Oxford University Press.
5. Houlihan D, Boujard T, Jobling M, editors (2001) Food intake in fish. Oxford: Blackwell Science Ltd. 418 p.
6. Hawkins CP (1986) Variation in individual growth rates and population densities of ephemerellid mayflies. Ecology 67: 1384–1395.
7. Bohlin T, Sundström LF, Johnsson JI, Höjesjö J, Pettersson J (2002) Density-dependent growth in brown trout: effects of introducing wild and hatchery fish. Journal of Animal Ecology 71: 683–692.
8. Sundström LF, Bohlin T, Johnsson JI (2004) Density-dependent growth in hatchery-reared brown trout released into a natural stream. Journal of Fish Biology 65: 1385–1391.
9. Mobæk R, Mysterud A, Holand Ø, Austrheim G (2012) Temporal variation in density dependent body growth of a large herbivore. Oikos 122: 421–427.
10. Loman J, Lardner B (2009) Density dependent growth in adult brown frogs *Rana arvalis* and *Rana temporaria* – A field experiment. Acta Oecologica 35: 824–830.
11. Lima SL, Dill LM (1990) Behavioral decisions made under the risk of predation – a review and prospectus. Canadian Journal of Zoology 68: 619–640.
12. McCue MD (2010) Starvation physiology: Reviewing the different strategies animals use to survive a common challenge. Comparative Biochemistry and Physiology – Part A: Molecular & Integrative Physiology 156: 1–18.

13. Ali M, Nicieza AG, Wootton RJ (2003) Compensatory growth in fishes: a response to growth depression. Fish and Fisheries 4: 147–190.
14. Wilson PN, Osbourn DF (1960) Compensatory growth after undernutrition in mammals and birds. Biological Reviews 35: 324–361.
15. Arendt JD (1997) Adaptive intrinsic growth rates: an integration across taxa. Quarterly Review of Biology 72: 149–177.
16. Hornick JL, Van Eenaeme C, Gérard O, Dufrasne I, Istasse L (2000) Mechanisms of reduced and compensatory growth. Domestic Animal Endocrinology 19: 121–132.
17. Johnsson JI, Bohlin T (2005) Compensatory growth for free? A field experiment on brown trout (*Salmo trutta*). Oikos 111: 31–38.
18. Johnsson JI, Bohlin T (2006) The cost of catching up: increased winter mortality following structural growth compensation in the wild. Proceedings of the Royal Society B: Biological Sciences 273: 1281–1286.
19. Álvarez D, Nicieza A (2005) Compensatory response 'defends' energy levels but not growth trajectories in brown trout, *Salmo trutta* L. Proceedings of the Royal Society B: Biological Sciences 272: 601–607.
20. van Dijk P, Staaks G, Hardewig I (2002) The effect of fasting and refeeding on temperature preference, activity and growth of roach, *Rutilus rutilus*. Oecologia 130: 496–504.
21. Bohlin T, Dellefors C, Faremo U (1996) Date of smolt migration depends on body-size but not age in wild sea-run brown trout. Journal of Fish Biology 49: 157–164.
22. Bohlin T, Dellefors C, Faremo U (1993) Optimal time and size for smolt migration in wild sea trout (*Salmo trutta*). Canadian Journal of Fisheries and Aquatic Sciences 50: 224–232.
23. Dellefors C, Faremo U (1988) Early sexual maturation in males of wild sea trout, *Salmo trutta* L., inhibits smoltification. Journal of Fish Biology 33: 741–749.
24. Quinn G, Keough MJ (2002) Experimental Design and Data Analysis for Biologists. Cambridge, UK: Cambridge University Press.
25. Seber GAF, Le Cren ED (1967) Estimating population parameters from catches large relative to the population. The Journal of Animal Ecology 36: 631–643.
26. Ayllón D, Almodóvar A, Nicola GG, Elvira B (2010) Ontogenetic and spatial variations in brown trout habitat selection. Ecology of Freshwater Fish 19: 420–432.
27. Carlson SM, Hendry AP, Letcher BH (2004) Natural selection acting on body size, growth rate and compensatory growth: an empirical test in a wild trout population. Evolutionary Ecology Research 6: 955–973.
28. Xu CL, Letcher BH, Nislow KH (2010) Size-dependent survival of brook trout *Salvelinus fontinalis* in summer: effects of water temperature and stream flow. Journal of Fish Biology 76: 2342–2369.
29. Sundström LF, Lõhmus M, Devlin RH (2005) Selection on increased intrinsic growth rates in coho salmon *Oncorhynchus kisutch*. Evolution 59: 1560–1569.
30. Biro PA, Abrahams MV, Post JR (2007) Direct manipulation of behaviour reveals a mechanism for variation in growth and mortality among prey populations. Animal Behaviour 73: 891–896.
31. Petranka JW, Sih A (1986) Environmental instability, competition, and density-dependent growth and survivorship of a stream-dwelling salamander. Ecology 67: 729–736.
32. Vøllestad LA, Quinn TP (2003) Trade-off between growth rate and aggression in juvenile coho salmon, *Oncorhynchus kisutch*. Animal Behaviour 66: 561–568.
33. Nephew BC, Romero LM (2003) Behavioral, physiological, and endocrine responses of starlings to acute increases in density. Hormones and Behavior 44: 222–232.
34. Nicieza AG, Metcalfe NB (1999) Costs of rapid growth: the risk of aggression is higher for fast-growing salmon. Functional Ecology 13: 793–800.

4

Longevity in Bovids Is Promoted by Sociality, But Reduced by Sexual Selection

Jakob Bro-Jørgensen*

Mammalian Behaviour & Evolution Group, Department of Evolution, Ecology & Behaviour, Institute of Integrative Biology, Faculty of Health and Life Sciences, University of Liverpool, Neston, United Kingdom

Abstract

Selection on intrinsic lifespan depends on both external factors affecting mortality and inherent tradeoffs in resource allocation between viability traits and other fitness-related traits. Longevity is therefore likely to vary between species in a sex-specific manner due to interspecific and intersexual differences in behavioural ecology. Here I focus on the bovid family to test two central hypotheses on longevity selection using the comparative method: firstly, that a reduction of extrinsic mortality in social species strengthens selection on intrinsic lifespan, and secondly, that mortality costs associated with intense sexual selection lead to shorter intrinsic lifespan. The results show that longevity (*i*) increases with sociality in both sexes and (*ii*) decreases with male-biased sexual size-dimorphism, but in males only. These discoveries suggest that sociality, a key ungulate strategy to reduce predation-related mortality, selects for inherently longer-lived organisms, and that strong sexual selection, which is known to compromise survival rates in the wild, can constrain also intrinsic lifespan. The contrasting results for males and females indicate that selection on longevity in the two sexes is partly uncoupled.

Editor: Jane M. Waterman, University of Manitoba, Canada

Funding: This author has no support or funding to report.

Competing Interests: The author has declared that no competing interests exist.

* E-mail: bro@liv.ac.uk

Introduction

The extent to which intrinsic lifespan is visible to selection depends on extrinsic mortality rates due to environmental conditions [1]. Where mortality from external factors, such as predation, is high, intrinsic lifespan is less likely to be expressed and thus subjected to selection; by contrast, low extrinsic mortality intensifies selection on longevity [2–4]. A key strategy to reduce predation in many taxa is the formation of groups, which may both enhance vigilance against predators ('detection effect' [5]) and reduce the probability of being preyed upon, should a predator attack ('dilution effect' [6]). Yet, where group-living has evolved as an antipedator strategy, the association between sociality and predation rate at the interspecific level is not straightforward, and empirical investigation is therefore essential to establish how longevity relates to sociality. Two scenarios as to how sociality and longevity are linked can be envisaged. On one hand, selection for longevity may be strongest in social species if the antipredator benefits of group-living result in lower extrinsic mortality compared to less social species. On the other hand, sociality may have evolved primarily where predation rates are high, and if the antipredator benefits of group-living only partly offset the higher extrinsic mortality, positive selection on longevity may be less intense in social species.

Intrinsic lifespan may also depend crucially on the intensity of sexual selection [7]. In natural environments, sexually selected traits can increase vulnerability to extrinsic mortality factors directly, e.g. where increased conspicuousness or reduced manoeuvrability enhances predation risk, and this may attenuate selection on intrinsic lifespan as described above. Also, investment in sexually selected traits may reduce longevity due to inherent tradeoffs in resource allocation against investment in longevity-promoting traits. Studies have found that in species where males are under stronger sexual selection than females, they exhibit higher rates of senescence in certain traits in the wild [8,9] and also invest less in morphological traits linked to viability [10,11]. These findings highlight the possibility of a general link between sexual selection and shorter intrinsic lifespan.

I here adopt the comparative method with phylogenetic control to investigate the impact of sociality and sexual selection on the evolution of intrinsic lifespan, using the family Bovidae as a model taxon. Bovids were chosen as the study system not only because they provide a classical example of a taxon where sociality has evolved due to anti-predator benefits [12,13], but also because this taxon is speciose and demonstrates pronounced interspecific variation in both sociality and the intensity of sexual selection.

First, I test the two hypotheses on the association between sociality and longevity. If antipredator benefits of group-living lead to positive selection on intrinsic lifespan in social species due to reduced extrinsic mortality, I predict that longevity will increase with group size. However, if sociality evolves primarily under intense predation and remains associated with high extrinsic mortality in spite of antipredator benefits from group-living, I predict that longevity will decrease with group size. Next, I investigate the hypothesis that sexual selection has a negative impact on intrinsic lifespan. As a proxy measure of the intensity of sexual selection, I use sexual body size dimorphism (SSD), a key sexually selected trait in bovids, whose evolution is linked to the advantage that large body size confers in contests over mates [14–16]. Thus, if viability costs of sexual selection constrain the

Table 1. Model of longevity in male bovids (59 species, 48 contrasts; $r^2=0.344$, F = 12.04, df 2,46, P<0.001).

	B (±SE)	β	df	t	P	VIF
Sexual body size dimorphism	−0.954± 0.219	−0.550	2,46	−4.356	<0.001	1.120
Group size	0.179± 0.050	0.449	2,46	3.553	0.001	1.438
Male body mass	0.018± 0.066	0.038	3,45	0.269	0.789	1.338
Interaction group size:male body mass	0.0351± 0.0318	-	4,44	1.105	0.275	-

Table 2. Model of longevity in female bovids (87 species, 66 contrasts; $r^2=0.245$, F = 10.4, df 2,64, P<0.001).

	B (±SE)	β	df	t	P	VIF
Female body mass	0.089±0.032	0.331	2,54	2.75	0.008	1.226
Group size	0.046±0.022	0.253	2,54	2.105	0.039	1.260
Sexual body size dimorphism	−0.196±0.136	−0.158	3,53	−1.443	0.154	1.037
Interaction group size:female body mass	−0.0068±0.0174	-	4,52	−0.390	0.698	-

evolution of intrinsic lifespan, I predict that longevity will decrease with increasing sexual size dimorphism. To disentangle the sex-specific evolutionary pressures, I conduct the analyses on the two sexes separately. In the analyses, I control for the effect of body size on longevity [17]. In particular, when investigating the link between sociality and longevity, I also take into account that the benefit of sociality as an antipredator strategy is thought to depend on body size: whereas larger species typically live in open habitats and rely on safety in numbers, minimal group size is believed to be advantageous in smaller species, which usually live in closed habitats and hide from predators [12,13].

Materials and Methods

Data

Focusing on all extant bovid species, data was collected from the literature on the maximum recorded lifespans of males and females. These records invariably came from captive populations and were assumed to broadly reflect intrinsic longevity, i.e. independent of ecologically induced mortality due to predation, food limitation, mate competition and disease transmission [18]. Data was also compiled on average body masses of adult males and females, as well as on average breeding group size. Since data was lacking for some species, the final data set was restricted to 100 out of a total of 136 species. SSD was calculated as the Lovich & Gibbons ratio (i.e. M/F if M>F and 2-F/M if F>M, where M and F denote male and female body mass, respectively [19]). All variables were $\log_{10}$-transformed to make them suitable for regression analysis. The data on maximum longevity was obtained primarily from [20], and supplemented from [21–24]. The data on the morphological and behavioural variables was copied from [16,25], and supplemented from [13,26–28]. The data set is presented in Table S1.

Comparative Analysis

The comparative method with phylogenetic control was used to identify predictors of longevity [29]. Evolutionarily independent contrasts were calculated in R [30] using the packages caper [31] and ape [32]. Phylogenetic relatedness was controlled for by assuming a gradual model of evolution based on the phylogenetic supertree and branch lengths in [33]. The independent contrasts were computed using the crunch-algorithm which assumes that trait evolution can be modelled as a random walk process [34]. To test this assumption, the estimated contrasts were regressed against their estimated nodal values [31]. I also tested the success of the contrast standardization procedure by regressing the absolute values of the standardized contrasts against the estimated standard deviation at each node, thereby assessing the heterogeneity of variance in the residuals and the suitability of the data for regression analysis [31]. As expected given the assumptions, no significant associations were found in any of these tests, which were all performed in R.

Contrasts in both male and female longevity followed a normal distribution, and I performed separate multivariate regression analysis on the two sexes in PASW Statistics 17.0.2. I tested the predictive power of the following independent variables: sex-specific body mass, group size, and SSD. Because antipredator strategy in ungulates depends on body size (see Introduction), the interaction between body mass and group size was also included as an independent variable in order to assess whether any effect of sociality on longevity was size-dependent. All regressions were through the origin. Non-significant variables were removed by backward elimination. Due to correlations between the independent variables, I assessed the potential effect of collinearity on the results of the multivariate analyses by calculating the variance inflation factors. In all cases, the variance inflation factors were <1.5 and thus well below the commonly accepted critical thresholds for significant collinearity of 5–10 [35]. Performing the analysis using the phylogenetic generalized least squares (PGLS) method yielded similar results [36]. Also, restricting the analyses to the subset of 46 species for which longevity records on both sexes were available revealed the same patterns, except that the positive association between body mass and longevity in females was not significant.

Results

Longevity in both males and females was positively correlated with group size across the bovid species (Tables 1 and 2, Figure 1); however, no effect was found of the interaction between group size and body size in either sex (Tables 1 and 2). Sexual size dimorphism showed an independent, negative, correlation with longevity in males but not in females (Tables 1 and 2, Figure 2). In females, longevity was positively related to body size, a pattern not found in males (Tables 1 and 2, Figure 3). In 24% of species, a male rather than a female held the longevity record, and these species were significantly less size-dimorphic (SSD where the record holder was male: 1.10±0.06, as compared to female: 1.39±0.04, mean±SE; $t=4.26$, df 98, P<0.001; Figure 4).

Discussion

This study reveals both sociality and sexual selection (as measured by SSD) as independent predictors of interspecific

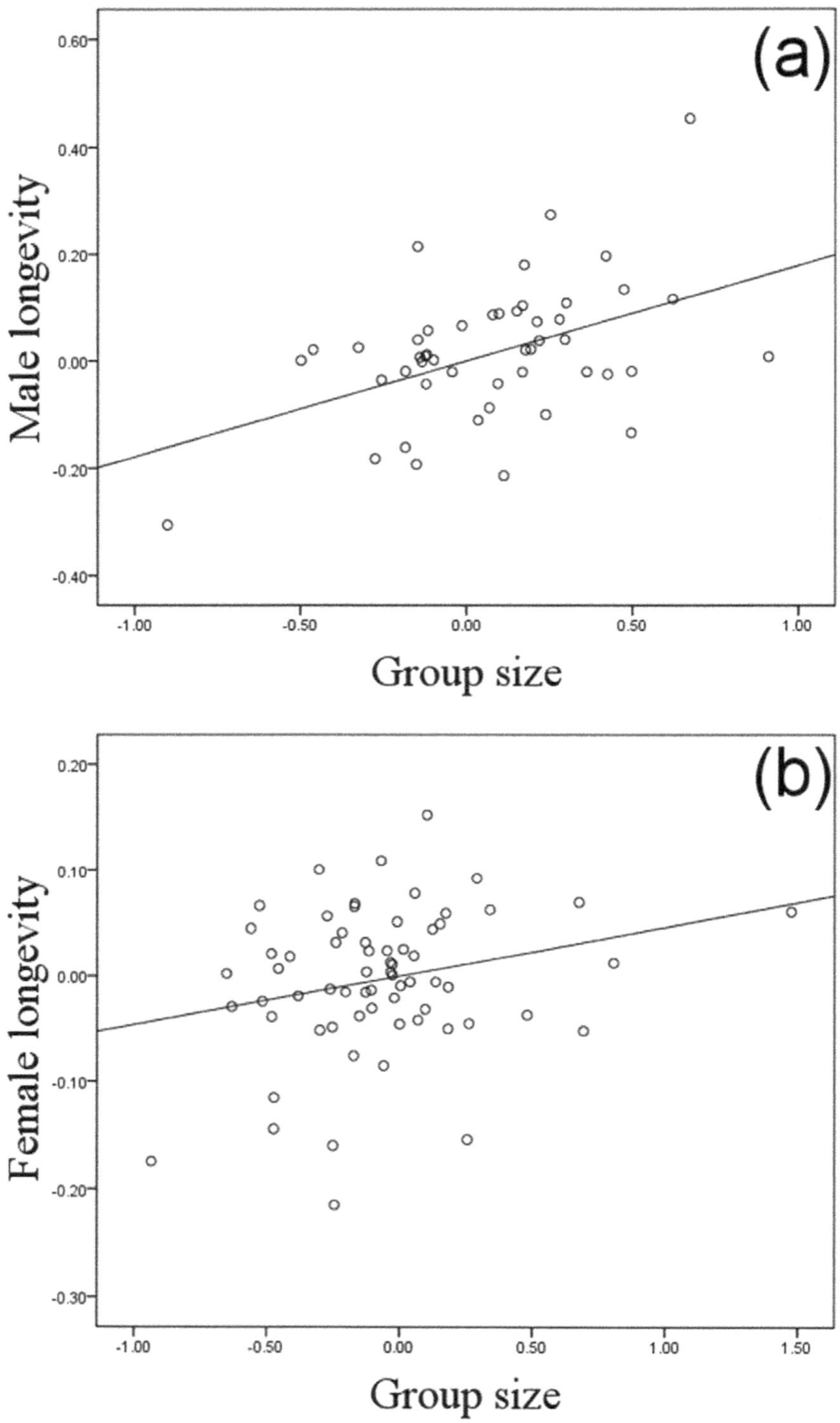

Figure 1. Partial regression plots of contrasts in longevity against contrasts in group size. (a) Males. (b) Females. The lines show the significant slopes from linear regression through the origin.

variation in longevity in bovids. Although the intensity of sexual selection is partly dependent on sociality [16], the multivariate approach taken here allows the distinct, oppositely directed effects of these two factors to be teased apart by analysing sex-specific relationships.

The negative relationship between male longevity and SSD suggests that in species where the fitness of males is strongly linked to their success in mate competition, viability costs associated with the development of sexually selected traits constrain selection on male lifespan. This interpretation is consistent with studies of wild populations, which show male survival rates to be compromised by intense mate competition [37–40]: such higher mortality is indeed expected to attenuate selection for long intrinsic lifespan. Further evidence that relatively short lifespan is not a general male

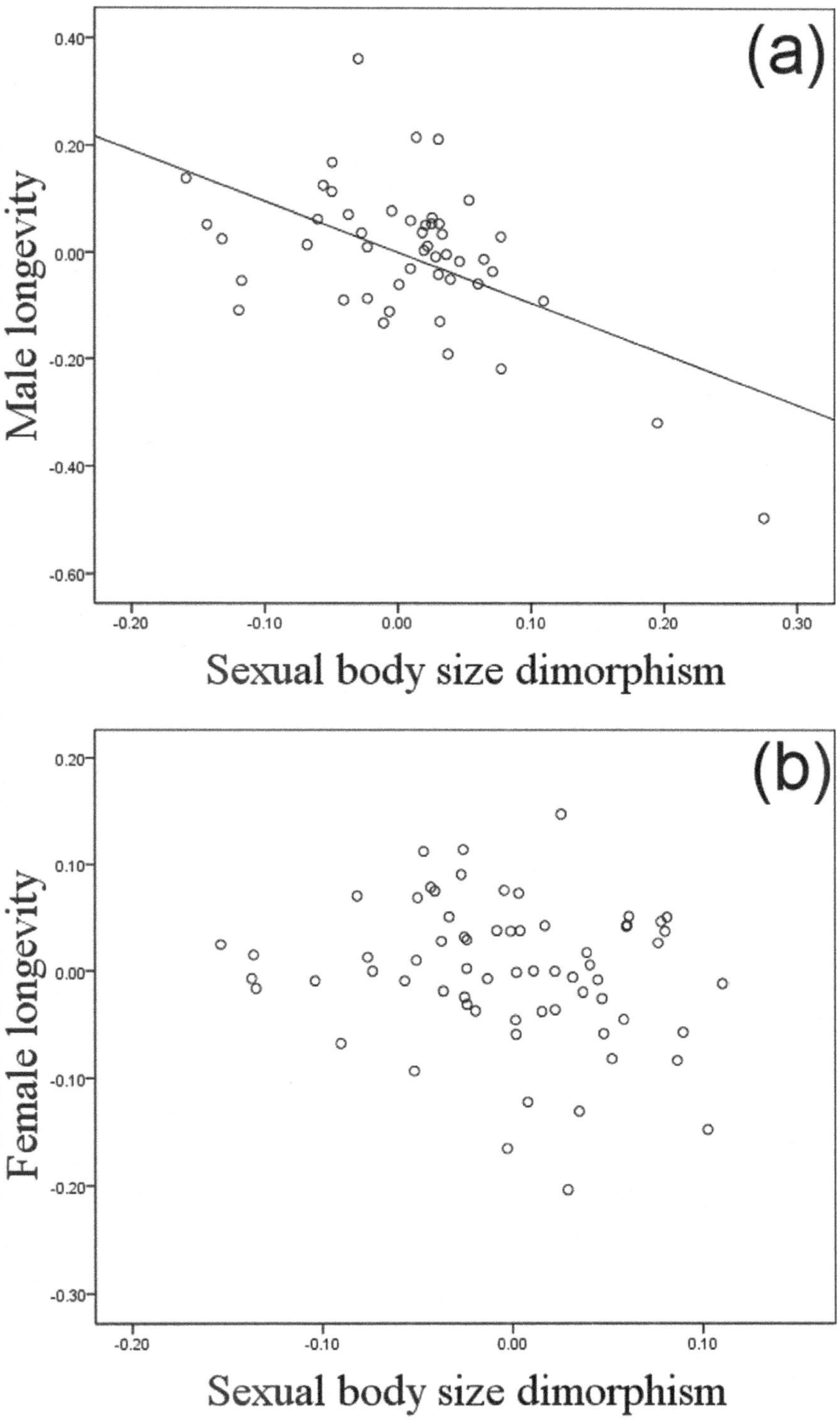

Figure 2. Partial regression plots of contrasts in longevity against contrasts in sexual body size dimorphism. (a) Males. (b) Females. The line shows the significant slope from linear regression through the origin.

characteristic but depends specifically on the intensity of sexual selection on males, is the finding that it is males rather than females, who tend to hold the longevity record in species where size dimorphism is female-biased. That the longevity of males not only equals, but generally supersedes that of females in this case, suggests that selection on lifespan in these species is more constrained by the costs of maternal investment than by the costs of reproductive competition. Again, this is consistent with survival rates in the wild as mortality in monogamous mammals is reported to be biased towards females in general [37].

The positive link between longevity and group size supports the hypothesis that selection for long intrinsic lifespan is stronger in social species due to lower extrinsic mortality from predation as a result of safety in numbers [1,3]. Apart from the classical detection

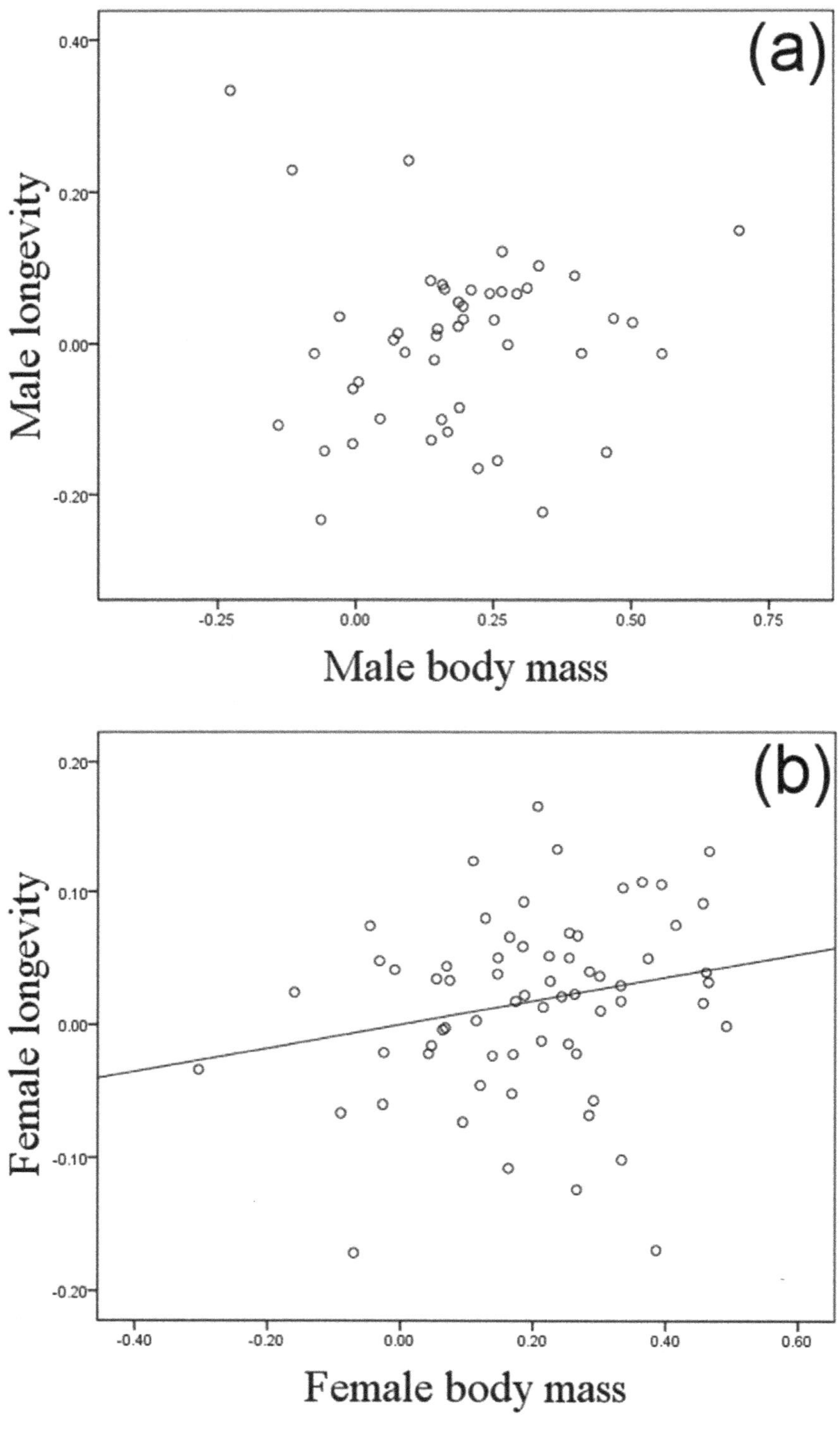

Figure 3. Partial regression plots of contrasts in longevity against contrasts in sex-specific body mass. (a) Males. (b) Females. The line shows the significant slope from linear regression through the origin.

and dilution benefits of group living, it is conceivable that the positive link between sociality and longevity is promoted also by a general link between large brain size and sociality [1], which has been reported specifically in ungulates [41,42]. The primary driver of large brain size evolution in social species is thought to be the need to cope with more complex inter-individual relations [41], but larger brains may also result in reduced mortality by enabling organisms to respond more appropriately to dangerous situations. A role of increased brain size in mediating the positive link between sociality and longevity could explain the somewhat surprising finding that the longevity-promoting effect of sociality was independent of body size: if group-living selects for increased longevity solely owing to detection and dilution effects, the link would be expected mainly in larger species since smaller ungulates

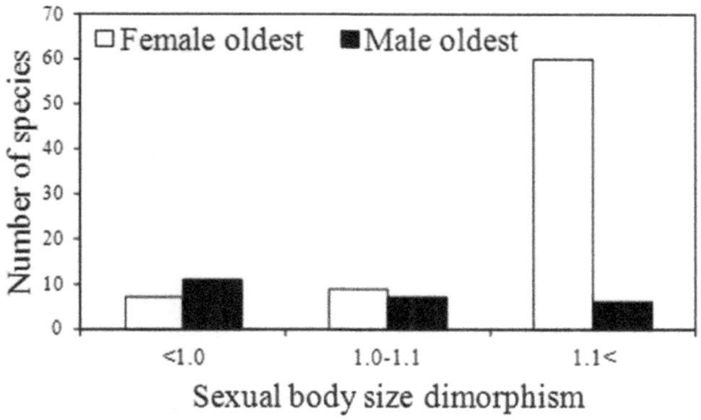

Figure 4. The number of species in which a male respectively a female holds the longevity record; shown in relation to the degree of sexual body size dimorphism.

typically rely on concealment in groups of minimal size to avoid predation. However, further investigation is required to disentangle the exact causal relationships between sociality, predation rates and longevity.

The finding that social life is associated with increased longevity in bovids contrasts with a recent study, which as part of a broader investigation examined non sex-specific longevity in a subset of Artiodactyl species (71 of 233 spp., i.e. 30%) [43]. Contrary to their expectation, the authors found a negative rather than a positive association. I propose that the discrepancy between the two studies can be explained by (*i*) the use of a more representative and comprehensive data set in the present study (100 of 136 bovid spp., i.e. 74%), and (*ii*) correction of errors in group sizes in the data set used in the former analysis (e.g. the present study uses a group size of 2 rather than 6.25 for the monogamous Guenther's dikdik *Madoqua guentheri*, and 6 rather than 1 for the social dama gazelle *Gazella dama*; see Table S1 for further details).

In summary, the positive link between longevity and sociality, a key antipredator strategy in ungulates, points to lower extrinsic mortality in social species as a factor selecting for inherently longer lifespan. At the same time, the negative link between male longevity and SSD, a key sexually selected trait, suggests that investment in reproductive fitness components may not only compromise survival in the wild but also constrain selection on intrinsic longevity. Now further studies are needed to establish the critical conditions under which these selective forces have had a significant impact on longevity selection.

Acknowledgments

I am grateful to the Mammalian Behaviour & Evolution Group at Liverpool University, Marco Apollonio and an anonymous reviewer for their insightful comments.

Author Contributions

Conceived and designed the experiments: JBJ. Wrote the paper: JBJ. Compiled and analyzed the data: JBJ.

References

1. Williams PD, Day T, Fletcher Q, Rowe L (2006) The shaping of senescence in the wild. Trends Ecol Evol 21: 458–463.
2. Medawar PB (1952) An unsolved problem in biology. London: HK Lewis.
3. Williams GC (1957) Pleiotropy, natural selection, and the evolution of senescence. Evolution 11: 398–411.
4. Stearns SC, Ackermann M, Doebeli M, Kaiser M (2000) Experimental evolution of aging, growth, and reproduction in fruitflies. Proc Natl Acad Sci U S A 97: 3309–3313.
5. Pulliam HR (1973) On the advantages of flocking. J Theor Biol 38: 419422.
6. Hamilton WD (1971) Geometry for the selfish herd. J Theor Biol 31: 295–311.
7. Stearns SC (1992) The evolution of life histories. Oxford: Oxford University Press.
8. Nussey DH, Kruuk LEB, Morris A, Clements MN, Pemberton JM, et al. (2009) Inter- and intrasexual variation in aging patterns across reproductive traits in a wild red deer population. Am Nat 174: 342–357.
9. Preston BT, Saint Jalme M, Hingrat Y, Lacroix F, Sorci G (2011) Sexually extravagant males age more rapidly. Ecol Lett 14: 1017–1024.
10. Carranza J, Alarcos S, Sanchez-Prieto CB, Valencia J, Mateos C (2004) Disposable-soma senescence mediated by sexual selection in an ungulate. Nature 432: 215–218.
11. Carranza J, Pérez-Barbería FJ (2007) Sexual selection and senescence: Male size-dimorphic ungulates evolved relatively smaller molars than females. Am Nat 170: 370–380.
12. Jarman P (1974) The social organisation of antelope in relation to their ecology. Behaviour 48: 215–267.
13. Brashares JS, Garland T, Arcese P (2000) Phylogenetic analysis of coadaptation in behaviour, diet, and body size in the African antelope. Behav Ecol 11: 452–463.
14. Darwin C (1874) The descent of man, and selection in relation to sex, 2nd edition. New York: Crowell.
15. Jarman P (1983) Mating system and sexual dimorphism in large, terrestrial, mammalian herbivores. Biol Rev Camb Phil Soc 58: 485–520.
16. Bro-Jørgensen J (2007) The intensity of sexual selection predicts weapon size in male bovids. Evolution 61: 1316–1326.
17. Austad SN, Fischer KE (1991) Mammalian aging, metabolism, and ecology: evidence from the bats and marsupials. J Geront 46: B47–B53.
18. Carey JR (2003) Longevity: The biology and demography of life span. Princeton: Princeton University Press.
19. Lovich JE, Gibbons JW (1992) A review of techniques for quantifying sexual size dimorphism. Growth, Development and Aging 56: 269–281.
20. Weigl R (2005) Longevity of mammals in captivity; from the living collections of the world. Stuttgart: Schweizerbart.
21. Jones ML (1982) Longevity of captive mammals. Zool Garten NF Jena 52: 113–123.
22. Jones ML (1993) Longevity of ungulates in captivity. Int Zoo Yb 32: 159–169.
23. Carey JR, Judge DS (2000) Longevity records: Life spans of mammals, birds, amphibians, reptiles, and fish. Odense: Odense University Press.
24. de Magalhães JP, Costa J (2009) A database of vertebrate longevity records and their relation to other life-history traits. J Evol Biol 22: 1770–1774. (Available as 'The Animal Ageing & Longevity Database' from: http://genomics.senescence.info/species/. Accessed 2012 April 26).
25. Bro-Jørgensen J (2008) Dense habitats selecting for small body size: a comparative study on bovids. Oikos 117: 729–737.
26. Mead JI (1989) *Nemorhaedus goral*. Mammal Spec 335: 1–5.
27. Kingdon J (1997) The Kingdon field guide to African mammals. San Diego: Academic Press.
28. Polak J, Frynta D (2009) Sexual size dimorphism in domestic goats, sheep, and their wild relatives. Biol J Linn Soc 98: 872–883.
29. Harvey PH, Pagel MD (1991) The comparative method in evolutionary biology. Oxford: Oxford University Press.
30. R Development Core Team (2011) R: A language and environment for statistical computing. Vienna: R Foundation for Statistical Computing. Available: http://www.R-project.org/. Accessed 2011 July 1.
31. Orme D, Freckleton R, Thomas G, Petzoldt T, Fritz S, et al. (2011) Caper: Comparative analyses of phylogenetics and evolution in R. R package version 0.4. Available: http://CRAN.R-project.org/package=caper. Accessed 2011 July 1.
32. Paradis E, Claude J, Strimmer K (2004) Ape: Analyses of phylogenetics and evolution in R language. Bioinformatics 20: 289–290.
33. Fernández MH, Vrba ES (2005) A complete estimate of the phylogenetic relationships in Ruminantia: A dated species-level supertree of the extant ruminants. Biol Rev Camb Phil Soc 80: 269–302.
34. Felsenstein J (1985) Phylogenies and the comparative method. Am Nat 125: 1–15.
35. McClave JT, Sincich T (2003) Statistics. Englewood Cliffs: Prentice Hall.
36. Rohlf FJ (2001) Comparative methods for the analysis of continuous variables: Geometric interpretations. Evolution 55: 2143–2160.
37. Promislow D (1992) Costs of sexual selection in natural populations of mammals. Proc R Soc Lond B 247: 203–210.

38. Mysterud A, Yoccoz NG, Stenseth NC, Langvatn R (2001) Effects of age, sex and density on body weight of Norwegian red deer: Evidence of density-dependent senescence. Proc R Soc Lond B 268: 911–919.
39. Promislow D (2003) Mate choice, sexual conflict, and evolution of senescence. Behav Genet 33: 191–201.
40. Liker A, Szekely T (2005) Mortality costs of sexual selection and parental care in natural populations of birds. Evolution 59: 890–897.
41. Shultz S, Dunbar RIM (2006) Both social and ecological factors predict ungulate brain size. Proc R Soc Lond B 273: 207–215.
42. Pérez-Barbería FJ, Shultz S, Dunbar RIM (2007) Evidence for coevolution of sociality and relative brain size in three orders of mammals. Evolution 61: 2811–2821.
43. Kamilar JM, Bribiescas RG, Bradley BJ (2010) Is group size related to longevity in mammals? Biol Lett 6: 736–739.

5

Disproportional Representation of Primates in the Ecological Literature

Eckhard W. Heymann[1]*, Dietmar Zinner[2], Jörg U. Ganzhorn[3]

1 Abt. Verhaltensökologie & Soziobiologie, Deutsches Primatenzentrum, Göttingen, Germany, **2** Abt. Kognitive Ethologie, Deutsches Primatenzentrum, Göttingen, Germany, **3** Abt. Tierökologie, Zoologisches Institut, Universität Hamburg, Hamburg, Germany

Abstract

We address the question why papers dealing with the ecology of primates are so sparsely represented in the general ecological literature. A literature analyses based on entries in Web of Science and PrimateLit reveals that despite a large number of papers published on primates in general and on the ecology of primates, only a very small fraction of these papers is published in high-ranking international ecological journals. We discuss a number of potential reasons for the disproportion and highlight the problems associated with experimental research on wild primates and constraints on sample size as major issues.

Editor: Joanna E. Lambert, The University of Texas at San Antonio, United States of America

Funding: The authors have no support or funding to report.

Competing Interests: The authors have declared that no competing interests exist.

* E-mail: eheyman@gwdg.de

Introduction

Primatological studies have played a major role in the development of conceptual frameworks in evolutionary ecology. The comparative approach of J. H. Crook on the social organization of birds and primates was instrumental to the development of the fields of "socio-biology" and "behavioral ecology" in general [1,2]. His work stimulated a number of primatological papers that contributed to the development of general ecological questions (e.g., [3,4]). Since then, the comparative approach seems to have shifted away from primate ecology, and primate ecology seems to have lost importance for the academic community at large, while analyses of social systems still remain prominent (e.g., [5]). Many ecological studies in primatology are designed or considered "case studies" rather than studies of general interest that contribute to the understanding of ecological and evolutionary questions in general. This perception or reality has important consequences as, first, academic funding agencies are reluctant to fund case studies, and second, the scientific community and primatologists themselves are unaware of the wealth of information for many distinct species (probably unmatched by any other taxon) and the potential to contribute to the understanding of general evolutionary questions in ecology. With the present analysis we wanted to test whether this perception on the lack of importance of primatological studies in the ecological literature is justified.

This perceived lack of importance is even more intriguing when considering that the number of publications on primate ecology is very high when compared to other taxa with a much larger number of species. A search in Web of Knowledge^{SM} v.5.9 (search date: 18 June 2013) with terms "primate* and ecology", "mammal* and ecology", "bird* and ecology" and "insect* and ecology" yielded 1646, 5947, 8979 and 9749 hits, respectively. Comparing these figures with the number of species (primates: 479, mammals: ~6000, birds: ~10.000, insects: >1.000.00) reveals an obvious disproportion: in relation to the number of species, there are many more papers on primates than on the other taxa.

Therefore, in this paper we examine the following questions:

(1) Which fraction of papers published on primates (independent of the type of publication, i.e. journal articles, book chapters etc.) deals with aspects of ecology?

(2) Which fraction of papers on primate ecology is published in high-ranking ecological journals?

(3) Which fraction of papers published in high-ranking ecological journals deals with primates and how does this compare with other taxa?

We then discuss potential reasons for the underrepresentation of primate papers in ecological journals.

Methods

To examine the significance of ecological research questions in primatology, we searched in PrimateLit (http://primatelit.library.wisc.edu) with "keyword = ecology and feral". The term "feral" was used to exclude papers dealing with "lab ecology", a keyword used in PrimateLit for papers dealing with housing of and environmental enrichment for captive primates. We restricted this search to the period 1980–2009 for two reasons: (a) some of the currently high-ranking ecological journals only started in the 80 s or 90 s; (b) PrimateLit ceased to be updated after November 2010. We plotted 5-year running means of the number of hits per year against the time in years.

To examine the representation of papers dealing with primate ecology in ecological journals, we performed two different analyses. First, we searched in PrimateLit, with "keyword = ecology and feral" and "publication = <journal name>" and restricted this search to the period 1980–2009. We used the following journals on the basis of their scope (which should make a journal potentially feasible for submitting primate ecology papers) and their impact factors: Ecology, Ecology Letters, Functional Ecology, Journal of Animal Ecology, Journal of Ecology, Oecologia, Oikos. We then determined the total number of articles dealing with any taxon or topic in the journals mentioned above by searching the Web of Knowledge[SM] [v.5.9] with the journal name in the field "publication name" (search date: 15 April 2013). On the basis of these search results, we calculated the percentage of papers from the search in PrimateLit that were published in the above mentioned journals. We also examined, on the basis of values summed up over 5-year periods, whether there is a temporal trend in the representation of primate ecology papers in ecological journals.

Second, we compared the primate data with other taxa, namely insects, all classes of vertebrates, and major orders of mammals. We searched the Web of Knowledge Knowledge[SM] v.5.9 (search date: 18 June 2013) first with the search term "<taxon> and ecology" (e.g. "bird* and ecology") in the field "topic" and then with this term plus the journal name in the field "publication name". The search period comprised 1945–2013. We then calculated, for each taxon and over the entire search period, the percentage of papers in the above mentioned ecological journals. To estimate how primate papers perform in comparison to other taxa, we calculated the mean percentage and SD over all taxa except primates and compared this mean to the value for primates with a single sample t-test in Statistica 10.0 [6].

Results

Significance of ecological research questions in primatology

Over the period 1980–2009, 7.6% of papers listed in PrimateLit dealt with aspects of primate ecology. The percentage almost doubled between 1980 and 2009, with a steep increase starting around 1990 (Fig. 1).

Representation of primate papers in ecological journals

Papers dealing with aspects of primate ecology are only marginally represented in high-ranking ecological journals, accounting on average for only 0.3% of all papers in these journals (Fig. 2). For the period 2010–2013, this remains unchanged (0.3% of a total of 4980 papers).

The representation of papers dealing with primate ecology in ecological journals dropped drastically from >3% in 1980–84 to the currently low value with some fluctuation over time, despite the strong increase in the total number of papers dealing with primate ecology (Fig. 3). The absolute number of primate papers in ecological journals varied relatively little between 1990 and 2004 (10–20 papers/5 year period), and increased to 33 in the period 2005–2009. The total number of primate papers with "keyword = ecology and feral" increased from 644 (1990–94) to 3027 (2005–09). Over the entire 30-year period, the percentage of primate ecology papers published in ecological journals is below 1%.

Comparison with other taxa

Compared to other taxa, primates are poorly represented in ecological journals. The proportion of papers that matched the search term "primate* and ecology" (1.5%) was lower than for most other taxa except for cetaceans that had an even lower value (1.4; Fig. 4). The mean percentage ± SD over all taxa is 5.1±2.8, and only primates and cetaceans lie outside one SD from the mean. The percentage for primates (1.5) is significantly smaller than the mean percentage over all other taxa (5.4±2.7; $t = 4.878$, d.f. $= 11$, $p = 0.000487$).

Discussion

Our analyses revealed that despite a strong increase in the number of papers dealing with primate ecology over the last decades, there is a disproportionately low representation of such papers in general ecological journals. The general increase in primatological papers was not matched by a similar increase in the number of primate papers published by ecological journals. Their percentage actually dropped (Fig. 3). Furthermore, the representation is similar to the representation of cetaceans and considerably lower than for most other taxa. There are several potential explanations for this which we discuss below:

(a) "Bad ecologist" hypothesis

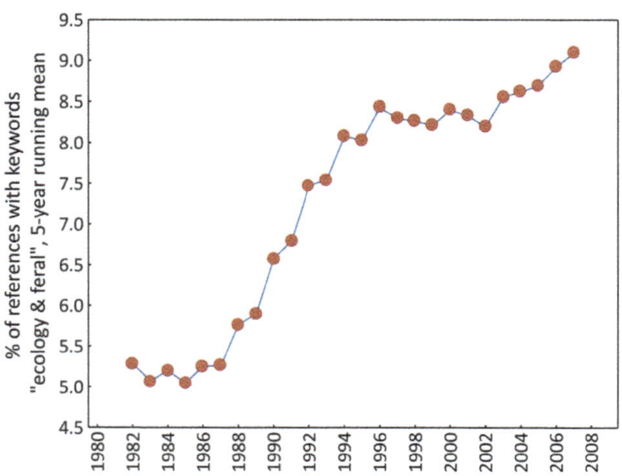

Figure 1. Temporal trend in the percentage of primatological publications dealing with aspects of primate ecology over the period 1980–2009. Dots represent 5-year running means; therefore, 1980, 1981, 2008 and 2009 are not represented by a dot.

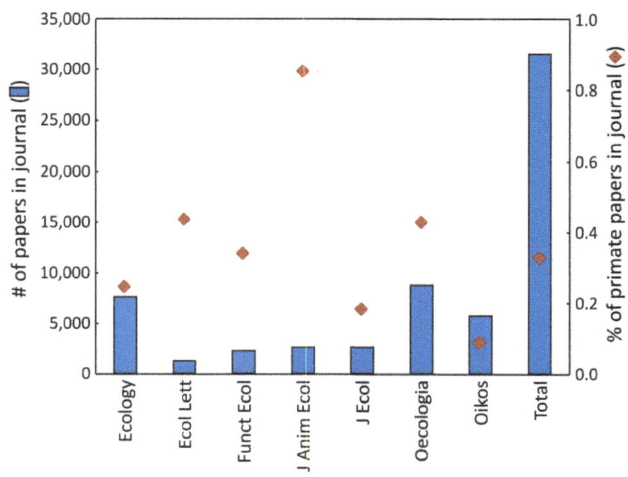

Figure 2. Total number of papers published in different ecological journals and percentage of these papers that relate to primates. Ecol Lett = Ecology Letters, Funct Ecol = Functional Ecology, J Anim Ecol = Journal of Animal Ecology, J Ecol = Journal of Ecology.

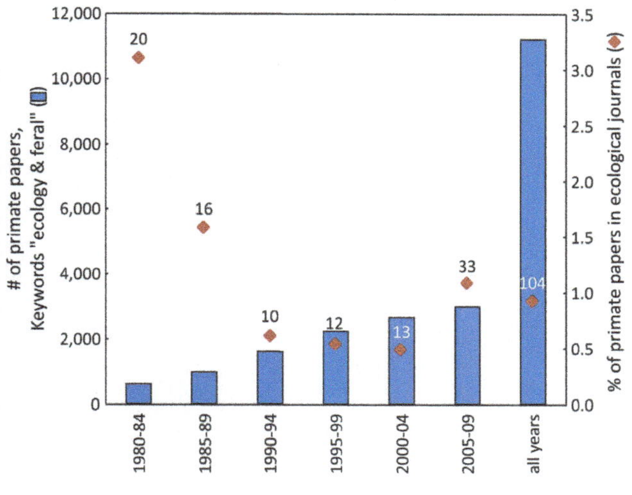

Figure 3. Number of primate papers over 5-year periods and percentage of primate papers in ecological journals. Numbers above diamonds indicate the total number of primate papers in ecological journals in the respective period.

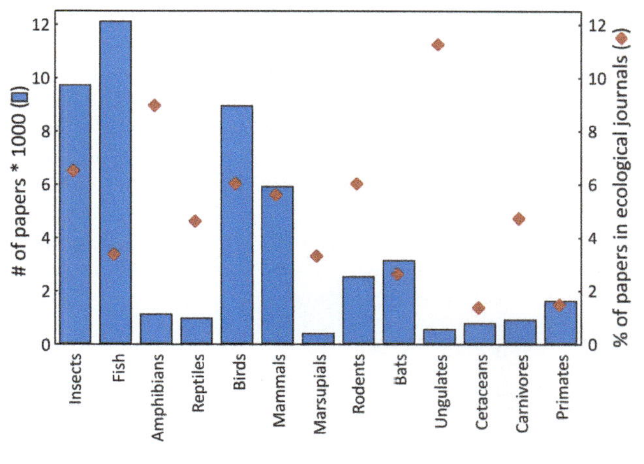

Figure 4. Number of papers on the ecology of insects, different classes of vertebrates and major orders of mammals, and the percentage of these papers that are published in high-ranking ecological journals.

Many primatologists may be interested in primates per se rather than in ecological problems and lack a sound theoretical background in general ecology. Thus, they may either not attempt publication outside the primatological circle, or they may not ask the "right" questions, or they cannot put their findings into a perspective that makes the results of their research interesting for a broader ecological readership. This is perhaps related to the history of primate ecological research. As Clutton-Brock [7, p. vii] stated in the preface to one of the first books dedicated to primate ecology, "interest in primate ecology has been stimulated by investigation of the adaptive significance of social systems". Furthermore, as Janson [8, p. 521] pointed out, "field primatology emerged largely from a descriptive anthropological tradition" which led to primate field studies focusing on a more descriptive primate ecology ("The diet of the ...", "Habitat use in a population of ..." etc.). This is not necessarily "bad ecology", but it clearly makes findings less appealing to a wider, ecologically interested audience. The relatively minor fluctuation in the total number of primate papers in ecological journals over the years, despite an almost 5-fold increase in the number of primate papers with keyword "ecology" and "feral" from 1990 to 2009, probably reflects this rather descriptive approach in primatology.

Carmel et al. [9] recently analysed trends in ecological research by examining the topics and approaches of papers published in all ecological journals and in "core ecological journals" (partially overlapping with our journal list) between 1980 and 2010. They found that a majority of studies deals with single species (71% and 66% of papers in all journals and core journals, respectively) and use observational approaches (59% and 45%). Therefore, the principal setup of most primate studies – observational on single species – should not preclude publication in ecological journals (although the percentage of observational studies in core ecological journals is considerably lower than in all ecological journals).

(b) Lack of experimental confirmation

Many research questions in ecology require experimental work to confirm observational findings and to test hypotheses. Lack of experimental confirmation of findings may result in lower acceptance of primate ecological research to ecologists who advocate more rigid experimental approaches. However, many experiments are impossible with primates on ethical and/or on practical grounds.

Ethical limitations

Primatologists have used playbacks of conspecific or allospecific vocalizations [10–12], platforms with food and/or sensory stimuli [13,14], cognitive test apparatuses [15,16], and provisioning with tools [17,18] to manipulate the behaviour and study the cognitive and sensory capabilities of primates. These kinds of experiments raise few ethical concerns, if any. However, many ecological experiments may require much stronger manipulation. E.g., to test whether a species exerts a competitive effect on another species or whether the lack of pollinators or seed dispersers affects plant populations requires the removal of complete groups or populations. For primate ecology studies, this would be absolutely unacceptable for ethical reasons.

As a proxy, pseudo-experimental situations where primate species have been extinguished, occur in reduced population densities due to hunting, or are absent for other reasons can provide relevant information. A comparison of hunted and non-hunted sites in Amazonia revealed competitive effects of large-sized on medium-sized primate species [19]. Similarly, a comparison of sites with low and high hunting pressure demonstrated that the lack of primate seed dispersal affects patterns of plant recruitment [20]. The absence of spider monkeys resulted in higher fruit set in a population of *Symphonia globulifera* compared to a population where these primates heavily preyed upon flowers of this plant [21].

Given that the majority of primates are group-living [5] experiments that require the temporal removal or isolation of individuals from a group are also ethically problematic (apart from practical problems). Experiments like those of Kawai [22] where entire groups were captured and released at a different place would no longer receive approval. Consequently, ecological experiments that require temporal removal of animals from their natural setting have been performed with solitary foragers like mouse lemurs, *Microcebus* [23,24], but not with group-living primates.

Practical limitations

(i) Sample size limitations

The majority of primates live in groups that range together and show a high degree of co-ordination and synchronization of activities [25,26]. Individuals within such groups thus cannot be considered as statistically independent samples. Observing a sufficiently large number of different groups to obtain high statistical power is in most cases practically impossible (due to limitations of funding, time constraints, limited size of study areas etc.).

Sample size can also be limited due to the high behavioural and ecological flexibility and high intra-specific and intra-population variability of most primates (see (ii)).

(ii) Flexibility and variability

Most primates are behaviourally and ecologically highly flexible [27–29]. This requires either many replica or the control of a large number of potentially or actually confounding variables; something that is rarely done in primate ecology [8]. Most primates also have slow life histories and long generation times [30] which can introduce an additional major source of variation [31]. Together with the issue of sample size limitations discussed above, this results in low statistical power.

(iii) Primate intelligence and cognition

Primates – by virtue of their higher cognitive capabilities compared to most other animals – can rapidly habituate to experimental stimuli and settings. They may even manipulate or destroy the experimental setup (see [32] for an example from captive primates). Even if planned very well and based on previous knowledge and experience, experiments may fail, simply because primates respond in an unanticipated manner. Janson [8] provides first-hand examples on failures of ecological and behavioural field experiments with primates. E.g., a setup that had worked very well with some groups, did not work with another group, or experiments turned out not "to fit within the lifestyle" [8, p. 531] of the animals. Experimental feeding platforms may not be attractive enough if alternative food resources become available, a factor that may be uncontrollable in the often large home ranges and complex tropical habitats of primates.

From all other taxa, practical problems, particularly for experimentation, and ethical concerns are probably most similar in cetaceans. The low representation of this taxon in high-ranking ecological journals supports our hypotheses outlined above.

There are a few other conceivable hypotheses that may apply.

(a) "Boring primates" hypothesis: Most primates live in tropical forest and occupy only a few, relatively similar niches. Therefore, ecological questions are similar for different primate species.

(b) "Geographic origin of researcher" hypothesis: Many researcher study ecology of species living in their home countries. Since still most researchers come from northern industrialized countries, species living in these regions could be overrepresented in the ecological literature.

(c) "Inbreeding" hypothesis: Primatologists may tend to publish their results – even if they could be of more general interest – in taxon-specific journals. Conversely, editors of general-interest ecology journals may be more willing to decide that 'there are other more appropriate journals for this manuscript' simply because in fact there are many primate journals relative to any other comparable order of animals (e.g. there is no 'Journal of Woodpeckerology'). However, the latter seems to be unlikely. This can be illustrated by the case of the representation of papers on birds and rodents. Both taxa have a similar representation in ecological journals (Fig. 4), but while there are 130 or more ornithological journals [33], we are aware of only one journal dedicated to rodents.

Conclusions

We have identified some potential causes for the low representation of primate papers in the general ecological literature, but do not pretend that this is an exhaustive listing. Conceptual problems are probably the easiest to overcome by modifying the curricula in primatology classes. Most practical problems could be overcome in an ideal world with less restricted funding and less time pressure for publications. However, in the real world, these problems constrain ecological research on primates and consequently may lead to results of primate ecological research being less attractive for a general ecological audience. Nevertheless, we hope that this paper stimulates discussions, both within the primatological community and between primate ecologists and general ecologists. Primate ecology has the potential to make important contributions to general ecology, particularly with regard the long-term ecological variation [34] and consequences of personality for ecological interactions and processes [35].

Acknowledgments

We thank Dr. Charlie Janson and two anonymous reviewers for commenting and very helpful suggestions on an earlier version of this manuscript.

Author Contributions

Analyzed the data: EWH. Wrote the paper: EWH DZ JUG.

References

1. Crook JH (1965) The adaptive significance of avian social organization. Symposia of the Zoological Society of London 14: 181–218.
2. Crook JH, Gartlan JS (1966) Evolution of primate societies. Nature 210: 1200–1203.
3. Clutton-Brock TH (1974) Primate social organisation and ecology. Nature 250: 539–542.
4. Milton K, May ML (1976) Body weight, diet, and home range area in primates. Nature 259: 459–462.
5. Mitani JC, Call J, Kappeler PM, Palombit RA, Silk JB (2012) The evolution of primate societies. Chicago: University of Chicago Press. 730 p.
6. StatSoft, Inc. (2011). STATISTICA für Windows (Software-System für Datenanalyse), Version 10.0. Available: www.statsoft.com.
7. Clutton-Brock TH (1977) Primate ecology. Studies on feeding and ranging behaviour in lemurs, monkeys and apes. London: Academic Press.
8. Janson CH (2012) Reconciling rigor and range: observations, experiments, and quasi-experiments in field primatology. International Journal of Primatology 33: 520–541.
9. Carmel Y, Kent R, Bar-Massada A, Blank L, Liberzon J, et al. (2013) Trends in Ecological Research during the Last Three Decades – a systematic review. PLoS ONE 8(4): e59813.
10. Robinson JG (1979) Vocal regulation of use of space by groups of titi monkeys *Callicebus moloch*. Behavioral Ecology and Sociobiology 5: 1–15.
11. Raemaekers JJ, Raemaekers P (1985) Field playback of loud calls to gibbons (*Hylobates lar*): territorial, sex-specific and species-specific response. Animal Behaviour 33: 481–493.
12. Zuberbühler K (2000) Interspecies semantic communication in two forest primates. Proceedings of the Royal Society of London B 267: 713–718.
13. Janson CH (1998) Experimental evidence for spatial memory in foraging wild capuchin monkeys, *Cebus apella*. Animal Behaviour 55: 1229–1243.
14. Bicca-Marques JC, Garber PA (2004) Use of spatial, visual and olfactory information during foraging in wild nocturnal and diurnal anthropoids: a field experiment comparing *Aotus*, *Callicebus*, and *Saguinus*. American Journal of Primatology 62: 171–182.

15. Pesendorfer MB, Gunhold T, Schiel N, Souto A, Huber L, et al. (2009) The maintenance of traditions in marmosets: individual habit, not social conformity? A field experiment. PLoS ONE 4: e4472.
16. Schnoell A, Fichtel C (2012) Wild redfronted lemurs (*Eulemur rufifrons*) use social information to learn new foraging techniques. Animal Cognition 15: 505–516.
17. Matsuzawa T (1994) Field experiments on use of stone tools by chimpanzees in the wild. In: Wrangham RW, McGrew WC, De Waal FBM, Heltne PG, editors. Chimpanzee cultures. Cambridge: Harvard University Press. pp. 351–370.
18. Fragaszy D, Pickering T, Liu Q, Izar P, Ottoni E, et al. (2010) Bearded capuchin monkeys' and a human's efficiency at cracking palm nuts with stone tools: field experiments. Animal Behaviour 79: 321–332.
19. Peres CA, Dolman PM (2000) Density compensation in neotropical primate communities: evidence from 56 hunted and nonhunted Amazonian forests of varying productivity. Oecologia 122: 175–189.
20. Effiom EO, Nuñez-Iturri G, Smith HG, Ottosson U, Olsson O (2013) Bushmeat hunting changes regeneration of African rainforests. Proceedings of the Royal Society B: Biological Sciences 280: 20130246.
21. Riba-Hernández P, Stoner KE (2005) Massive destruction of *Symphonia globulifera* (Clusiaceae) flowers by Central American spider monkeys (*Ateles geoffroyi*). Biotropica 37: 274–278.
22. Kawai M (1960) A field experiment on the process of group formation in the Japanese monkey (Macaca fuscata), and the releasing of the group at Ôhirayama. Primates 2: 181–255.
23. Goerlitz HR, Siemers BM (2007) Sensory ecology of prey rustling sounds: acoustical features and their classification by wild grey mouse lemurs. Functional Ecology 21: 143–153.
24. Thorén S, Linnenbrink M, Radespiel U (2011) Different competitive potential in two coexisting mouse lemur species in northwestern Madagascar. American Journal of Physical Anthropology 145: 156–162.
25. King AJ, Cowlishaw G (2009) All together now: behavioural synchrony in baboons. Animal Behaviour 78: 1381–1387.
26. Boos M, Kolbe M, Kappeler PM, Ellwart T (2011) Coordination in human and primate groups. Heidelberg: Springer. 288 p.
27. Chapman CA, Fedigan LM (1990) Dietary differences between neighboring *Cebus capucinus* groups: local traditions, food availability or responses to food profitability? Folia Primatologica 54: 177–186.
28. Jones CB (2005) Behavioral flexibility in primates: causes and consequences. New York: Springer.
29. Codron D, Lee-Thorp JA, Sponheimer M, de Ruiter D, Codron J (2006) Inter- and intrahabitat dietary variability of chacma baboons (Papio ursinus) in South African savannas based on fecal δ13C, δ15N, and %N. American Journal of Physical Anthropology 129: 204–214.
30. Martin RD (1990) Primate origins and evolution. London: Chapman and Hall. 804 p.
31. Nowell AA, Fletcher AW (2008) The development of feeding behaviour in wild western lowland gorillas (*Gorilla gorilla gorilla*). Behaviour 145: 171–193.
32. Vitale AF, Visalberghi E, De Lillo C (1991) Responses to a snake model in captive crab-eating macaques (*Macaca fascicularis*) and captive tufted capuchins (*Cebus apella*). International Journal of Primatology 12: 277–286.
33. Wikipedia website. Available: http://en.wikipedia.org/wiki/List_of_ornithology_journals. Accessed: 25 Sep 2013.
34. Kappeler PM, Watts DP (2012) Long-term field studies of primates. Heidelberg: Springer. 495 p.
35. Wolf M, Weissing FJ (2012) Animal personalities: consequences for ecology and evolution. Trends in Ecology & Evolution 27: 452–461.

6

Breeding Biology and Variable Mating System of a Population of Introduced Dunnocks (*Prunella modularis*) in New Zealand

Eduardo S. A. Santos[1,2]*, Shinichi Nakagawa[1]

1 Department of Zoology, University of Otago, Dunedin, New Zealand, **2** Departamento de Ecologia, Universidade de São Paulo, São Paulo, Brazil

Abstract

Species with variable mating systems provide a unique opportunity to investigate whether females receive direct fitness benefits from additional male partners. The direct benefits provide an obvious explanation for why females would breed polyandrously, in a situation where males clearly do not attain their optimal reproductive success. Evidence for these direct benefits is, however, mixed. Here, we present a detailed study of the breeding biology of the dunnock, *Prunella modularis*, which inform an investigation into the effects of the social mating system on the reproductive success in a population of dunnocks in Southern New Zealand. We studied 80 different social groups over the course of three breeding seasons. Dunnocks in our population presented a variable mating system, with socially monogamous (45%), socially polyandrous (54%) and socially polygynandrous (1%) groups being observed in the same breeding season. We did not observe any polygynous social units in our study period although polygyny exists in the population. We found little difference in the numbers of eggs laid, and egg volume between monogamous and polyandrous nests. However, polyandrous groups had better hatching and fledging success than monogamous groups (composite $d = 0.385$, 95% CI: 0.307 to 0.463). Overall our results support the notion that polyandry is beneficial for females.

Editor: Csaba Moskát, Hungarian Academy of Sciences, Hungary

Funding: The Association for the Study of Animal Behaviour (ASAB) and the Department of Zoology of the University of Otago for research grants. The University of Otago Postgraduate Scholarship to ESAS. The funders had no role in study design, data collection and analysis, decision to publish, or preparation of the manuscript.

Competing Interests: The authors have declared that no competing interests exist.

* E-mail: e.salves@gmail.com

Introduction

Usually, there is little intra-specific variation in social breeding systems, with one form (e.g., monogamy) being commonly the only one observed in a population. This is especially the case for bird species, although, in some instances, the genetic breeding system could provide another dimension of variation [1,2]. When individuals engage in multiple copulations, be it with different social partners or with extra-pair mates, it is thought that they seek additional fitness benefits. While it is clear that males benefit directly from breeding outside of their pairs (i.e. by increasing the number of offspring produced), the search for female fitness benefits of multiple paternity has been more challenging [3–5]. In many species, females do not receive obvious direct benefits (e.g., increases in reproductive success of the individual or its offspring), and thus, it is predicted that they should receive indirect genetic benefits [3].

Nevertheless, some species exhibit variable breeding systems, in which two or more types of social units occur during the reproductive season (e.g., pukeko, *Porphyrio porphyrio*, [6]). In these species, the outcome of sexual conflict over mating optima is more readily observed. For instance, some females might breed in socially polyandrous groups, gaining direct fitness benefits conferred by additional paternal care. Other females, in contrast, might be constrained to breeding with a single partner in monogamy. However, studies that investigated the direct benefits of additional male help in species with variable mating system have provided mixed findings. There is little evidence that the additional help provided in polyandrous groups of the ground tit, *Pseudopodoces humilis* [7], and of the moustached warbler, *Acrocephalus melanopogon* [8] is converted into fitness benefits for these polyandrous females. On the other hand, in another species with a variable mating system, the dunnock, *Prunella modularis*, socially polyandrous females fledge more offspring than socially monogamous ones [9]. More investigations are clearly needed to gain better understanding of what benefits, if any, females in variable mating system species gain from breeding polyandrously. Especially, a replication using the same species in a different population is likely to provide us with fresh insights.

In this study, we describe the breeding ecology of dunnocks in one population of these introduced birds in southern New Zealand (NZ). Using this information, we then investigate whether the type of social mating system affects breeding parameters (e.g., clutch size, egg volume, territory size), and hatching and fledging success of dunnocks. Our study provides important information for this species in a non-native range. Moreover, we have a unique opportunity to investigate the fitness benefits of polyandry for a species that has already been extensively studied.

Methods

Ethics statement

All the research reported here was conducted with the approval of the Animal Ethics Committee of the University of Otago (permit no. 08/09) and birds were banded under the New Zealand National Bird Banding Scheme Institutional Permit to Band Birds no. 2008/075. We thank the Dunedin Botanic Garden and its staff for allowing us to conduct this research on their grounds.

Study species

In-depth studies of dunnocks have been conducted in its native range in Europe (e.g., [10–16]). These studies have identified that dunnocks present a variable mating system within populations. Dunnocks form breeding groups that are composed of unrelated adults and can range from monogamy, polyandry (one female associated with two or three males concurrently), polygyny (one male associated with two or more females) and polygynandry (two or more females paired with two or more males) (see [10,17,18]). Although dunnocks have been well studied in Europe and considered to be one of the classic model species in the study of behavioural ecology, no study of its breeding ecology and behavioural ecology has been conducted in its introduced range in New Zealand (but see [8] for a study of the dunnock's diet). Thus, little is known of the effects that the release of the species in a new environment could have had on its breeding ecology and behaviour. A better understanding of the dunnock's breeding biology in New Zealand would provide important information for future studies of its behaviour from a comparative point of view with the previous European studies.

The dunnock is abundant in New Zealand, due to their deliberate introduction by English settlers in the 19th century [19–21], with approximately 640 dunnocks released at five different locations during this period [19,20,22]. Data from a dunnock population in Southern New Zealand suggest little genetic and morphological divergence from a native population in England [23].

Study area

This study was conducted at the Dunedin Botanic Garden (45.87° S, 170.50° E), Dunedin, New Zealand. We focused on a 7.1 ha area of the Botanic Garden that contained a diversity of human-made habitats, including open lawns, hedgerows, flower-beds, and open woodland. The study was conducted between July 2009 and January 2012, comprising three breeding seasons: 2009–2010, 2010–2011 and 2011–2012.

Nest monitoring and general field methods

In order to obtain breeding ecology data, we searched for dunnock nests during spring and summer, from September to January, either by walking through the study area and checking the vegetation for potential nests, or by following adults carrying nesting materials. We usually located nests before egg laying, as we were able to observe nest-building adults. Once we found a nest, we recorded its height above the ground, and for a subset of nests we measured four structural variables (inside cup diameter and cup depth, and outside cup diameter and cup height; ±0.01 cm). We used these data to describe the general strata of the vegetation that dunnocks use for nesting, and also the basic structural characteristics of the nests.

To collect breeding parameter data we monitored nests as follows. From the moment we found a nest, we checked it every two days until the first egg was laid (we considered nests to be active when we found either eggs or nestlings in them). During the egg laying stage, we checked nests daily in order to record the laying order. We measured each egg with dial callipers (length and width; ±0.1 mm), weighed on an electronic scale (±0.01 g), and sequentially numbered with a permanent felt-tip marker to indicate the position in the laying order. We used the length and width of eggs to estimate their volume (mm^3) based on Hoyt's formula: $0.51\times$ length $\times$ width^2 [24]. During the incubation period, we visited nests every two days until the first egg hatched. Most eggs in a nest hatched on the same day, and in most cases we were not able to match nestlings to the eggs. We also recorded the fate of each egg (disappeared, unhatched, hatched) and nestling (died in nest, disappeared, fledged). We marked all nestlings at hatching by uniquely clipping their superciliary feathers [25] and by applying a unique combination of dots with a permanent felt-tip marker to their nape for individual recognition. Before nestlings fledged, we colour banded them at nine days of age (age at hatching: zero day old) with a combination of three colour and one numbered aluminium bands.

Colour banding and mating system assessment

We captured adult dunnocks throughout our study site during both the non-breeding season and the breeding season by placing mist nets at frequently used flight paths (i.e. usually at territory borders and around nesting sites). We banded each dunnock with a unique combination of three colour and one numbered aluminium bands for individual recognition.

We characterized the social mating system of each breeding group by visiting nests at least five times during the nest cycle and recording the number of unique adults present in the territory that interacted with each other in social behaviours (e.g., foraging jointly, joint territory defence, mating, parental care). Moreover, to ascertain the social mating system, we also used the identity of adults that provided parental care to nestlings (i.e. through video recordings of parental behaviour). We did not characterize the genetic mating system of the groups in this study although we plan to do so in a future study.

We recorded GPS relocations of breeding colour-banded adults in order to estimate territory sizes. We estimated territory size per group using the minimum convex polygon method [26] implemented in the *adehabitatHR* package [27] in *R* (version 2.13.0; [28]). In our estimates of territory size, we used the locations where we saw all birds from a group engaging in the following behaviours: singing, fighting, patrolling, and/or foraging.

We estimated the density of breeding males using the number of males assigned to a breeding territory during each breeding season (when males had multiple breeding attempts, they were included only once in the estimate). We present the average density of breeding males per hectare over the three breeding seasons. Furthermore, we estimated the adult sex-ratio of the population as the proportion of breeding unique males to unique females in each breeding season.

Statistical analyses

To investigate the effect of the social mating system on the different breeding parameters, we used Bayesian mixed-effect models (BMM) implemented using the *MCMCglmm* package [29] in *R* (version 2.13.0; [28]). We built BMMs for the following breeding parameters as response variables: territory size (hectares; Gaussian link function), laying date of nests (Julian date; Gaussian link function), number of eggs laid (clutch size; Poisson link function), egg volume (mm^3; Gaussian link function), proportion of eggs that hatched (Binomial link function), and proportion of chicks that fledged (Binomial link function). We fitted all BMMs with the social mating system as a categorical fixed effect with two levels (polyandrous and monogamous; we did not include polygynandrous mating system in this analysis due to low sample size [one group and four nesting attempts]). We also included the breeding season in which the data was collected as a categorical fixed effect with three levels in all models (Season A for nests found in 2009–2010, Season B for nests found in 2010–2011 and Season C for nests found in 2011–2012). We removed the breeding season fixed effect from final reported models if its effect was non-significant for a given response variable (in all occasions where it was non-significant, the Deviance Information Criterion (DIC) supported the most parsimonious model) [30]. In the BMMs, we included the breeding group identity as a random effect to account for the lack of independence between measures. Moreover, we inspected plots of the models' residuals for heteroscedasticity [31]. When there was evidence for heteroscedasticity in the data (e.g., the variance in territory size of polyandrous groups was greater than that of monogamous groups), we accounted for it by modelling the residual variance structure to allow the variance in the effects to be different for each level of the fixed effect. We used a sample of 1,000 iterations from which to estimate the BMM model estimates. In order to obtain this sample for each model, we used the following range of model parameters: number of iterations = 1,000,000–2,000,000; burn-in = 900,000–1,000,000; and thinning of 100 or 1,000. We used an Inverse Gamma prior for all models.

To estimate nest daily survival rates (DSR), we used the package *RMark* (version 2.06; [32] for *R* (version 2.13.0; [28]). We used the method developed by [33] to evaluate variation in DSR as a function of the type of social mating system that nests belonged to. We used Pearson's correlation to evaluate the magnitude of the association between nest structural characteristics. All descriptive statistics are presented as means (standard deviations) and sample size, unless specified otherwise. Model estimates (slopes of the differences between factor levels) are presented as posterior means (standard deviations) and 95% credible intervals (95% CI). We considered parameters with 95% CIs not spanning zero to be statistically significant. For each result investigating the effect of social mating system on a breeding parameter, we present a standardized effect size statistic. We use Cohen's d as the effect size metric in order to present results in an 'effective' thinking framework [34]. Cohen proposed that estimates of d of 0.2, 0.5 and 0.8 are considered 'small', 'medium' and 'large', respectively [35]. We use these values as benchmarks in order to assess the magnitude of our findings. Finally, we use a within-study meta-analysis approach [36] to calculate a composite Cohen d value for hatching and fledging success, which is a measure of the overall effect of social mating system on reproductive success.

Results

Breeding season phenology and nest characteristics

We found and monitored 42 active nests in 2009–2010, 62 in 2010–2011 and 54 in 2011–2012 (henceforth, Season A, Season B and Season C, respectively; Figure 1). Breeding groups began to build nests in the first week of October in Season A, the third week of September in Season B, and again in the third week of September in Season C. Behavioural observations of colour-banded individuals during the nest-building stage revealed that only females built nests. These observations also showed that females built nests quickly (mean number of days to build nest: Season A: 7.8 (2.77) days, $n = 5$; Season B: 5.5 (1.71) days, $n = 10$; Season C: 5.6 (1.50) days, $n = 11$). The period of egg laying extended for 86 days in Season A (from the second week of October 2009 to the first week of January 2010), 101 days in Season B (from the last week of September 2010 to the second week of January 2011) and 88 days in Season C (from the last week of September 2011 to the last week of December 2011). Groups frequently re-nested in the same breeding season after depredation of the nest (up to three times), and also after successfully fledging young. Fifty percent (21/42) of the nesting territories of Season A were re-used in Season B, and 46.7% of those (29/62) from Season B were reused in Season C; all re-nesting events (both within- and between seasons) by the same group took place in immediately adjacent nests to the old ones (i.e., no nests were re-used).

Nests were open-cup structures of sticks, lined mainly with mosses and feathers. Nest height ranged from 0.16 to 3 m above the ground (mean height: 1.21 (0.67) m, $n = 100$). We measured a subset of nest for their structural dimensions ($n = 10$): inside cup diameter (mean: 3.75 (0.40) cm); inside cup depth (mean: 6.55 (0.93) cm); outside cup diameter (mean: 12.63 (1.58) cm); outside cup height (mean: 7.52 (1.22) cm). All the nest dimensions were positively associated ($0.34 < r < 0.74$).

Social mating system, territories, and adult sex-ratio

We captured and colour banded 126 adults and 202 nestlings during all three breeding seasons. We confidently assessed the social mating system of 85% of the active nests (134/158), as colour-banded individuals attended them. A total of 80 unique breeding groups attended this subset of nests. We observed the following frequency of social mating systems in our population: 45.0% (36/80) of the breeding groups were monogamous pairs, 53.7% (43/80) were polyandrous groups composed of one female and two or three males, and 1.3% (1/80) was a polygynandrous group composed of two females and two males. We did not observe any polygynous breeding group. Group composition persisted throughout the year (i.e. we did not observe any individuals switching groups outside of the breeding season).

Dunnocks in our population are year-round residents and defend territories throughout the year. Territories encompassed lawns, flowerbeds, and a variety of gardens containing shrubs, herbs and trees. For most groups, the territory borders made contact with at least another territory, indicating that all suitable nesting habitat was occupied. The effect size of mating system on territory size was 'small' ($d = 0.008$) and the difference was not statistically significant between polyandrous groups and monogamous groups (Table 1; Figure 2). There was, however, a significant negative effect of the breeding season on territory size, with territories in Season B being smaller than those of both Season A and Season C for both polyandrous and monogamous groups (Gaussian BMM: $\beta_{\text{breeding season B}}$: −0.202 (0.052), 95% CI: −0.302 to −0.097; $\beta_{\text{breeding season C}}$: 0.078 (0.046), 95% CI: −0.002 to 0.176; Figure 2). Over the course of both seasons, our study site had an average density of 5.047 (0.861) breeding males per hectare, and the adult breeding population was male-biased (average proportion of adult males, ASR = 0.569 (0.009)).

Social mating systems and breeding parameters

Eggs that successfully hatched were laid over an average period of 2.87 (1.254) days (one egg was laid daily until clutch completion), incubated on average for 12.62 (1.624) days, and chicks fledged on average 12.17 (1.617) days after hatching. Completed clutches ranged in size from one to five eggs. The effect sizes of mating system on laying date of the first egg, clutch size, and egg volume were 'small' (all $d < 0.3$; Table 1; Figure 3), and the differences between polyandrous and monogamous groups were statistically non-significant. The estimate of d was positive for the difference in the proportion of eggs that hatched between polyandrous and monogamous nests, indicating that polyandrous

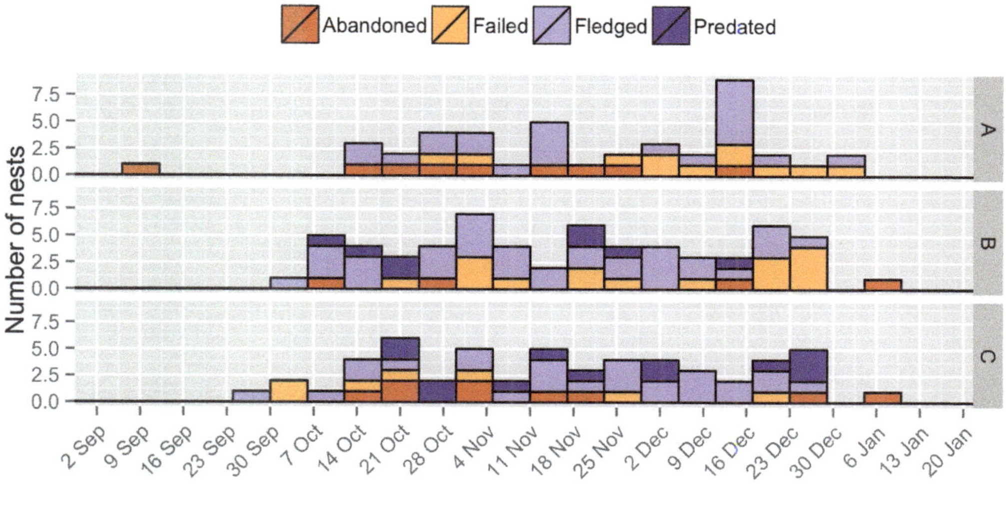

Figure 1. The timing of breeding in dunnocks. The timing of breeding by a population of dunnocks in Dunedin, New Zealand during the breeding seasons of (a) 2009–2010, (b) 2010–2011, and (c) 2011–2012. Bars summarise the number of active nests per week, based on the date that the first egg was laid. When the date of the first egg was unknown, it was estimated by backtracking from the average length of the incubation or nestling periods. For nests that never hatched, we used the date in which the nest was encountered as a proxy for the first egg date. The different fillings of the bars represent the outcome of the nests (i.e., abandoned, failed [e.g. unfertilized eggs, dead nestlings due to starvation or extreme weather], fledged, or predated).

nests hatched a greater proportion of eggs per clutch than monogamous nests ($d = 0.367$; Table 1; Figure 3). This d estimate can be considered 'small' to 'medium' [35], which suggests that the effect of polyandry on hatching success is weak. Among the known reasons associated with eggs failing to hatch were: nest desertion, depredation, infertility, and embryo mortality (Table 2). The probability that eggs failed to hatch due to infertility presented a negative, 'medium' to 'large', d estimate, indicating that it was higher in monogamous nests than in polyandrous ones ($d = -0.642$; Table 2) although the difference was not statistically significant. All other causes showed 'small' to 'medium' differences and were not significantly different between the two mating systems (Table 2).

Overall, 50% of nests (81/162) successfully fledged at least one offspring. The d estimate was positive for the difference in the proportion of chicks that fledged in polyandrous nests in relation to monogamous nests ($d = 0.404$; Table 1; Figure 3). This d estimate is considered 'medium' [35], which suggests that the effect of the social mating system on fledging success is moderate. The composite d estimate of hatching and fledging success was considered 'medium' and its 95% CI did not overlap zero ($d = 0.385$, 95% CI: 0.307 to 0.463), which indicates that overall socially polyandry has a moderate effect on the reproductive success.

Of the nests that failed to fledge at least one chick, 28 failed to hatch any eggs, 10 were abandoned, and 20 were predated. We estimated the DSR (daily survival rate) of dunnock nests as a

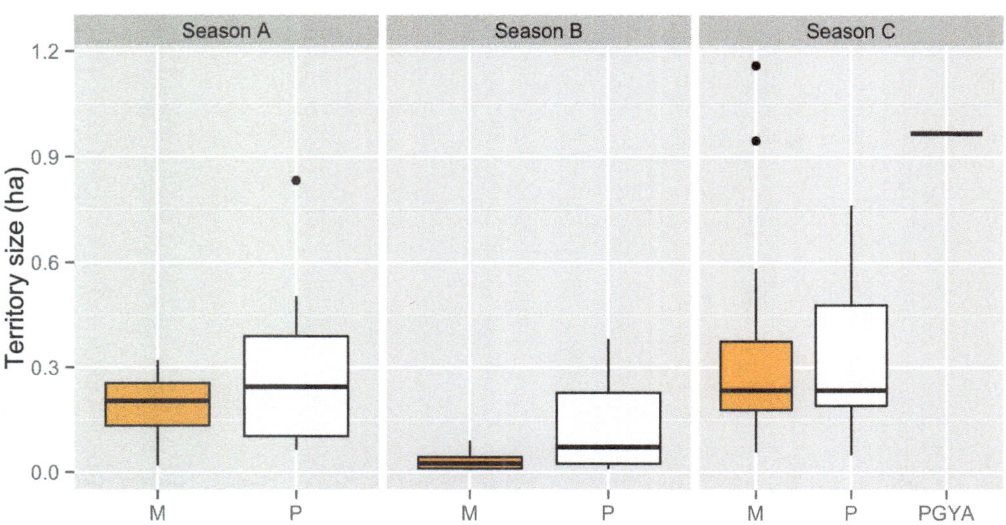

Figure 2. Territory size in dunnocks by type of mating system. Territory size in hectares of monogamous (M), polyandrous (P), and polygynandrous (PGYA) social groups of dunnocks in Dunedin, New Zealand during the 2009–2010 (Season A), 2010–2011 (Season B) and 2011–2012 (Season C) breeding seasons.

Table 1. Breeding parameters by the type of social mating system in dunnocks.

	Polyandrous	Monogamous	Effect size (β (SD))	95% CI	Cohen's d (SE), 95% CI
Territory size (ha)	0.246 (0.216), n = 32	0.244 (0.258), n = 30	0.050 (0.040)	−0.029 to 0.127	0.008 (0.254), −0.489 to 0.506
Laying date (days since the beginning of the breeding season)	44.375 (25.13), n = 40	38.43 (26.221), n = 41	5.286 (5.916)	−6.083 to 16.804	0.229 (0.222), −0.207 to 0.666
Clutch size (eggs)	3.576 (0.648), n = 61	3.507 (0.841), n = 69	0.029 (0.092)	−0.152 to 0.205	0.090 (0.175), −0.253 to 0.435
Egg volume (mm^3)	2076 (165), n = 198	2062 (169), n = 229	9.908 (29.25)	−51.223 to 61.940	0.083 (0.097), −0.106 to 0.273
Proportion of hatched eggs	0.714 (0.311), n = 55	0.588 (0.364), n = 64	0.166 (0.619)	−0.061 to 0.345	0.367 (0.185), 0.004 to 0.731
Proportion of fledglings	0.518 (0.385), n = 61	0.363 (0.377), n = 69	0.988 (0.627)	−0.251 to 2.254	0.404 (0.177), 0.056 to 0.752

Descriptive statistics are reported as means (SD) and sample sizes for each type of mating system. Effect sizes (β) and associated 95% credible intervals (CI) are shown for the difference between types of social mating system in relation to the breeding parameters. We present Cohen's d effect sizes (with associated standard error (SE) and 95% confidence intervals (95% CI)) for the difference between polyandrous and monogamous nests calculated from the raw data.

function of the type of social mating system. Nests attended by socially polyandrous pairs had higher DSR than those attended by socially monogamous groups in the first two seasons, but in season C the DSR of monogamous nests was higher than that of polyandrous (Season A: DSR polyandrous = 0.992 (0.005); DSR monogamous = 0.982 (0.006); Season B: DSR polyandrous = 0.989 (0.003); DSR monogamous = 0.973 (0.006); and Season C: DSR polyandrous = 0.971 (0.007); DSR monogamous = 0.979 (0.002)).

Discussion

In this study, we described the previously unavailable breeding ecology of dunnocks in New Zealand. More importantly, we investigated whether the types of social mating system affected breeding parameters and nest survival in our population. Our results indicate that social polyandry is beneficial for females given that polyandrous nests hatch more eggs and fledge more nestlings than socially monogamous nests. In the following paragraphs, we will discuss these results and their implications.

We observed socially polygynandrous, monogamous and polyandrous mating systems, but did not observe any cases of polygyny. The fact that we only observed these three types of mating system differs from European studies of dunnocks, especially the well-studied population in Cambridge (see [2] for a summary of this study), where all types of social mating systems have been observed. Nevertheless, it is important to note that even in the studies that observed polygynous breeding groups [10,12,18], polygyny accounted for only 4 to 8% of the observed breeding systems. More recently, polygynous breeding groups were observed in our New Zealand population (Holtmann, pers. comm.). During this breeding season, polygynous groups accounted for 3% of the breeding groups, and the adult sex ratio was of 0.569 (Holtmann, pers. comm.). All these populations have male-biased adult sex ratios that range from 0.545 to 0.583 (males/all adults), suggesting that males can only achieve polygyny, the breeding system that potentially yields the highest fitness payoffs to males, under very particular scenarios. It seems plausible that the adult sex ratio could potentially explain the difference in the frequency and types of mating systems between dunnock

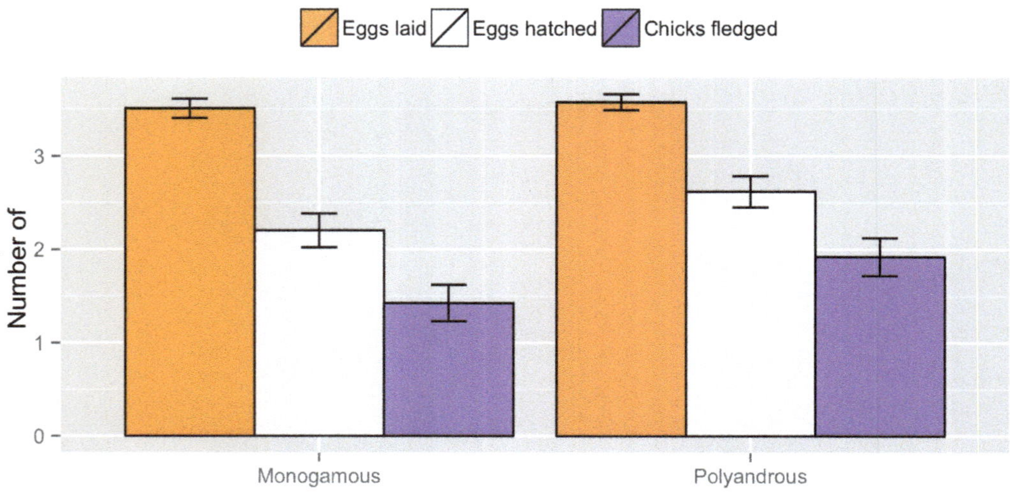

Figure 3. Estimates of breeding parameters by mating system in dunnocks. The mean number of eggs laid, hatched, and chicks fledged per nest by type of social mating system at 158 active dunnock nests in Dunedin, New Zealand (2009–2012). The error bars represent standard errors.

Table 2. Causes of egg hatching failure by the type of social mating system in dunnocks.

	Polyandrous (%)	Monogamous (%)	Effect size (β (SD))	95% CI	Cohen's d (SE), 95% CI
Desertion	10 (18.18)	11 (12.64)	0.378 (0.515)	−0.542 to 1.406	0.307 (0.439), −0.554 to 1.169
Depredation	14 (25.45)	13 (14.94)	0.701 (0.498)	−0.261 to 1.660	0.525 (0.392), −0.243 to 1.294
Infertility	12 (21.81)	32 (36.78)	−0.858 (0.444)	−1.719 to 0.033	−0.642 (0.345), −1.319 to 0.035
Embryo mortality	8 (14.54)	14 (16.09)	−0.180 (0.530)	−1.239 to 0.849	−0.144 (0.443), −1.014 to 0.725
Unknown	11 (20.0)	17 (19.54)	−0.029 (0.489)	−0.957 to 0.977	−0.022 (0.386), −0.780 to 0.736

The number of eggs and percentages are shown for each category of hatching failure for polyandrous and monogamous nests. Effect sizes (β) and associated 95% credible intervals (CI) are shown for the difference between types of social mating system in relation to the kind of hatching failure. We present Cohen's d effect sizes (with associated standard error (SE) and 95% confidence intervals (95% CI)) for the difference between polyandrous and monogamous nests calculated from the *z*-value of the models and sample sizes from raw data.

populations. It has been recently shown, across bird species, that adult sex ratio is associated with social mating systems, in such a way that when the sex ratio is male biased, female polygamy is more frequent [37]. Given their variable mating system, dunnocks provide a rare opportunity to test this prediction intra-specifically. A direct consequence of the variable mating system in male-biased populations, is the occurrence of polyandrous breeding systems in which females can count with paternal care provided by two or more males.

It has been previously shown that female dunnocks breeding in polyandrous groups have higher productivity than monogamous females [9]. We found little evidence for differences in clutch size, and egg volume between polyandrous or monogamous nests. These results indicate that polyandrous and monogamous female dunnocks have similar initial investment in the egg stage, and are not allocating resources differentially to eggs according to expectations of future help (see [38,39] for discussions of load lightening and differential allocation). Nevertheless, polyandrous groups hatched and fledged more chicks than monogamous pairs, as suggested by the moderate positive composite *d* estimate, demonstrating that having more adults attending the nest is beneficial. The benefits of having more males in a breeding groups in dunnocks is similar to the effect of having more helpers-at-the-nest in classic cooperative breeding systems (see [40–44]). This finding was supported by the higher daily survival rate of polyandrous nests than that of monogamous ones. This evidence further demonstrates (see [2]) that in the dunnock, breeding in a polyandrous social group provides direct benefits to breeding females. It will be interesting in the future to investigate whether indirect (genetic) benefits also occur as a consequence of breeding polyandrously [5], or whether female age can affect the mating system in which she breeds or how much investment she provides at the egg stage.

While in the dunnock, two studies have found evidence of direct fitness benefits of social polyandry (current study and [9]), studies of other bird species with variable mating system have failed to do so. Both the ground tit [7], and the moustached warbler [8] exhibit variable mating systems, with some females breeding with more than one partner. However in these latter species, the formation of the polyandrous group differs from that of the dunnock. While in dunnocks, the polyandrous males generally form the bond with the female at the beginning of the breeding season, in ground tits and moustached warblers, polyandrous group formation occurs after the female has initiated incubation [7,8]. It seems reasonable that differences in adult sex ratio among these species (see discussion above) create scenarios in which males: (i) are constrained into forming breeding coalitions (e.g., as in dunnocks) and thus contribute more to paternal care; and (ii) have opportunities to breed monogamously (e.g., as in ground tits and moustached warblers), but when faced with perturbations, such as nest failures, can join another pair and try to secure residual reproductive success. Under scenario (ii), joining males would not be compelled to supply high levels of paternal care, given the low certainty of paternity from joining the pair at such late stage (for a discussion of when males should reduce paternal care, see [45]).

The results of this study allow us to describe, for the first time, important reproductive aspects of the dunnock, one of the most common introduced species of birds in New Zealand. The dunnock has been and is the focus of extensive behavioural ecology studies that investigate aspects of parental care and sexual conflict (e.g., current study, and see [46,47]). The data presented here represent the first part of results from on-going studies of the genetic mating system, parental investment, and genetic and morphological divergence between NZ and UK dunnocks. These data will provide fruitful contrasts between dunnocks in their native environment with a population introduced to a new environment, providing a rare replication of behavioural ecological studies. Finally, the variable mating system of dunnocks provides opportunities for females to breed polyandrously, which lead to direct fitness benefits.

Acknowledgments

We thank Raphael I. Dias, Wolfgang Forstmeier and an anonymous reviewer who provided many useful suggestions that improved the quality of the manuscript. For help with fieldwork, we thank Patrick Crowe, James Coats, Delphine Scheck, Sam van der Horst, Sophie Gibson and Luana Santos.

Author Contributions

Conceived and designed the experiments: ESAS SN. Performed the experiments: ESAS. Analyzed the data: ESAS. Contributed reagents/materials/analysis tools: ESAS. Wrote the paper: ESAS SN.

References

1. Griffith S, Owens I, Thuman K (2002) Extra pair paternity in birds: a review of interspecific variation and adaptive function. Mol Ecol 11: 2195–2212.
2. Westneat DF, Stewart I (2003) Extra-pair paternity in birds: causes, correlates, and conflict. Annu Rev Ecol Evol Syst 34: 365–396.
3. Jennions MD, Petrie M (2000) Why do females mate multiply? A review of the genetic benefits. Biol Rev 75: 21–64.

4. Arnqvist G, Kirkpatrick M (2005) The evolution of infidelity in socially monogamous passerines: The strength of direct and indirect selection on extrapair copulation behavior in females. Am Nat 165: S26–S37.
5. Slatyer RA, Mautz BS, Backwell PRY, Jennions MD (2012) Estimating genetic benefits of polyandry from experimental studies: a meta-analysis. Biol Rev 87: 1–33. doi:10.1111/j.1469-185X.2011.00182.x.
6. Jamieson IG, Quinn JS, Rose PA, White B (1994) Shared paternity among non-relatives is a result of an egalitarian mating system in a communally breeding bird, the pukeko. Proc R Soc Lond B Biol Sci 257: 271–277.
7. Du B, Lu X (2009) Bi-parental vs. cooperative breeding in a passerine: fitness-maximizing strategies of males in response to risk of extra-pair paternity? Mol Ecol 18: 3929–3939. doi:10.1111/j.1365-294X.2009.04336.x.
8. Blomqvist D, Fessl B, Hoi H, Kleindorfer S (2005) High frequency of extra-pair fertilisations in the moustached warbler, a songbird with a variable breeding system. Behaviour 142: 1133–1148.
9. Davies NB, Houston AI (1986) Reproductive success of dunnocks, *Prunella modularis*, in a variable mating system. II. Conflicts of interest among breeding adults. J Anim Ecol: 139–154.
10. Birkhead ME (1981) The social behaviour of the dunnock *Prunella modularis*. Ibis 123: 75–84.
11. Karanjia WK (1982) The biology of the dunnock *Prunella modularis occidentalis* with special emphasis to its breeding biology Oxford: University of Oxford.
12. Snow BK, Snow DW (1982) Territory and social organization in a population of dunnocks *Prunella modularis*. J Yamashina Inst Ornithol 14: 281–292.
13. Davies NB (1983) Polyandry, cloaca-pecking and sperm competition in dunnocks. Nature 302: 334–336.
14. Tuomenpuro J (1989) Habitat preferences and territory size of the dunnock *Prunella modularis* in Southern Finland. Ornis Fenn 66: 133–141.
15. Hatchwell BJ, Davies NB (1990) Provisioning of nestlings by dunnocks, *Prunella modularis*, in pairs and trios: compensation reactions by males and females. Behav Ecol Sociobiol 27: 199–209.
16. Langmore NE, Davies NB (1997) Female dunnocks use vocalizations to compete for males. Anim Behav 53: 881–890.
17. Burke T, Davies NB, Bruford M, Hatchwell BJ (1989) Parental care and mating behaviour of polyandrous dunnocks *Prunella modularis* related to paternity by DNA fingerprinting. Nature 338: 249–251.
18. Davies NB (1992) Dunnock behaviour and social evolution. Oxford: Oxford University Press.
19. Thomson GM (1922) The naturalisation of plants and animals in New Zealand. Cambridge: Cambridge University Press.
20. Long JL (1981) Introduced birds of the World. London: David & Charles.
21. Robertson CJR, Society of New Zealand O (2007) Atlas of bird distribution in New Zealand 1999–2004. Wellington: Ornithological Society of New Zealand.
22. Santos ESA (2012) Discovery of previously unknown historical records on the introduction of dunnocks (*Prunella modularis*) into Otago, New Zealand during the 19th century. Notornis 59: 79–81.
23. Santos ESA, Jamieson IG, Santos LLS, Nakagawa S (2013) Low genetic and morphological differentiation between an introduced population of dunnocks in New Zealand and an ancestral population in England. Biol Invasions 15: 185–197. doi:10.1007/s10530-012-0278-x.
24. Hoyt D (1979) Practical methods of estimating volume and fresh weight of bird eggs. Auk 96: 73–77.
25. Zimmerling J, Craigie G, Robinson A (2004) A comparison of techniques for marking passerine nestlings. Wilson Bull 116: 240–245.
26. Mohr CO (1947) Table of equivalent populations of North American small mammals. Am Midl Nat 37: 223–249.
27. Calenge C (2006) The package "adehabitat" for the R software: A tool for the analysis of space and habitat use by animals. Ecol Modell 197: 516–519. doi:10.1016/j.ecolmodel.2006.03.017.
28. R Development Core Team (2012) R: a language and environment for statistical computing. Vienna: R Foundation for Statistical Computing. Available: http://www.R-project.org/. Accessed 2013 Jun 16.
29. Hadfield JD (2010) MCMC methods for multi-response generalized linear mixed models: the MCMCglmm R package. J Stat Softw 33: 1–22.
30. Lunn D, Thomas A, Best N, Spiegelhalter D (2000) WinBUGS - A Bayesian modelling framework: Concepts, structure, and extensibility. Stat Comput 10: 325–337.
31. Cleasby IR, Nakagawa S (2011) Neglected biological patterns in the residuals. Behav Ecol Sociobiol 65: 2361–2372. doi:10.1007/s00265-011-1254-7.
32. Laake J (2011) RMark: R Code for MARK analysis.
33. Dinsmore SJ, White GC, Knopf FL (2002) Advanced techniques for modeling avian nest survival. Ecology 83: 3476–3488.
34. Nakagawa S, Cuthill IC (2007) Effect size, confidence interval and statistical significance: a practical guide for biologists. Biol Rev 82: 591–605. doi:10.1111/j.1469-185X.2007.00027.x.
35. Cohen J (1992) A power primer. Psychol Bull 112: 155–159.
36. Nakagawa S, Santos ESA (2012) Methodological issues and advances in biological meta-analysis. Evol Ecol 26: 1253–1274. doi:10.1007/s10682-012-9555-5.
37. Liker A, Freckleton RP, Székely T (2013) The evolution of sex roles in birds is related to adult sex ratio. Nat Commun 4: 1587. doi:10.1038/ncomms2600.
38. Russell AF, Langmore NE, Cockburn A, Astheimer L, Kilner RM (2007) Reduced egg investment can conceal helper effects in cooperatively breeding birds. Science 317: 941–944.
39. Santos ESA, Macedo RH (2011) Load lightening in southern lapwings: group-living mothers lay smaller eggs than pair-living mothers. Ethology 117: 547–555. doi:10.1111/j.1439-0310.2011.01905.x.
40. Brown JL, Brown ER, Brown SD, Dow DD (1982) Helpers: Effects of experimental removal on reproductive success. Science 215: 421–422.
41. Emlen ST, Wrege PH (1991) Breeding biology of white-fronted bee-eaters at Nakuru: the influence of helpers on breeder fitness. J Anim Ecol 60: 309–326.
42. Komdeur J (1994) Experimental evidence for helping and hindering by previous offspring in the cooperative-breeding Seychelles warbler *Acrocephalus sechellensis*. Behav Ecol Sociobiol 34: 175–186.
43. Mumme RL (1992) Do helpers increase reproductive success? An experimental analysis in the Florida scrub jay. Behav Ecol Sociobiol 31: 319–328.
44. Hatchwell BJ, Russell AF, Maccoll ADC, Ross DJ, Fowlie MK, et al. (2004) Helpers increase long-term but not short-term productivity in cooperatively breeding long-tailed tits. Behav Ecol 15: 1–10.
45. Griffin AS, Alonzo SH, Cornwallis CK (2013) Why do cuckolded males provide paternal care? PLoS Biol 11: e1001520. doi:10.1371/journal.pbio.1001520.
46. Langmore NE (1996) Territoriality and song as flexible paternity guards in dunnocks and alpine accentors. Behav Ecol 7: 183–188.
47. Davies NB, Hatchwell BJ (1992) The value of male parental care and its influence on reproductive allocation by male and female dunnocks. J Anim Ecol 61: 259–272.

7

SEARCH: Spatially Explicit Animal Response to Composition of Habitat

Benjamin P. Pauli[1]*, Nicholas P. McCann[1,2], Patrick A. Zollner[1], Robert Cummings[3], Jonathan H. Gilbert[4], Eric J. Gustafson[5]

1 Department of Forestry and Natural Resources, Purdue University, West Lafayette, Indiana, United States of America, **2** Minnesota Zoo, Apple Valley, Minnesota, United States of America, **3** Rancho Deluxe Consulting, Stevens Point, Wisconsin, United States of America, **4** Great Lakes Indian Fish and Wildlife Commission, Odanah, Wisconsin, United States of America, **5** USDA Forest Service, Northern Research Station, Rhinelander, Wisconsin, United States of America

Abstract

Complex decisions dramatically affect animal dispersal and space use. Dispersing individuals respond to a combination of fine-scale environmental stimuli and internal attributes. Individual-based modeling offers a valuable approach for the investigation of such interactions because it combines the heterogeneity of animal behaviors with spatial detail. Most individual-based models (IBMs), however, vastly oversimplify animal behavior and such behavioral minimalism diminishes the value of these models. We present program SEARCH (Spatially Explicit Animal Response to Composition of Habitat), a spatially explicit, individual-based, population model of animal dispersal through realistic landscapes. SEARCH uses values in Geographic Information System (GIS) maps to apply rules that animals follow during dispersal, thus allowing virtual animals to respond to fine-scale features of the landscape and maintain a detailed memory of areas sensed during movement. SEARCH also incorporates temporally dynamic landscapes so that the environment to which virtual animals respond can change during the course of a simulation. Animals in SEARCH are behaviorally dynamic and able to respond to stimuli based upon their individual experiences. Therefore, SEARCH is able to model behavioral traits of dispersing animals at fine scales and with many dynamic aspects. Such added complexity allows investigation of unique ecological questions. To illustrate SEARCH's capabilities, we simulated case studies using three mammals. We examined the impact of seasonally variable food resources on the weight distribution of dispersing raccoons (*Procyon lotor*), the effect of temporally dynamic mortality pressure in combination with various levels of behavioral responsiveness in eastern chipmunks (*Tamias striatus*), and the impact of behavioral plasticity and home range selection on disperser mortality and weight change in virtual American martens (*Martes americana*). These simulations highlight the relevance of SEARCH for a variety of applications and illustrate benefits it can provide for conservation planning.

Editor: Matteo Convertino, University of Florida, United States of America

Funding: Funding was provided by Purdue University and a USFS Chief's Award given to Pat Zollner via USFS RJVA 06-JV-11231300-0361. The funders had no role in study design, data collection and analysis, decision to publish, or preparation of the manuscript.

Competing Interests: The authors have the following interests: Co-author RC is employed by Rancho Deluxe Consulting. There are no patents, products in development or marketed products to declare. This does not alter the authors' adherence to all the PLOS ONE policies on sharing data and materials.

* E-mail: bpauli@purdue.edu

Introduction

Individual-based (or agent-based) modeling is established as a valuable approach in disciplines such as landscape ecology and conservation biology for cases where individual variation and behavior are important drivers of system behavior [1], [2]. The conceptual basis for individual-based modeling is that the behavior of higher-level aggregations (i.e. populations, communities, ecosystems) can be simulated through the mechanistic behavior of the individuals that comprise that system [3]. In such systems the population-level attributes of simulated species emerge from the behavior and interaction of individuals that act according to detailed mechanistic rules [1], [4]. This bottom-up approach offers the opportunity for the marriage of traditional behavioral ecology (which focuses on the small-scale behavior and individual variation) and population and community ecology (which focus on large-scale dynamics of higher-order groups) through the field of behavioral landscape ecology [5–7].

Early IBMs used simplified behavioral rules for simulated animals in order to reduce computing time and reduce model complexity [8]. Unfortunately, oversimplified behavioral rules have persisted in many modern models that are less constrained by computing power. IBMs typically still include simplifications of animal behavior such as fixed dispersal distances, omniscient dispersers, or purely random walks [5], [9], [10]. The omission of behavioral complexity can have important implications. Research has shown that complex behavioral decisions drive patterns of animal movement [5]. Empirical studies show that movements based on behavioral decisions drive population expansion [10], species invasion success [11], and animal response to changing landscapes [12]. Further, ecological modeling has demonstrated that complex behavioral decisions can dramatically affect the viability of populations [13] and metapopulations [14], [15].

There is a need to incorporate more behavioral complexity into IBMs to understand the main drivers of animal dispersal and population dynamics [16–19]. Models that include too much detail, however, are in danger of becoming overly convoluted and difficult to interpret [20]. Therefore, models should strive for the optimal degree of complexity [20]. Because the appropriate degree of complexity is difficult to assess and may vary with the research goal, models with user-controlled levels of complexity are ideal. Individual-based models that are able to include sufficient behavioral complexity allow for the creation of a virtual environment in which a wide variety of questions related to population ecology could be investigated at relatively low cost and considerably lower effort than large scale field experiments.

One area of research that is well suited for individual-based modeling is animal dispersal. Natal dispersal [21–23] is vital for the maintenance of viable populations because it is associated with reductions in inbreeding [24], expansion of population range [25], the "rescue" of metapopulation patches [26], reduction in intraspecific competition for resources and mates [27], and the ability of wild populations to respond to dynamic landscapes [28]. Dispersal constitutes a complex interaction between landscape characteristics and animal behavior. Due to the rarity of dispersal events and the challenges associated with observing it, empirical research on animal dispersal is difficult [29], [30]. However, individual-based modeling offers a method for investigating the process and effect of animal dispersal [31].

Here we present program SEARCH (Spatially Explicit Animal Response to Composition of Habitat), a spatially explicit, individual-based model that incorporates a great degree of behavioral complexity. SEARCH simulates the dispersal of animals across a virtual landscape comprised of vector-based Geographic Information System (GIS) maps that determine the movement, foraging, mortality and spatial arrangement of animals (referred to as movement, food, risk and suitability maps, respectively). These maps can represent real areas such as those generated from remotely sensed data or theoretical landscapes with particular characteristics like those created based upon habitat metrics. These maps are vector-based rather than raster or regular geometry to reduce the potential for bias from spatial representation [32]. As animals move they respond to local stimuli such as habitat boundaries and are capable of changing behavior states (e.g. foraging vs. searching) in response to their experience. Simulations in SEARCH may employ temporally dynamic landscapes, such that the environment to which animals respond changes during the simulation. Thus, simulations can model events such as land cover change, succession, seasonal shifts in food availability, or diurnal patterns in predation pressure. As SEARCH is a population model and incorporates breeding stochastically, simulations may span many years and population-level trends emerge as a result of the behavior of the individuals. SEARCH also has flexibility in that many components can be easily turned on or off allowing for variation in model complexity. All code for SEARCH was written in the C# language utilizing the.NET framework and employs ArcGIS (Environmental Systems Research Institute, Redlands, California, USA) for map manipulation procedures. SEARCH is freely available in many versions (as a graphic user interface or command-line application) at http://code.google.com/p/paz-search.

SEARCH incorporates concepts and features from numerous other individual-based and spatially explicit models. The modeling work of Gustafson and Gardner [33] and Gardner and Gustafson [25], which simulates animal movement along with energetics and predation of dispersers, provide an important conceptual foundation for this model. SEARCH also incorporates aspects and concepts of other simulation models including (but not limited to) the dynamic landscapes of ALMaSS [34] and BACH-MAP [35], vector-based movement of Vuilleumier and Metzger [30], behavioral state changes of Morales et al. [36], boundary permeability of HexSim [37] and habitat selection rules similar to Kramer-Schadt et al. [38] and Wiegand et al. [39]. SEARCH incorporates many features of the models listed above (and others) into a single population model which allows researchers to include a high degree of behavioral complexity and landscape dynamics. Most features in the model are optional and most users would not use all capabilities of SEARCH in a single simulation, rather, model complexity would be driven by the research question. In fact, SEARCH incorporates flexibility to the degree that it could be considered a modeling framework in which alternative features can be implemented to create particular models.

To illustrate the ways in which the added behavioral complexity of the program may be utilized, we present examples of the application of SEARCH to three study populations. We demonstrate map-swapping capabilities of SEARCH and illustrate its use to investigate the impact of seasonally dynamic food resources on raccoon (*Procyon lotor*) dispersers. Similarly, we demonstrate the impacts of temporally variable predation risk on eastern chipmunk (*Tamias striatus*) dispersal. Additionally, the effects of behavioral state changes on weight distribution and disperser mortality were investigated using American martens (*Martes americana*). Finally, we simulated virtual martens that were given a range of home-range choice rules that varied the importance of food resources, safety, proximity, and search time when selecting a home-range center to determine their effect on settlement time, dispersal distance, and disperser mortality. Through these simulations we were able to highlight some of the capabilities of SEARCH and underscore the importance of incorporating an appropriate but flexible degree of behavioral complexity into individual-based models.

Model Overview

The following model description follows the Overview – Design Concepts – Details (ODD) protocol for describing individual-based models [40], [41]. Therefore, the "Model Overview" section describes the model in broad terms, the "Design Concepts" section discusses certain design concepts and the degree to which they are employed in SEARCH, the "Details" section and the supplementary materials give detailed descriptions of the processes and algorithms used in the model and the remaining sections describe case studies conducted to illustrate the capabilities of SEARCH.

Purpose

SEARCH simulates the dispersal and home-range establishment of animals across a virtual landscape. Animals respond to four vector GIS layers that contain values used by the rules for animal movement, foraging, risk of predation, and the suitability of habitat for home-range establishment and the configuration of areas occupied by established resident animals. Thus, users are able to investigate such factors as the potential impacts of landscape change, habitat permeability, and energetic budgets on animal populations. Output from SEARCH provides information on both the characteristics of animal dispersal as well as the associated emergent population-level attributes. SEARCH can be used to simulate a variety of species and utilizes research from disparate fields such as animal movement, foraging ecology, and physiology to parameterize the model.

State Variables and Scales

Individuals within SEARCH can be one of two classes – juvenile dispersers or adult residents. Dispersers are characterized by a unique number, sex, weight, perception, activity mode, behavioral state, and location. State variables for residents include animal number, sex, and home-range location. Population characteristics such as age structure, sex ratio, and mortality rates can be derived from these model outputs.

Dispersers interact with the environment by responding to habitat characteristics represented in four vector GIS maps. Polygon values contained in these maps drive animal movement, foraging, mortality and home-range establishment by dictating correlated random walk movement parameters, the probability of acquiring prey and size of prey, the probability of being killed, and the suitability for home-range establishment and whether or not a location is currently occupied by a resident animal, respectively (Table 1).

Time periods in SEARCH are discrete. The user defines the time-step length (≥ 1 minutes), dispersal season length (≥ 1 day), and the number of years of simulation (≥ 1 year). The period of the year outside of the dispersal season (i.e. the inter-dispersal period) is modeled as a single, discrete time period. Spatial scales employed by SEARCH follow the resolution and extent of user-input GIS maps. The extents for case studies presented below are 0.25 km^2–660 km^2 but larger areas can be used [42].

Process Overview and Scheduling

In SEARCH, the user parameterizes the model to emulate the movement, foraging ecology, and habitat use of the study species (Table 2). Additionally, the default behavior of individuals can be modified based upon sex, behavioral state, and time (Table 3).

During SEARCH simulations, animals traverse a virtual landscape comprised of four GIS maps that each contains multiple parameter values that are used to model virtual animal behavior (Table 1). These maps are user created and can represent biologically relevant landscape features such as habitat type, land use, land cover, or topology. The initial landscape is populated by adult residents and/or released juveniles from the social and release maps, respectively. The initial population is input by the user depending on the scenario that best models the system under study. As animals move throughout the landscape, all four maps are queried during each time-step by each animal. After each animal movement segment, the location, energetic reserves, and behavioral states are updated for that individual. Animals move, forage, die and establish home ranges according to habitat and species parameters. During a time-step, each animal completes every action for that time-step in sequential order based on animal number assigned geographically at the beginning of each year.

Once a user-defined threshold for number of steps taken or number of sites visited is exceeded, animals select a site for a possible home-range center from areas searched during dispersal. Virtual animals then move to that site, and attempt to establish a home range but may continue dispersing if they fail to locate a site with unoccupied suitable habitat. Animals become residents once a home range is established, but die if a home range is not established during the dispersal period. During the inter-dispersal period, residents are subject to random mortality and females stochastically give birth to young which disperse the following dispersal season.

Design Concepts

Objectives

In SEARCH each disperser's objective is to establish a home range before the end of the dispersal period. The behavioral traits of virtual animals are expected to affect dispersal success. Animals can be parameterized to remain in habitat that is of higher quality than adjacent areas. Since habitat quality in simulations is typically parameterized to correlate with increased safety and foraging opportunities (although it need not be) dispersers that remain in higher quality areas are typically less susceptible to predation or starvation. Similarly, animals that switch behavior due to perceived danger or low energy reserves behave in a way that minimizes risk of predation or starvation. Finally, implicit in many of the home-range selection criteria is the assumption that animals choosing home-range locations of higher quality (i.e. better foraging opportunities and/or lower mortality risks) will have offspring that are less likely to succumb to predation or starvation and will therefore have increased fitness.

Adaptation

Simulated animals make decisions in response to the environment and change behavior based on their individual experiences. Such dispersal behaviors are expected to change dispersal success. For example, animals calculate whether to cross a habitat boundary during dispersal by comparing the relative habitat value

Table 1. Landscape parameter maps and field definitions input by the user to reflect variation in animal behavioral or physiological responses to different GIS classifications.

Map type	Field	Definition	Range
Movement	Tortuosity	Tortuosity of movement drawn from a wrapped Cauchy distribution where 0 produces a purely random walk and 1 produces a linear movement	0–1
	Step length	Mean step length (m) per time-step; includes a field for standard deviation (±SD)	≥ 0
	Energy use	Energy used per time-step	≥ 0
	Crossing	Rank of location quality	≥ 0
	Perceptual window modifier	Modifies distance of perceptual window	≥ 0
Food	Probability	Probability of capturing a prey item per time-step	0–1
	Size	Mean size energy gain of captured prey; includes a field for standard deviation (±SD)	≥0
Risk	Probability	Probability of mortality due to depredation per time-step	0–1
Suitability	Suitability	Suitability of habitat for home-range establishment	0 or 1
	Occupancy	If occupied by a male	0 or 1
	Occupancy	If occupied by a female	0 or 1
Release		Location, number, and gender of animals upon initialization	na

Table 2. Animal parameter values input by the user for the temporal aspects of the simulation along with basic attributes of virtual animals.

Parameter	Description
Dispersal season dates	Start and end dates for dispersal season each year and the number of years to conduct simulation runs
Time-step resolution	Number of minutes between time-steps
Start time	Time of day that dispersal begins
Activity and resting periods	Hours (±SD) of activity and rest from start time
Home-range center selection threshold	Number of steps or suitable and unoccupied sites traversed before selecting a home-range center
Minimum home-range size	Gender-specific minimum area required for a home range
Distance weighting factor	Gender specific coefficient that modifies the effect of proximity on home-range center selection
Energy	Initial, minimum and maximum energy allowable for each animal. Death occurs below the minimum value
Search/forage trigger	Energy threshold below which animals switch behaviors from primarily searching to primarily foraging
Perception window	Distance at which habitat suitability and occupancy are perceived during dispersal
Safe to risky	Probability of switching from safe behavior to risky behavior
Risky to safe	Probability of switching from risky behavior to safe behavior

at their current location with adjacent locations based upon values in the movement map. Implied in this decision making is a simple predictive model that assumes that remaining in areas of higher habitat quality will increase their chance of home-range establishment. Animals also evaluate the probability of successfully capturing prey/forage when selecting a location for a home range. In addition, animal activity mode may change based on a user-defined energy threshold and the behavior of active animals can also be affected by their perceived risk of predation. In these ways, virtual animals respond to perceived danger or low energy reserves by changing their behavioral state to respond to conditions.

Sensing

Animals in SEARCH are able to detect information in their environment and respond to that information accordingly. During movement, virtual animals sense the predation pressure, foraging resources, habitat quality and the suitability of habitat for home range establishment along with the presence of resident animals around them. These attributes are detected by virtual animals at a distance dependent upon the perceptual window of that animal (see S.1.4 of Text S1 for detail). The characteristics detected by virtual animals can be influenced by their behavioral states, energetic reserves, gender, time of day and season such that animals maintain a memory of the habitat as they perceived it during dispersal. If animals revisit a location, the most recent memory is stored and the totality of this memory is used when animals select potential home range centers.

Interaction

SEARCH incorporates little interaction between individuals and no direct interaction between dispersers. Indirect interaction between individuals is modeled through the restriction of non-overlapping home ranges between same-sex animals. This implies some form of interactive exclusion between individuals. Similarly, inter-dispersal reproduction implies male-female interaction though this is not modeled explicitly in SEARCH.

Stochasticity

Nearly every action taken by animals in SEARCH is probabilistic, thus stochasticity is critical. Animals in SEARCH move based upon a correlated random walk where path tortuosity can vary from completely random (mean vector length = 0) to completely straight (mean vector length = 1) based upon a value drawn from a wrapped Cauchy distribution [43], [44] that the virtual animal obtains from the polygons in the movement map. In addition, when dispersing individuals calculate the probability of crossing a habitat boundary they do so by comparing the quality values of adjacent polygons in the movement map to a random number to determine whether to cross a habitat boundary.

Animal foraging uses stochasticity by assigning a probability of successfully foraging to each polygon to determine if an animal

Table 3. Modifiers of animal behavior employed in SEARCH which allow the user to modify habitat values by multiplying them by a real number to reflect variation caused by gender, activity mode, vigilance mode, time of day, and date.

Modifier	Parameter modified
Prey acquisition probability	Probability of acquiring prey in a given GIS classification
Predation probability	Probability of mortality due to predation in a given GIS classification
Energy use	The energy used in each GIS classification
Movement speed	The movement distance per time-step in each GIS classification
Movement tortuosity	The movement tortuosity in each GIS classification
Perception	Animal perception distance in each GIS classification

gains energy. The amount of energy gained during successful foraging bouts is drawn from a normal distribution based on a user-specified mean and standard deviation derived from a map polygon from the food map.

The likelihood of an animal dying during a time-step is assigned based on the animal's location, the time of day, season, and animal parameters. Stochasticity is incorporated through a randomly drawn number that determines if that animal dies. Resident mortality is calculated in a similar way, although resident morality probabilities are aspatial and occur during each time-step of the dispersal season as well as once during the inter-dispersal season.

The vigilance mode of dispersing animals is determined by a stochastic perception of risk. In this way, animals switch between risky and safe behaviors. Animal activity bouts (i.e. active/resting) are also determined by a stochastic process. The duration of each active and resting period is drawn from a normal distribution based on a user-defined mean and standard deviation of time.

Home-range center selection is weighted toward higher quality sites but incorporates stochasticity in selecting a location. Animals create home range polygons by generating a set of points around a potential home-range center with random orientation and with distances that are drawn from a distribution based on the minimum home-range area for each sex to delineate the vertices of that home-range polygon.

Breeding incorporates a number of stochastic processes. Whether a resident female becomes pregnant during an inter-dispersal period is based upon the probability of a female breeding. For those female residents that do become pregnant, the number of offspring produced is drawn from a normal distribution based on a user-specified mean and standard deviation of litter size. The sex of each offspring is determined probabilistically according to the sex ratio parameter.

Emergence

Population-level patterns emerge based upon the behavior and interactions of the individual animals in response to the spatio-temporal configuration of habitat and conspecific residents. These emergent properties develop from the interaction of baseline behavioral parameters, individual variation (due to gender, activity mode, etc.), interaction between individuals and stochasticity. Such higher-order emergent properties can include population density and spatial configuration, animal weight distribution, mortality rates by source (starvation vs. predation), and mortality locations.

Observation

During SEARCH simulations, data on the behavioral state (energetic, vigilance mode, etc.) and fate of each individual are produced during every time-step. This output includes a GIS polygon map that depicts the animal movement and perception during dispersal. Following each dispersal season, a landscape map is also produced that depicts all existing home ranges. Other population-level attributes (such as annual survivorship, population density, habitat selection, etc.) can be calculated from the individual and population output.

Details

Initialization

SEARCH can be initialized to reflect one of three scenarios. First, the simulation can begin without existing home ranges and all dispersers can be created based upon a map of releases. Such a simulation could model the reintroduction of a species into an area from where it had been extirpated. Second, the simulation can begin with established home ranges throughout the landscape without any released animals. Thus, all dispersers would be the result of reproduction of established females (chosen randomly based upon user-input parameters). This scenario would best reflect the population dynamics of an established population. Finally, simulations can implement a combination of resident reproduction and release of individuals. In this way an augmentation or supplementation of an existing population can be modeled [42].

At the beginning of each simulation an empty memory map is created for each individual. This map reflects the area perceived by an animal during dispersal and the occupancy and suitability of all areas observed. During each time-step the memory map is updated for every animal to reflect the area perceived during dispersal. When animals begin selecting potential home-range centers, this map is used to eliminate all points in areas perceived as unsuitable or occupied. For each animal a text file is also created that records the animal's conditions for all state variables (i.e. location, energy level, etc.) during each time-step.

Input

Time Parameters. SEARCH incorporates flexibility in the temporal scale and extent of simulations. The user inputs the start date of the simulation, the number of years to simulate (≥ 1), the number of days in the dispersal season (≥ 1), the start time of day 1 of the dispersal season (0–23 hours) and the length of each time-step in minutes (≥ 1).

Maps. At the most basic level, virtual animals in SEARCH respond to the user-specified parameters of five GIS maps (Table 1). These map types include 4 polygon maps (movement, food, risk and suitability) and 1 point map (assigning location and number of released animals). The landscape in SEARCH is dynamic in that any of the 5 GIS layers can be replaced at any time during the simulation with a map with different parameters and/or spatial configuration. Such map swapping can be employed to simulate habitat change or management, landscape disturbance or simply the variation in animal response to habitats at different times of day or seasons (see case studies for examples).

Species Attributes. Energy parameters (with no units) represent the initial energetic reserves of each disperser, the minimum allowable energy level (below which animals die of starvation) and the maximum possible energy level (Table 2). During each time-step, energy is lost based upon the energetic cost associated with a particular habitat as defined in the movement map. Energy is gained if prey or forage is acquired based upon the probability of foraging success and the amount gained is a function of the size of the item on the food map.

Active dispersers move across the landscape relative to the various parameters on the movement map (resting animals remain static). The mean durations of active and resting periods are assigned by the user (along with a standard deviation) that applies to all dispersers. Activity periods of individual animals, however, may diverge based upon the stochasticity in period length due to variance around the mean (described by the standard deviation) so that animal activity and rest cycles need not be synchronized with one another. Animals may have many active and resting periods within a single day but must begin each year active. The mean values of all active and rest periods must sum to 24. Variability around mean active and resting periods may cause animals to have activity periods that do not exactly follow a 24-hour cycle.

Within SEARCH, active, dispersing animals can be either foraging or searching. The particular behavior of each activity mode can be specified by the user through the parameterization of modifiers (see "Modifiers" in "Details" section; Table 3). The user could parameterize SEARCH so that the probability of an animal successfully capturing a forage or prey item, for instance, would be higher during foraging activity but lower when in searching mode. All animals begin in searching mode but switch to foraging if their energetic reserves fall below a user-defined threshold. Note both searching and foraging animals are both capable of foraging or establishing home ranges.

As with activity modes, virtual animals can also exhibit one of two possible vigilance modes at any time. Individuals can either be in safe mode or risky mode. The modifiers (as defined by the user; Table 3) that affect animal behavior can differ based upon an animal's vigilance mode. The risk of an animal being killed, for instance, could be decreased when an animal is in safe mode (reflecting higher vigilance, for example) relative to risky mode. Animals begin each dispersal season in risky mode but will change to safe mode during any time-step if a randomly drawn number falls within a user specified interval (see section S.7.2 of Text S1 for details). Animals in safe mode change back to risky mode if a randomly drawn number falls within the user specified interval (see section S.7.2 of Text S1 for details).

The individual memory of each animal in SEARCH is retained explicitly through the use of a memory map. An animal's perceptual range is the distance at which it can perceive and respond to landscape features [45]. SEARCH, however, utilizes a perceptual window that includes both perceptual range and small scale wandering of animals during a time-step (similar to assessment corridors of Doerr and Doerr [46], ellipses of Bélisle et al. [47], and circle-ellipses of Selonen et al. [48]. In SEARCH, this perceptual window is the area (with radius in meters) animals perceive during dispersal. Within this memory map, animals retain a complete record of the suitability and occupancy of perceived areas. If a specific location or area is revisited, the most recent suitability and occupancy status is overwritten on the memory map. The perceptual distance of an animal can be modified based upon that habitat type in which an animal occurs, the time of day, or the season. For example, the user can reduce the animal-perception window during low moonlight relative to the full moon perception window [49].

Home-Range Attributes. Animals in SEARCH can be triggered to begin selecting home-range centers based upon one of two user-selected criteria – either a user-specified number of time-steps have elapsed since the inception of dispersal or the animal has visited a user-specified number of suitable and unoccupied sites. Once this trigger is surpassed, animals choose their preferred home-range center based on user-specified criteria including the factors that most influence patch selection [50], [51]. These criteria consider either (1) the proximity of the site ("closest"), (2) the proximity and food availability of a site ("food"), (3) the proximity and risk of mortality at a site ("risk"), or (4) the proximity, food and risk of a site ("integrated"). The user chooses one of these four criteria for each run. The user also inputs home-range establishment requirements including the minimum area for a home range for each sex and the relative importance of site proximity in home-range center selection for each sex (Table 2). It is possible for users to effectively negate the effect of proximity in home site selection by choosing a large value for the distance weighting factor (Table 2).

Modifiers. In SEARCH, the user can modify the baseline parameters for many behavioral traits of animals to reflect the variability in behavior as a result of an animal's gender, behavioral state, the time of day, or the season (Table 3). Modifiers can be created for both sexes (male and female), all four behavioral states (risky-searching, risky-foraging, safe-searching, and safe-foraging), and any number of temporal modifiers at two scales (hourly, daily). For instance, the perceptual distance of an animal that relies on vision may be increased during the day relative to night [49].

Resident Attributes. Residents are assigned a single, user-specified probability of mortality during each time-step (independent of location) but during the inter-dispersal period are subject to a single mortality probability. Additionally, a proportion of randomly selected females give birth to a user-defined (mean±SD) number of young (when this integer is negative, 0 is used) that have a sex ratio based on the defined probability of female offspring. Young begin dispersal in the subsequent season at the center of the mother's home range.

Submodels

See Supplementary material for details including submodel descriptions (Text S1), technical documentation (Text S2) and process flow diagrams (Figure S1).

Illustrative Case Studies

To demonstrate the functionality of map swapping, behavioral states and home range selection in SEARCH, we provide a number of case studies here. These case studies are meant to be illustrative of the capabilities of the model rather than complete studies of model performance. These examples highlight some of the possible applications of SEARCH to real-world scenarios and demonstrate SEARCH's applicability to a variety of systems.

Case Studies: Food Map Swapping

SEARCH is unique among individual-based models in that it allows the user to exchange any of the four GIS maps to which virtual animals respond (i.e. movement, risk, food, or social) at a daily, seasonally, or yearly time scale allowing temporally dynamic parameters. To illustrate the functionality and implications of these map swapping capabilities we simulated two study systems (raccoons in Indiana and chipmunks in Wisconsin).

In the first study system, the effects of seasonally dynamic food resources on raccoon population dynamics were investigated. Raccoons in north-central Indiana experience dramatic shifts in forage availability throughout the year due to corn maturation and the subsequent superabundance of food resources [52], [53]. This temporal shift in food availability was modeled in SEARCH by swapping food maps during corn maturation in the summer to reflect the increase in both the likelihood of successful foraging and the amount of energy gained during a foraging bout in an agricultural polygon. Because agriculture was the dominant cover type in the area and virtual raccoons were expected to forage heavily there, we predicted that raccoons that experienced temporally dynamic food resources would have dramatically different changes in mass compared to animals exposed to static foraging resources.

Methods. Raccoon simulations were conducted for a single year with a dispersal season of 150 days (the maximum observed raccoon colonization time in one study – Beasley unpublished data) and 1 hour time-steps on GIS layers digitized from USGS aerial photos of a 4 km×4 km area of the upper Wabash River basin in north-central Indiana (for more detail see: [53] and [54]; Supplementary material, Table S1) simulations began with resident raccoons present on the landscape based on estimated raccoon density in the study area [55] and all dispersers were the product of resident reproduction. For virtual raccoons, energy parameters were used as a surrogate of mass. To approximate raccoon weight limits observed in field studies, dispersers began with a mass of 3750 g, had a maximum mass of 10000 g and died of starvation if their body mass fell below 1800 g [56], [57],

(Beasley unpublished data). Dispersers were active for 12 h during the night (1800 until 600) and rested for 12 h during the day [58], [59], [60]. Translocated raccoons typically establish home ranges after 2 weeks of dispersing [61] so this value (168 active steps) was used as the trigger time for virtual raccoons to begin choosing home-range centers. Animals used the integrated criteria for home-range center selection and had minimum home-range sizes of 0.29 km^2 and 0.12 km^2 for males and females, respectively [53]. Resident virtual raccoons were subject to a mortality probability each time-step (8.34×10^{-5}) and during the inter-dispersal season (0.194) based on published raccoon survival estimates [62–66]. Surviving resident females had a 90% pregnancy probability during the inter-dispersal season [67], [68], Beasley unpublished data) and produced young based on a litter size with mean 3.5 and standard deviation of 1 [56], [69–72], Beasley unpublished data). To simulate corn maturation, virtual animals were exposed to maps with low food resources in agricultural areas for 37 days, superabundant food resources in these areas for 76 days and then low resources for another 37 days. The weight distribution of successful dispersers in simulations with dynamic food maps was compared to that of the null model with a constant intermediate forage probability and energy gain. All modifiers were set to 1 to effectively eliminate variability due to gender, time, or behavior.

For each scenario, 10 replicates were conducted and the weight change of every successful disperser in each simulation was recorded. Because the weight changes of dispersers within a simulation were not independent, we used a mixed modeling approach to avoid pseudoreplication [73], [74]. Each replicate simulation was nested within the corresponding scenario (static or dynamic food resources) and a nested ANOVA was conducted [75] to determine if the mean weight change of virtual raccoons differed among the scenarios.

Results/Discussion. In simulations of raccoons, the weight distributions of animals differed significantly between simulations with and without temporally dynamic food resources ($t_{18} = -20.78$, $p<0.0001$). Virtual raccoons that experienced static values for energetic gain from agricultural habitats gained an average of 475 g (SD = 230, n = 58). Virtual raccoons subjected to temporally dynamic agricultural food resources (with relative scarcity followed by superabundance and then scarcity again) lost an average of 473 g (SD = 228, n = 54) during the simulations (Figure 1).

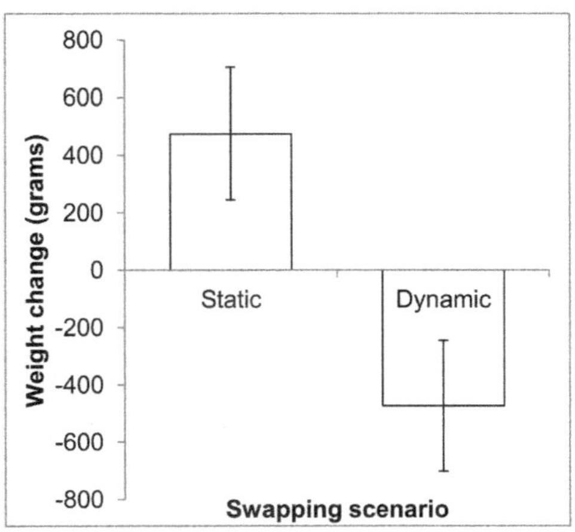

Figure 1. Raccoon weight change. Mean (± 1 SD) change in weight of virtual raccoons for simulations with static and temporally dynamic food maps. Weight change values are for animals that successfully established home ranges during the simulation (Base n = 58; Swap n = 54).

The observed differences in virtual raccoon weight changes seem to be due to the amount of time animals dispersed before settling. Nearly all (96%) successful dispersers in the temporally dynamic simulation settled before agricultural areas produced superabundant food resources. These animals, therefore, were only subjected to food scarcity in agricultural areas and, not surprisingly, all lost weight during the simulation. The two dispersers that settled after the food map swap either settled one day after the food switch (losing 1108 g) or settled well after the superabundant food emergence and gained 701 g during the simulation. Therefore, temporally dynamic food resources appear to have a substantial effect on virtual animal weight distribution but this effect is dependent on the animal's dispersal time.

Simulations of raccoon weight changes in response to dynamic food resources underscore the impact dynamic landscapes can have on animal populations. Animal populations that exploit seasonal food resources (including anthropogenic or naturally ephemeral sources) can be easily simulated in SEARCH. This capability allows researchers to include more temporal complexity in foraging resources than would be possible in simulations with static food availability.

Case Studies: Risk Map Swapping and Behavioral Response

An appreciation for the importance of spatial heterogeneity in predation pressure on animal populations (i.e. the 'landscape of fear'; [76] has gained recognition within the ecological community. Numerous animal populations are impacted by and respond to spatial variation in mortality risk [77–81]. The importance of temporal heterogeneity in predation pressure has been studied less frequently, however. Temporal variation in mortality risk has been shown to have significant impact on survival and behavior in a few populations [82–84].

We investigated the potential population-level impacts of temporal variation in predation and the impact of behavioral response to such variation by modeling the dispersal of chipmunks in northern Wisconsin (Supplementary material, Table S2). Over the entirety of each simulation run virtual chipmunks were exposed to 1) a homogeneous static risk of mortality, 2) a spatially heterogeneous but temporally static predation risk or 3) different diurnal and nocturnal predation probabilities that were both spatially heterogeneous. Additionally we simulated virtual chipmunks that responded to variation in predation risk through variable permeability of habitat boundaries. Animals either had no response to habitat boundaries, or preferentially remained in habitat with lower predation risk (independent of time), lower predation risk based upon current time or lower predation risk based upon predicted future risk.

Methods. Chipmunks are exposed to different predators at different times of day, such as raptors during daylight and mustelids at night [85], [86], (Zollner unpublished data). Furthermore, the timing of chipmunk activity in conjunction with predators has been shown to affect survival in field experiments [87]. In SEARCH, predation risk was modeled using a habitat-specific, empirically derived index of predation pressure where motion sensor cameras observed relative predation intensity of taxidermied chipmunks at different times in varying habitats (Zollner unpublished data). This was combined with published annual mortality values of chipmunks [88] to estimate simulation predation rates. Data used to calculate predation probabilities were combined, segregated spatially or segregated spatially and temporally to create predation maps that were aspatial, spatially heterogeneous or spatially and temporally heterogeneous, respectively. Temporally dynamic landscapes were simulated by swapping risk maps that represented daytime (6:00 – 18:00) and nighttime (18:00 – 6:00) mortality risk. Alternatively, chipmunks were simulated that had a constant predation risk with either aspatial or spatially variable risk.

Simulation scenarios contrasted the empirical conditions described above with a range of responses to predation risk by virtual chipmunks. Animals without response to habitat boundaries used crossing values that were identical for all habitat types except for those areas into which animals never entered (cross. value none in Supplementary material, Table S2). Virtual animals that responded to spatial variation in predation risk had crossing values that scaled boundary permeability to the relative predation risk of habitat types independent of time (cross. value base in Supplementary material, Table S2). Temporally responsive animals had crossing values that determined boundary crossing relative to time-dependent predation risk either currently (cross. value day and night relative to risk map swapping in Supplementary material, Table S2) or predictively (cross. value day and night 1 hour prior to risk map swapping in Supplementary material, Table S2).

All combinations of predation variation and behavioral response were simulated for a total of 12 simulation scenarios. Therefore, animals in particular simulations 1) responded at a coarser scale than risk was simulated (under-response), 2) responded at the same scale as risk was simulated (correct-response) or 3) responded at a finer scale than risk was simulated (over-response).

Simulation output consisted of both overall and predation-specific (e.g. only animals that died of predation rather than from a failure to establish a home range by the end of the dispersal season) mortality rates. Both rates were compared among populations of all scenarios to determine if variation in predation risk and animal response to predation pressure impacted survival of the population. We predicted that both types of mortality rate in each population would scale inversely with the degree of response of the virtual chipmunks (i.e. under-response > correct response > over-response).

All chipmunk simulations were run for 2 years with a 30 day dispersal season and 5 min time-steps on GIS maps 500 m×500 m (derived from digitized aerial photos). All virtual dispersers were produced through reproduction of resident animals present on the landscape at the beginning of the simulation based on estimated chipmunk density in the study area (Zollner unpublished data). Dispersing animals were active for 4 h beginning at 9 AM [89] and had a baseline perceptual window of 120 m [90]. Animals were triggered to begin choosing home-range centers after 1000 active steps or roughly 3 weeks of dispersal [91]. Virtual chipmunks used the closest site criteria and had minimum home-range requirements of 0.002 km^2 for both sexes [89]. All resident females became pregnant and produced litters (mean = 4.5, SD = 0.5) and had an equal chance of male and female young [89]. We eliminated energy as a limiting factor due to insufficient data by creating ideal foraging conditions (i.e. 100% success) and no energy loss during movements. Similarly, resident mortality was eliminated and all multiplicative modifiers were set as 1 to turn off variability due to gender, time and behavior.

For each scenario, 10 replicate simulations were conducted and the overall and predation-specific mortality rates of the dispersers were determined. The data were then power transformed (raised to 0.75) in order to satisfy assumptions of equal variance and normalized residuals (Shapiro-Wilk, W = 0.985073, p = 0.2086) and an ANOVA was conducted to compare overall mortality and predation rates based upon simulated risk and animal behavior. Contrasts were then used to compare the average predation rates of simulations with varying degrees of animal responsiveness (i.e. under, correct or over-response; all tests conducted in SAS 9.3; [75]).

Results/Discussion. Simulations of virtual chipmunks suggested that the effect of predation was compensatory in the overall mortality rate of the population. Overall mortality rates did not differ among virtual dispersing chipmunks in the 12 simulated scenarios ($F_{11,\ 108} = 0.85$, p = 0.59) nor did overall mortality differ between the three levels of responsiveness of animals to their experienced predation (all pairwise p>0.25; Figure 2a). Predation-specific mortality, however, was significantly different between scenarios ($F_{11,\ 108} = 2.08$, p = 0.0277). In addition, contrasts of predation rate between levels of responsiveness differed with greater simulated mortality rates for under-responsive populations as compared to the others (Table 4, Figure 2b).

The overall mortality rates were the same for all simulations despite differences in predation-specific mortality. Thus, it appears that predation, as a result of variation in probabilities of mortality and animal response, is only a compensatory factor within overall mortality of simulated eastern chipmunks. Therefore, factors other than predation, such as competition for space, seem to be driving mortality. Interestingly, the compensatory effect of predation has also been suggested for eastern chipmunks based on the results of field studies [92].

As expected, chipmunks that were under-responsive to the variability in predation risk exhibited the highest predation rates. Interestingly, virtual animals that over-responded to predation variability had the same predation rates as those that responded at the appropriate scale. This suggests that while over-responding animals did not gain any advantage through their behavior they also suffered no mortality cost from their over-responsiveness.

These simulations highlight the capabilities and possible implications of fine-scale, temporally dynamic predation risk in conjunction with variable behavioral response in SEARCH. Most models use coarse, static mortality risk to model predation [82], [83]. SEARCH simulations with virtual chipmunks have demonstrated that the inclusion of temporally and spatially variable risk, when combined with various degrees of behavioral response, can dramatically affect predation mortality. While the overall mortality was unaffected (thus the population dynamics nearly identical) by the inclusion of this added complexity, predation-specific mortality differed greatly. Therefore, research concerned with cause-specific mortality would benefit from the fine-scale temporal component of predation available in SEARCH.

Case Studies: Behavioral States

Many spatially explicit individual-based models are behaviorally minimalistic and assume static behavior for mathematical convenience or due to lack of empirical data [5]. Empirical research, however, has shown the behavioral states of dispersers to have dramatic effects on population dynamics [36], [93], [94]. SEARCH allows users to provide virtual animals with greater behavioral complexity by defining the conditions under which animals switch activity or behavioral state.

We conducted SEARCH simulations to investigate the impact of increasing behavioral complexity on dispersal characteristics of simulated American martens. We simulated martens with varying degrees of responsiveness to behavioral triggers (low energy reserves and narrow predator escapes) and measured the weight change of individual virtual martens along with the disperser mortality rates under each scenario.

Methods. Simulations were conducted for one year on GIS maps of Wisconsin (derived from data from the U.S. Forest Service

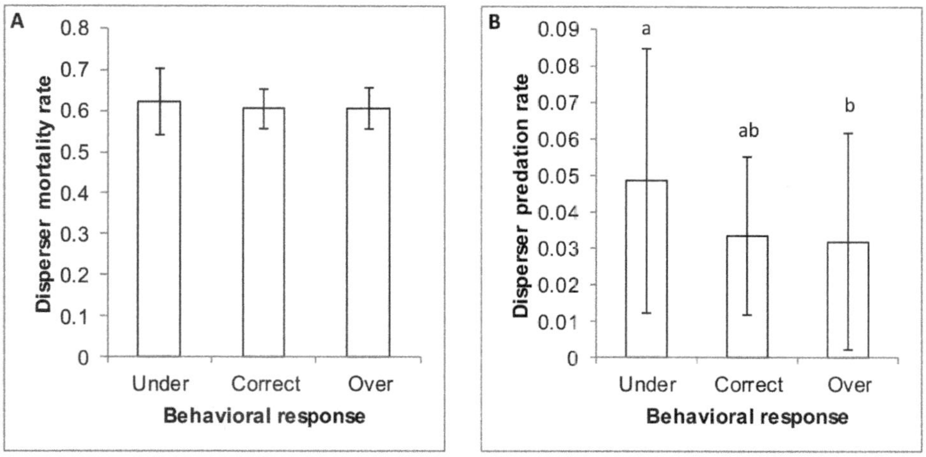

Figure 2. Eastern chipmunk mortality rate. Mean (± 1 SD) A) overall mortality rate and B) predation-specific mortality rate for virtual chipmunks with varying levels of response to simulated predation risk.

– Chequamegon-Nicolet National Forest Combined Data Systems STAND data and the Wisconsin Department of Natural Resources' WISCLAND Level 3 GIS data [95]. Initial residents (17 male and 25 females) were created using the output of a separate simulation (the closest home-range selection of final case study) to approximate the distribution of martens following the four years of releases from 1987–1990 [96]. Additionally, the population was augmented with a release of 14 animals to simulate a portion of the marten releases in the study area in 2008 [97], [98].

Spatial parameters for movement were derived from marten snow backtracking data, foraging parameters were based on small mammal trapping data, and predation risk parameters were based on predator indices (Zollner unpublished data; Supplementary material, Tables S3, S4 and S5). Habitat suitability on the social map was based on Dumyahn et al. [99] and Wright [100]. The dispersal season was 60 d [101] with 1 h time-steps. Animals began active behavior at 4 am with alternating activity and rest periods of 4.5 h and 7.5 h (all with SD = 8) [102]. Marten energy values were based on conversions of body mass to kilocalories. Animals dispersed with 4548 units of initial energy and had minimum and maximum energy limits of 3866 and 5003 units, respectively [102]. Virtual martens had a baseline perceptual window of 100 m (from perceptual range of Gardner and Gustafson [25]. A baseline value of 270 active steps, or an average of 30 days, was used for a trigger value after which individuals began establishing home ranges. Virtual martens had minimum home ranges of 4.25 km^2 and 2.32 km^2 for males and females, respectively [99]. Residents were subject to a 5×10^{-5} time-step mortality probability and a 0.17 inter-dispersal mortality [103]. Surviving resident females had a 74.4% likelihood of becoming pregnant [104], [105], had litter sizes with a mean of 3 and standard deviation of 1 [105] with a balanced sex ratio [104]. Gender and temporal modifiers were set to 1 with the exception of the male risk modifier which was 0.7632 to model the low mortality risk of male martens compared to females [103].

We simulated virtual martens with different sensitivities to low energy reserves. Martens that fell below the energy threshold switched activity mode from searching to foraging behavior (and vice versa). Martens in foraging mode had an increased likelihood of capturing prey and a decrease in energy use, movement speed, mean vector length and perceptual window distance as compared to searching martens (Supplementary material, Table S6). The baseline, reduced and increased threshold levels for simulations were set at 4250, 4000 and 4500, respectively. We predicted a positive relationship between energetic threshold level and animal weight change due to those animals' ability to respond to their level of energetic reserves. We predicted no effect of threshold level on disperser mortality compared to simulations with behaviorally static animals, however, due to the low likelihood of animal starvation.

Similarly, we simulated martens with one of three levels of response to perceived mortality risk (animals switching from risky to safe behavior and vice versa). The baseline response was a 1%

Table 4. Contrasts of average predation-specific mortality between simulations of virtual chipmunks with varying degrees of responsiveness to simulated predation pressure.

Contrast	Difference	Standard error	Degrees of freedom	t-value	p-value
Over vs. Under	−0.4425	0.1829	108	−2.42	0.0172
Over vs. Correct	−0.1280	0.2240	108	−0.57	0.5687
Correct vs. Under	−0.2772	0.1530	108	−1.81	0.0728
Over vs. Pooled Others	−0.5706	0.3421	108	−1.67	0.0982
Correct vs. Pooled Others	−0.0373	0.0818	108	−0.46	0.6493
Under vs. Pooled Others	0.2399	0.1002	108	2.40	0.0183

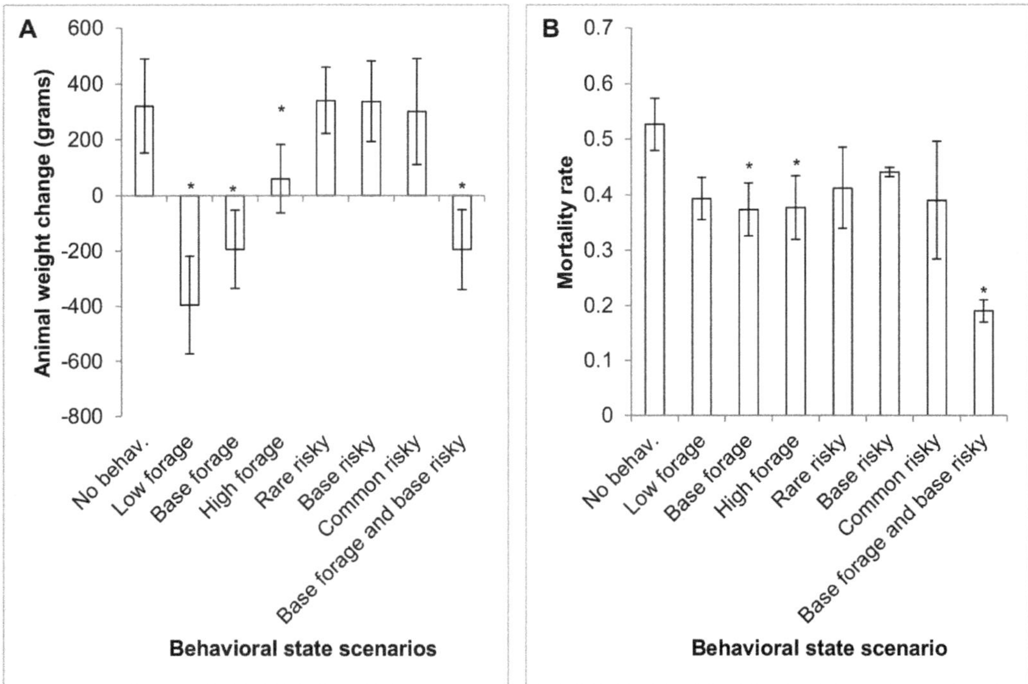

Figure 3. Effect of behavioral state switching on American martens. Mean (± 1 SD) marten A) weight change (pooled for all replicates) and B) mortality rate for eight behavioral state scenarios (three replicates per scenario). Simulations consisted of animals with no behavioral state changes (no behav.), a search-forage threshold of 4000 (low forage), a search-forage threshold of 4250 (base forage), a search-forage threshold of 4500 (high forage), a risky-safe proability of 0.001 (rare risky), a risky-safe probability of 0.01 (base risky), a risky-safe probability of 0.1 (common risky) or a search-forage threshold of 4250 and a safe-risky probability of 0.01 (base forage and base risky). Asterisk denotes scenarios that were significantly different from no behavior simulations.

probability of switching vigilance modes during a single time-step. Animals with increased responsiveness had a 10% likelihood of changing and animals with reduced response had a 0.1% chance of changing behavior. Animals displaying safe behavior represented animals with increased vigilance and thus had an increased perceptual window [106] but decreased speed (due to vigilance pauses; [107], [108]) and mortality risk [109] compared to those exhibiting risky behavior (Supplementary material, Table S6). Because the safe-risky trigger responds only to perceived predation risk, we predicted that animals with varying levels of responsiveness would have different levels of disperser mortality but no differences in mean animal weight change over the course of the simulation as compared to virtual animal without behavioral state changes.

Finally, a scenario was conducted where marten had both a baseline level of danger response (1% probability of behavioral switch) and a baseline level of response to low energy (4250 units). Modifiers for animals in each of the four possible behavioral states were the product of the values for the activity and vigilance modes (Supplementary material, Table S6). We predicted that virtual animals with both predation and energetic behavioral responses would have different weight changes and mortality rates compared to the null model where animal behavior was constant.

Three replicates of each scenario were conducted in SEARCH. A single disperser mortality rate per simulation was recorded. Because these data satisfied the assumptions of normality (Shapiro-Wilk, $W = 0.983142$, $p = 0.9459$) and homogeneous variances (Levene, $F = 2.35$, $p = 0.0745$) the survival values for the 8 behavioral states were compared using an ANOVA. A post hoc Dunnett's comparison [110] of the control with the 7 experimental setups (three levels of risk response, three levels of energetic response, one level of both responses; simultaneous $\alpha = 0.05$) was conducted.

Because the dispersal characteristics of individual animals in the same simulation were not independent (i.e. fate of one animal affected that of another such as an animal settling precluding another individual from establishing a home range in an area), animal weights were considered multiple measurements from a single replicate to avoid pseudoreplication [73], [74]. Because these data failed to satisfy both the normality and equality of variances assumptions, these data were rank transformed in order to conduct non-parametric tests [111]. The means of the rank transformed data were compared using a nested ANOVA. A post hoc Dunnett's comparison was conducted to compare the mean values of simulation runs to those virtual martens that had no behavioral modifiers relative to weight distribution (all tests conducted in SAS 9.3; [75]).

Results/Discussion. The inclusion and degree of sensitivity of animal behavioral state changes had significant effects on dispersal mortality and weight distribution. The mean ranked weight distributions of animals differed ($F_{7, 16} = 280.57$, $p < 0.0001$) and was the response variable most affected by the various behavioral states. The differences in weight changes of animals appeared to be driven primarily by the search-foraging threshold (Figure 3a). All four conditions that used the search-forage trigger (low threshold, base threshold, high threshold, base foraging threshold combined with base risky) had significantly lower weight changes compared to animals that had no behavioral state changes (Dunnett's test, $p < 0.05$, $df_{error} = 874$, $MS_E = 22968$, $t_{crit} = 2.6$). Mean weight change had a positive relationship with forage-search threshold level where the lowest threshold level resulted in animals with the greatest weight loss. The risky-safe vigilance modes, however, had no significant effect on animal weight compared to animals without behavioral state changes (all $p>0.05$). There was an apparent positive correlation between the variance in animal weight and the probability of vigilance mode change though this was not tested explicitly.

As expected, the search-forage threshold had a much stronger effect on virtual marten weights than the risky-safe probability. Animals subjected to differences in foraging and searching behavior had weights approaching the threshold for behavioral state change. In these situations, animals in the searching mode lost weight until they fell below the threshold, switched to foraging mode and began gaining weight. This resulted in animal weights that commonly oscillated around the threshold value. The risky-search probability, on the other hand, was stochastic and independent of an animal's energy reserves. All levels of this parameter, therefore, had little direct effect on weight change. Instead the probability of switching vigilance modes seemed to have more influence on the variance of animal weight changes than on the mean population weight change.

Simulations including behavioral state changes differed significantly in terms of disperser mortality ($F_{7,\ 16}=8.10$, $p=0.0003$). Simulations with the high and baseline forage-search threshold as well as simulations with both the base threshold and base risky had significantly lower disperser mortality than simulations with virtual animals with static behavior (using a simultaneous $\alpha=0.05$). Most dramatically, simulations with base levels of both vigilance mode and activity mode changes had less than 40% the mortality rate of simulations with static behavior (Figure 3b).

Both types of behavioral complexity reduced mortality of virtual martens with the lowest mortality rates associated with animals with both types of behavioral response. Inclusion of the search-foraging threshold allowed animals to avoid starvation by responding to low energy reserves and changing behavior to maximize energetic gain. Similarly, the risky-safe probability allowed virtual animals to react to predation escapes and respond with safer behavior but was independent of observed mortality risk and was, therefore, purely stochastic. Thus, the forage-search threshold more dramatically affected mortality as it responded to the systematic threat of starvation while the safe-risky behavior responded to a stochastic probability of a predation escape.

Behavioral variability in SEARCH resulted in animals that behaved differently than would have been possible in simulations without this flexibility. This added behavioral complexity significantly affected dispersal characteristics of virtual martens. Therefore, the dynamics of animals that utilize different behavioral states or strategies could be dramatically impacted by the inclusion or exclusion of behaviorally complexity in simulation modeling. SEARCH allows researchers to evaluate whether this increased complexity affects the population under study and incorporates it when it is found to be necessary to accurately model the system.

Case Studies: Home-Range Trigger and Decision Criteria

SEARCH employs a number of options for home-range selection. These selection criteria differentially prioritize sites based on the factors most associated with animal habitat selection (i.e. foraging opportunities, [112]; predation risk, [113]; proximity, [114]). Empirical and modeling studies have shown that dispersal and space use of animals is affected by how they weigh the potential costs and benefits of sites when selecting from a number of potential home range locations [39], [115], [116].

We studied the impact of varied prioritization of these costs and benefits for particular locations on dispersal distance, settlement time, and disperser mortality of virtual American martens. Dispersers in SEARCH select home-range locations based on a user-specified prioritization of attributes. We investigated the effect of the different home-range selection criteria on the dispersal of American marten. Simulations consisted of virtual animals that chose home-range locations based on only proximity (to current location), proximity and food availability, proximity and mortality risk, or proximity and food and risk together. For each scenario the dispersal distance, time to settlement and mortality of dispersers was measured. We predicted that the different home range selection criteria would result in differences in dispersal distance but not settlement time or disperser mortality because we expected animals to travel further to find appropriate home sites when using more restrictive selection criteria.

Methods. Marten simulations (with same parameterization as 6.3 except where specified) were run for 4 years on 33.4 km×28.7 km GIS map layers (same sources as "Behavioral States" case study). The simulation began without any resident individuals present in the area and animals were added to the simulation through releases every year that corresponded with actual releases of American martens in Wisconsin [96] as well as reproduction of successful dispersers. For each scenario, three replicates were simulated.

To satisfy the assumptions of normality and equal variances, some of the data were transformed. Dispersal distances were square-root transformed (Shapiro-Wilk, $W=0.9979$, $p=0.2604$; Levene, $F=1.19$, $p=0.2920$), settlement times were rank transformed and unmodified mortality values were used (Shapiro-Wilk, $W=0.8750$, $p=0.0757$; Levene, $F=1.07$, $p=0.4162$). Dispersal distances and times contained pseudoreplication due to the fact that the fate of an individual could be affected by that of another animal in the same simulation. Therefore, a nested ANOVA was conducted to detect differences among scenarios. Because a single mortality rate was measured for each simulation, a standard ANOVA was conducted to determine if scenarios differed in respect to disperser mortality (all tests conducted in SAS 9.3; [75]).

Results/Discussion. The criteria used for home-range center selection had little effect on any of the response variables of virtual animals. Mean dispersal distances were nearly identical for all four criteria ($F_{3,\ 8}=1.31$, $p=0.338$; Figure 4a). Similarly, settlement times were fairly constant across the home-range selection scenarios ($F_{3,\ 8}=1.69$, $p=0.2450$; Figure 4b). Finally, the mortality rates for simulations with the four criteria were nearly identical ($F_{3,\ 8}=0.56$, $p=0.654$; Figure 4c).

Overall, virtual martens in SEARCH simulations exhibited the same dispersal characteristics (distance, time and mortality rate) in response to a variety of home range selection rules. At first, these negative results appear inconsequential and such non-significant results are often overlooked [117]. This case study, however, highlights one of the major advantages of simulation models that have the capability of flexible levels of complexity. Models with features that can be turned on or off allow researchers to experimentally test the level of model complexity needed to adequately simulate the species in question [1]. In our case we found that more complex home-range selection criteria had no effect on the dispersal characteristics of virtual martens in Wisconsin. Therefore, future research on this study system could use simplified home-range selection rules (primarily the 'closest' home-range criterion) allowing for a simulation structure that only includes necessary complexity [19]. Of course this particular form of model simplification would not pertain to all cases, but the principle of refinement of model application through experimentation is a valuable asset that could be used in many implementations of SEARCH.

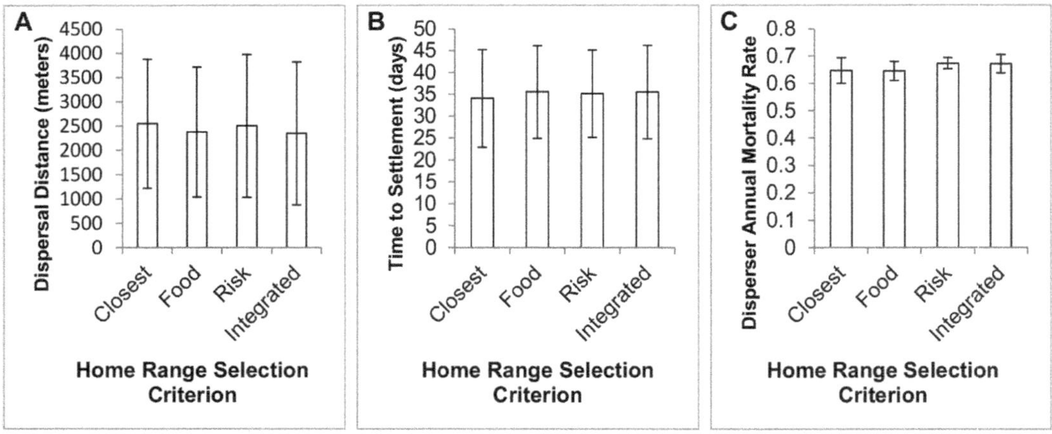

Figure 4. Effect of home range selection criterion on American martens. Mean($\pm$ 1 SD) A) dispersal distance B) settlement time and C) annual mortality for virtual marten home range selection scenarios. Simulations consisted of martens that selected home range locations based upon proximity (closest), proximity and food availability (food), proximity and predation risk (risk), or proximity, predation risk and food availability (integrated).

Summary

We present SEARCH, a newly developed, spatially explicit, individual based model. SEARCH incorporates a high degree of behavioral complexity and allows for temporally dynamic landscapes. SEARCH is parameter-intensive which allows researchers to utilize all available data in parameterizing the model. However, SEARCH has the flexibility to allow users to "turn off" functions in the model when data for parameterization are unavailable (as would be the case for some component of the model in nearly every case). This functionality enables users to investigate when added behavioral complexity results in quantitatively different model outcomes. Thus, users can investigate the model's sensitivity to added complexity and evaluate the benefits and costs of incorporating behavioral complexity. Users are therefore able to optimize model functionality for the research question and population under study.

SEARCH is applicable to a number of species in a wide variety of systems though probably best suited for solitary mammals. It is a model that is ideal for simulating behaviorally complex populations with small abundances in a conservation setting. Furthermore, SEARCH allows researchers to simulate habitat and population manipulations that would be impractical in a field setting and offers that ability to project population dynamics into the future. There are a number of limitations to SEARCH, however. For example, the interaction between individuals in SEARCH is fairly rudimentary and the dynamic aspects of maps in SEARCH must be determined *a priori* (rather than as a response to model behavior). Furthermore, the breeding algorithms in SEARCH are not spatially explicit nor are they responsive to the state of the individual animal. Despite such shortcomings, SEARCH offers researchers a tool for investigating animal dispersal (and the subsequent population dynamics) that is not species specific but is capable of incorporating behavioral complexity not found in most comparable models. Thus the use of this tool has the potential to offer valuable insight into the role of the interplay between complex behavior and landscape configuration to animal population dynamics and management.

Supporting Information

Figure S1 SEARCH process schematic. Process flow of SEARCH simulation (left) with detailed schematic of animal processes during dispersal (right).

Table S1 Spatial parameters of raccoon simulations with values for each habitat type corresponding to the movement map, the risk map and the food map.

Table S2 Spatial parameters of chipmunk simulations with values for each habitat type corresponding to the movement map, the social map and the risk map.

Table S3 Spatial parameters of movement map for American marten simulations.

Table S4 Spatial parameters of food map for American marten simulations.

Table S5 Spatial parameters of risk map for American marten simulations.

Table S6 Modifier values for behavioral states of American martens with various levels of behavioral switching.

References S1 Citations for materials referenced in supporting information documents.

Text S1 Submodel descriptions.

Text S2 Technical documentation.

Acknowledgments

We thank Tim Van Deelen, Ron Moen and Karl Malcolm for thoughts on model design. Dave McCabe, Bryan Pijanowski, David Braun and the Purdue Rosen Center for Advanced Computing provided computational assistance. We thank Gene Rhodes and Jim Beasley for sharing data and providing insight into raccoon ecology. Barny Dunning, Bill Peterman, Gene Rhodes, Julia Earl, Ray Semlitsch, Steven Railsback and Vanessa Quinn contributed valuable feedback on manuscript drafts.

Author Contributions

Conceived and designed the experiments: BPP NPM PAZ RC. Performed the experiments: BPP. Analyzed the data: BPP PAZ. Contributed reagents/materials/analysis tools: BPP PAZ RC JHG EJG. Wrote the paper: BPP NPM PAZ JHG EJG.

References

1. Grimm V, Railsback SF (2005) Individual-based modeling and ecology. Princeton, NJ: Princeton University Press. 428 p.
2. McLane AJ, Semeniuk C, McDermid GJ, Marceau DJ (2011) The role of agent-based models in wildlife ecology and management. Ecol Model 222: 1544–1556.
3. Łomnicki A (1992) Population ecology from the individual perspective. In: DeAngelis DL, Gross LJ, editors. Individual-based models and approaches in ecology: populations, communities and ecosystems. New York: Chapman and Hall. pp. 3–17.
4. Railsback SF (2001) Concepts from complex adaptive systems as a framework for individual-based modelling. Ecol Model 139: 47–62.
5. Lima SL, Zollner PA (1996) Towards a behavioral ecology of ecological landscapes. Trends in Ecol Evol 11: 131–135.
6. Bélisle M (2005) Measuring landscape connectivity: the challenge of behavioral landscape ecology. Ecology 86: 1988–1995.
7. Knowlton JL, Graham CH (2010) Using behavioral landscape ecology to predict species' responses to land-use and climate change. Biol Conserv 143: 1342–1354.
8. Topping CJ, Dalkvist T, Forbes VE, Grimm V, Sibly RM (2009) The potential for the use of agent-based models in ecotoxicology. In: Devillers J, editor. Ecotoxicology modeling. Dordrecht:Springer. pp. 205–236.
9. South A (1999) Dispersal in spatially explicit population models. Conserv Biol 13: 1039–1046.
10. Morales JM, Ellner SP (2002) Scaling up animal movements in heterogeneous landscapes: the importance of behavior. Ecology 83: 2240–2247.
11. Holway DA, Suarez AV (1999) Animal behavior: an essential component of invasion biology. Trends Ecol & Evol 14: 328–330.
12. Baguette M, Van Dyck H (2007) Landscape connectivity and animal behavior: functional grain as a key determinant for dispersal. Landsc Ecol 22: 1117–1129.
13. Gosling LM (2003) Adaptive behavior and population viability. In: Festa-Bianchet M, Apollonio M, editors. Animal Behavior and Wildlife Conservation. Washington DC:Island Press. pp. 13–30.
14. Woodroffe R (2003) Dispersal and conservation: a behavioral perspective on metapopulation persistence. In: Festa-Bianchet M, Apollonio M, editors. Animal behavior and wildlife conservation. Washington DC:Island Press. pp. 33–48.
15. Heinz SK, Wissel C, Frank K (2006) The viability of metapopulations: individual dispersal behavior matters. Landsc Ecol 21: 77–89.
16. Stephens PA, Frey-Roos F, Arnold W, Sutherland WJ (2002) Model complexity and population predictions: the alpine marmot as a case study. J Anim Ecol 71: 343–361.
17. Jepsen JU, Baveco JM, Topping CJ, Verboom J, Vos CC (2005) Evaluating the effect of corridors and landscape heterogeneity on dispersal probability: a comparison of three spatially explicit modelling approaches. Ecol Model 181: 445–459.
18. Patterson TA, Thomas L, Wilcox C, Ovaskainen O, Matthiopoulos J (2008) State-space models of individual animal movement. Trends in EcolEvol 23: 87–94.
19. Hudgens BR, Morris WF, Haddad NM, Fields WR, Wilson JW, et al. (2012) How complex do models need to be to predict dispersal of threatened species through matrix habitats? Ecol Appl 22: 1701–1710.
20. Grimm V, Revilla E, Berger U, Jeltsch F, Mooij WM, et al. (2005) Pattern-oriented modeling of agent-based complex systems: lessons from ecology. Science 310: 987–991.
21. Howard WE (1960) Innate and environmental dispersal of individual vertebrates. American Midland Naturalist 63: 152–161.
22. Greenwood PJ (1980) Mating systems, philopatry and dispersal in birds and mammals. Animal Behavior 28: 1140–1162.
23. Belthoff JR, Ritchison G (1990) Natal dispersal: Greenwood (1980) revisited. Condor 92: 803–804.
24. Bollinger EK, Harper SJ, Barrett GW (1993) Inbreeding avoidance increases dispersal movements of the meadow vole. Ecology 74: 1153–1156.
25. Gardner RH, Gustafson EJ (2004) Simulating dispersal of reintroduced species within heterogeneous landscapes. Ecol Model 171: 339–358.
26. Brown JH, Kodric-Brown A (1977) Turnover rates in insular biogeography: effect of immigration on extinction. Ecology 58: 445–449.
27. Dobson FS, Jones WT (1985) Multiple causes of dispersal. Am Nat 126: 855–858.
28. Travis JMJ, Dytham C (1998) The evolution of dispersal in a metapopulation: a spatially explicit, individual-based model. Proc Biol Sci 265: 17–23.
29. Koenig WD, Van Vuren D, Hooge PN (1996) Detectability, philopatry, and the distribution of dispersal distances in vertebrates. Trends Ecol Evol 11: 514–517.
30. Vuilleumier S, Metzger R (2006) Animal dispersal modelling: handling landscape features and related animal choices. Ecol Model 190: 159–170.
31. Mooij WM, DeAngelis DL (2003) Uncertainty in spatially explicit animal dispersal models. Ecol Appl 13: 794–805.
32. Holland EP, Aegerter JN, Dytham C, Smith GC (2007) Landscape as a model: the importance of geometry. PLoS Comput Biol 3: 1979–1992.
33. Gustafson EJ, Gardner RH (1996) The effect of landscape heterogeneity on the probability of patch colonization. Ecology 77: 94–107.
34. Topping CJ, Hansen TS, Jensen TS, Jepsen JU, Nikolajsen F, et al. (2003) ALMaSS, an agent-based model for animals in temperate European landscapes. Ecol Model 167: 65–82.
35. Pulliam HR, Dunning JB Jr, Liu J (1992) Population dynamics in complex landscapes: a case study. Ecol Appl 2: 165–177.
36. Morales JM, Haydon DT, Frair J, Holsinger KE, Fryxell JM (2004) Extracting more out of relocation data: building movement models as mixtures of random walks. Ecology 85: 2436–2445.
37. Schumaker NH (2011) HexSim (Version 2.3). U.S. Environmental Protection Agency, Environmental Research Laboratory. 237 p. Available: http://www.epa.gov/hexsim.Accessed 12 January 2013.
38. Kramer-Schadt S, Revilla E, Wiegand T, Breitenmoser U (2004) Fragmented landscapes, road mortality and patch connectivity: modeling influences on the dispersal of Eurasian lynx. J Appl Ecol 41: 711–723.
39. Wiegand T, Moloney KA, Naves J, Knauer F (1999) Finding the missing link between landscape structure and population dynamics: a spatially explicit perspective. Am Nat 154: 605–627.
40. Grimm V, Berger U, Bastiansen F, Eliassen S, Ginot V, et al. (2006) A standard protocol for describing individual-based and agent-based models. Ecol Model 198: 115–126.
41. Grimm V, Berger U, DeAngelis DL, Polhill JG, Giske J, et al. (2010) The ODD protocol: a review and first update. Ecol Model 221: 2760–2768.
42. McCann NP (2011) American marten survival and movement in the presence of fishers. PhD dissertation. West Lafayette, Indiana. Purdue University.
43. Batschelet E (1965) Statistical methods for the analysis of problems in animal orientation and certain biological rhythms. Washington DC :American Institute of Biological Sciences. 57 p.
44. Haefner JW, Crist TO (1994) Spatial model of movement and foraging in harvester ants (*Pogonomyrmex*) (I): the role of memory and communication. J Theor Biol 166: 299–313.
45. Zollner PA, Lima SL (1997) Landscape-level perceptual abilities in white-footed mice: perceptual range and the detection of forested habitat. Oikos 80: 51–60.
46. Doerr ED, Doerr VAJ (2005) Dispersal range analysis: quantifying individual variation in dispersal behavior. Oecologia 142: 1–10.
47. Bélisle M, Desrochers A, Fortin MJ (2001) Influence of forest cover on the movements of forest birds: a homing experiment. Ecology 82: 1893–1904.
48. Selonen V, Hanski IK, Desrochers A (2010) Measuring habitat availability for dispersing animals. Landsc Ecol 25: 331–335
49. Zollner PA, Lima SL (1999) Illumination and the perception of remote habitat patches by white-footed mice. Anim Behav 53: 489–500.
50. Sih A (1980) Optimal behavior: can foragers balance two conflicting demands? Science 210: 1041-1043.
51. Holbrook SJ, Schmitt RJ (1988) The combined effects of predation risk and food reward on patch selection. Ecology 69: 125–134.
52. Rivest P, Bergeron JM (1981) Density, food habits, and economic importance of raccoons (*Procyon lotor*) in Quebec agrosystems. Can J Zool 59: 1755–1762.
53. Beasley JC, DeVault TL, Retamosa M, Rhodes OE Jr (2007) A hierarchical analysis of habitat selection by raccoons in northern Indiana. J Wildl Manage 71: 1125–1133.
54. Retamosa MI, Humberg LA, Beasley JC, Rhodes OE Jr (2008) Modeling wildlife damage to crops in northern Indiana. Human-Wildlife Conflicts 2:225–239.
55. Beasley JC, Rhodes OE Jr (2008) Relationship between raccoon abundance and crop damage. Human-Wildlife Conflicts 2: 248–259.
56. Stuewer FW (1943) Raccoons: their habitats and management in Michigan. Ecol Monogr 13: 203-257.
57. Mech LD, Barnes DM, Tester JR (1968) Seasonal weight changes, mortality and population structure of raccoons in Minnesota. J Mamm 49: 63–73.

58. Sharp WM, Sharp LH (1956) Nocturnal movements and behavior of wild raccoons at a winter feeding station. J Mamm 37: 170–177.
59. Bider JR, Thibault P, Sarrazin R (1968) Schèmes dynamiques spatio-temporales de l'activité de *Procyon lotor* en relation avec le comportement. Mammalia 32: 137–163.
60. Lotze JH, Anderson S (1979) *Procyon lotor*. Mamm Species 119: 1–8.
61. Mosillo M, Heske EJ, Thompson JD (1999). Survival and movements of translocated raccoons in northcentral Illinois. J Wildl Manage 63: 278–286.
62. Fritzell EK, Greenwood RJ (1984) Mortality of raccoons in North Dakota. Prairie Nat 16: 1–4
63. Glueck TF, Clark WR, Andrews RD (1988) Raccoon movement and habitat use during the fur harvest season. Wildl Soc B 16: 6–11.
64. Clark WR, Hasbrouck JJ, Kienzler JM, Glueck TF (1989) Vital statistics and harvest of an Iowa raccoon population. J Wildl Manage 53: 982–990.
65. Hasbrouck JJ, Clark WR, Andrews RD (1992) Factors associated with raccoon mortality in Iowa. J Wildl Manage 56: 693–699.
66. Gehrt SD, Fritzell EK (1999) Survivorship of a nonharvested raccoon population in south Texas. J Wildl Manage 63: 889–894.
67. Fritzell EK, Hubert GF, Meyen BE, Sanderson GC (1985) Age-specific reproduction in Illinois and Missouri raccoons. J Wildl Manage 49: 901–905.
68. Hasbrouck JJ (1991) Demographic responses of raccoons to varying exploitation rates. PhD dissertation. Ames, Iowa,Iowa State University.
69. Sagar RG (1956) A study of factors affecting raccoon reproduction in Ohio. MS thesis. Columbus, Ohio, The Ohio State University.
70. Sanderson GC, Hubert GF (1981) Selected demographic characteristics of Illinois (USA) raccoons (*Procyon lotor*). In: Chapman JA, Pursley D, editors. Worldwide Furbearer Conference Proceedings. Frostburg: Worldwide Furbearer Conference, Inc. pp. 487–513
71. Ritke ME (1990) Quantitative assessment of variation in litter size of the raccoon *Procyon lotor*. Am Midl Nat 123: 390–398.
72. Rees EE, Pond BA, Phillips JR, Murray D (2008) Raccoon ecology database: a resource for population dynamics modeling and meta-analysis. Ecol Inform 3: 87–96.
73. Hurlbert SH (1984) Pseudoreplication and the design of ecological field experiments. Ecol Monogr 54: 187–211.
74. Millar RB, Anderson MJ (2004) Remedies for pseudoreplication. Fish Res 70: 397–407.
75. SAS Institute Inc. (2011) SAS/STAT® 9.3 User's Guide. SAS Institute Inc. 8621 p.
76. Laundré JW, Hernández L, Altendorf KB (2001) Wolves, elk, and bison: reestablishing the "landscape of fear"in Yellowstone National Park, U.S.A. Can J Zool 79: 1401–1409.
77. Altendorf KB, Laundré JW, López González CA, Brown JS (2001) Assessing effects of predation risk on foraging behavior of mule deer. J Mamm 82: 430–439.
78. Brown JS, Kotler BP (2004) Hazardous duty pay and the foraging cost of predation. Ecol Lett 7: 999–1014.
79. Hernández L, Laundré JW (2005) Foraging in the 'landscape of fear' and its implications for habitat use and diet quality of elk *Cervus elaphus* and bison *Bison bison*. Wildl Biol 11: 215–220.
80. Van der Merwe M, Brown JS (2008) Mapping the landscape of fear of the cape ground squirrel (*Xerus inauris*). J Mamm 89: 1162–1169.
81. Laundré JW, Hernández L, Ripple WJ (2010) The landscape of fear: ecological implications of being afraid. Open Ecol J 3: 1–7.
82. Lima SL, Bednekoff PA (1999) Temporal variation in danger drives antipredator behavior: the predation risk allocation hypothesis. Am Nat 153: 649–659.
83. Sih A, Harding KC (2000) New insights on how temporal variation in predation risk shapes prey behavior. Trends in Ecol Evol 15: 3–4.
84. Ferrari MCO, Messier F, Chivers DP (2008) Larval amphibians learn to match antipredator response intensity to temporal patterns of risk. Behav Ecol 19: 980–983.
85. Allen EG (1938) The habits and life history of the eastern chipmunk, *Tamias striatus lysteri*. New York State Museum Bulletin 314: 1–122.
86. Yahner RH (1978) The adaptive nature of the social system and behavior in the eastern chipmunk, *Tamias striatus*. Behav Ecol Sociobiol 3: 397–427.
87. DeCoursey PJ, Walker JK, Smith SA (2000) A circadian pacemaker in free-living chipmunks: essential for survival? J Comp Physiol A 186: 169–180.
88. Tryon CA, Snyder DP (1973) Biology of the eastern chipmunk, *Tamias striatus*: life tables, age distributions, and trends in population numbers. J Mamm 54: 145–168.
89. Snyder DP (1982) *Tamias striatus*. Mamm Species 168: 1–8.
90. Zollner PA (2000) Comparing the landscape level perceptual abilities of forest sciurids in fragmented agricultural landscapes. Landsc Ecol 15: 523–533.
91. Loew SS (1999) Sex-biased dispersal in eastern chipmunks, *Tamias striatus*. Evol Ecol 13: 557–577.
92. Nupp TE, Swihart RK (1998) Effects of forest fragmentation on population attributes of white-footed mice and eastern chipmunks. J Mamm 79: 1234–1243.
93. Frair JL, Merrill EH, Visscher DR, Fortin D, Beyer HL, et al. (2005) Scales of movement by elk (*Cervus elaphus*) in response to heterogeneity in forage resources and predation risk. Landsc Ecol 20: 273–287.
94. Delgado MM, Penteriani V (2008) Behavioral states help translate dispersal movements into spatial distribution patterns of floaters. Am Nat 172: 475–485.
95. Reese HM, Lillesand TM, Nagel DE, Stewart JS, Goldmann RA, et al. (2002) Statewide land cover derived from multiseasonal Landsat TM data: a retrospective of the WISCLAND project. Remote Sens Environ 82: 224–237.
96. Williams BW, Gilbert JH, Zollner PA (2007) Historical perspective on the reintroduction of the fisher and American marten in Michigan and Wisconsin. Newton Square: USDA, Forest Service. 29 p.
97. Woodford JE, Dumyahn SL (2011) Conservation and management plan for American martens in Wisconsin. Madison:Wisconsin Department of Natural Resources. 43 p.
98. Powell RA, Lewis JC, Slough BG, Brainerd SM, Jordan NR, et al. (2012) Evaluating translocations of martens, sables, and fishers: testing model predictions with field data. In: Aubry KB, Zielinski WJ, Raphael MG, Proulx G, Buskirk SW, editors. Biology and conservation of martens, sables, and fishers: a new synthesis. Ithaca:Cornell University Press. pp. 93–137.
99. Dumyahn JB, Zollner PA, Gilbert JH (2007) Winter home-range characteristics of American marten (*Martes americana*) in northern Wisconsin. Am Midl Nat 158: 382–394.
100. Wright JL (1999) Winter home range and habitat use by sympatric fishers (*Martes pennanti*) and American martens (*Martes americana*) in northern Wisconsin. MS thesis. Stevens Point, Wisconsin, University of Wisconsin-Stevens Point.
101. Bull EL, Heater TW (2001) Home range and dispersal of the American marten in northeastern Oregon. Northwest Nat 82: 7–11.
102. Gilbert JH, Zollner PA, Green AK, Wright JL, Karasov WH (2009) Seasonal field metabolic rates of American martens in Wisconsin. Am Midl Nat 162: 327–334.
103. McCann NP, Zollner PA, Gilbert JH (2010) Survival of adult martens in northern Wisconsin. J Wildl Manage 74: 1502–1507.
104. Strickland MA, Douglas CW (1987) Marten. In: Novak M, Baker JA, Obbard ME, Malloch B, editors. Wild furbearer management and conservation in North America.Toronto: Ontario Trappers Association. pp. 531–546.
105. Aune KE, Schladweiler P (1997) Age, sex structure, and fecundity of the American marten in Montana. In: Proulx G, Bryant HN, Woodard PM, editors.*Martes*: taxonomy, ecology, techniques, and management.Edmonton:-:Provincial Museum of Alberta. pp. 61–77.
106. Lima SL, Bednekoff PA (1999) Back to the basics of anitpredatory vigilance: can nonvigilant animals detect attack? Anim Behav 58: 537–543.
107. McAdam AG, Kramer DL (1998) Vigilance as a benefit of intermittent locomotion in small mammals. Anim Behav 55: 109–117.
108. Zollner PA, Lima SL (2005) Behavioral trade-offs when dispersing across a patchy landscape. Oikos 108: 219–230
109. Lima SL (1996) The influence of models in the interpretation of vigilance. In: Bekoff M, Jamieson D, editors.Readings in Animal Cognition. Cambridge:-:Massachusetts Institute of Technology Press. pp. 201–216.
110. Dunnett CW (1955) A multiple comparison procedure for comparing several treatments with a control. J Am Stat Assoc 50: 1096–1121.
111. Conover WJ, Iman RL (1981) Rank transformations as a bridge between parametric and nonparametric statistics. Am Stat 35: 124–133.
112. Dussault C, Ouellet JP, Courtois R, Huot J, Breton L, et al. (2005) Linking moose habitat selection to limiting factors. Ecography 28: 619–628.
113. Festa-Bianchet M (1988) Seasonal range selection in bighorn sheep: conflicts between forage quality, forage quantity, and predator avoidance. Oecologia 75: 580–586.
114. Rhodes JR, McAlpine CA, Lunney D, Possingham HP (2005) A spatially explicit habitat selection model incorporating home range behavior. Ecology 86: 1199–1205.
115. Doligez B, Danchin E, Clobert J (2002) Public information and breeding habitat selection in a wild bird population. Science 297: 1168–1170.
116. Stamps JA, Krishnan VV, Reid ML (2005) Search costs and habitat selection by dispersers. Ecology 86: 510–518.
117. Csada RD, James PC, Espie HM (1996) The "file drawer problem" of non-significant results: does it apply to biological research? Oikos 76: 591–593.

8

Harvester Ant Colony Variation in Foraging Activity and Response to Humidity

Deborah M. Gordon*, Katherine N. Dektar[¤a], Noa Pinter-Wollman[¤b]

Department of Biology, Stanford University, Stanford, California, United States of America

Abstract

Collective behavior is produced by interactions among individuals. Differences among groups in individual response to interactions can lead to ecologically important variation among groups in collective behavior. Here we examine variation among colonies in the foraging behavior of the harvester ant, *Pogonomyrmex barbatus*. Previous work shows how colonies regulate foraging in response to food availability and desiccation costs: the rate at which outgoing foragers leave the nest depends on the rate at which foragers return with food. To examine how colonies vary in response to humidity and in foraging rate, we performed field experiments that manipulated forager return rate in 94 trials with 17 colonies over 3 years. We found that the effect of returning foragers on the rate of outgoing foragers increases with humidity. There are consistent differences among colonies in foraging activity that persist from year to year.

Editor: Corrie S. Moreau, Field Museum of Natural History, United States of America

Funding: The work was funded by NSF grant IOS-0718631 to DMG, and by a grant from the CleanSlate program at Stanford University. The funders had no role in study design, data collection and analysis, decision to publish, or preparation of the manuscript.

Competing Interests: One of the authors, Katherine N. Dektar, is currently employed at Google. She did the work reported in this manuscript while she was an undergraduate at Stanford and the work is not related in any way to her work at Google. This does not alter the authors' adherence to all the PLOS ONE policies on sharing data and materials. Deborah M. Gordon is a member of the PLOS ONE editorial board. This does not alter the authors' adherence to all the PLOS ONE policies on sharing data and materials.

* E-mail: dmgordon@stanford.edu

¤a Current address: Google, Mountain View, California, United States of America
¤b Current address: University of California San Diego, BioCircuits Institute, La Jolla, California, United States of America

Introduction

Collective behavior arises from the local interactions of individuals. Bird flocks and fish schools turn, termites build nests, and wildebeest cross the plains, because individuals respond to what others are doing nearby. The first question about collective behavior is how the actions of individuals add up to the dynamic behavior we observe. The recent explosion of work on collective behavior in many different animal groups (e.g. [1–5]) seeks to describe how local interactions among individuals produce a certain collective outcome.

Evolutionary questions about behavior require us to think about variation, and there is growing interest in individual variation in behavior [6]. Consistent differences among individuals in behavior [7], recently called personality [8–9], temperament [10] or behavioral syndromes [11], may be heritable reaction norms like those created by phenotypic plasticity in any other trait [12].

Recent work shows variation among social insect colonies in behavior. Honeybee colonies differ in pollen and nectar collection [13] and in foraging behavior [14]. Ant colonies differ in foraging activity [15–18], and in the aggression levels of colonies [19] and individuals [20–21]. Colonies function as reproductive individuals; the colony produces reproductives, who mate with the reproductives of other colonies and then form offspring colonies. When the collective behavior of a colony is heritable and ecologically important, selection may act on variation among colonies in collective behavior.

Here we examine differences among colonies of the red harvester ant, *Pogonomyrmex barbatus*, in the collective behavior that regulates foraging in response to food availability. An inactive forager is stimulated to leave the nest on its next trip by the return of foragers with food [17,22–24]. Interactions between returning and outgoing foragers consist of brief antennal contacts inside the nest entrance as the ants come in and out. During an antennal contact a forager detects the task-specific cuticular hydrocarbon profile of the other [25] and whether it is carrying food [26]. Forager return rate reflects food availability because the duration of a foraging trip depends on search time [27]. The more food is available, the less time is needed to search and the more quickly a forager returns with food. Thus the overall rate of return of successful foragers reflects the availability of food on that day.

Harvester ants foraging in hot, dry conditions lose water, but obtain water from metabolizing fats in the seeds that they eat [28–29]. Positive feedback on foraging activity, from returning foragers with food, allows the colony to regulate its foraging activity according to the current costs of desiccation and the benefits based on current food availability.

In many harvester ant species, foraging behavior is influenced by the weather [30–31]. For example, in the ant *Messor andrei*, recruitment to food bait is higher in more humid conditions [18]. Both humidity and food availability are affected by day to day changes in weather conditions. Food is distributed by wind and flooding and rain uncovers seeds in the top layer of the soil [32–33]. In *P. barbatus*, daily changes in conditions such as humidity and food availability produce strong daily trends in the foraging activity of all colonies [24,34].

Colonies may vary in the relation between humidity and foraging activity. Previous work shows that colonies differ consistently from year to year in how often they forage at all [16]. Most colonies forage on days with high humidity and high food availability, such as those just after a rain when flooding has exposed a layer of seeds in the soil. Few colonies forage on very dry days. Previous work also showed that colonies differ in how likely they are to adjust the rate of outgoing foragers to the rate of forager return [17]. While all colonies tend to adjust outgoing foraging rate closely when conditions are good, only some colonies do so in poor conditions.

Here we asked whether the response of outgoing foragers to returning foragers depends on humidity, and whether the magnitude of response predicts the level of foraging activity. We then considered whether colonies differ in foraging activity, and whether colony differences in foraging behavior persist from year to year.

Methods

We examined foraging behavior repeatedly in the same colonies over several years and over a range of naturally occurring conditions, so as to encompass the diverse factors that influence colony foraging decisions. To examine how the rate of outgoing foragers depends on the rate of returning foragers, we conducted experiments that manipulated forager return rate. Experiments were performed in August 2009, August-September 2010, and August-September 2011, at the site of a long-term study since 1985 of a population of *P. barbatus* near Rodeo, New Mexico, USA. No permits were required for the field work; permission was granted by Stanford University. Over the course of 3 years we conducted a total of 94 trials with 17 colonies. Three of the 17 colonies were measured in all 3 years, and an additional 2 colonies were measured in 2 of the 3 years. In 2009 there were 32 trials in 8 colonies on 10 days; in 2010 there were 29 trials in 8 colonies on 5 days; in 2011 there were 33 trials in 10 colonies on 12 days (Table S1). All colonies were mature, more than 5 years old (ages determined by yearly census; methods in Gordon and Kulig 1996), except for one colony, 112, which was measured in all 3 years and was 4 years old in 2009.

Returning foragers were prevented from returning to the nest in minutes 4–7 of a 20-min observation period in 2009 and 2010, and a 14-min observation period in 2011. The observation period was reduced because the results from 2009 and 2010 showed that foraging rates did not change much after about 3 minutes following the removal of returning foragers. After the 3 minutes of removals, from minute 8 to the end of the trial, other returning foragers went back to the nest undisturbed, while the ants that had been collected were not released until after the trial. Methods were the same as in [24] and [17]. Rates of returning and outgoing foragers crossing an imaginary line along the trail were measured from video film using an image analysis system developed by Martin Stumpe (http://www.antracks.org). Methods for obtaining the data from image analysis results are described in Prabhakar et al. [35].

As a measure of foraging rate, we used the mean rate of returning foragers per sec. We used the rate of forager return because it reflects both the rate at which foragers went out and food availability. Foraging rate was measured when colonies were undisturbed, before any removals were made. For each trial, the foraging rate was the mean rate of returning foragers per sec from 1 to 240 sec.

To characterize differences among colonies in response to the rate of returning foragers, we used a stochastic version of a threshold model (e.g. [13]). In the model, the probability that foragers leave the nest depends on the rate of forager return. We estimated the colony-specific value of the parameter c in a modified version of an algorithm [35] that predicts the flow of outgoing foragers (α_n):

$$\alpha_n = max(\alpha_{n-1} - qD_{n-1} + cA_n, \alpha_0 = 0.01 \quad (1)$$

$$D_n \sim Poisson(\alpha_n)D \quad (2)$$

A_n is the observed number of returning food-bearing foragers at time n. To find the value of c that best fit the data for each trial, we kept all other parameters fixed at the values that, in previous work, provided a good fit to data from forager removal experiments, predicting the number of outgoing foragers at time n, D_n, from the observed rate of returning foragers. The baseline value at which outgoing foragers leave the nest if no foragers return was $\alpha_0 = 0.01$, based on previous data from 5 trials of a similar experiment with each of 14 colonies in field experiments in 2008 [17]. If at a given time step very few ants were returning and $(\alpha_{n-1} - qD_{n-1} + cA_n) < \alpha_0$, the value of α_n was set to α_0. The parameter q, which represents the reduction in availability of outgoing foragers due to the departure of foragers, was set at 0.05, based on observations inside the nest in field colonies (Pinter-Wollman et al. in review). To estimate c for each trial, we used observed rates of returning foragers throughout the trial (A_n), from 0 to 1200 seconds for the 20-min trials from 2009 and 2010, and 0 to 840 sec for the 14-min trials in 2011, and the algorithm from equations 1) and 2) above [35] to simulate the rate of outgoing foragers. For each trial we found c by testing a range of values from 0.01 to 0.25, using the same model-fitting procedures as in [35]: to set a value of c for each trial, we compared simulated and observed rates of outgoing foragers in 200 simulations of each value, and chose the value of c that gave the lowest root mean square error (RMSE) [36] of the difference between the simulated and observed rates.

We examined whether c, the effect of returning foragers on the rate of outgoing foragers, is associated with current humidity. As a measure of humidity we used dew point, the temperature at which water vapor condenses. The higher the humidity, the higher the temperature at which water vapor condenses, and thus dew point is an absolute measure of humidity. Measures of dew point for a given day were obtained from http://www.wunderground.com/ showing weather data for San Simon, which tends to have similar conditions as the study area because it is in the same floodplain and at a similar elevation, about 50 km from the study area. We used linear mixed-effects models to test whether foraging rate depends on dew point, and whether the effect of returning foragers on the rate of outgoing foragers (c) depends on dew point. Foraging rate and c were the dependent variables in two separate models. In each model, dew point was the main fixed effect and, to account for variation among colonies, we included colony as a random effect. To examine the relation of c and foraging rate, we performed a linear regression of c on foraging rate, using the mean values of c and foraging rate for each colony within each year.

To examine whether colony differences in foraging activity persist from year to year, we first considered the data for 5 colonies for which removal experiments were performed in both 2010 and 2011 (Table S1). We compared the values for 2010 and 2011 for each colony with 2 repeated-measures ANOVAs, either with foraging rate or *c* as dependent variables. Three of these 5 colonies were also tested in 2009. For these 3 colonies we then used 2 different repeated measures ANOVAs with year within colony (2009, 2010, 2011) as main effect and either foraging rate or c as dependent variables.

We then examined variation among colonies in *c* and foraging rate. To test for colony differences, we performed 2 nested ANOVAs with colony, year, and day within year as main effects, and either foraging rate or c as dependent variables.

All statistical tests were conducted using the software R (version 2.12.1).

Results

The effect of returning foragers on the rate of outgoing foragers, *c*, depends on humidity. The higher the humidity, the higher the value of *c*, or the more each returning forager stimulates other workers to go out and forage. There was a positive association between the effect of returning foragers and dew point (data for all 94 trials of 17 colonies in 3 years; linear mixed-effects model: $p<0.0001$, Fig. 1A).The increased response to returning foragers when humidity is high is not always sufficient to raise foraging rates. There was no significant relation of foraging rate and dew point (linear mixed-effects model, dew point: $p=0.49$, Fig. 1B), and no relation of the effect of returning foragers on outgoing foragers and foraging rate (Fig. 1C) (linear regression: $r2=0.02$, $p=0.48$).

The effect of each returning forager on the rate of outgoing foragers, *c*, appears to shift somewhat from year to year, presumably reflecting changes in conditions. For all 5 colonies tested both in 2010 and 2011, there was no significant difference from 2010 to 2011 in the effect of returning foragers (Repeated measures ANOVA: $F=0.07$; $df=1$; $p=0.8$, Fig. 2A), but over 3 years, for the three colonies that were tested in 2009, 2010, and 2011, there were significant differences among years in the effect of returning foragers (Repeated measures ANOVA: $F=12.4$, $df=2$, $p=0.02$). Foraging rate varied from year to year for the 5 colonies tested in 2010 and 2011 (Repeated measures ANOVA: $F=15.3$; $df=1$; $p=0.02$), and for the three colonies that were tested in 2009, 2010, and 2011 (Repeated measures ANOVA: $F=9.1$; $df=2$; $p=0.03$).

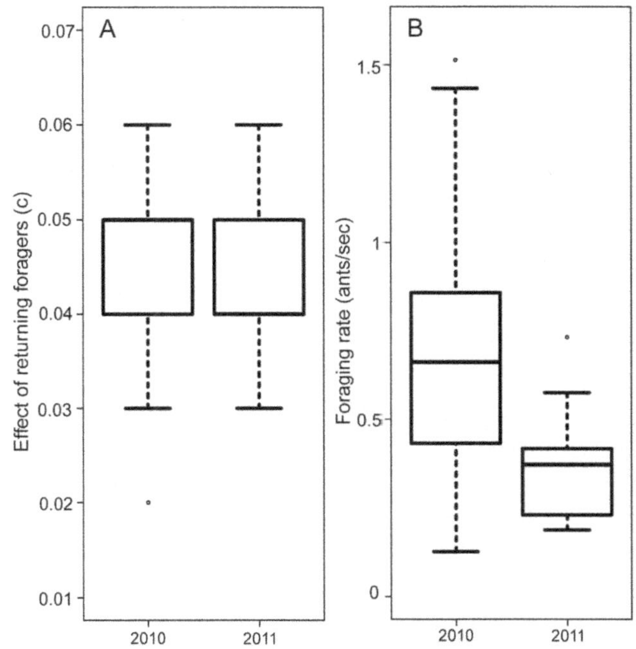

Figure 2. Differences between 2010 and 2011 for 5 colonies. A. The effect of returning foragers on the rate of outgoing foragers (*c*). B. Foraging rate. Boxes indicate the lower and upper quartiles; horizontal lines within boxes indicate the median, whiskers extend to 1.5 interquartile range from the box, and points indicate outliers.

Colonies differed consistently in foraging rate, the rate of returning foragers per second (Fig. 3). There was a significant effect of colony on foraging rate (Table 1), despite significant differences among years (e.g. Fig. 2B) and from day to day. Colonies did not differ in the effect of returning foragers on the rate of outgoing foragers (*c*) (Table 1).

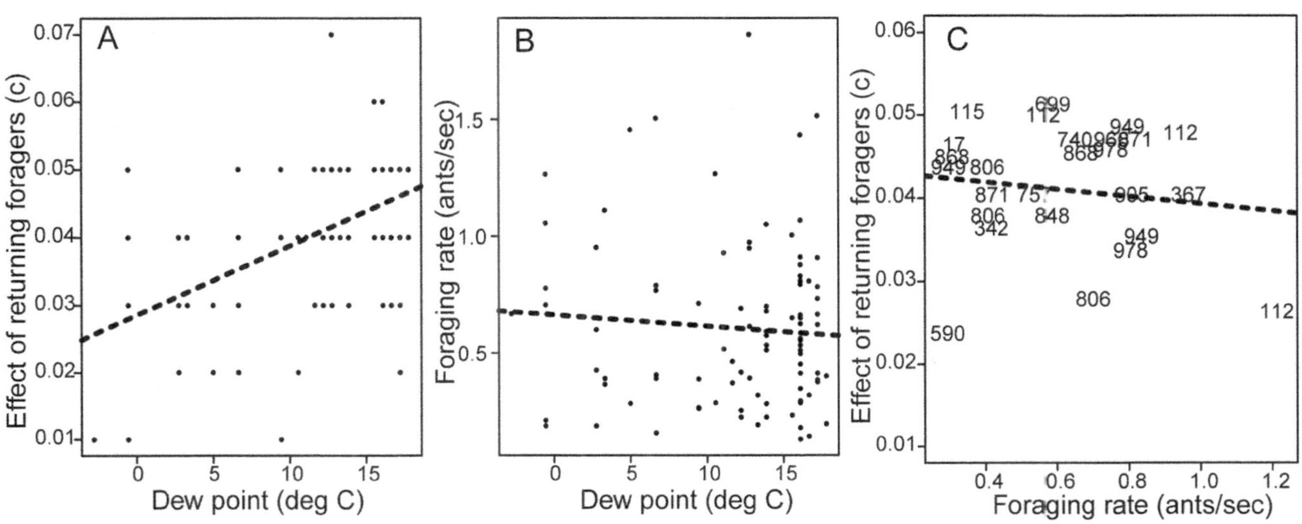

Figure 1. Relation of foraging activity and dew point, a measure of humidity. A. Relation of the effect of returning foragers on the rate of outgoing foragers (*c*) and dew point. B. Relation of foraging rate and dew point. The dotted lines show the least squares fit. C. Relation of effect of returning foragers and forager rate.

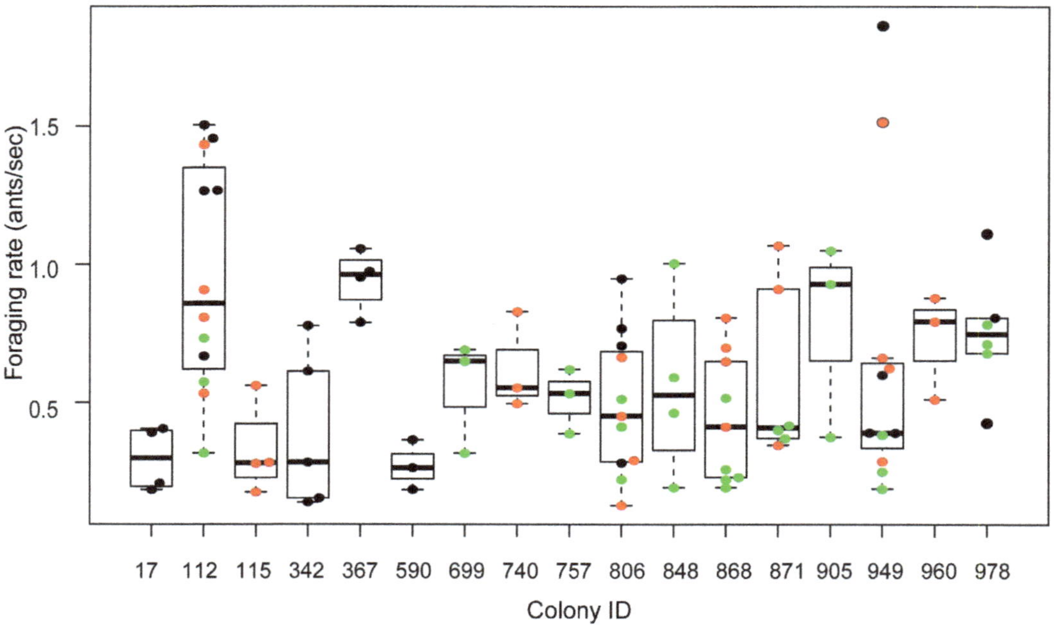

Figure 3. Variation among colonies in foraging rate. Numbers on the x-axis are the colony IDs. Boxes indicate the lower and upper quartiles; horizontal lines within boxes indicate the median, whiskers extend to 1.5 interquartile range from the box. Points represent the values of foraging rate for each trial. Black dots are for trials conducted in 2009, red dots for 2010, and green dots for 2011. Overlapping points are slightly offset along the x axis.

Discussion

Colonies differ consistently in foraging activity (Fig. 3), and such differences persist from year to year. It is important to note that colony differences in foraging rate do not arise simply from differences in the numbers of ants available to forage. Colony behavior depends on colony size in many ant species (e.g. [37–40]). However, differences among colonies are also due to differences in how colonies regulate the activity of their available foragers. For example, some colonies with a low average foraging rate clearly have enough foragers to go out at high rates on some days (e.g. colonies 342 and 949, Fig. 3). Colony variation in foraging rate was also observed in previous work done in a single year [18], and was correlated with variation among colonies in their response to patrollers early in the morning.

Further work is needed to determine why colonies differ in foraging activity. In harvester ants, olfactory cues are used in interactions [25,41], and there may be variation among colonies in the action of olfactory receptors [42–43]. In *Camponotus* ants, workers react to an olfactory stimulus from 5 minutes earlier [44]; such olfactory responses change in the course of an ant's development [45] and this process may vary among colonies. Finally, differences among colonies in nest structure (e.g. [18,46]) may influence the strength of chemical cues that ants exchange [47].

Humidity influences the response of outgoing foragers to returning foragers. It is possible that humidity affects the behavior of returning foragers in some way that increases the extent to which returning foragers stimulate those waiting inside the nest entrance. It is also possible that humidity influences the perception of chemical cues in interactions between returning and outgoing forager. Further work is needed to investigate these possibilities. However, humidity does not in itself predict foraging activity. The lack of relation between dew point and foraging rate suggests that within the range of humidity conditions on the days we observed foraging, food availability, as well as any other factors that diminish forager return rate, had a stronger influence on foraging activity than humidity. Though in general, returning foragers tend to stimulate the rate of outgoing foragers more when humidity is higher (Fig. 1A), this does not lead to a significant effect of dew point on foraging rate (Fig. 1C). This suggests that the negative effect of low food availability can override the positive effect of dew point: if food availability is low enough on a humid day, then the rate of forager return will be very low. Even if the effect of each

Table 1. Results of ANOVAs on effect of returning foragers and foraging rate.

	Overall model			Main effects								
				Colony			Year			Day (nested in year)		
	Adjusted r^2	F	P	DF	F	P	DF	F	P	DF	F	P
Effect of returning foragers (*c*)	0.39	2.4	**0.001**	16	1.14	0.35	2	13.63	**<0.0001**	24	1.86	**0.03**
Foraging rate (ants/sec)	0.47	2.98	**<0.0001**	16	4.78	**<0.0001**	2	11.98	**<0.0001**	24	1.81	**0.04**

returning forager is high, if the foragers return very infrequently, the rate at which outgoing foragers leave the nest will remain low, which perpetuates a low rate of forager return. Experiments that manipulate food supply in different humidity conditions would help to determine how colony response to food availability and humidity each contribute to foraging rates.

Many factors affect foraging rate at any time, such as temperature [31] and the vegetation ants have to travel through to reach the food [48], while other processes such as predation [49] and amount of stored food [50], probably influence whether a colony forages at all on a given day. Colonies compete with neighbors for foraging area, and colony decisions about interactions with the foragers of neighboring colonies depend on the ages of the colonies involved [32,51–52], so foraging activity also depends on a colony's age and the age distribution of neighboring colonies. Combining all of these factors, colonies show consistent differences in foraging activity. The question remains open whether colony differences in foraging activity are associated with variation in colony fitness [53].

Supporting Information

Table S1 Forager removal experiments by colony and year. Bold indicates colonies for which trials were performed in all 3 years; italics indicates colonies for which trials were performed in 2 years.

Acknowledgements

We thank Ximena Ampuero, Sam Crow, Jordan Haarmsa, James Hickman, Mattias Lanas, Doo Young Lee, LeAnn Howard, Andrew Merrell, Evie Pless, Tom Yue, Uyhun Ung, and Claire Zabel for assistance with field work. We thank Martin Stumpe for developing the image analysis software used to measure foraging rates. Discussion with Balaji Prahabkar contributed greatly to the work. We thank William Flesch, Edith Katsnelson, Andrew Merrell, Lis Nelis, and Zhiyuan Song for comments on the manuscript.

Author Contributions

Conceived and designed the experiments: DMG. Performed the experiments: DMG KND NPW. Analyzed the data: DMG KND NPW. Contributed reagents/materials/analysis tools: DMG. Wrote the paper: DMG.

References

1. Couzin ID, Krause J, James R, Ruxton GD, Franks NR (2002) Collective memory and spatial sorting in animal groups. Journal of Theoretical Biology 218: 1–11.
2. Torney C, Neufeld Z, Couzin ID (2009) Context-dependent interaction leads to emergent search behavior in social aggregates. Proceedings of the National Academy of Sciences of the United States of America 106: 22055–22060.
3. Buhl J, Deneubourg JL, Grimal A, Theraulaz G (2005) Self-organized digging activity in ant colonies. Behavioral Ecology and Sociobiology 58: 9–17.
4. Pratt SC, Sumpter DJT (2006) A tunable algorithm for collective decision-making. Proceedings of the National Academy of Sciences of the United States of America 103: 15906–15910.
5. Sumpter DJT (2010) Collective Animal Behavior. Princeton NJ: Princeton University Press.
6. Gordon DM (2011) The fusion of behavioral ecology and ecology. Behavioral Ecology 22: 225–230.
7. Lott DF (1991) Intraspecific variation in the social systems of wild vertebrates: Cambridge University Press.
8. Wilson DS, Clark AB, Coleman K, Dearstyne T (1994) Shyness and boldness in humans and other animals. Trends in Ecology & Evolution 9: 442–446.
9. Gosling SD (2001) From mice to men: What can we learn about personality from animal research? Psychological Bulletin 127: 45–86.
10. Reale D, Reader SM, Sol D, McDougall PT, Dingemanse NJ (2007) Integrating animal temperament within ecology and evolution. Biological Reviews 82: 291–318.
11. Sih A, Bell A, Johnson JC (2004) Behavioral syndromes: an ecological and evolutionary overview. Trends in Ecology & Evolution 19: 372–378.
12. Dingemanse NJ, Kazem AJN, Reale D, Wright J (2010) Behavioural reaction norms: animal personality meets individual plasticity. Trends in Ecology & Evolution 25: 81–89.
13. Page RE, Erber J, Fondrk MK (1998) The effect of genotype on response thresholds to sucrose and foraging behavior of honey bees (Apis mellifera L.). Journal of Comparative Physiology a-Neuroethology Sensory Neural and Behavioral Physiology 182: 489–500.
14. Wray MK, Mattila HR, Seeley TD (2011) Collective personalities in honeybee colonies are linked to colony fitness. Animal Behaviour 81: 559–568.
15. Cole BJ, Smith AA, Huber ZJ, Wiernasz DC (2010) The structure of foraging activity in colonies of the harvester ant, *Pogonomyrmex occidentalis*. Behavioral Ecology 21: 337–342.
16. Gordon DM (1991) Behavioral flexibility and the foraging ecology of seed-eating ants. American Naturalist 138: 379–411.
17. Gordon DM, Guetz A, Greene MJ, Holmes S (2011) Colony variation in the collective regulation of foraging by harvester ants. Behavioral Ecology 22: 429–435.
18. Pinter-Wollman N, Gordon DM, Holmes S (2012) Nest site and weather affect the personality of harvester ant colonies. Behavioral Ecology 23: 1022–1029.
19. Crosland MWJ (1990) Variation in ant aggression and kin discrimination ability within and between colonies. Journal of Insect Behavior 3: 359–379.
20. Modlmeier AP, Foitzik S (2011) Productivity increases with variation in aggression among group members in *Temnothorax* ants. Behavioral Ecology 22: 1026–1032.
21. Foitzik S, Backus VL, Trindl A, Herbers JM (2004) Ecology of *Leptothorax* ants: impact of food, nest sites, and social parasites. Behavioral Ecology and Sociobiology 55: 484–493.
22. Gordon DM (2002) The regulation of foraging activity in red harvester ant colonies. American Naturalist 159: 509–518.
23. Schafer RJ, Holmes S, Gordon DM (2006) Forager activation and food availability in harvester ants. Animal Behaviour 71: 815–822.
24. Gordon DM, Holmes S, Nacu S (2008) The short-term regulation of foraging in harvester ants. Behavioral Ecology 19: 217–222.
25. Greene MJ, Gordon DM (2003) Cuticular hydrocarbons inform task decisions. Nature 423: 32–32.
26. Greene MJ, Pinter-Wollman N, Gordon DM (2013) Interactions with combined chemical cues inform harvester ant foragers' decisions to leave the nest in search of food. PLoS One 8: e52219.
27. Beverly BD, McLendon H, Nacu S, Holmes S, Gordon DM (2009) How site fidelity leads to individual differences in the foraging activity of harvester ants. Behavioral Ecology 20: 633–638.
28. Lighton JRB, Bartholomew GA (1988) Standard energy-metabolism of a desert harvester ant, *Pogonomyrmex rugosus* - effects of temperature, body-mass, group-size, and humidity. Proceedings of the National Academy of Sciences of the United States of America 85: 4765–4769.
29. Lighton JRB, Feener DH (1989) Water-loss rate and cuticular permeability in foragers of the desert ant Pogonomyrmex-rugosus. Physiological Zoology 62: 1232–1256.
30. Brown MJF, Gordon DM (2000) How resources and encounters affect the distribution of foraging activity in a seed-harvesting ant. Behavioral Ecology and Sociobiology 47: 195–203.
31. Azcarate FM, Kovacs E, Peco B (2007) Microclimatic conditions regulate surface activity in harvester ants *Messor barbarus*. Journal of Insect Behavior 20: 315–329.
32. Gordon DM, Kulig AW (1996) Founding, foraging and fighting: colony size and the spatial distribution of harvester ant nests. Ecology 77: 2393–2409.
33. Gordon DM (1993) The spatial scale of seed collection by harvester ants. Oecologia 95: 479–487.
34. Gordon DM (1991) Behavioral flexibility and the foraging ecology of seed-eating ants. American Naturalist 138: 379–411.
35. Prabhakar B, Dektar KN, Gordon DM (2012) A Poisson model of the regulation of ant colony foraging activity without spatial information. PLoS Computational Biology 8(8): e1002670.
36. Papoulis A (1965) Probability, Random Variables, and Stochastic Processes. New York.: McGraw-Hill.
37. Tschinkel WR (1993) Sociometry and sociogenesis of colonies of the fire ant *Solenopsis invicta* during one annual cycle. Ecological Monographs 63: 425–457.
38. Bourke AFG (1999) Colony size, social complexity and reproductive conflict in social insects. Journal of Evolutionary Biology 12: 245–257.
39. Thomas ML, Elgar MA (2003) Colony size affects division of labour in the ponerine ant *Rhytidoponera metallica*. Naturwissenschaften 90: 88–92.
40. Gordon D (2010) Ant encounters : interaction networks and colony behavior. Princeton, N.J.: Princeton University Press. vi.i, 167 p. p.
41. Greene MJ, Gordon DM (2007) Interaction rate informs harvester ant task decisions. Behavioral Ecology 18: 451–455.

42. Yamagata N, Nishino H, Mizunami M (2006) Pheromone-sensitive glomeruli in the primary olfactory centre of ants. Proceedings of the Royal Society B-Biological Sciences 273: 2219–2225.
43. Muscedere ML, Johnson N, Gillis BC, Kamhi JF, Traniello JFA (2012) Serotonin modulates worker responsiveness to trail pheromone in the ant Pheidole dentata. Journal of Comparative Physiology a-Neuroethology Sensory Neural and Behavioral Physiology 198: 219–227.
44. Dupuy F, Sandoz JC, Giurfa M, Josens R (2006) Individual olfactory learning in Camponotus ants. Animal Behaviour 72: 1081–1091.
45. Muscedere ML, Traniello JFA (2012) Division of labor in the hyperdiverse ant genus Pheidole is associated with distinct subcaste- and age-related patterns of worker brain organization. PLoS ONE 7: e31618.
46. Mailleux AC, Buffin A, Detrain C, Deneubourg JL (2011) Recruitment in starved nests: the role of direct and indirect interactions between scouts and nestmates in the ant Lasius niger. Insectes Sociaux 58: 559–567.
47. Burd M, Shiwakoti N, Sarvi M, Rose G (2010) Nest architecture and traffic flow: large potential effects from small structural features. Ecological Entomology 35: 464–468.
48. Crist TO, Wiens JA (1994) Scale effects of vegetation on forager movement and seed harvesting by ants. Oikos 69: 37–46.
49. Munger JC (1984) Long-term yield from harvester ant colonies - implications for horned lizard foraging strategy. Ecology 65: 1077–1086.
50. Cassill DL, Tschinkel WR (1995) Allocation of liquid food to larvae via trophallaxis in colonies of the fire ant, *Solenopsis invicta*. Animal Behaviour 50: 801–813.
51. Gordon DM (1992) How colony growth affects forager intrusion between neighboring harvester ant colonies. Behavioral Ecology and Sociobiology 31: 417–427.
52. Adler FR, Gordon DM (2003) Optimization, conflict, and nonoverlapping foraging ranges in ants. American Naturalist 162: 529–543.
53. Ingram KK, Pilko A, Heer J, Gordon DM (2013) Colony life history and lifetime reproductive success of red harvester ant colonies. Journal of Animal Ecology, in press.

9

Syndromes or Flexibility: Behavior during a Life History Transition of a Coral Reef Fish

James R. White[1]*, Mark I. McCormick[1], Mark G. Meekan[2]

1 School of Marine and Tropical Biology, James Cook University, Townsville, Queensland, Australia, 2 Australian Institute of Marine Science, Botany Building, The University of Western Australia, Crawley, Western Australia, Australia

Abstract

The theory of behavioral syndromes focuses on quantifying variation in behavior within and among individual organisms and attempts to account for the maintenance of differences in behavior that occur in a consistent manner among individuals. Behavioral syndromes have potentially important ecological consequences (e.g. survivorship tradeoffs) and can be shaped by population dynamics through selective mortality. Here, we search for any evidence for consistency of behavior across situations in juveniles of a common damselfish, *Pomacentrus amboinensis* (Pomacentridae) at the transition between larval habitats in the plankton and juvenile habitats on the reef. Naïve fish leaving the pelagic phase to settle on reefs were caught by light traps and their behaviors observed using similar methods across three different situations (small aquaria, large aquaria, field setting); all of which represent low risk and well-sheltered environments. Seven behavioral traits were compared within and among individuals across situations to determine if consistent behavioral syndromes existed. No consistency was found in any single or combination of behavioral traits for individuals across all situations. We suggest that high behavioral flexibility is likely beneficial for newly-settled fish at this ontogenetic transition and it is possible that consistent behavioral syndromes are unlikely to emerge in juveniles until environmental experience is gained or certain combinations of behaviors are favored by selective mortality.

Editor: Allan V. Kalueff, Tulane University Medical School, United States of America

Funding: This work was supported by the ARC Centre of Excellence for Coral Reef Studies. The funders had no role in study design, data collection and analysis, decision to publish, or preparation of the manuscript.

Competing Interests: The authors have declared that no competing interests exist.

* E-mail: james.ryan.white@gmail.com

Introduction

The study of individual consistency in animal behavior has attracted recent attention from researchers because of the potential for this variability to reflect underlying processes influencing an animal's responses to a range of situations and for it to have significant effects on fitness [1,2]. Because variation in behavioral traits is so widespread taxonomically, an understanding of flexibility in these traits has important implications for the ecology and evolution of different species [1,3–6] and for predictions of how they will respond to environmental and ecological shifts [1,7].

Sih and Bell (2008) define within-individual consistency as occurring when an individual behaves in a consistent way over time or across situations (e.g. absence vs. presence of predators) [8]. Between-individual consistency occurs when there is a correlation between behavioral traits (e.g. boldness and aggression) among individuals. Having behavioral consistency either within- and between-individuals is termed a behavioral syndrome [1,8]. Individual consistency in one behavioral trait (e.g. activity rate in different situations) is referred to as a behavioral type [1].

Most previous research has focused on explaining why individual variation in behavioral traits exists and why there should be consistency within an individual over time. The former question deals with factors maintaining variation in behavior among individuals from the same population, while the latter focuses on factors that maintain stability in behavioral traits [2]. However, inconsistent use of terminology and methodology in previous work has led to similar behavioral traits being defined differently or distinct traits defined as equivalent across various animal behavioral studies, making comparisons difficult [9,10]. Furthermore, there is a lack of fish studies that demonstrate consistent patterns of individual behavior by showing that multiple behavioral traits are correlated across multiple situations (especially in laboratory vs. natural settings). Additionally, few studies have used identical measures of behavior across situations. Perhaps unsurprisingly, these studies conclude that their target species showed little consistency in behavioral traits [11,12]. Finally, a majority of fish behavioral studies have been conducted in the laboratory [10] on captive or captive-bred populations [13]. Although this is done to control for factors that could potentially confound results, it is extraordinarily difficult to create environments that approximate natural situations in the laboratory. Consequently, such studies assume that behavior of an animal in an artificial setting will be representative of its natural state. This assumption is rarely tested in the field [14].

Behavior can be influenced selectively by a wide variety of abiotic (temperature, illumination, habitat; [14–16] or biotic factors (hunger, thirst, stress; [14,15]) that can vary significantly through ontogeny [17]. These factors, along with experience gained as an individual grows, can lead to situations where it is beneficial for an individual to change their behavior to adjust to conditions (behavioral flexibility) or risk elimination from the population (selective mortality) [14]. Group behavior also can influence changes in individual behavior. For example, bold rainbow trout (*Onchorhyncus mykiss*) reduced boldness after observing shy conspecifics [15]. The outcomes of research on associations between behavioral traits and behavioral flexibility [13] have been inconsistent, although some studies have suggested that a tendency to display bold behavior increases an individual's ability to solve novel tasks [18,19], while others have shown that individuals that are more shy and unaggressive have more behavioral flexibility [20].

Maintaining a certain degree of behavioral flexibility to suit changing environmental conditions is likely to be necessary for animals with complex life cycles that undergo life history transitions. The transition between these planktonic and benthic environments is a major developmental landmark for most coral reef fishes and makes these ideal organisms on which to investigate the relationship between adaptive behavioral traits and biotic factors. Furthermore, young reef fish arriving from the plankton into benthic habitats have no experience of their new environment. As such they make a useful model organism because these naïve juveniles enable researchers to control for learning behaviors through experience and examine behavioral consistency precisely at the time of settlement, which is a critical ontogenetic boundary and mortality bottleneck [21–25]. In this phase of their life cycle, reef fishes typically experience high mortality [26], with rates within the first 48 hours averaging 57% [26,27] but can be >90% [23,28]. Behavioral decisions at early settlement can thus influence survival and possibly the structure of reef communities [23,25]. Thus, we expect juvenile reef fish to quickly adopt consistency in behaviors that are likely to influence survival at this life stage (e.g. boldness and aggression; [21]).

Our study examines whether naïve juveniles of the Ambon damselfish (*Pomacentrus amboinensis*) display consistent behavioral traits across three low-risk situations (i.e. whether they possessed behavioral syndromes). Because the same behavioral measures are used across different-sized laboratory arenas and within the field, this is one of the first studies to compare behaviors of fish observed in the laboratory with those in a natural setting. Furthermore, we investigated the role of an environmental factor (temperature) in influencing flexibility in behavioral traits. The relationships among behavioral traits across situations were also examined. If newly-settled fish display behavioral syndromes, then theory would predict that individuals should maintain a similar ranking of behavioral types (e.g. boldness) among situations.

Methods

Ethics Statement

This study was carried out in strict accordance with the recommendations under James Cook University (JCU) ethics protocols and approved by the JCU Animal Ethics Committee (Permit Number: A1067). All efforts were made to minimize animal handling and stress. Fish and coral collection was permitted by the Great Barrier Reef Marine Park Authority (Permit Number: G10/33784.1). Fish were collected using light traps and kept in flow through aquaria for the duration of the study where they were observed visually and subsequently returned to the field upon completion.

Study Site and Species

Field experiments were conducted in the shallow coral habitats (2–4 m depths) at the back-reef of Lizard Island in the northern Great Barrier Reef (GBR) (14°40′S, 145°28′E). Juvenile *P. amboinensis* settle from the plankton at night to a variety of habitats in the northern GBR [29] with the greatest densities found on small reef patches at the base of shallow (<10 m depth) reefs. *P. amboinensis* has a pelagic larval duration of 15–23 days and settles from 10.3–15.1 mm standard length [30]. The juvenile body is mostly complete at settlement; however fish go through a rapid change in body pigmentation in less than 12 hours after settlement [31]. Previous studies have shown *P. amboinensis* is relatively site-attached [32] and moves only small distances (<1 m) during the first few months after settlement. Also, these damselfish can be collected immediately prior to the end of their larval phase before settling on the reef and thus are largely naïve to reef-based predators and behaviors learned after settlement. As experience can influence the behavioral phenotypes a fish will exhibit, using reef-naïve individuals reduces the variability that may result from markedly different experience histories. Due to its high abundance, small size, rapid development, and sedentary nature after settlement, *P. amboinensis* is an ideal model species for field observations and laboratory manipulations [24].

Similar studies with *P. amboinensis* and *P. wardi* found individuals displayed consistent behaviors over multiple sampling periods across short time scales after settlement in the field (1 to 3 d; White et al., *in prep.*, [23]). This study examines if the behavioral consistency remains when individuals are observed within different situations.

Collection

We collected late-stage pelagic larvae of *P. amboinensis* using moored light traps (see small light trap in Figure 1 of [33]for design) during January 2010. Fish caught in similar traps have been used in numerous published behavioral studies and individuals characteristically display considerable among-individual variability in behavioral traits such as boldness [21]. Traps were anchored approximately 100 m from the nearest reef in ~16 m of water at dusk and left overnight. Catches were emptied from the traps the next morning at 0730–0800 hours. Fish collected from the traps were transported to the laboratory where *P. amboinensis* was separated from all other species and maintained in a 30 L aquarium of aerated seawater. Individual fish were then assigned a label and placed in separate small aquaria (13 L) for 24 h to acclimatize to laboratory conditions before experiments began during which time they were fed *Artemia* nauplii twice during daylight hours 1.5 mL^{-1}; ~ 768 nauplii per fish per feeding). Fish were fasted for 12 h prior to initial observations. Seawater was unfiltered, therefore potentially contained food, although no feeding behaviors were observed outside of standard feeding times.

Situations

Standard behavioral observations (below) were conducted on each individual fish under three different environmental situations: small aquaria (13 L, 20 cm water depth), large aquaria (285 L, 25 cm water depth) and shallow reef (field). All were established so they represented a simple low-risk, resource-rich environment with no competitors.

Firstly, behaviors of focal fish were observed in small aquaria (13 L) after 24 h acclimatization. Short pieces of PVC tubing (approx. 4 cm diameter) in each tank provided a refuge for each fish. Observers viewed fish through slits cut into a blind (black plastic sheet) to reduce any effect of observer presence. Barriers of black plastic sheeting prevented individuals from seeing fish in neighboring tanks.

After behavioral observations in the small tanks, individual fish were then transferred to the center of large circular tanks (285 L) and given at least a 20 min acclimation period before another series of behavioral observations. A short piece of PVC tubing

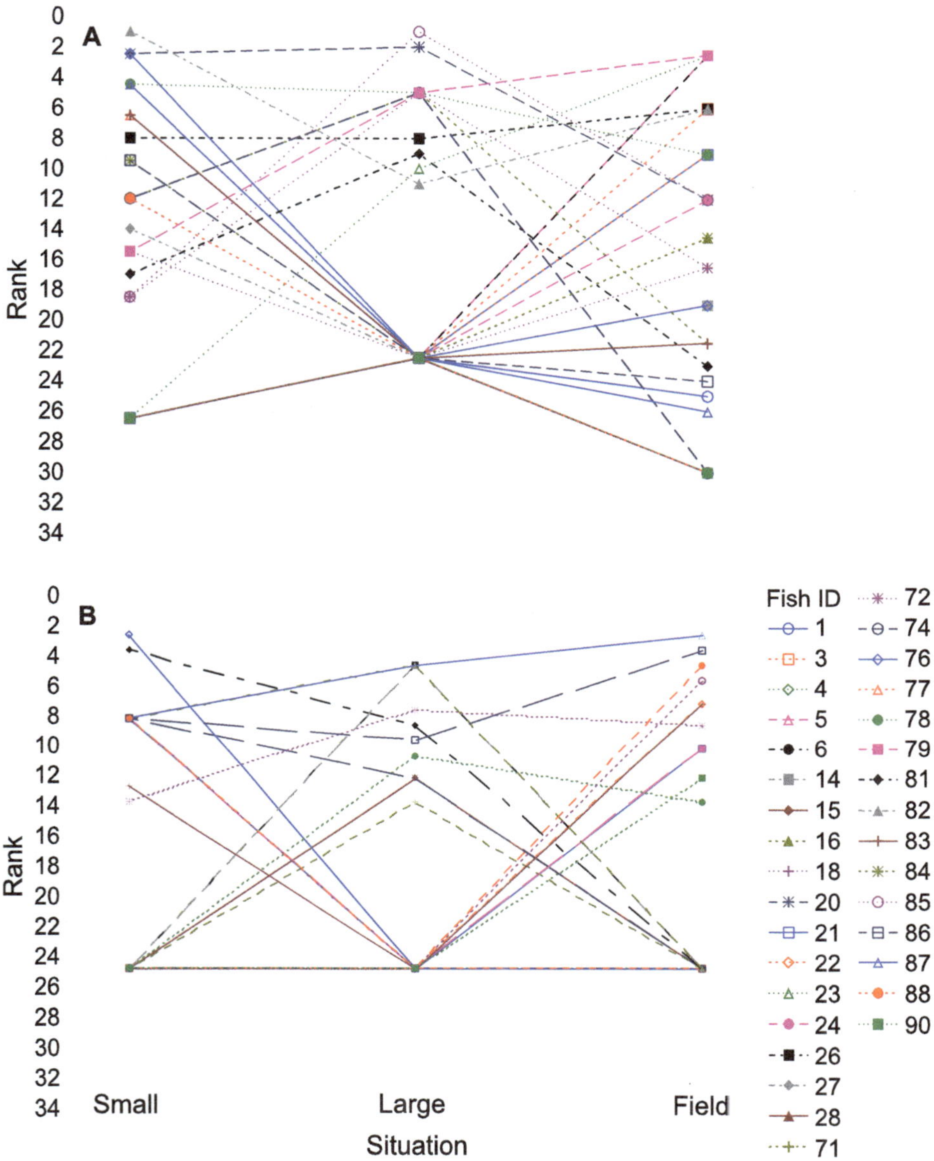

Figure 1. Rank order of boldness and height across situations. A) Ranking of boldness and B) ranking of height across small tanks, large tanks and field site). Each line and unique symbol represents individual fish (N = 33). Individuals were ranked sequentially according to the individual's observed behavioral trait (1 = highest recorded value). Average ranking scores were assigned when multiple fish had a tie in values.

(approx. 4 cm diameter) and replica low complexity artificial coral (white molded resin branching coral, item no. 21505; Wardleys/TFH, Sydney; dimensions 14×11.5×5 cm) provided refuge for each fish. Observations were made from a distance of at least 1.5 m from the tank and fish took no apparent notice of observers.

A final series of observations of the same fish was made in the field. Each *P. amboinensis* was placed into a labeled 2 L clip-seal plastic bag containing aerated seawater and transported to the field. One fish was released onto each reef of an array of small (30×30×30 cm) patch reefs of live bushy hard coral (*Pocillopora damicornis*) on a shallow sand flat. Reefs were positioned 5 m apart and ~20 m from the nearest area of natural reef to avoid the re-dispersal of fish among reefs or colonization from natural reef. *P. amboinensis* naturally settles to this habitat. Prior to release of focal fish, patch reefs were cleared of all resident fishes using hand nets. Once fish were released, small wire cages (about 40×40×40 cm, 12 mm mesh size) were placed over the patch to allow the fish to acclimate to the new surroundings while being protected from predation. Cages were left a minimum of 20 min and removed (carefully lifted with slow movements) immediately before observations. Divers conducted observations with the aid of a 2× magnifying glass from at least 1 m away to avoid any observer affects that may have been caused by close proximity to the target fish. This protocol has been used in other behavioral studies and has not been found to disturb fish or alter their behavior in any significant way [21,23,24,34].

While there were differences in habitat (e.g. variation in refuge size in different-sized arenas) and acclimation times among situations (small and large tanks and the field) these were mostly unavoidable trade-offs associated with logistical and efficiency issues typical of any empirical study. Once given time to de-stress after initial capture, the juvenile fish are generally quick to adapt to new situations, hence the shorter acclimation times for the large tank and field situations. Each situation resulted in different conditions for the subject, but all provided a living space, shelter, and an absence of competition and larger predators. Also, the same testing stimuli were presented to the focal fish in all settings. The way the fish responds to the stimulus will be a product of how it behaves in its environment (its perception of risk is likely to be different between situations) and this will influence how the fish responds to the stimulus. There are many things that are associated with the situations that are different that may affect fish behavior. But, this is the main aim of this study; given the difference in the environments, do fishes respond in consistent ways.

In addition to behavioral data, water temperature in the field and laboratory was recorded every 20 min using calibrated data loggers (32K StowAway Tidbit).

Behavioral Observations

Identical measurements of behavior were made on each fish under the three settings (small & large aquaria, field). Fish were fasted for 12 hours before initial observations and were observed in all three situations on the same day in order to minimize the behavioral influence of individual metabolism or food availability. We recorded: bite rate (the number of strikes towards objects floating in the water column during 3 min); distance moved (the total distance covered in cm during 3 min); distance ventured as the percentage of time spent at various distances from refuge (e.g. 90% of time spent 2 cm away, 10% of time 5 cm away etc.); height from substrate (an estimate of percentage of time spent in either the upper, middle, or lower third of the coral patch); aggression recorded as mirror strike rate (after the initial 3 min observation a mirror was carefully placed in front of the fish and the combined number of strikes, tail whips, or aggressive displays made toward their reflection over 3 min was recorded [35,36]; boldness (see below); latency (see below) to emerge from shelter after stimulus by a novel object (see below). The initial 7 behaviors were recorded during a 3 min observation, aggression was recorded in a separate 3 min trial, while boldness and latency were recorded in a separate 10 minute trial.

Boldness was defined as a continuous variable on a 0–3 scale, where 0 was hiding in a refuge before or immediately after introduction of a novel object and seldom emerging afterwards; 1 was retreating to refuge after a threat and taking more than 5 sec to re-emerge, then tentatively striking at food; 2 was retreating to refuge after a threat but emerging quickly and tenaciously striking at food; 3 was not hiding but continuing to explore or strike at food aggressively. This scale was similar to other measures of boldness used in earlier studies and has been shown to be consistent over short time periods (hours) within individuals, normally distributed, and related to survival [21,23,24,34]. The boldness score basically establishes a spectrum of behavioral reaction to a stimulus and records where an individual lies within this spectrum.

Latency to emerge from hiding was recorded as the amount of time it took the fish to leave their refuge after introduction of a novel object. This variable was limited to a 10 min observation time. In both boldness and latency measures, the novel object was a lead bean sinker weight that was tied to clear fishing line and dropped from 1 m above the refuge in each setting. In the small and large aquariums, short sections of PVC pipe were suspended above each tank in order to consistently guide the lead weights over the center of each refuge. In the field setting, a PVC frame held a guiding section of pipe, to standardize and center the weight to a 1 m drop over the patch reef. The length of each line was calibrated to prevent the weight from hitting the substrate in order to reduce auditory or vibration cues. In the field setting, the weight was dropped while underwater, so it lacked the auditory and visual cues from the weight breaking the surface of the water. However, all fish in the field responded in some way to the weight drop. Observers triggered the release from either behind a blind (small aquaria setting) or from at least 1.5 m away (large aquaria and field settings).

Distances were estimated visually. A pilot study revealed visual estimates were within 10% of actual distances measured with a ruler. Two observers collected the data used in this study; both were trained by a more experienced observer. During training, inter-observer variation was less than 10% for all behavioral measures. Observations of behavior were aided by the use of a magnifying glass (2×). An observation time of 3 min was used to assess behavior since McCormick and Meekan (2010) found that this period produced low coefficients of variation (0–0.15) in behavioral observations [23]. Their study demonstrated consistency in behaviors among three consecutive 3 min observations, and White et al. (*in prep.*) demonstrated behavioral consistency of individuals (using the variables measured in our study) in a field situation over 2–3 days.

Data Analysis

Height from substrate was recorded as the percentage of time an individual fish spent on the bottom (B), middle (M), or top (T) portions of the coral or PVC refuge. These values were transformed to a single, continuous mean variable using the formula: $(B \times 0/100)+(M \times 5/100)+(T \times 10/100)$. Similarly, distance ventured was recorded as the percentage of time spent 0 (A), 2 (B), 5 (C), or 10 (D) cm away from refuge. Data was transformed to a mean distance ventured using the formula: $(A \times 0/100)+(B \times 2/100)+(C \times 5/100)+(D \times 10/100)$.

To examine whether individual fish changed behaviors across situations, but did so consistently, individuals were ranked and plotted for each behavioral trait and situation. Person's product-moment correlations were compared across situations for each trait and were also used to identify relationships between specific traits, as well as behavioral traits and water temperature for each situation.

Principal components analysis (PCA) was used to examine the inter-relationships of behaviors (bite rate, distance moved, distance ventured, height, boldness, latency, and aggression) and individual behavioral consistency across each situation. Parallel analysis was used to determine the number of factors to be extracted (using permutations of 1000 parallel generated datasets) [37]. With the correct number of factors determined by the parallel analysis, principal component loadings were calculated using a correlation matrix with Direct Oblimin rotation [37]. Hierarchical agglomerative cluster analysis [38] for the 7 behavioral variables was overlaid with the PCA in order to determine if fish behaved similarly within each situation. Euclidean distance and unweighted pair-group method using arithmetic averages (UPGMA) were used to calculate clusters. For the PCA and Pearson's product-moment correlations, the seven traits were $\log_{10}(x+1)$ transformed to improve normality. Analyses used SPSS [39] software.

Results

A pilot study revealed that fish in the field began to explore their environment and feed within 1 minute of release onto patch reefs. Fish released into aquaria needed 20 mins to a few hours before exhibiting similar behavior (large and small aquaria respectively). The quicker acclimation time suggested that fish were less stressed and naturally inclined to start exhibiting "normal" behaviors in the field. There was no effect of time of day or observer on observed behavioral measures. All behavioral traits were highly variable both within and among individuals among settings.

Consistency among Situations

There was no clear pattern of rankings of individuals for any of the behaviors such as boldness, distance moved and bite rate across situations (Fig. 1). A fish ranked highly for these traits in the small tank was just as likely to be medium or low ranked in the large tank and field situations. There was a significant positive correlation in height scores between small tanks and the field ($r = 0.48$, $p = 0.004$).

Situation-dependent Relationships among Behaviors

Principal component analysis showed distance moved and bite rate had similar loadings across PC1 for each situation (Fig. 2; Table S1). Other relationships between behavioral traits differed among situations (Figs 2 & 3). In small tanks, boldness and latency had nearly opposite loadings over PC2 (Fig. 2a), which differed across situations. In large tanks, distance moved, distance ventured, and boldness scores had similar correlations with PC1. Height rank and latency also showed similar loadings over PC2 (Fig. 2b). In the field, boldness and distance ventured had similar loadings on both PC1 and PC2 (Fig. 2c). Height rank and latency had similar loadings, yet with different strengths of correlation with PC1 and PC2 for observations within large tanks and in the field (Figs. 2b & c).

Hierarchical cluster analysis created overlapping groups when superimposed on the principal components analysis. Groups did not separate clearly by situation (Fig. 3a). We did not find any consistency in behavioral traits when comparing a single situation at a time (Fig. 2) or across all situations at once (Fig. 3b). Correlation analysis showed a pattern of an increasing number of significant correlations among measured behavioral traits from small tanks, field site, to the large tanks (Table S2). In the small tank and field situations, there were strong significant correlations between bite rate and distance moved. In large tanks, distance moved was correlated with distance ventured. Also, boldness was positively correlated with distance ventured. While in the field, bite rate was positively related to distance moved. Fish which ventured a greater distance also tended to have higher scores of boldness.

Effects of Temperature

There were no consistent correlations across situations for the relationship between behavioral traits and temperature. There was a positive relationship between temperature and bite rate, and height rank in the small tanks. Aggression was negatively related to water temperature in the large tanks while distance ventured and maximum distance ventured were negatively correlated in the field. However, none of the relationships were significant after Holm's sequential Bonferonni adjustment [38]. Overall, temperatures averaged 29°C, ranging between 24–33.6°C, however this varied slightly among situations (Table 1).

Discussion

A behavioral syndrome across individuals can appear as consistent trends in the direction of loadings in a principal component analysis and can also be demonstrated among individuals by multiple significant correlations among the same behavioral traits across situations [1]. *P. amboinensis* did not show any evidence of a behavioral syndrome (i.e. a suite of correlated behaviors across situations) based on these analyses, although the lack of behavioral stability is not necessarily surprising.

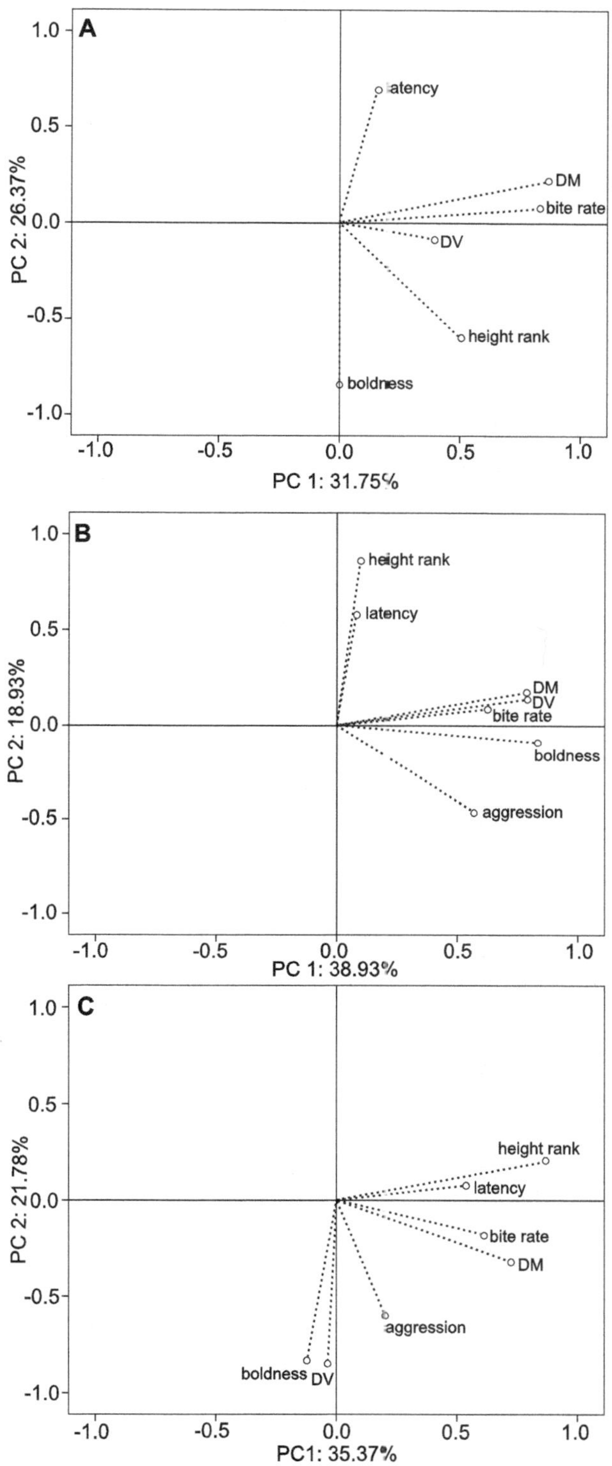

Figure 2. Principal component analysis of relationships between 7 behavioral traits. Traits include: bite rate, distance moved, distance ventured, height, boldness, latency and aggression in *P. amboinensis* in A) small tanks, B) large tanks and C) field situations.

An important result of our study was the lack of consistency in the rank order of all behavioral traits for individual *P. amboinensis* across the three situations (small tanks, large tanks, field site). The definition of behavioral syndromes accounts for this type of

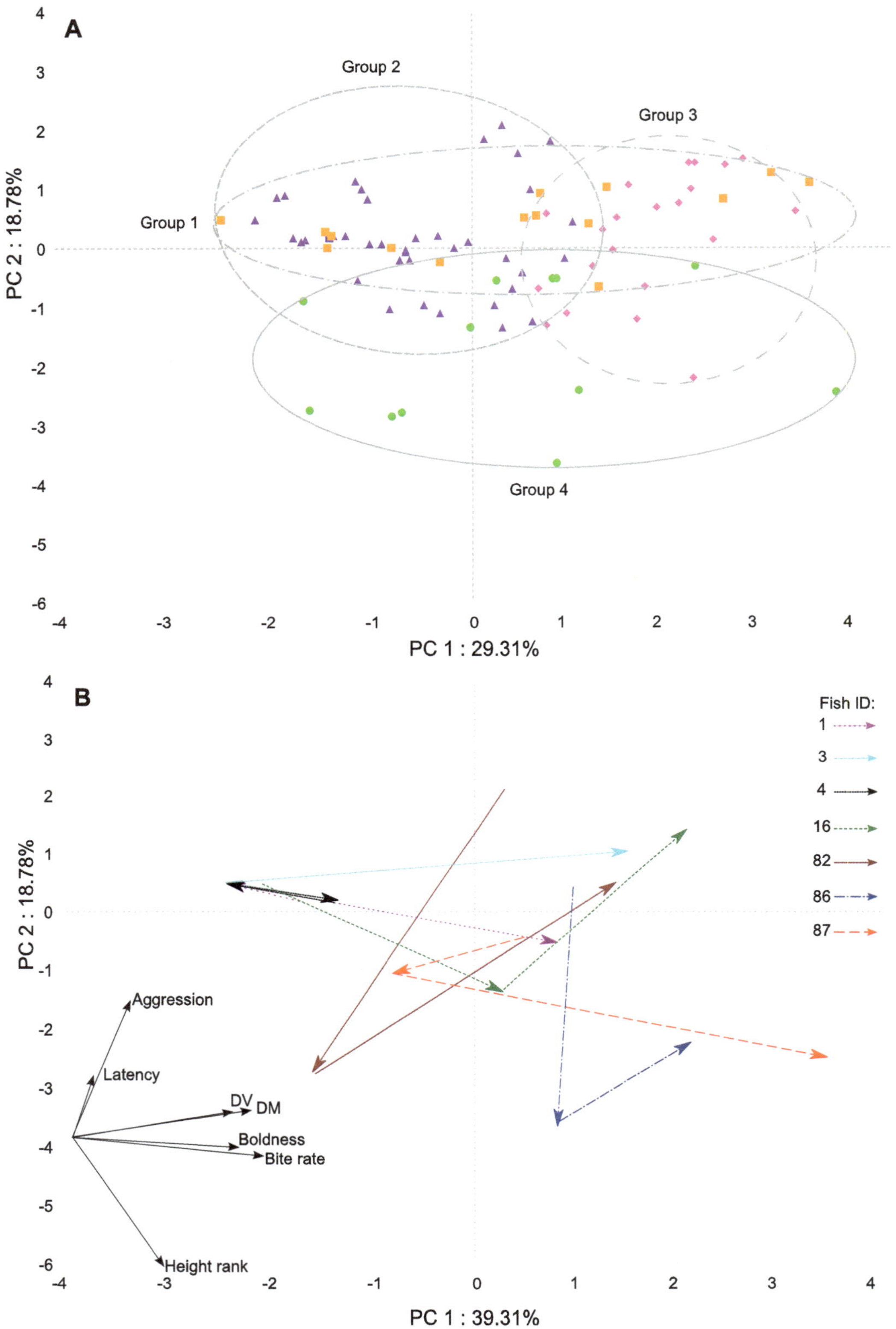

Figure 3. Principal component analysis of 7 behavioral traits for individual fish. A) PCA traits include: bite rate, distance moved, distance ventured, height, boldness, latency and aggression for individual fish (N = 33) across small tanks, large tanks, and field site (Total N = 99). Factor loadings of these traits represented by arrows in lower left corner. Ovals represent groupings created by clustering analysis. 67% of replicates in group 1 (square symbol) were from observations made in large tanks. 56% of replicates in group 2 (triangle symbol) were from small tanks. Group 3 (diamond symbol) is comprised of 76% of large tank observations. Group 4 (circle symbol) is composed of 42% large tank observations. Group 4 and the combination of groups 1, 2, and 3 represent the first split in the hierarchy. B) Identical principal components analysis with fish plots removed. Arrows represent factor loading patterns for seven randomly chosen fish from small tanks, large tanks, and field site.

Table 1. Average temperature ranges during behavioral assessments within each situation (small tank, large tank, field situation).

	Temperature (°C)			
Situation	**Mean**	**Min.**	**Max.**	**Range**
Small tank	29.1	24.1	33.0	8.9
Large tank	29.4	24.6	33.6	9.0
Field	29.0	28.6	30.5	1.9

†Values of P do not control for multiple testing of the same data (*P<0.05; **P<0.01; ***P<0.001). Only values printed in bold are significant after Holm's sequential Bonferonni adjustment of experimental error rates (Quinn and Keough 2002).

flexibility across situations [8] yet the premise of behavioral syndromes suggests some limitation of flexibility of behavioral responses [40]. Our results suggest that at this life history stageit is advantageous to remain highly flexible [1,23] in behavior, rather than to develop syndromes. Young fish at settlement undergo high rates of mortality (averaging ~ 60% within 48 hours, [26]) due to their small size and relatively poor competitive abilities [41] and they must be prepared to adapt rapidly to novel conditions. For these reasons, the ability to alter behavior to suit the new challenges they face may be key to survival.

White et al. (*in prep.*) recorded consistent individual rankings in scores of boldness, distance moved and occupancy height of reefs for newly-settled *P. amboinensis* over a three day period in the field. This finding suggests that behavioral patterns can be highly variable across different situations, yet at the same time show consistency within a single situation. Similarly, Coleman and Wilson (1998) found consistent individual boldness scores in two different contexts, but no correlation across contexts in juvenile pumpkinseed sunfish (*Lepomis gibbosus*) [42]. This implies that behavioral studies may have limited predictive ability when expanded to other situations; a finding that may be particularly relevant to laboratory-based work. Artificial environments can introduce variation in behavior due to confounding factors such as handling stress or experiences gained from life in captivity [14]. For example, farmed fish that live in an environment of high competition and no predation pressure are often bolder, more aggressive and take more risks than their wild counterparts [43,44]. This idea was also supported by Wilson et al. (1993) who found individual boldness to be stable in nature but absent in the laboratory for juvenile pumpkinseed sunfish (*L. gibbosus*). They argued local environmental conditions maintain differences between individual behavioral phenotypes [45]. Our pilot study revealed that fish in the field have a reduced acclimation time compared to those held in aquaria, suggesting they are less stressed and naturally inclined to start exhibiting "normal" behaviors more quickly in the field. Studies in the field also have the added benefit of incorporating realistic environmental and ecological factors that may influence behavior (e.g. quantifying the ecological trade-offs of individual variation in behavior).

There were no significant correlations between behavioral traits and temperature for each of the three situations after correcting for multiple comparisons. Thus, despite large fluctuations in daily water temperature (up to 9°C and 2°C in the lab and field respectively), temperature did not meaningfully affect behavioral traits in any consistent manner. In a laboratory study using 6 L tanks filled halfway, Biro *et al.* (2010) found average values for activity, boldness and aggressiveness to increase by a factor of 2.5 to 6 in two species of damselfish in response to daily water temperature fluctuations of 3°C or less [16]. McCormick and Meekan (2010) also found a significant positive relationship between activity and temperature in the field for *P. amboinensis* [23]. Metabolic rate has been shown to increase exponentially with temperature in other ectotherms [46], and individual differences in metabolism are thought to be contribute to individual differences in behavioral traits for these animals [6,47]. In our study, bite rate showed a positive correlation (however this was not significant after Bonferonni adjustment) with temperature within small tanks (over twice as large as the aquaria used by Biro *et al.* 2010), which would seem to agree with Biro *et al.*'s (2010) findings. Perhaps if behavioral observations in the present study were conducted following the protocol of Biro *et al.* (2010) (only one situation, in very small aquaria), our results might concur. In any event, it is clear that correlations developed from laboratory studies require validation in field environments in order to confirm that they have real ecological meaning.

The most consistent relationships among behaviors across situations were the close positive relationship between bite rate and distance moved across both the principal components (PC1 & PC2). Similarly, the correlation analysis showed positive correlations between bite rate and distance moved across small tank and field situations. These relationships are intuitive because individuals move more often and at greater distances when in pursuit of planktonic food. These relationships probably reflect the active foraging style of damselfish. All other relationships between behavioral traits were inconsistent across all situations. This result agrees with previous work on sticklebacks, which found no consistency in boldness, aggression, and activity across different situations between two populations [17]. This suggests behavioral syndromes do not always fit within the theory of the "constraint hypothesis" [48] which states behavioral syndromes are derived from a shared link between behaviors and assumes that the decoupling mechanisms underlying correlated behavioral traits do not evolve readily, because they would require changes in hormonal machinery [17]. The different relationships of behavioral traits among situations suggests either that there are biotic or abiotic factors that influence certain traits in different situations [17] or that the traits themselves can represent different things in different situations (e.g. the same measurement of boldness, or propensity to take risks, across different situations may in fact represent different traits due to how the individual perceives risk across the different situations).

It has been assumed that consistent individual differences in behavioral traits can occur due to differences in underlying physiological, behavioral, or morphological characteristics (i.e. state variables; [2,49]) and that these variables establish the efficiency of certain types of behavior [5]. For example, if predation risk is a function of body size, and since body size is stable over short time scales (daily), animals of different body sizes should differ consistently with respect to their tendency to take risks while foraging. Therefore, theory predicts behavioral patterns related to body size should also be stable over the same time frame [2,50]. The same logic applies to any other behavioral pattern linked to underlying state variables that are stable over time but vary among individuals [5]. While state variables are likely important in establishing stable behaviors, individual fish can rapidly respond to environmental factors that influence the behavioral patterns displayed. Thus, environmental factors are likely to be just as important in developing or maintaining stable behavioral syndromes. Behavioral flexibility is likely necessary for fish to quickly adjust to a completely different environment once they leave their pelagic larval phase. Bell and Sih (2007) found only populations of sticklebacks raised under strong predation pressure developed a correlative relationship between aggressiveness and boldness [51], demonstrating selective mortality and/or experience can help shape and establish behavioral patterns. Coral reef fishes may need some exposure to predators and environmental experience before developing a consistent behavioral syndrome.

In this study, naïve newly settled juveniles of *P. amboinensis* were found to lack consistency in: i) the rank order of behaviors across situations; ii) the relationship between behaviors and water temperature; iii) correlations in behaviors across situations. Given our results, it is interesting to note the lack of multi-situation or setting comparisons in previous studies that are often cited as evidence for the existence of behavioral syndromes. There have been a small number of mark-recapture [52] or short term (days) repeated studies (White et al., *in prep.*) that have properly met the definition of behavioral syndromes; yet the vast majority of fish studies only measure one behavioral trait, or show correlation under a single situation/context. Evidence for behavioral syndromes and personality in fish is weakened by the lack of multi-situation or multi-context comparisons. For example, boldness is defined as risk-taking and aggression as agonistic interactions between individuals and correlations between these two traits are often cited as a demonstration of behavioral syndromes. However, the operational definition and methods used to assess boldness are quite varied [53]. In some instances, aggressive interactions could also be considered risky. In some previous studies, this relationship could be an artifact of overlapping behavioral traits, where boldness and aggression are characteristics of a single behavioral trait rather than two distinct traits, or alternatively inadequate measures, if methods used to assess boldness actually assess both boldness and aggression. These issues cannot be resolved from observations made in a single situation. Comparisons across multiple situations (ideally including a natural setting) are necessary to establish within-individual or between-individual behavioral syndromes that by Sih et al.'s (2004) definition suggest some degree of situation or context independence.

Supporting Information

Table S1 Factor components from PCA of Behaviors in each situation.

Table S2 Correlations between 8 behavioral traits of *P. amboinensis* in A) small tanks, B) large tanks, and C) field site. Only significant values are presented ($p<0.05$). DM = distance moved, DV = distance ventured. † Values of P do not control for multiple testing of the same data (*$P<0.05$; **$P<0.01$; ***$P<0.001$). Only values printed in bold are significant after Holm's sequential Bonferonni adjustment of experimental error rates (Quinn and Keough 2002).

Acknowledgments

We would like to thank T. Bodine for assistance in the field as well as the staff of Lizard Island Research Station (Australian Museum) for logistical support. Comments and statistical advice from M. Ferrari improved the manuscript. J. H. Christ provided valuable insight on earlier versions of this manuscript.

Author Contributions

Conceived and designed the experiments: JRW MIM. Performed the experiments: JRW. Analyzed the data: JRW MIM MGM. Contributed reagents/materials/analysis tools: JRW MIM MGM. Wrote the paper: JRW MIM MGM.

References

1. Sih A, Bell AM, Chadwick Johnson J (2004) Behavioral syndromes: an ecological and evolutionary overview. Trends Ecol Evol 19: 372–378. doi:10.1016/j.tree.2004.04.009.
2. Stamps JA (2007) Growth-mortality tradeoffs and "personality traits" in animals. Ecol Lett 10: 355–363. doi:10.1111/j.1461-0248.2007.01034.x.
3. Wilson DS, Clark AB, Coleman K, Dearstyne T (1994) Shyness and boldness in humans and other animals. Trends Ecol Evol 9: 442–446. doi:10.1016/0169-5347(94)90134-1.
4. Wilson DS (1998) Adaptive individual differences within single populations. Philos Trans R Soc Lond B Biol Sci 353: 199–205. doi:10.1098/rstb.1998.0202.
5. Dall SRX, Houston AI, McNamara JM (2004) The behavioural ecology of personality: consistent individual differences from an adaptive perspective. Ecol Lett 7: 734–739. doi:10.1111/j.1461-0248.2004.00618.x.
6. Careau V, Thomas D, Humphries MM, Réale D (2008) Energy metabolism and animal personality. Oikos 117: 641–653. doi:10.1111/j.0030-1299.2008.16513.x.
7. Réale D, Reader SM, Sol D, McDougall PT, Dingemanse NJ (2007) Integrating animal temperament within ecology and evolution. Biol Rev 82: 291–318. doi:10.1111/j.1469-185X.2007.00010.x.
8. Sih A, Bell AM (2008) Insights for Behavioral Ecology from Behavioral Syndromes. Adv Study Behav 38: 227–281. doi:10.1016/S0065-3454(08)00005-3.
9. Gosling SD (2001) From Mice to Men: What Can We Learn About Personality From Animal Research? Psychol Bull 127: 45–86. doi:10.1037/0033-2909.127.1.45.
10. Toms CN, Echevarria DJ, Jouandot DJ (2010) A Methodological Review of Personality-Related Studies in Fish: Focus on the Shy-Bold Axis of Behavior. Int J Comp Psychol 23: 1–25.
11. Sinn DL, Moltschaniwskyj NA (2005) Personality traits in dumpling squid (*Euprymna tasmanica*): context-specific traits and their correlation with biological characteristics. J Comp Psychol 119: 99–110. doi:10.1037/0735-7036.119.1.99.
12. Wilson ADM, Stevens ED (2005) Consistency in context-specific measures of shyness and boldness in rainbow trout, *Oncorhynchus mykiss*. Ethology 111: 849–862. doi:10.1111/j.1439-0310.2005.01110.x.
13. Adriaenssens B, Johnsson JI (2011) Shy trout grow faster: exploring links between personality and fitness-related traits in the wild. Behav Ecol 22: 135. doi:10.1093/beheco/arq185.
14. Brown C, Jones F, Braithwaite V (2005) In situ examination of boldness-shyness traits in the tropical poeciliid, *Brachyraphis episcopi*. Anim Behav 70: 1003–1009. doi:10.1016/j.anbehav.2004.12.022.
15. Frost AJ, Winrow-Giffen A, Ashley PJ, Sneddon LU (2007) Plasticity in animal personality traits: does prior experience alter the degree of boldness? Proc R Soc B Biol Sci 274: 333–339. doi:10.1098/rspb.2006.3751.
16. Biro PA, Beckmann C, Stamps JA (2010) Small within-day increases in temperature affects boldness and alters personality in coral reef fish. Proc R Soc B Biol Sci 277: 71–77. doi:10.1098/rspb.2009.1346.
17. Bell AM (2005) Behavioural differences between individuals and two populations of stickleback (*Gasterosteus aculeatus*). J Evol Biol 18: 464–473. doi:10.1111/j.1420-9101.2004.00817.x.
18. Dugatkin LA, Alfieri MS (2003) Boldness, behavioral inhibition and learning. Ethol Ecol Evol 15: 43–49. doi:10.1080/08927014.2003.9522689.
19. Sneddon LU (2003) The bold and the shy: individual differences in rainbow trout. J Fish Biol 62: 971–975. doi:10.1046/j.1095-8649.2003.00084.x.
20. Koolhaas JM, Korte SM, De Boer SF, Van Der Vegt BJ, Van Reenen CG, et al. (1999) Coping styles in animals: current status in behavior and stress-physiology. Neurosci Biobehav Rev 23: 925–935. doi:10.1016/S0149-7634(99)00026-3.
21. Fuiman LA, Meekan MG, McCormick MI (2010) Maladaptive behavior reinforces a recruitment bottleneck in newly settled fishes. Oecologia 164: 99–108. doi:10.1007/s00442-010-1712-3.
22. Holmes TH, McCormick MI (2010) Size-selectivity of predatory reef fish on juvenile prey. Mar Ecol Prog Ser 399: 273–283.
23. McCormick MI, Meekan MG (2010) The importance of attitude: the influence of behaviour on survival at an ontogenetic boundary. Mar Ecol Prog Ser 407: 173–185. doi:10.3354/meps08583.
24. Meekan MG, von Kuerthy C, McCormick MI, Radford B (2010) Behavioural mediation of the costs and benefits of fast growth in a marine fish. Anim Behav 79: 803–809. doi:10.1016/j.anbehav.2009.12.002.
25. Lönnstedt OM, McCormick MI, Meekan MG, Ferrari MCO, Chivers DP (2012) Learn and live: the role of predator experience in influencing prey behaviour and survival. Proc R Soc B Biol Sci. doi:10.1098/rspb.2001.2516.

26. Almany GR, Webster MS (2006) The predation gauntlet: early post-settlement mortality in reef fishes. Coral Reefs 25: 19–22. doi:10.1007/s00338-005-0044-y.
27. Doherty PJ, Dufour V, Galzin R, Hixon MA, Meekan MG, et al. (2004) High mortality during settlement is a population bottleneck for a tropical surgeonfish. Ecology 85: 2422–2428. doi:10.1890/04-0366.
28. Gosselin LA, Qian P (1997) Juvenile mortality in benthic marine invertebrates. Mar Ecol Prog Ser 146: 265–282. doi:10.3354/meps146265.
29. Pitcher CR (1988) Validation of a technique for reconstructing daily patterns in the recruitment of coral reef damselfish. Coral Reefs 7: 105–111. doi:10.1007/BF00300969.
30. Kerrigan BA (1996) Temporal patterns in size and condition at settlement in two tropical reef fishes (Pomacentridae: *Pomacentrus amboinensis* and *P. nagasakiensis*). Mar Ecol Prog Ser 135: 24–41. doi:10.3354/meps135027.
31. McCormick MI, Makey L, Dufour V (2002) Comparative study of metamorphosis in tropical reef fishes. Mar Biol 141: 841–853. doi:10.1007/s00227-002-0883-9.
32. McCormick MI, Makey LJ (1997) Post-settlement transition in coral reef fishes: overlooked complexity in niche shifts. Mar Ecol Prog Ser 153: 247–257. doi:10.3354/meps153247.
33. Meekan MG, Wilson SG, Halford A, Retzel A (2001) A comparison of catches of fishes and invertebrates by two light trap designs, in tropical NW Australia. Mar Biol 139: 373–381. doi:10.1007/s002270100577.
34. McCormick M (2009) Behaviourally mediated phenotypic selection in a disturbed coral reef environment. PLoS ONE 4: e7096. doi:10.1371/journal.pone.0007096.
35. Gerlai R (2003) Zebrafish: an uncharted behavior genetic model. Behav Genet 33: 461–468.
36. Marks C, West T, Bagatto B, Moore F (2005) Developmental environment alters conditional aggression in zebrafish. Copeia: 901–908.
37. Budaev SV (2010) Using Principal Components and Factor Analysis in Animal Behaviour Research: Caveats and Guidelines. Ethology 116: 472–480. doi:10.1111/j.1439-0310.2010.01758.x.
38. Quinn GP, Keough MJ (2002) Experimental Design and Data Analysis for Biologists. Cambridge: Cambridge University Press. 537 p.
39. IBM SPSS Statistics for Windows (2011). Armonk, NY: IBM Corp.
40. Conrad JL, Weinersmith KL, Brodin T, Saltz JB, Sih A (2011) Behavioural syndromes in fishes: a review with implications for ecology and fisheries management. J Fish Biol 78: 395–435. doi:10.1111/j.1095-8649.2010.02874.x.
41. Munday P, Jones G, Caley M (2001) Interspecific competition and coexistence in a guild of coral-dwelling fishes. Ecology 82: 2177–2189. doi:10.1890/0012-9658(2001)082[2177:ICACIA]2.0.CO;2.
42. Coleman K, Wilson DS (1998) Shyness and boldness in pumpkinseed sunfish: individual differences are context-specific. Anim Behav 56: 927–936. doi:10.1006/anbe.1998.0852.
43. Sundström LF, Petersson E, Höjesjö J, Johnson JI, Jarvi T (2004) Hatchery selection promotes boldness in newly hatched brown trout (*Salmo trutta*): implications for dominance. Behav Ecol 15: 192–198. doi:10.1093/beheco/arg089.
44. Dingemanse NJ, Réale D (2005) Natural selection and animal personality. Behaviour 142: 1159–1184. doi:10.1163/156853905774539445.
45. Wilson DS, Coleman K, Clark AB, Biederman L (1993) Shy-bold continuum in pumpkinseed sunfish (*Lepomis gibbosus*): An ecological study of a psychological trait. J Comp Psychol 107: 250–260. doi:10.1037/0735-7036.107.3.250.
46. Clarke AB, Johnston NM (1999) Scaling of metabolic rate with body mass and temperature in teleost fish. J Anim Ecol 68: 893–905. doi:10.1046/j.1365-2656.1999.00337.x.
47. Biro P, Stamps J (2008) Are animal personality traits linked to life-history productivity? Trends Ecol Evol 23: 361–368. doi:10.1016/j.tree.2008.04.003.
48. Stamps JA (1991) Why evolutionary issues are reviving interest in proximate behavioral mechanisms. Am Zool 31: 338–348. doi:10.1093/icb/31.2.338.
49. Houston AI, McNamara JM (1999) Models of adaptive behaviour. Cambridge: Cambridge University Press.
50. Ambrose R, Strimling P (2006) How noisy information and individual asymmetries can make "personality" an adaptation: a simple model. Anim Behav 72: 1135–1139. doi:10.1016/j.anbehav.2006.04.001.
51. Bell AM, Sih A (2007) Exposure to predation generates personality in threespined sticklebacks (*Gasterosteus aculeatus*). Ecol Lett 10: 828–834. doi:10.1111/j.1461-0248.2007.01081.x.
52. Wilson AD., Godin JG. (2009) Boldness and behavioral syndromes in the bluegill sunfish, *Lepomis macrochirus*. Behav Ecol 20: 231.
53. White JR, Meekan MG, McCormick MI, Ferrari MCO (2013) A comparison of measures of boldness and their relationships to survival in young fish. PLoS ONE 8. Available: http://dx.plos.org/10.1371/journal.pone.0068900.

Joint Effects of Asymmetric Payoff and Reciprocity Mechanisms on Collective Cooperation in Water Sharing Interactions: A Game Theoretic Perspective

Cho Nam Ng[1], Raymond Yu Wang[1]*, Tianjie Zhao[2]

1 Department of Geography, The University of Hong Kong, Pokfulam, Hong Kong, **2** Institute of Remote Sensing and Digital Earth, Chinese Academy of Sciences, Beijing, China

Abstract

Common-pool resource (CPR) dilemmas distinguish themselves from general public good problems by encompassing both social and physical features. This paper examines how a physical mechanism, namely asymmetric payoff; and a social mechanism, reciprocity; simultaneously affect collective cooperation in theoretical water sharing interactions. We present an iterative N-person game theoretic model to investigate the joint effects of these two mechanisms in a linear fully connected river system under three information assumptions. From a simple evolutionary perspective, this paper quantitatively addresses the conditions for Nash Equilibrium in which collective cooperation might be established. The results suggest that direct reciprocity increases every actor's motivation to contribute to the collective good of the river system. Meanwhile, various upstream and downstream actors manifest individual disparities as a result of the direct reciprocity and asymmetric payoff mechanisms. More specifically, the downstream actors are less willing to cooperate unless there is a high probability that long-term interactions are ensured; however, a greater level of asymmetries is likely to increase upstream actors' incentives to cooperate even though the interactions could quickly end. The upstream actors also display weak sensitivity to an increase in the total number of actors, which generally results in a reduction in the other actors' motivation for cooperation. It is also shown that the indirect reciprocity mechanism relaxes the overall conditions for cooperative Nash Equilibrium.

Editor: Angel Sánchez, Universidad Carlos III de Madrid, Spain

Funding: Financial support is appreciated from the General Research Fund, Research Grants Council of Hong Kong (No.HKU750911H), as well as the Short Stay Fellowship for Ph.D. candidates of Utrecht University. The funders had no role in study design, data collection and analysis, decision to publish, or preparation of the manuscript.

Competing Interests: The authors have declared that no competing interests exist.

* E-mail: wangyuray@connect.hku.hk

Introduction

The emergence and evolution of collective cooperation in common-pool resource (CPR) dilemmas have fascinated scholars from various disciplines [1–8]. The fundamental puzzle lies in the mechanisms that facilitate costly cooperative and altruistic behavior of individuals who interact with other competitors in rigorous environments. Previous literature on social theory and public good problem has shown that some social mechanisms such as direct reciprocity (repeated interactions) and indirect reciprocity (reputation) can augment the level of collective cooperation [8–15]. Yet as much as the underlying structure of CPR dilemmas might be analogous to general public good problems, it is worth noting that a CPR system is unique in the sense of involving both physical and social attributes. These two mutually interconnected dimensions constitute the context in which human beings interact with nature as well as with each other. Therefore, to better understand the dynamics of collective action in governing CPRs, we argue that it is important to simultaneously examine the social and physical characteristics of CPR systems.

One of the main features of CPR systems is individuals' heterogeneities that are attributed to their physical geographies. For instance, different upstream and downstream actors are heterogeneous in terms of their influences to a river system. They are also diverse in the sense of being victims or beneficiaries who are dependent on other actors' behavior. However, most general theoretic studies of collective action in CPR dilemmas have been established on an assumption that all actors share symmetric access and position with regard to the commons [2,4,16–19]. It is worth noting that individual asymmetries might be able to substantially change the structures and results of previous theoretic models. Therefore, it warrants further study on the elements that constitute the asymmetries as well as the internal and external asymmetric mechanisms under which certain regularities might hold.

In this paper, we focus on surface water, which is a controversial CPR that flows across physical boundaries. With a lack of theoretical studies on the asymmetric gains and losses associated with the geographical locations and actions of different upstream and downstream actors, we aim to establish a formal model to investigate how a physical mechanism (asymmetric payoff) and a social mechanism (reciprocity) jointly affect collective cooperation in water sharing interactions. Built upon Raub and Weesie's [20] model regarding reputation and efficiency in social interactions, our study adds to previous literature in four main aspects. Firstly, we apply a game theoretic model to analyze a specific linear configuration which resembles a river system in the real world. Secondly, we incorporate multi-players into the game theoretic model to demonstrate a more complex group interaction of sharing limited water resources. Thirdly, asymmetric payoffs are integrated with the model to reflect heterogeneous features that are attached to different upstream and downstream actors. Finally, the study simultaneously analyzes the effects of two mechanisms which involve both physical and social attributes of a river system.

This paper first presents an iterative N-person Prisoner's Dilemma Game (PDG) which enables the direct reciprocity mechanism under which peer punishment could be enforced in future encounters between any two actors. Then with a simple evolutionary approach, we introduce the indirect reciprocity mechanism under which every actor in the game could respond to other actors' behavior based on the information one has received through a linear system. Also, we incorporate the asymmetric payoff mechanism into the model by modifying the standard PDG payoff matrix. Moreover, we quantitatively address conditions for cooperative Nash Equilibrium (NE) under three different information scenarios. Lastly, we perform numerical simulations in Matlab to illustrate the effects of each independent variable under the equilibrium conditions and provide intuitive explanations on the implications of the asymmetric payoff and reciprocity mechanisms.

Methods

Consider a river system that consists of a finite number of n actors ($A_1, A_2 \ldots A_i \ldots A_n$) who are located along the river in a fixed sequence as shown in Figure 1. The subscripts denote geographical locations of the actors. We refer to A_1 as the head-end actor and A_n as the tail-end actor. Being confined by the boundaries of the catchment, interactions between all water users are assumed to occur in an abstract linear system. Also, the water users along the river are institutional actors who have to engage in activities of utilizing water resources on a regular basis. Thus they are assumed to be unable to refuse participation; however, there is a probability that the interactions may end, i.e. the decision-makers of the institutional actors might be replaced or purposely move out of the river system. Follow the rational choice theory, the actors are assumed to be selfish who aim at maximizing their payoffs.

We assume that all actors are playing the PDGs on a discrete time scale ($t = 1, 2, 3\ldots$) till an indefinite end. The event continues with a probability $0<\beta<1$. Within each event moment t, every actor must make a binary choice between contributing to preserve the river environment (C) and inconsiderately exploiting the water resources (D). The choice of every actor is made simultaneously. We assume every actor is fully informed on both alternatives (C or D). We also refer to C as cooperation and D as defection in some of the following analysis. Once the action is made, any actor A_i cannot behave differently towards others within a single event moment. It implies that actor A_i actually plays a large game which is composed of n-1 pairwise PDGs (supergames) against all other actors within every event moment. By allowing the game to be indefinitely repeated, the direct reciprocity mechanism is thus established in the way that any actor is able to reciprocate with others who had cooperated or defected against himself in potential future interactions.

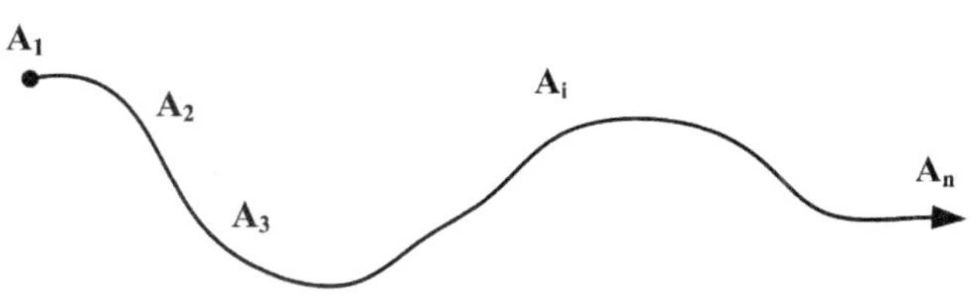

Figure 1. Geographic distribution of water users.

Then we introduce the indirect reciprocity mechanism by adding an information set to the model. The information set I_i is a profile of other actors' behavior that any actor A_i obtained. It is a critical element of the indirect reciprocity mechanism; because, in an interconnected social structure, every actor's present behavior might not only influence his present utility, it also generates a reputation which influences other participants' actions towards him in their potential future interactions. Therefore, we assume that every actor's choice of action in each event moment is dependent on the evolving information set that they have obtained during the course of the game.

Next, we introduce the asymmetric payoff mechanism by adding a parameter α to the standard PDG payoff matrix as follows,

		j: C	j: D
i	C	$2R\alpha^i_{i,j}, 2R\alpha^j_{i,j}$	$2S\alpha^i_{i,j}, 2T\alpha^j_{i,j}$
	D	$2T\alpha^i_{i,j}, 2S\alpha^j_{i,j}$	$2P\alpha^i_{i,j}, 2P\alpha^j_{i,j}$

$$T>R>0>P>S;\ 2R>T+S$$

where $\alpha^i_{i,j}=\frac{i^\varphi}{i^\varphi+j^\varphi}$, $\alpha^j_{i,j}=\frac{j^\varphi}{i^\varphi+j^\varphi}$, $\alpha^i_{i,j}+\alpha^j_{i,j}=1$; φ is a positive real number which indicates the degree of asymmetries; $1\le i<j\le n$, $i,j \in Z^+$. For later reference, we define $\delta(x)=\alpha^i_{i,i-x}+\alpha^i_{i,i+x}$ i.e. $j=i-x$ or $i+x$, $\alpha^i_{i,j}=0$ if $j>n$ or $j<1$. We also define $\gamma=(T-R)/(T-P)$, where $0<\gamma<1$ and γ is a constant under the given payoff matrix. Basically, γ is an indicator of actors' short-term incentive for defection.

This modification to the payoff matrix corresponds to the basic physical feature of the river system. That is, any actor's cooperative behavior will produce a public good, i.e. ecological service to the river system. The public good is shared by all actors, among whom the relative downstream actors benefit more than the upstream ones. Similarly, any actor's defective behavior will generate a loss to the public good, i.e. environmental degradation. This negative outcome is also shared by all actors, among whom the relative downstream actors lose more than the upstream ones. One might have a mistaken belief that the downstream actors' behavior cannot affect the upstream actors. As a matter of fact, a river system is a complete ecological unit in which one actor's behavior affects all other actors one way or another. For instance, the water quality at the river mouth influences fishes that migrate back upstream to mate and reproduce. The sediment deposition and wetlands at the very downstream of the river are also of great ecological value to upstream areas. Therefore, we argue it is safe to assume, as we presented in the payoff matrix, that the relative downstream actors are more vulnerable to the defection made by the upstream actors than the other way around.

It is worth noting that every pairwise interaction is essentially a PDG although the shares that are obtained by every actor are asymmetric. Here $\alpha^i_{i,j}$ and $\alpha^j_{i,j}$ denote the share of utilities that are allocated to actor A_i and actor A_j respectively in their interaction. The asymmetric parameter α is adjusted by relative geographical distribution of any two actors A_i and A_j, as well as the exploitation variable φ. Here we assume $i<j$ and thus actor A_j is located downstream actor A_i. The exploitation variable φ represents the degree to which one actor is differentiated from another in terms of utilities allocation. The larger φ is the greater asymmetric payoff effect is produced on every actor in the game. In consistent with the ecological characteristics, notice that the tail-end actor is the one that most exposed to the behavior of others, whereas the potential ramifications of other actors' behavior gradually decreases the further upstream the actors are located.

With the asymmetric payoff mechanism, the long-term expected utilities of any actor is not only dependent on the strategies of their own and the strategies of other actors, it is also dependent on their geographical locations and the asymmetric parameter α. Accordingly, we can address the utility function for any actor A_i at a particular moment t as follows:

$$U_{it}=f(V_t,\alpha^i_{i,j})=\sum_j^{V_t}2R\alpha^i_{i,j}+\sum_j^{N-V_t-i}2S\alpha^i_{i,j}\text{ if }s_{it}=C \quad (1)$$

$$U_{it}=g(V_t,\alpha^i_{i,j})=\sum_j^{V_t}2T\alpha^i_{i,j}+\sum_j^{N-V_t-i}2P\alpha^i_{i,j}\text{ if }s_{it}=D \quad (2)$$

where U_{it} denotes the utility of A_i at moment t, s_{it} denotes the action of A_i at moment t. $j\in V_t$, V_t = the numbers which denote the locations of actors who choose C apart from the focal actor A_i at moment t. $\boldsymbol{N}$ = the numbers which denote the locations of all actors in the game. $\boldsymbol{N}=(1, 2,\ldots,n), V_t\subseteq N$. Thus the total utility $\overline{U_i}$ that any actor A_i could receive during the entire game is:

$$\overline{U_i}=\sum_{t=1}^{\infty}\beta^{t-1}U_{it} \quad (3)$$

where $\overline{U_i}$ is the exponentially discounted utility sum of A_i from $t=1$ till the indefinite end of the game.

As the game is repeated, we will analyze the game from a simple evolutionary perspective. A desired outcome, from the view of social efficiency, is an equilibrium in which collective cooperation is established. In the scenario of water sharing interactions, actors cannot be 'dead' in a biological sense, instead, we assign supergame strategies which allow every actor to update their behavior based on the information they obtain during the course of the game. It is well known that pure strategy C is strictly dominated by D in the PDG. It is also known that unconditional cooperative supergame strategy, which means an actor always chooses C in a supergame regardless of other actors' choices, is not individually rational because it is strictly dominated by unconditional defective supergame strategy [21,22]. However, conditional cooperation could be a rational strategy if the choice of C by all actors in all their interactions is the best strategy to use against each other. In other words, collective cooperation might be established when all actors reach an NE in which C is chosen by every actor throughout the game at each moment and no actor can be strictly better off by switching one or more of their supergame strategies given the remaining actors stick with their supergame strategies [20]. There exist countless conditional cooperative supergame strategies as the game is indefinitely repeated. For analytical simplicity, we adopt a classical conditional strategy called "trigger", which suggests an actor to enter the game with C and then always play D once notified with any defection. The "trigger" is a non-forgiving strategy. One might consider using other conditional strategies that are less strict. Surely more sophisticated defined supergame strategies will enrich the study. However, they will meanwhile extensively increase the complexity of the model. Most importantly, the purpose of this paper is to focus on the effects of asymmetric payoff and reciprocity mechanisms and our following analysis remains valid if other strategies are used. Hence the application of alternative supergame strategies can be discussed in future studies.

Results

The model is analyzed in three scenarios in which information diffuses through the linear system at different rates. This section aims at addressing conditions for NE in which collective cooperation might be respectively achieved under the three information scenarios. The asymmetric payoff and reciprocity mechanisms are analyzed simultaneously.

Atomized interactions

We begin with the simplest scenario. Assuming that the information any actor A_i can possibly receive is only from actors who are located adjacent to him (A_{i+1} and A_{i-1}). This information is assumed to be received right after an action is committed by A_{i+1} and A_{i-1} at moment t. Under this specific assumption, interactions are atomized in the sense that an actor only gets information from nobody else but his contiguous actors.

Assumption 1. Each actor A_i receives information on the history of their contiguous actors A_{i+1} and A_{i-1}, that is for all i and t,

$$I_i(t)=H_{i-1}(t)\cup H_{i+1}(t)$$

where $I_i(t)$ denotes actor A_i's information set at moment t, $H_i(t)$ denotes the history of A_i's actions at moment t, $H_i(t)=\varphi$ if $i\leq 0$ or $i>n$ or $t\leq 1$.

Based on non-cooperative game theory, it is already known that either ALL-C (always play C or any other strategy that never initiates a D) or ALL-D (always play D) is a best-response strategy should all other actors use "trigger" [20,23]. Therefore, to address the necessary and sufficient conditions under which collective cooperation is in NE, one needs to compare any actor A_i's expected utilities of using ALL-C with his expected utilities of using ALL-D. The conditions for cooperative NE is equivalent to the conditions under which the following inequality stands for any $i=1,2,\ldots,n$.

$$E(\overline{U_i}|ALL-D)\leq E(\overline{U_i}|ALL-C) \quad (4)$$

With the asymmetric payoff matrix, obviously each actor receives different utilities in each event. Yet the expected utilities for any actor A_i of using ALL-C against all other actors who stick with "trigger" is only dependent on their spatial locations regardless of their information situations; because, C will be chosen by every actor throughout the game as no actor will ever initiate a defection according to their supergame strategy (ALL-C or trigger). According to Equation (1), any actor A_i's expected utilities of using ALL-C is,

$$E(\overline{U_i}|ALL-C)=\frac{2R}{1-\beta}\left(\sum_{j=1}^{i-1}\alpha^i_{i,j}+\sum_{j=i+1}^{n}\alpha^i_{i,j}\right) \quad (5)$$

It is, however, complex to calculate actor A_i's expected utilities of using ALL-D. The results vary significantly under different information assumptions. Due to the asymmetric payoff mechanism, the results are also greatly dependent on the actors' geographical locations. We provide a brief summary on the calculation and then derive the conditions for cooperative NE in each information scenario.

Assumption 1 illustrates a situation in which actors are poorly connected by information. For example, if actor A_i initiates a defection in the game at moment *1*, then only actors A_{i-1} and A_{i+1} will be aware of the defection and start to defect at moment *2*, the rest of the actors will still choose C. Likewise, only actors A_{i-2} and A_{i+2} will realize the defection of A_{i-1} and A_{i+1}, and then start to defect at moment *3*, and so forth. In general, defection can only diffuse through contiguity in the atomized system towards both upstream and downstream directions. As actor A_i's geographical location affects the evolution of the game, the calculation is carried out in two scenarios in which A_i is located either in relative upstream ($i \leq \lfloor n/2 \rfloor$) or downstream ($i > \lfloor n/2 \rfloor$) the river.

(1) If actor A_i uses ALL-D and $i \leq \lfloor n/2 \rfloor$, then actor A_n will be the last one who realizes that another actor had defected before. Thus actor A_n will start to defect at moment *n-i+1*, from which all actors will always defect afterwards. We divide the total expected utility of A_i into two parts as shown in Equation (6). The first part calculates actor A_i's utilities before actor A_n turns to defection; the second part is actor A_i's utilities when there is no cooperative behavior exists in the game.

$$E(\overline{U_i}|ALL-D) = \sum_{t=1}^{\infty} \beta^{t-1} U_{it} = \sum_{t=1}^{n-i} \beta^{t-1} U_{it} + \sum_{t=n-i+1}^{\infty} \beta^{t-1} U_{it} \quad (6)$$

For part one,

$$\begin{aligned}
\sum_{t=1}^{n-i} \beta^{t-1} U_{it} = {} & 2T\Big(\sum_{j=1}^{i-1} \alpha_{i,j}^{i} + \sum_{j=i+1}^{n} \alpha_{i,j}^{i}\Big) + \\
& \beta\Big[2T\Big(\sum_{j=1}^{i-2} \alpha_{i,j}^{i} + \sum_{j=i+2}^{n} \alpha_{i,j}^{i}\Big) + 2P\Big(\sum_{j=i-1}^{i-1} \alpha_{i,j}^{i} + \\
& \sum_{j=i+1}^{i+1} \alpha_{i,j}^{i}\Big)\Big] + \beta^{2}\Big[2T\Big(\sum_{j=1}^{i-3} \alpha_{i,j}^{i} + \sum_{j=i+3}^{n} \alpha_{i,j}^{i}\Big) + \\
& 2P\Big(\sum_{j=i-2}^{i-1} \alpha_{i,j}^{i} + \sum_{j=i+1}^{i+2} \alpha_{i,j}^{i}\Big)\Big] + \cdots\cdots + \\
& \beta^{n-i-2}\Big[2T\Big(\sum_{j=1}^{2i-n+1} \alpha_{i,j}^{i} + \sum_{j=n-1}^{n} \alpha_{i,j}^{i}\Big) + \\
& 2P\Big(\sum_{j=2i-n+2}^{i-1} \alpha_{i,j}^{i} + \sum_{j=i+1}^{n-2} \alpha_{i,j}^{i}\Big)\Big] + \\
& \beta^{n-i-1}\Big[2T\Big(\sum_{j=1}^{2i-n} \alpha_{i,j}^{i} + \sum_{j=n}^{n} \alpha_{i,j}^{i}\Big) + \\
& 2P\Big(\sum_{j=2i-n+1}^{i-1} \alpha_{i,j}^{i} + \sum_{j=i+1}^{n-1} \alpha_{i,j}^{i}\Big)\Big]
\end{aligned}$$

Multiply β on each side of the equation and with some basic algebra we will have

$$\begin{aligned}
(1-\beta)\sum_{t=1}^{n-i} \beta^{t-1} U_{it} = {} & 2T\Big(\sum_{j=1}^{i-1} \alpha_{i,j}^{i} + \sum_{j=i+1}^{n} \alpha_{i,j}^{i}\Big) + \\
& 2(P-T)[\beta(\alpha_{i,i-1}^{i} + \alpha_{i,i+1}^{i}) + \beta^{2}(\alpha_{i,i-1}^{i} + \alpha_{i,i+1}^{i}) + \cdots + \\
& \beta^{n-i-1}(\alpha_{i,2i-n+1}^{i} + \alpha_{i,n-1}^{i})] - \beta^{n-i}\Big[2T\Big(\sum_{j=1}^{2i-n} \alpha_{i,j}^{i} + \\
& \sum_{j=n}^{n} \alpha_{i,j}^{i}\Big) + 2P\Big(\sum_{j=2i-n+1}^{i-1} \alpha_{i,j}^{i} + \sum_{j=i+1}^{n-1} \alpha_{i,j}^{i}\Big)\Big]
\end{aligned} \quad (7)$$

For part two,

$$\sum_{t=n-i+1}^{\infty} \beta^{t-1} U_{it} = \frac{\beta^{n-i} 2P\big(\sum_{j=2i-n}^{i-1} \alpha_{i,j}^{i} + \sum_{j=i+1}^{n} \alpha_{i,j}^{i}\big)}{1-\beta} \quad (8)$$

Combining Equation (7) & (8) and we can solve Equation (6). Put them back to Inequality (4) with Equation (5) we will have,

$$\gamma \leq \frac{\beta\delta(1) + \beta^{2}\delta(2) + \cdots + \beta^{n-i-1}\delta(n-i-1) + \beta^{n-i}\delta(n-i)}{\sum_{j=1}^{i-1} \alpha_{i,j}^{i} + \sum_{j=i+1}^{n} \alpha_{i,j}^{i}} \quad (9)$$

(1) If actor A_i uses ALL-D and $i > \lfloor n/2 \rfloor$, then actor A_1 will be the last one who realizes another actor had defected before. Thus actor A_1 will start to defect at moment *i*, from which all actors will always defect afterwards. Likewise, we divide the total expected utilities of A_1 into the following two parts.

$$E(\overline{U_i}|ALL-D) = \sum_{t=1}^{\infty} \beta^{t-1} U_{it} = \sum_{t=1}^{i-1} \beta^{t-1} U_{it} + \sum_{t=i}^{\infty} \beta^{t-1} U_{it} \quad (10)$$

Without repeating a similar calculation as the one presented above, we will have the following equilibrium condition.

$$\gamma \leq \frac{\beta\delta(1) + \beta^{2}\delta(2) + \cdots + \beta^{i-2}\delta(i-2) + \beta^{i-1}\delta(i-1)}{\sum_{j=1}^{i-1} \alpha_{i,j}^{i} + \sum_{j=i-1}^{n} \alpha_{i,j}^{i}} \quad (11)$$

Summarizing the results of actor A_i's expected utilities when he is either located upstream or downstream, there is a condition for equilibrium outcome which suggests collective cooperation might be reached under Assumption 1.

Condition 1. If every actor uses a trigger against each other in an N-person PDG under Assumption 1, the system is in cooperative Nash Equilibrium if and only if

$$\gamma \leq \frac{\beta\delta(1) + \beta^{2}\delta(2) + \cdots \beta^{\max\{n-i,i-1\}}\delta[\max\{n-i,i-1\}]}{\sum_{j=1}^{i-1} \alpha_{i,j}^{i} + \sum_{j=i+1}^{n} \alpha_{i,j}^{i}}$$

Perfectly embedded interactions

In this section we relax the first assumption by introducing a system which allows each actor to obtain information from all other actors rather than his neighbors. Besides, we assume the information is perfectly embedded in the system. It implies that every actor receives full information about all other actors' behaviors immediately after an action is made.

Assumption 2. Each actor A_i receives information on the history of all actors in the game, that is for all i and t.

$$I_i(t) = H_1(t) \cup H_2(t) \cup \cdots \cup H_n(t)$$

Under assumption 2, the expected utilities of actor A_i if he uses ALL-D are simple to calculate. Because, any defective actor only has one-shot opportunity to abuse other actors' cooperation and all actors would immediately realize the existence of a defective actor once the defection is made. Then the entire game would turn into full defection. Thus the expected utilities of actor A_i when he chooses ALL-D against trigger are,

$$E(\overline{U_i}|ALL-D) = \sum_{t=1}^{\infty} \beta^{t-1} U_{it} = 2T\left(\sum_{j=1}^{i-1} \alpha^i_{i,j} + \sum_{j=i+1}^{n} \alpha^i_{i,j}\right) + \frac{2P\beta\left(\sum_{j=1}^{i-1} \alpha^i_{i,j} + \sum_{j=i+1}^{n} \alpha^i_{i,j}\right)}{1-\beta} \quad (12)$$

Put Equations (12) & (5) back to Inequality (4), we have the equilibrium conditions for collective cooperation under Assumption 2.

Condition 2. If every actor uses a trigger strategy against each other in an N-person PDG under Assumption 2, the system is in cooperative Nash Equilibrium if and only if

$$\gamma \leq \beta$$

Imperfectly embedded interactions

Either atomized or perfectly embedded interactions represent a relatively extreme situation of information exchange. In the following section we introduce a more realistic assumption under which information is partly or imperfectly informed to all actors. In particular, for any actor A_i, it is still assumed that information can be immediately received from his contiguous actors. Meanwhile, he can also obtain information about the behavior of any other actor A_j ($i \neq j$), but only after a certain time lag π_{ij}, which increases with the distance between actors A_i and A_j.

Assumption 3. Each actor A_i immediately receives information on the history of his contiguous actors A_{i+1} and A_{i-1}, and is informed with a time delay $\pi_{ij}>0$ on the history of all other actors, that is for all i and t.

$$I_i(t) = H_1(\max(t-\pi'_{i1},0)) \cup \cdots \cup H_{i-1}(t) \cup H_{i+1}(t) \cup \cdots \cup H_n(\max(t-\pi'_{in},0))$$

where π'_{ij} is the smallest integer that is strictly larger than π_{ij}.

Finding the conditions for cooperative NE under Assumption 3 is more complex due to the new parameter π_{ij}. This assumption allows everyone to receive information from distant actors. The parameter π_{ij} determines how soon the information can be received. Clearly, the longer time it takes for information to transfer, the fewer actors could have been notified that defections have been made in earlier rounds. More generally, if a defection was made by actor A_i, at moment *1*, for any other actors A_j, $i \neq j$

If $\pi'_{ij} > |i-j|$, then actor A_j will defect at moment $|i-j|+1$;

If $\pi'_{ij} \leq |i-j|$, then actor A_j will defect at moment $\pi'_{ij}+1$;

It is important to note is that Assumption 1 & 2 could be interpreted as two extreme cases of Assumption 3. That is, when $\pi_{ij} \to \infty$, the information among distant actors travels so slow that the case is the same as atomized interaction; when $\pi_{ij} \to 0$, the information travels so fast that everyone will immediately be aware of the history of all other actors as in perfectly embedded interactions.

Although the information might travel at different rates under Assumption 3, to find the condition for cooperative NE, we only need to focus on a scenario in which actor A_i gains the highest expected utilities if he uses ALL-D against all other actors who use trigger strategy. Obviously, no rational actor would defect if his highest expected utilities are smaller than his expected utilities of using ALL-C.

To allow actor A_i the highest expected utilities of using ALL-D under Assumption 3, information should only travel slightly faster than that in atomized interactions; because, it will give actor A_i the highest short-term benefits before the whole game turns into universal defection. More specifically, it implies that if A_i initiates a defection at moment *1*, then actors A_{i-1} and A_{i+1}, being A_i's neighbors, will be aware of the defection and start to defect at moment *2*; besides, actors A_{i-2} and A_{i+2} will also receive the information about actor A_i and start to defect at moment *2*. Because, the information is better embedded in this scenario than it is in atomized interactions and actors A_{i-2} and A_{i+2} are geographically closer to A_i than other actors. Yet we consider A_{i-2} and A_{i+2} are the only two more actors who can receive the defective information so that actor A_i is ensured to gain the maximum benefits by using ALL-D under Assumption 3.

The calculation of actor A_i's expected utilities is consisted of four different scenarios in terms of the geographical location of A_i as well as the number of actors who are located in the upstream and downstream directions of A_i. We present the deduction for one scenario in which actor A_i is located in relative upstream ($i \leq \lfloor n/2 \rfloor$) and the total amount of downstream actors is an even number ($n-i = 2m$, $m \in \mathbb{Z}^+$). In this scenario, actor A_n will be the last one who realizes another actor had defected in earlier rounds. Actor A_n will start to defect at moment $(n-i)/2+1$, from which all actors will defect afterwards. Similar to the calculation under Assumption 1, we divide the total expected utilities of actor A_i into two parts as follows.

$$E(\overline{U_i}|ALL-D) = \sum_{t=1}^{\infty} \beta^{t-1} U_{it} = \sum_{t=1}^{\frac{n-i}{2}} \beta^{t-1} U_{it} + \sum_{t=\frac{n-i}{2}+1}^{\infty} \beta^{t-1} U_{it} \quad (13)$$

For part one,

$$\sum_{t=1}^{(n-i)/2}\beta^{t-1}U_{it}=2T(\sum_{j=1}^{i-1}\alpha_{i,j}^{i}+\sum_{j=i+1}^{n}\alpha_{i,j}^{i})+$$
$$\beta[2T(\sum_{j=1}^{i-3}\alpha_{i,j}^{i}+\sum_{j=i+3}^{n}\alpha_{i,j}^{i})+2P(\sum_{j=i-2}^{i-1}\alpha_{i,j}^{i}+$$
$$\sum_{j=i+1}^{i+2}\alpha_{i,j}^{i})]+\beta^{2}[2T(\sum_{j=1}^{i-5}\alpha_{i,j}^{i}+\sum_{j=i+5}^{n}\alpha_{i,j}^{i})+$$
$$2P(\sum_{j=i-4}^{i-1}\alpha_{i,j}^{i}+\sum_{j=i+4}^{i+2}\alpha_{i,j}^{i})]+\cdots\cdots+$$
$$\beta^{(n-i)/2-2}[2T(\sum_{j=1}^{2i-n+3}\alpha_{i,j}^{i}+\sum_{j=n-3}^{n}\alpha_{i,j}^{i})+$$
$$2P(\sum_{j=2i-n+4}^{i-1}\alpha_{i,j}^{i}+\sum_{j=i+1}^{n-4}\alpha_{i,j}^{i})]+$$
$$\beta^{(n-i)/2-1}[2T(\sum_{j=1}^{2i-n+1}\alpha_{i,j}^{i}+\sum_{j=n-1}^{n}\alpha_{i,j}^{i})+$$
$$2P(\sum_{j=2i-n+2}^{i-1}\alpha_{i,j}^{i}+\sum_{j=i+1}^{n-2}\alpha_{i,j}^{i})]$$

Multiply β on each side of the equation and with some basic algebra we will have

$$(1-\beta)\sum_{t=1}^{(n-i)/2}\beta^{t-1}U_{it}=2T(\sum_{j=1}^{i-1}\alpha_{i,j}^{i}+\sum_{j=i+1}^{n}\alpha_{i,j}^{i})+$$
$$2(P-T)[\beta(\alpha_{i,i-2}^{i}+\alpha_{i,i-1}^{i}+\alpha_{i,i+1}^{i}+\alpha_{i,i+2}^{i})+$$
$$\beta^{2}(\alpha_{i,i-4}^{i}+\alpha_{i,i-3}^{i}+\alpha_{i,i+3}^{i}+\alpha_{i,i+4}^{i})+\cdots+$$
$$\beta^{(n-i)/2-1}(\alpha_{i,2i-n+2}^{i}+\alpha_{i,2i-n+3}^{i}+\alpha_{i,n-3}^{i}+\alpha_{i,n-2}^{i})]$$
$$-\beta^{(n-i)/2}[2T(\sum_{j=1}^{2i-n+1}\alpha_{i,j}^{i}+\sum_{j=n-1}^{n}\alpha_{i,j}^{i})+$$
$$2P(\sum_{j=2i-n+2}^{i-1}\alpha_{i,j}^{i}+\sum_{j=i+1}^{n-2}\alpha_{i,j}^{i})] \quad (14)$$

For part two,

$$\sum_{t=(n-i)/2+1}^{\infty}\beta^{t-1}U_{it}=\frac{\beta^{(n-i)/2}2P(\sum_{j=2i-n}^{i-1}\alpha_{i,j}^{i}+\sum_{j=i+1}^{n}\alpha_{i,j}^{i})}{1-\beta} \quad (15)$$

Combining Equation (14) & (15) and we will solve Equation (13). Put them back to Inequality (4) with Equation (5) we will have,

$$\gamma\le\frac{\beta[\delta(1)+\delta(2)]+\beta^{2}[\delta(3)+\delta(4)]+\cdots+\beta^{\frac{n-i}{2}}[\delta(n-i-1)+\delta(n-i)]}{\sum_{j=1}^{i-1}\alpha_{i,j}^{i}+\sum_{j=i+1}^{n}\alpha_{i,j}^{i}} \quad (16)$$

The other three scenarios are $i\le\lfloor n/2\rfloor$, $n-i=2m-1$, $m\in\mathbb{Z}^{+}$; $i>\lfloor n/2\rfloor$, $i=2m-1$, $m\in\mathbb{Z}^{+}$; and $i>\lfloor n/2\rfloor$, i = 2m, $m\in\mathbb{Z}^{+}$. The deductions about these three scenarios are fairly the same as the one we presented above. There are only subtle differences in their results. Without repeating the same procedure, we have the equilibrium conditions for collective cooperation under Assumption 3.

Condition 3. If every actor uses a trigger strategy against each other in an N-person PDG under Assumption 3, the system is in cooperative Nash Equilibrium if and only if

when max{n−i, i} = 2m, m∈Z⁺,

$$\gamma\le\frac{\beta[\delta(1)+\delta(2)]+\beta^{2}[\delta(3)+\delta(4)]+\cdots+\beta^{\max\{\frac{n-i}{2},\frac{i}{2}\}}[\delta(\max\{n-i-1,i-1\})+\delta(\max\{n-i,i\})]}{\sum_{j=1}^{i-1}\alpha_{i,j}^{i}+\sum_{j=i+1}^{n}\alpha_{i,j}^{i}}$$

when max{n−i, i} = 2m−1, m∈Z⁺,

$$\gamma\le\frac{\beta[\delta(1)+\delta(2)]+\beta^{2}[\delta(3)+\delta(4)]+\cdots+\beta^{\max\{\frac{n-i+1}{2},\frac{i-1}{2}\}}[\delta(\max\{n-i,i-2\})+\delta(\max\{n-i+1,i-1\})]}{\sum_{j=1}^{i-1}\alpha_{i,j}^{i}+\sum_{j=i-1}^{n}\alpha_{i,j}^{i}}$$

Discussion

The conditions for cooperative NE under the three assumptions are very complex. It is thus difficult to get intuitive insights into the implications of the equilibrium conditions. However, we are able to find regularities from their mathematical expressions. In particular, all of the three conditions are inequalities which comprise of the "temptation to defect" γ on the left and a function on the right, which the latter is dependent on four variables n, β, φ and i. For analytical simplicity, we respectively define the right side function in condition 1, 2 and 3 as f_a, f_p and f_{im}. Therefore the conditions for cooperative NE could be translated into the following mathematical expressions,

$$\gamma\le V_a=f_a(n,\beta,\varphi,i) \quad (17)$$

$$\gamma\le V_p=f_p(\beta) \quad (18)$$

$$\gamma\le V_{im}=f_{im}(n,\beta,\varphi,i) \quad (19)$$

where V_a, V_p and V_{im} denote the values of f_a, f_p and f_{im}.

The three conditions basically imply that cooperative outcomes might be supported under each information assumption, providing that V_a, V_p and V_{im} are sufficiently large to outweigh the incentive for short-term defection γ. Apparently, the larger V_a, V_p and V_{im} is, the more likely cooperation is achieved. We should note that the condition for cooperative NE under perfect information is straightforward in the sense of being only dependent on β regardless of asymmetries in the system and the geographical location of an actor. Whereas in the other two scenarios, it is difficult to establish explicit understandings of the effects of n, β, φ and i; because, all of them affect V_a and V_{im} concurrently. Therefore, we carry out numerical simulations in Matlab to examine how V_a and V_{im} respond to changes of n, β, φ and i on a comparative basis. The simulations produce 142,560 data sets under circumstances when n varies from 3 to 50 at a 1 interval, β varies from 0.01 to 0.99 at a 0.01 interval, φ varies from 0.1 to 3 at a 0.1 interval and i varies from 1 to n at a 1 interval.

To better examine the effects of each independent variable in f_a and f_{im}, our analysis is designed to hold variables other than the focal variable constant in selective scenarios. We start with the total number of actor n. Then the variables β, φ and i are analyzed under the condition that the total number of actors is fixed to 20.

Figure 2 indicates how V_a and V_{im} would change as the total number of actors n increases. We considered three scenarios in which the level of asymmetry φ is set to 0.4, 1 and 3.We select the head-end, tail-end and midstream actors as examples to illustrate. In each subplot, we also compare V_a and V_{im} under two different continuing probability 0.3 and 0.9. In general, the results demonstrate a descending trend of V_a and V_{im} in most scenarios. The results correspond to Olson's influential argument about "the logic of collective action" which states that the larger a group is, the less likely they are to create social incentives which lead its members to provide collective goods [11]. Nonetheless, we also discover that the head-end actor's incentive to cooperate is hardly affected by the group size when a high level of asymmetries exists during the course of interactions. An intuitive explanation for this phenomenon is that the head-end actor has most control over his potential loss. Greater level of asymmetries reduces his dependence on other actors' behavior. Hence his risk of being defected by others does not increase with the number of actors involved.

Figure 3 indicates how V_a and V_{im} would change as an actor's geographical location i moves gradually from the source to the end of the river. We select three scenarios in which the continuing probability β is set to 0.3, 0.6 and 0.9. In each subplot, we also compare how each actor reacts to variations of the levels of asymmetric payoff under the circumstances when φ equals to 0.1, 0.5, 1, 2 and 3. Two remarks can be drawn from this figure. First, V_a and V_{im} increase with i at the beginning and then decrease after V_a and V_{im} reach their apexes. Although the position of the apexes varies with φ and β, it implies that relative up-midstream actors are more likely to cooperate than the others when the remaining variables are held invariant. Second, the upstream curves in each subplot become steeper as φ increases. It shows that greater individual differences exist among upstream actors with higher levels of asymmetries.

Figure 4 indicates how V_a and V_{im} would change as the level of asymmetries φ increases. Likewise, we select three scenarios in which the continuing probability β is set to 0.3, 0.6 and 0.9. We compare the motivation for cooperation of five different actors (head-end, mid-upstream, mid, mid-downstream, tail-end). It is shown from Figure 4 that V_a and V_{im} for the head-end and mid-upstream actors increase with φ, yet for downstream actors they decrease slightly with φ and tend to stabilize though φ continues to increase. It implies upstream actors are sensitive to the degree of asymmetries and more likely to cooperate than downstream actors when higher levels of asymmetries display. An intuitive explanation for the joint effect of φ and i is the actors' reactions to risks. As the level of asymmetries increases, the upstream actors have less reservation about the risk of being hurt by others and thus tend to cooperate; whereas the situation is reversed for the downstream

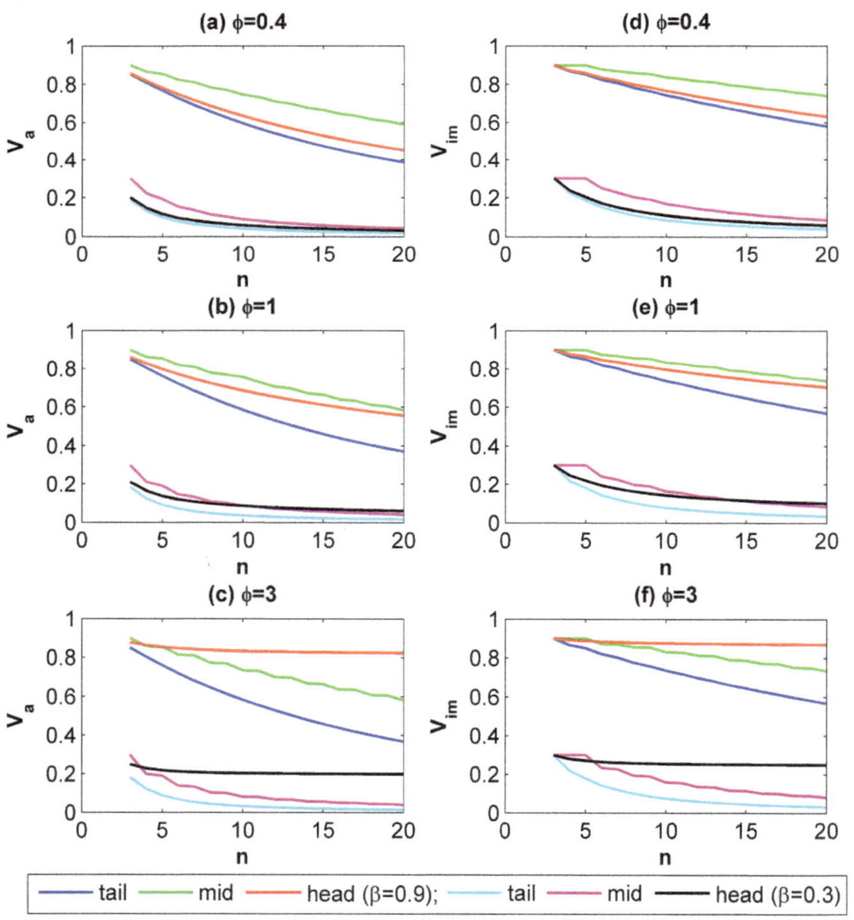

Figure 2. The effect of total number of actors n on the conditions for cooperative NE. (a, b, c) V_a for the head-end, mid-stream and tail-end actors in atomized interactions when $\varphi = 0.4$, 1 and 3 respectively; (d, e, f) V_{im} for the head-end, mid-stream and tail-end actors in imperfectly embedded interactions when $\varphi = 0.4$, 1 and 3 respectively. The curves provide each actor's general response to the increase of the group size. Each actor's motivation for cooperation is represented by two curves $\beta = 0.3$ and 0.9 in every subplot.

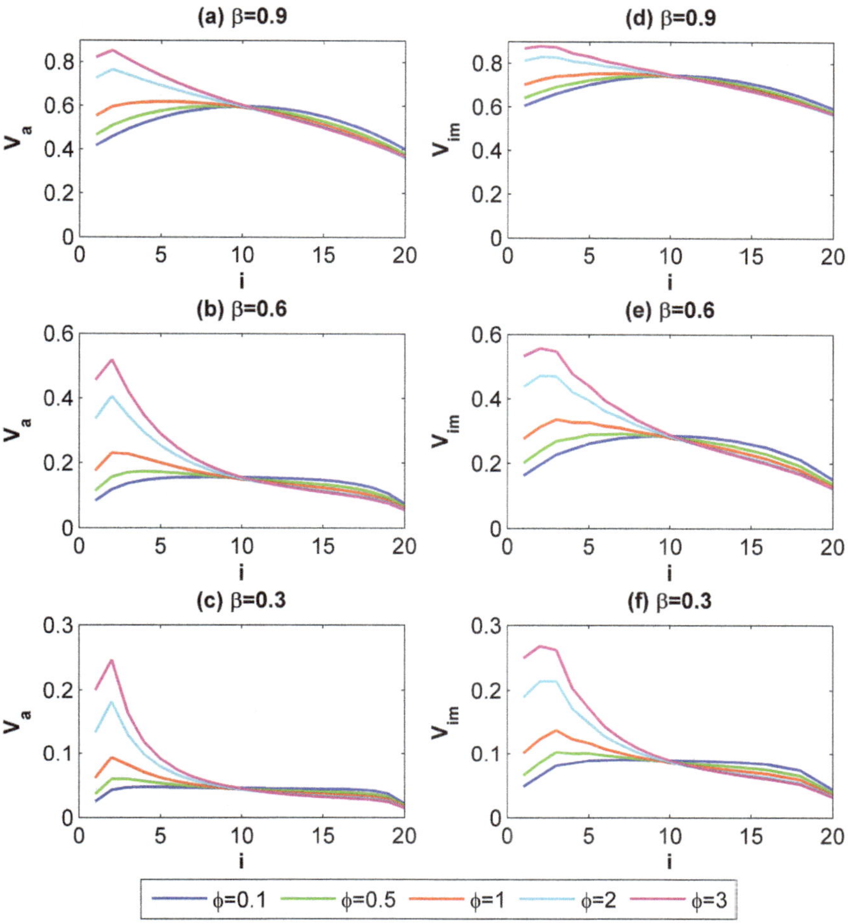

Figure 3. The effect of actors' geographical locations *i* on the conditions for cooperative NE when the group size is fixed to 20. (a, b, c) V_a for all of the 20 involved actors in atomized interactions when $\beta = 0.9, 0.6$ and 0.3 respectively; (d, e, f) V_{im} for all of the 20 involved actors in imperfectly embedded interactions when $\beta = 0.9, 0.6$ and 0.3 respectively. The curves provide a comparison in terms of motivation for cooperation for all actors who are located at different positions on the river. Each actor's motivation for cooperation is also compared in every subplot when the interactions take place under different levels of asymmetries $\varphi = 0.1$, 0.5, 1, 2 and 3.

actors who bear increasing risks thus their motivation for cooperation tend to maintain at a low level.

Figure 5 indicates how V_a and V_{im} would change as the continuing probability β increases. We consider three scenarios in which the level of asymmetries φ is set to 0.4, 1 and 3. In each subplot, we compare the motivation for cooperation of five different actors (head-end, mid-upstream, mid, mid-downstream, tail-end). The value of β is generally referred as "the shadow of the future" which indicates the possibility of future interactions between all involved actors. Hence, the larger β is the more likely the game can continue. We can draw two remarks from Figure 5. On one hand, for a particular actor, V_a and V_{im} increase with β when is φ constant. It confirms that an actor is more likely to cooperate when "the shadow of the future" is more significant. On the other hand, the slope of the curves for downstream actors becomes steeper than upstream actors when β is relatively large; the situation is reversed when β is relatively small. It implies that downstream actors' motivation for cooperation increase faster than upstream actors when there is a higher possibility that future interactions will continue taking place. To the contrary, upstream actors are more motivated to cooperate than downstream actors even when there is a greater chance that the game could quickly end. An intuitive explanation for the joint effects of β and i is actors' vision for their long-term interactions. For the downstream actors, assurance of future interactions will reduce their risks by giving them more control over other actors. They will therefore more likely to provide cooperation. This effect is amplified by the level of asymmetries. Whereas upstream actors are less exposed to others hence direct reciprocate behavior would not reduce their motivation for cooperation.

After analyzing the effect of each variable individually, in Figure 6 we present the overall conditions for cooperative NE under Assumption 1, 2 & 3 in three dimensional graphs. Instead of providing a condition for cooperation that is applicable to any actor A_i in the game, we located the minimum f_a and f_{im} with respect to i by screening our massive amount of simulation results. This approach helps us find the single one actor who has the least incentive for cooperation when other variables (n, β and φ) are held invariant. The results confirmed that the tail-end actor is the one that shows least willingness to choose C. In other words, the conditions for cooperative NE should always stand if we apply $i = n$ in all the three assumptions above. In that case, f_a and f_{im} become functions of n, β and φ. Consider the group size $n = 20$, we generate the overall cooperative NE conditions as shown in Figure 6. The comparison of cooperative NE conditions reveals the effect of indirect reciprocity mechanism. Specifically, the cooperative

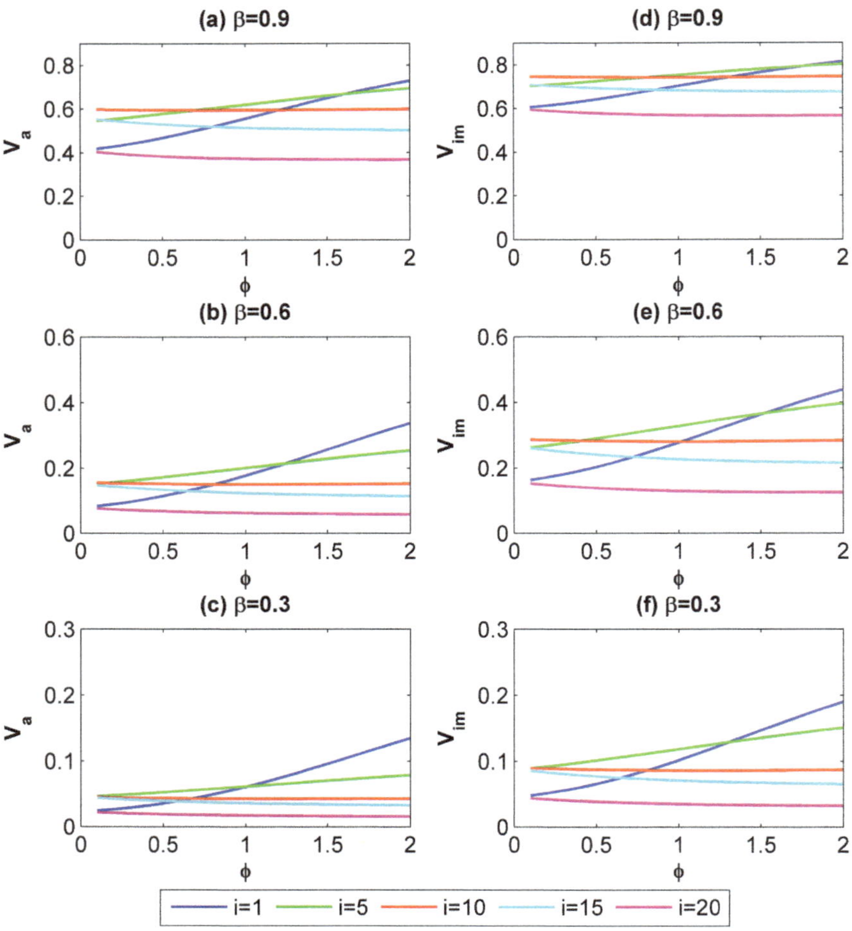

Figure 4. The effect of the level of asymmetries φ on the conditions for cooperative NE when the group size is fixed to 20. (a, b, c) V_a for five involved actors (head-end, mid-upstream, mid, mid-downstream, tail-end) in atomized interactions when $\beta = 0.3$, 0.6 and 0.9 respectively; (d, e, f) V_{im} for five involved actors (head-end, mid-upstream, mid, mid-downstream, tail-end) in imperfectly embedded interactions when $\beta = 0.3$, 0.6 and 0.9 respectively. The curves provide each actor's general response to the increase of the level of asymmetries.

NE condition in perfectly embedded interactions is only dependent on β and thus is a flat surface in Figure 6(b) regardless of other variables. The flat surface is always above the other two curved surfaces in Figure 6 (a) & (c), among which the former represents atomized interactions and the latter represents imperfectly embedded interactions. Figure 6(d) illustrates that the effect of indirect reciprocity mechanism and demonstrates that the conditions for collective cooperation become less restrict when information is better embedded in the game.

Conclusions

In this paper we simultaneously investigate the effects of asymmetric payoff and reciprocity mechanisms on collective cooperation in water sharing interactions. We establish a quantitative model of iterative N-person PDG and study the game as it evolves with all actors' actions which are conditional on their available information during the course of the game. Under different information scenarios, our analysis produces conditions for NE in which collective cooperation is likely to be established. The results suggest that the direct reciprocity, or put it poetically "the shadow of the future", can increase all actors' motivation to contribute to the collective good. Meanwhile, various upstream and downstream actors manifest individual disparities as a result of the direct reciprocity and asymmetric payoff mechanisms. More specifically, the downstream actors are less willing to contribute unless there is a high probability that long-term interactions are guaranteed; however, a greater level of asymmetries is more likely to increase upstream actors' incentives to cooperate even though the interactions could quickly end. The upstream actors also display weak sensitivity to an increase in the total number of actors, which generally results in a reduction in the other actors' motivation for cooperation. It is also shown that the indirect reciprocity mechanism relaxes the overall conditions for cooperative NE.

In general, our model is a preliminary theoretical attempt to connect the asymmetric payoff with reciprocity mechanisms. We endeavor to examine their joint effects on the collective behavior of heterogeneous selfish actors in a theoretical river system. We generate theoretical predictions based on the N-person iterative asymmetric PDG. In this paper, we do not intend to conclude with a deterministic argument about a causal relationship between the two mechanisms and collective cooperation in water governance. Obviously, a lot more theoretical models and empirical case studies remain to be conducted. However, we do expect to provide a more comprehensive perspective on the theory of collective cooperation, which emphasizes the integrity of CPR systems in the sense of including both physical and social characteristics.

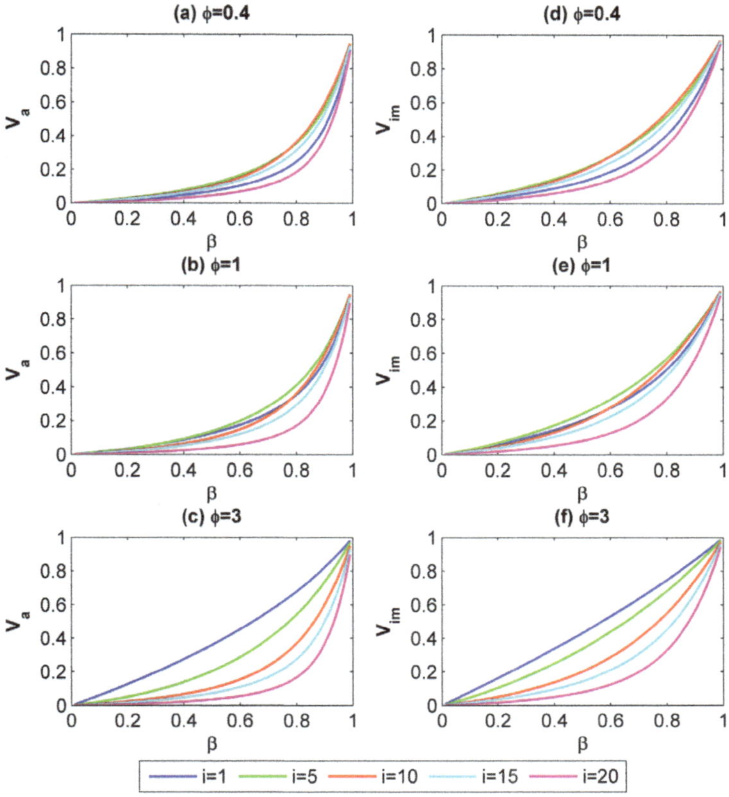

Figure 5. The effect of continuing probability β on the conditions for cooperative NE when the group size is fixed to 20. (a, b, c) V_a for five involved actors (head-end, mid-upstream, mid, mid-downstream, tail-end) in atomized interactions when $\varphi = 0.4$, 1 and 3 respectively; (d, e, f) V_{im} for five involved actors (head-end, mid-upstream, mid, mid-downstream, tail-end) in imperfectly embedded interactions when $\varphi = 0.4$, 1 and 3 respectively. The curves provide each actor's general response to the increase of the continuing probability β.

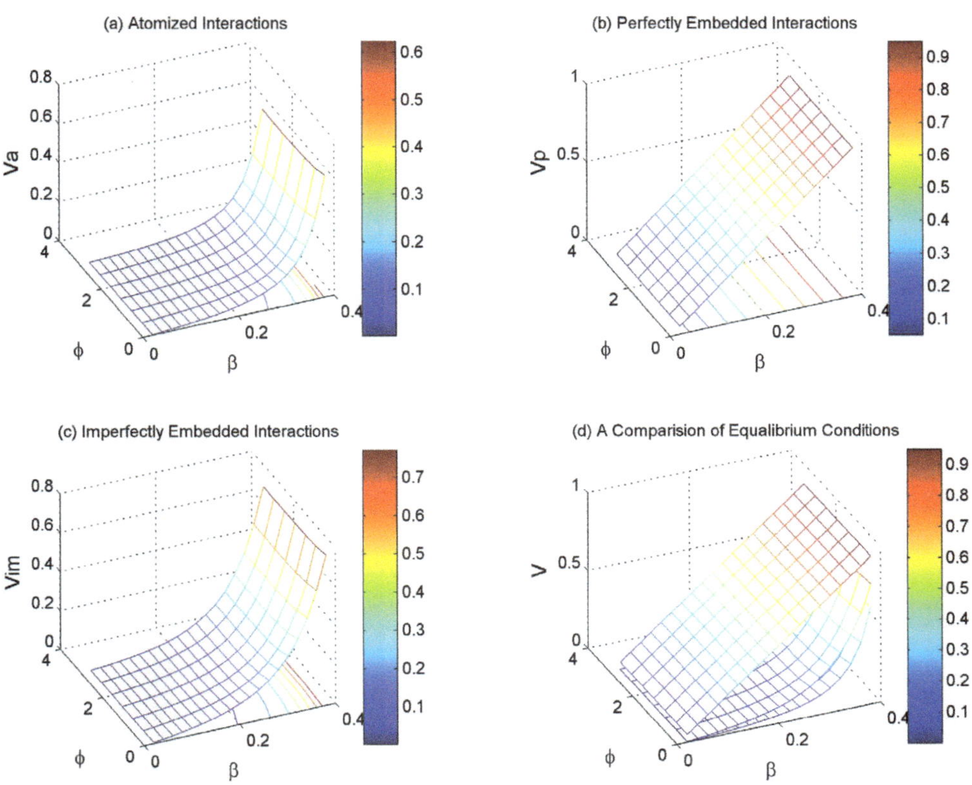

Figure 6. Overall conditions for cooperative NE when the group size is fixed to 20. (a) conditions for cooperative NE in atomized interactions; (b) conditions for cooperative NE in perfectly embedded interactions; (c) conditions for cooperative NE in imperfectly embedded interactions; (d) comparison of the conditions for cooperative NE under Assumption 1, 2 & 3.

Acknowledgments

The authors appreciate the comments made on this paper by Prof. Werner Raub and Prof. Vincent Buskens. Thanks also to Su Di for her help in Matlab programming and to James Lenzer for his excellent editing.

Author Contributions

Conceived and designed the experiments: CNN RYW. Performed the experiments: RYW CNN. Analyzed the data: CNN RYW TJZ. Contributed reagents/materials/analysis tools: RYW TJZ CNN. Wrote the paper: RYW CNN.

References

1. Ostrom E (2009) A General Framework for Analyzing Sustainability of Social-Ecological Systems. Science 325: 419–422.
2. Janssen MA, Holahan R, Lee A, Ostrom E (2010) Lab Experiments for the Study of Social-Ecological Systems. Science 328: 613–617.
3. Ostrom E (1990) Governing the commons: The evolution of institutions for collective action. New York: Cambridge University Press.
4. Ostrom E, Gardner R, Walker J (1994) Rules, games, and common-pool resources. Ann Arbor: University of Michigan Press.
5. Velez MA, Stranlund JK, Murphy JJ (2012) Preferences for government enforcement of a common pool harvest quota: Theory and experimental evidence from fishing communities in Colombia. Ecological Economics 77: 185–192.
6. Madani K, Dinar A (2012) Non-cooperative institutions for sustainable common pool resource management: Application to groundwater. Ecological Economics.
7. Rahman HMT, Hickey GM, Sarker SK (2012) A framework for evaluating collective action and informal institutional dynamics under a resource management policy of decentralization. Ecological Economics 83: 32–41.
8. Nowak MA (2012) Evolving cooperation. Journal of Theoretical Biology 299: 1–8.
9. Granovetter M (1985) Economic Action and Social Structure: the Problem of Embeddedness. American Journal of Sociology 91: 481–510.
10. Coleman JS (1990) Foundations of Social Theory. Cambridge, Massachusetts: Harvard University Press.
11. Olson M (1971) The logic of collective action: public goods and the theory of groups. Cambridge: Harvard University Press.
12. Raub W, Buskens V, van Assen MALM (2011) Micro-Macro Links and Microfoundations in Sociology. Journal of Mathematical Sociology 35: 1–25.
13. Krippner GR, Alvarez AS (2007) Embeddedness and the intellectual projects of economic sociology. Annual Review of Sociology 33: 219–240.
14. Nowak MA (2006) Five rules for the evolution of cooperation. Science 314: 1560–1563.
15. Hauert C, Traulsen A, Brandt H, Nowak MA, Sigmund K (2007) Via freedom to coercion: The emergence of costly punishment. Science 316: 1905–1907.
16. Janssen MA, Rollins ND (2012) Evolution of cooperation in asymmetric commons dilemmas. Journal of Economic Behavior & Organization 81: 220–229.
17. Janssen MA, Anderies JM, Cardenas JC (2011) Head-enders as stationary bandits in asymmetric commons: Comparing irrigation experiments in the laboratory and the field. Ecological Economics 70: 1590–1598.
18. Lindahl T (2012) Coordination problems and resource collapse in the commons — Exploring the role of knowledge heterogeneity. Ecological Economics 79: 52–59.
19. Ostrom E, Gardner R (1993) Coping with Asymmetries in the Commons - Self-Governing Irrigation Systems Can Work. Journal of Economic Perspectives 7: 93–112.
20. Raub W, Weesie J (1990) Reputation and Efficiency in Social Interactions - an Example of Network Effects. American Journal of Sociology 96: 626–654.
21. Axelrod R, Hamilton W (1981) The Evolution of Cooperation. Science 211: 1390–1396.
22. Nash J (1951) Non-Cooperative Games. The Annals of Mathematics 54: 286–295.
23. Friedman J (1977) Oligopoly and the Theory of Games. Amsterdam: North-Holland.

11

Human Adaptive Behavior in Common Pool Resource Systems

Gunnar Brandt[1]*, Agostino Merico[1,2], Björn Vollan[3,4], Achim Schlüter[3,5]

1 Systems Ecology, Leibniz Center for Tropical Marine Ecology, Bremen, Germany, **2** School of Engineering and Science, Jacobs University, Bremen, Germany, **3** Institutional & Behavioural Economics, Leibniz Center for Tropical Marine Ecology, Bremen, Germany, **4** Institute of Public Finance, University of Innsbruck, Innsbruck, Austria, **5** School of Humanities and Social Sciences, Jacobs University, Bremen, Germany

Abstract

Overexploitation of common-pool resources, resulting from uncooperative harvest behavior, is a major problem in many social-ecological systems. Feedbacks between user behavior and resource productivity induce non-linear dynamics in the harvest and the resource stock that complicate the understanding and the prediction of the co-evolutionary system. With an adaptive model constrained by data from a behavioral economic experiment, we show that users' expectations of future pay-offs vary as a result of the previous harvest experience, the time-horizon, and the ability to communicate. In our model, harvest behavior is a trait that adjusts to continuously changing potential returns according to a trade-off between the users' current harvest and the discounted future productivity of the resource. Given a maximum discount factor, which quantifies the users' perception of future pay-offs, the temporal dynamics of harvest behavior and ecological resource can be predicted. Our results reveal a non-linear relation between the previous harvest and current discount rates, which is most sensitive around a reference harvest level. While higher than expected returns resulting from cooperative harvesting in the past increase the importance of future resource productivity and foster sustainability, harvests below the reference level lead to a downward spiral of increasing overexploitation and disappointing returns.

Editor: Angel Sánchez, Universidad Carlos III de Madrid, Spain

Funding: This research was fully funded by the Leibniz Center for Tropical Marine Ecology (www.zmt-bremen.de). The funders had no role in study design, data collection and analysis, decision to publish, or preparation of the manuscript.

Competing Interests: The authors have declared that no competing interests exist.

* E-mail: gunnar.brandt@zmt-bremen.de

Introduction

Many social-ecological systems (SESs) that comprise a common pool resource (CPR) face the problem of overexploitation, because it is very costly, albeit not impossible, to exclude users from subtracting resource units [1–6]. Resource appropriation in such SESs often produces benefits for the individual, while all share the costs. This gives users an obvious incentive to maximize their harvest, thus, preventing cooperation and sustainability [7]. There is, however, compelling evidence both from economic experiments [8–11] and from real systems [12–15] that under certain conditions users may overcome the egoistic temptation of maximizing individual profits.

The decision to forgo part of a possible harvest from a renewable CPR is, particularly at low resource levels, an investment into future productivity at the cost of reduced short-term returns [16]. Because such an investment into the future always comes with uncertainties and because of the human preference for proximate returns, users discount potential future pay-offs [11,17,18]. According to the standard discounted utility model, rational users integrate all current and expected future returns after discounting them at a constant rate [17]. If discounted future benefits are large enough, users are willing to forego current benefits and cooperate [8,10,19].

Users of CPRs face highly uncertain decisions for two reasons. First, many ecological systems are characterized by high intrinsic variability, which complicates predictability [20]. Second, harvesting itself affects the stock and eventually the productivity of the resource. By this means, the harvest behavior may also influence the weight users assign to future resource productivity giving rise to a trade-off between the harvest behavior and the expected future returns [21]. The key to sustainability in many real SES is therefore to enhance the certainty of receiving the future benefits of cooperation, a goal achieved best by creating an institutional environment that is capable of accounting for the specific characteristics of the system under consideration [21–23]. Change, whether social or ecological, may, however, overstrain also robust institutions, when it is too rapid for successful adaptation [24].

Studies of real resource-user systems often disregard these close links between harvest behavior, resource dynamics, and future certainty and focus on either ecological or social aspects. The two sub-systems are described on different levels of detail, which hinders an integrated understanding of their coupled dynamics [25,26]. Moreover, observational or experimental data from CPR systems usually cover only a short period of time or comprise many confounding factors and hence do not allow observing and understanding temporal changes. The quantitative relationship and the feedbacks between resource productivity and user behavior remain largely unknown. Consequently, regime shifts or collapses observed in overexploited ecological resources are still not fully understood [27,28].

A novel approach for understanding the combined dynamics of a CPR and the users is to reduce the number of confounding factors by studying the system under controlled laboratory conditions. In a recent study, Janssen et al. [9] presented such a computer-based laboratory experiment, in which a group of five users could harvest continuously from a renewable CPR. Each group played six consecutive rounds of $240s$ with a change of the treatment after the third round. In three of the six rounds neither communication nor punishment were possible (a treatment labeled as NCP), while in the other three rounds users could coordinate resource extraction using either communication (C), punishment (P), or a combination of both (CP).

Although the composition of each group was fixed and all properties of the game except the treatments were identical, the experimental results revealed a great variability of harvest behaviors within rounds, between rounds with the same treatment, as well as between different treatments. Users realized highest total harvests (H_{tot}) when they cooperated and allowed the CPR to grow and to produce more resource units at the beginning of a round. Communication, punishment (albeit to a lesser extend), and the total harvest realized in previous rounds influenced the harvest behavior of the users and their returns [9].

We combine here Janssen et al.'s experimental data with an adaptive model to identify the main drivers of the co-evolution of the users' harvest behavior and the CPR. Our model is based on a mechanistic trade-off between the current harvest and the discount factor of future productivity. The trade-off accounts for the effect of resource exploitation on the certainty of future returns and reflects the central decision users face while harvesting from a renewable resource with a density-dependent growth. While maximizing the current harvest reduces the certainty of future returns, because the resource stock may decline significantly or even collapse as a consequence of intensive exploitation, lower current harvests enhance the chances of higher productivities and hence higher returns in the future.

Our model simulates a renewable CPR (R, Equation 1) with a fixed number of users (C), who realize a harvest (H, Equation 9) by adopting a variable harvest strategy. We define the harvest strategy as a continuous behavioral trait (x, Equation 5) that determines the harvest rate of the users. x adapts to changes in 1) the current harvest opportunity and 2) the discounted future productivity, which we consider to be equivalent to the potential future returns. The CPR grows at a density-dependent logistic growth rate (r_r, Equation 2) and users subtract variable amounts of resource units according to a Monod-type harvest rate (r_h, Equation 3). Changes in x alter r_h via the half-saturation constant (K_S, Equation 4), but also affect the discounted future productivity of the resource (r_f, Equation 7).

More precisely, harvesting becomes less intense with increasing x, while the discount factor for future productivity (w, Equation 8) rises. The discount factor w, which is a function of the time horizon and the harvest trait, can vary between 0 and the maximum discount factor (w^*). The parameter w^* sets an upper limit for the weight of future productivity and represents the maximum level of certainty that is sustained by the social system, i.e. by the rules of the game, the institutions, and the experiences of the users. It is constant on short time-scales, because rules and institutions usually change slower than the fastest processes in ecological or social systems [24], but may vary between simulations accounting for different institutional environments or different harvest experiences.

Users maximize the net present value, which is the sum of the current pay-off and all discounted future pay-offs for given maximum discount factor and CPR level. Following adaptation models of continuous traits [29–32], the temporal change of the harvest trait x is proportional to the gradient of the fitness function F, which is the sum of current and discounted future pay-offs ($r_h + r_f$) and the costs for optional punishment (r_p, Equation 6). By adjusting x, users change their harvest behavior to increase their fitness.

In behavioral economic experiments, user behavior is measured as cooperation. Cooperation is typically expressed as a dimensionless number between 0 and 1 and determined by the user's investment relative to a potential maximum value. Following this approach, we define the average cooperation of the group (φ) by the normalized foregone harvest (Equation 10), which is the amount of resource units that the users decide not to harvest divided by the maximum possible harvest.

In our model, users are not resolved as indidivuals. By contrast, the group of users is considered as a single adaptive entity, and the state variables of the model describe the dynamics of average group properties. The model hence corresponds to the typical resolution of observational data from real SESs and does not require detailed assumptions on the behavior of each individual in the system.

While Janssen et al. [9] focused on the statistical analysis of outcomes in terms of total harvest, our aim is to find a mechanistic explanation for the dynamics of the coupled system and the observed differences between rounds. Our major assumption is that the observed variability in harvest behavior and cooperation of users is caused by a trade-off between the current harvest and the discounted future productivity, which is mainly driven by differing maximum discount factors between rounds. Therefore, we 1) constrain the proposed trade-off by experimental data [9], 2) study the influence of the social environment, which is represented by a single parameter (the maximum discount factor w^*), on the co-evolution of the user-CPR system, and 3) assess the effect of previous experience on the users' perception of future certainty.

Results and Discussion

Trade-off between Harvest Behavior and Future Expectations

The expected future pay-offs of the CPR users were not measured directly during the experiments of Janssen et al. [9]. We instead use the relative resource productivity P^* as a qualitative indication of the group's expectations to constrain our model. P^* is defined as the cumulated resource productivity from the current time to the end of the experiment normalized to the resource level at the beginning of each round. In other words, P^* expresses the future productivity as a fraction of the initial resource level. Correspondingly, the inverse of the half-saturation constant Ks^{-1}, calculated from the experimental time-series of the harvest rate, is proportional to the resource affinity and used here equivalently. A high affinity value (or a low K_S) indicates aggressive harvesting already at low resource levels, while at low resource affinities users reach near maximum harvest rates only at high resource levels.

We discover a strong trade-off between P^* and Ks^{-1} and a high variability of these two variables in the experimental data (Figure 1C). More specifically, in the experiment, relative resource productivities significantly exceeding 1.0 only occur at low resource affinities ($K_S^{-1} < 0.005\text{N}^{-1}$). In contrast, high resource affinities ($K_S^{-1} > 0.005\text{N}^{-1}$) lead to large current harvests, while limiting the production of new resource units to values of $P^* < 0.5$. Similar to known relationships in real SESs (cf. Figure 2 in [26]), this trade-off is highly non-linear and introduces a tipping point to the system that clearly separates the effects of sustainable use from overexploitation. Users, thus, face the decision of increasing either short-term benefits or the long-term resource productivity [33].

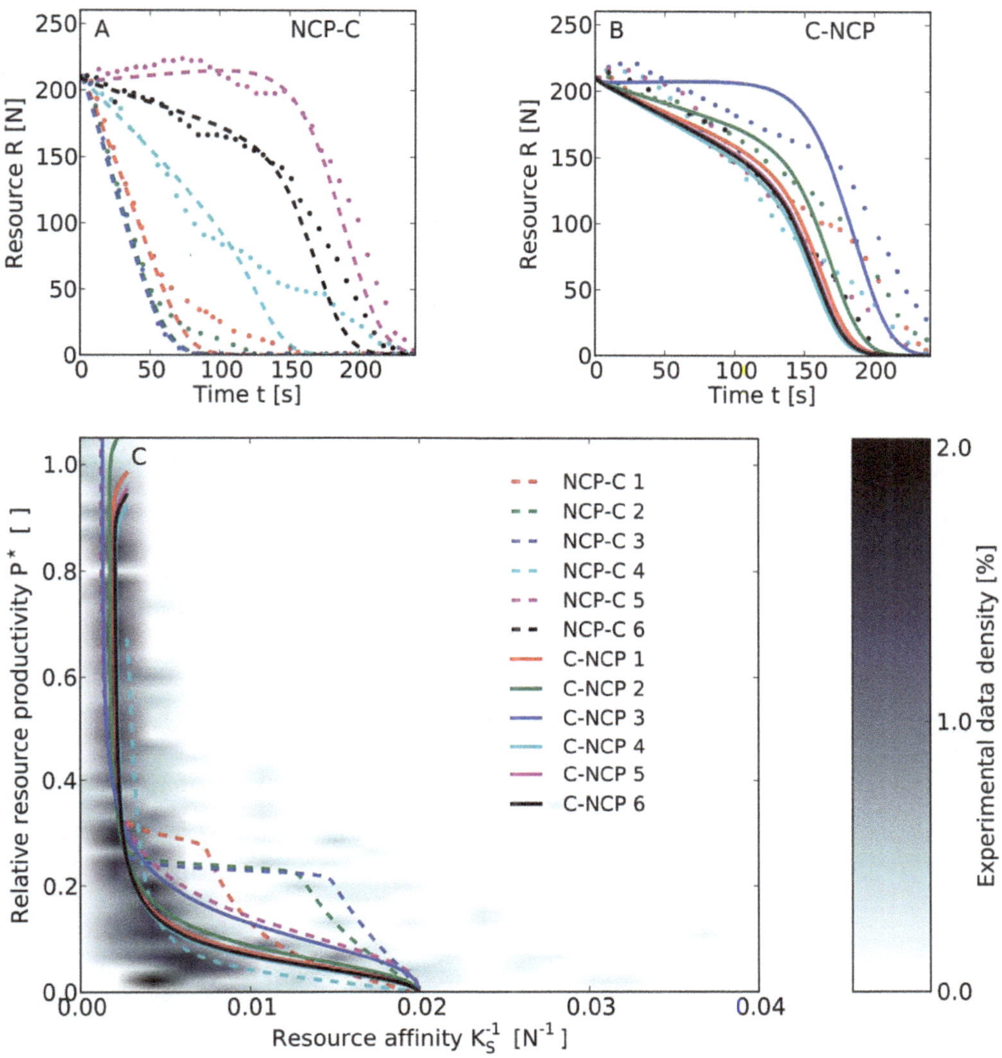

Figure 1. Trade-off between consumer's resource affinity and the productivity of a renewable resource. A–B, Resource level in six rounds of a computer-based laboratory game with five users and different treatments (dotted lines indicate experimental data [9]). While in the first three rounds of **A** neither communication nor punishment was possible (NCP-C 1–3), users could communicate in subsequent rounds (NCP-C 4–6). In the treatment C-NCP (**A**), three rounds with communication (C-NCP 1–3) were followed by three NCP rounds (C-NCP 4–6). Set-ups of model runs (dashed and solid lines in **A** and **B**, respectively) only differ in the the maximum discount factor $w^{\star}$ (see Equation 8, NCP-C: [4.4, 1.9, 1.4, 11.7, 32.5, 20.4], C-NCP: [18.0, 20.4, 29.3, 16.5, 17.3, 17.0]). **C**, Phase plot of the users' resource affinity, here defined as the inverse of the half-saturation constant K_S (see Equations 3 and 4), and the relative resource productivity $P^{\star}$, defined as the resource productivity from the current point of time to the end of a round normalized to the initial resource level. The shaded area shows the density distribution of the experimental data from the two treatments shown in **A** and **B**. Solid lines indicate the trade-off between resource affinity and potential future harvest from the resource system in corresponding model results.

We calibrated our model to match the distribution of the experimental data and adjusted only the parameter $w^{\star}$ between rounds (Figure 1A and B). The model trajectories reveal the continuous change of user behavior over the course of the different rounds (Figure 1C). Starting from low affinities all model simulations end with $P^{\star}=0$ and $Ks^{-1}=0.02\mathrm{N}^{-1}$, which represents the highest possible harvest rate and the complete exhaustion of the resource at the end of all rounds. While all rounds end similarly, they differ in the trajectories that lead to the exhaustion of the resource. When communication is possible (NCP-C 4–6, C-NCP 1–3), maximum $P^{\star}$ values are considerably higher than in NCP-rounds with no prior experience of communication (NCP-C 1–3). In contrast, users increase the resource affinity in NCP-C 1–3 right from the start (cf. Figure 1A and C) and by doing so avert high resource productivities.

Effect of Different $w^{\star}$ on the Temporal Dynamics of the User-CPR System

The different values of the maximum discount factor $w^{\star}$ can be attributed to the changes in the social environment of the users, because the simulated rounds differ only by the available treatments and the history of previous round, whereas the resource characteristics and the composition of the groups of users were identical.

If the future is irrelevant to users and future returns are disregarded ($w^{\star}=0$, Figure 2A and D), cooperation levels deteriorate within the first $30s$ of the simulation. In this case, the

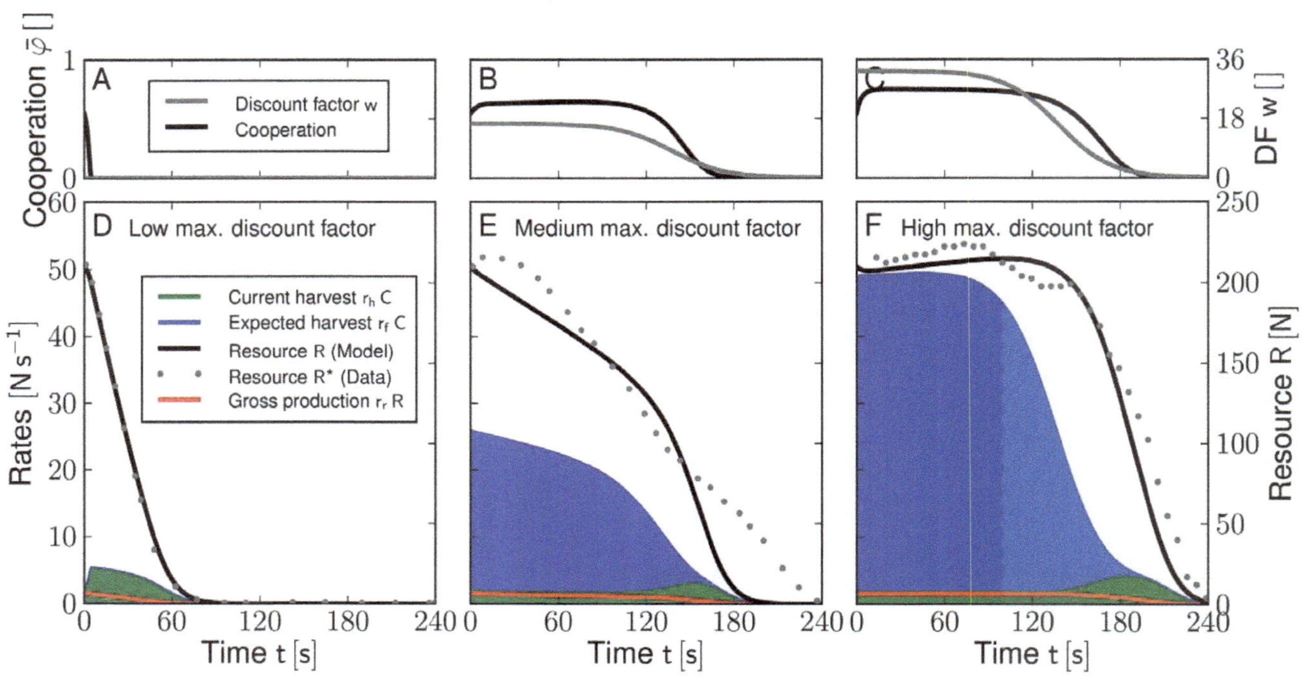

Figure 2. Time evolution of cooperation, harvest rates and a renewable resource for three different levels of future certainty. Increasing the maximum discount factor $w^\star$ (Equation 8) lowers the current harvest r_hC (Equation 3, lower panels **d–f**, green shaded area), but raises the future resource productivity r_fC that is considered by the users (Equation 7, blue shaded area). Cooperation φ decreases sharply when future pay-offs are ignored (**a**, $w^\star=0.0$) causing an immediate resource collapse (**d**, black solid line (model) and gray dots (experimental data from [9])). Larger values of $w^\star$ (**b**, $w^\star=16.5$ and **c**, $w^\star=32.5$) result in higher cooperation and reduce the current harvest as resource users account for a much higher proportion of future productivity (**e** and **f**). Resource collapse occurs later and the extended period of sustainable resource use leads to significantly higher total harvests (cf. Figures. S3a and S1). Red lines in panels **a–c** indicate the temporal evolution of the discount factor w.

current harvest rate significantly exceeds the growth rate of the CPR and the unsustainable use leads to a collapse within the first $100s$ and to a poor total harvest ($H_{tot}<300N$, Figure 3A). Increasing the importance of the future, that is increasing $w^\star$, results in higher cooperation over longer periods of time and slows down (Figure 2B and E) or even reverses overexploitation (Figure 2C and F). Towards the end of all simulations, however, w declines with the remaining time of the experiment causing an erosion of cooperation that eventually triggers the collapse of the resource (Figure 2A–C), because there is no potential future productivity to account for in a finite game.

Users continuously adjust their harvest strategy according to changing present harvest opportunities and expected future pay-offs. Therefore, cooperation and sustainable harvesting become rational when the discounted total harvest for one strategy is higher than for other strategies [34]. By treating resource users as an adaptive entity, our model unveils their great behavioral variability and the smooth transition from a sustainable to unsustainable resource use. These results support studies [35] that question stable norms of cooperation derived from "one-shot" field experiments [36–39]. The observed behavioral variability among CPR users with identical cultural background corroborates our assumption that users adapt their harvest behavior to the properties of the social and ecological environment [13,35,40].

Relation between the Maximum Discount Factor and the Total Harvest

The harvest behavior of CPR users and hence the outcome of the artificial commons vary greatly between rounds (cf. Figure S1 and S2 for the results of all rounds). A quantitative measure that may explain those differences of the total harvest H_{tot} is the maximum discount factor $w^\star$. The total harvest indicates the success of the group's behavior and is positively related to $w^\star$ (Figure 3A), because in the model the productivity increases with R for the range of resource levels observed in the experiment ($R<0.5K=420.5$N). In other words, a high $w^\star$ leads to harvest rates below productivity, i.e. sustainability, at the beginning of a round and eventually to high total harvests. Given the constraints of the experiment, the highest H_{tot} of $520N$ is therefore realized at the highest value of $w^\star=41.8$ (Figure 3A).

By only adjusting $w^\star$ to match the experimental results, we assume that differences between rounds are mainly caused by variations in the users' perception of future certainty. Allowing for variations also in the parameters α and β reduces the error between model and experimental data indicating that users also adjust the temporal dynamics of the harvest from round to round (Figure S4 and Table S1). However, the results of a systematic sensitivity analysis (Figure S5) confirm the high sensitivity of the model results to changes in $w^\star$ and corroborate the choice of $w^\star$ as the only free parameter explaining the observed differences between the rounds of the experiment.

Effect of Different Treatments on the Maximum Discount Factor

Our analysis reveals that the variations between rounds in the perception of future certainty, represented by $w^\star$ in the model, are determined by a combination of factors, including 1) treatment in the current round, 2) prior experience from rounds with the same treatment, and 3) possible exposure to a different treatment in the past.

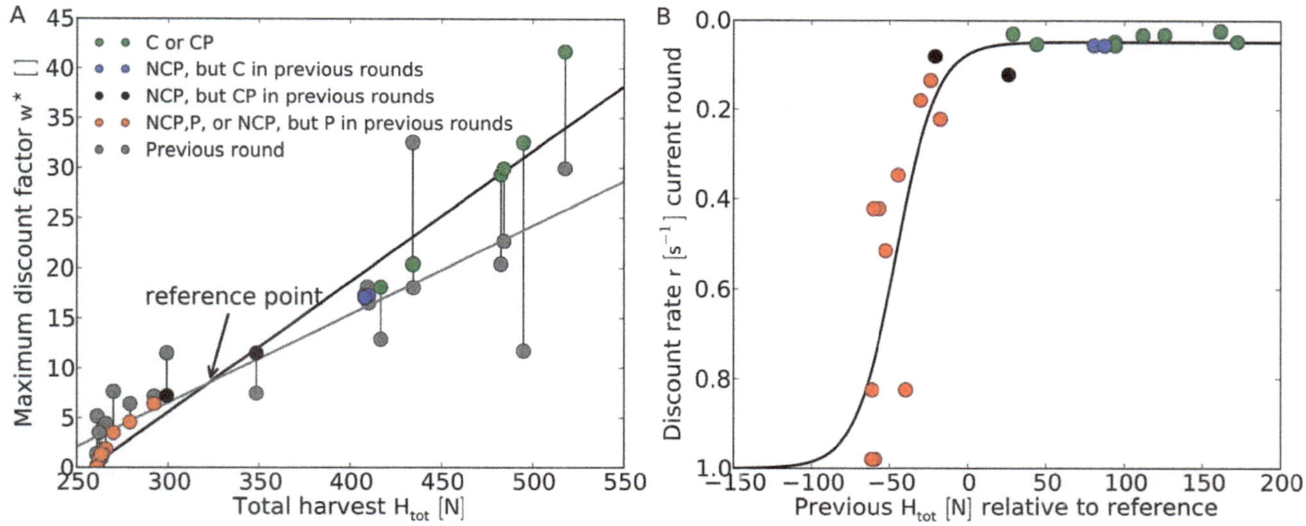

Figure 3. Outcomes of the common pool resource system. The total harvest H_{tot} is closely related to the certainty of the future, here expressed as the maximum discount factor $w^{\star}$ (Equation 8, colored dots). Only data from rounds that were preceded by a round with identical treatment (rounds 2, 3, 5, and 6) were included. **a,** Communication with or without punishment (CP and C, green dots) is essential to establish high $w^{\star}$ and to increase the H_{tot}. Punishment (P), even if experienced only in previous rounds, or the lack of communication and punishment (NCP), keep H_{tot} below the reference point (red dots). Blue and black dots indicate NCP rounds in which either communication or communication and punishment were available in previous rounds. The lines connecting two dots show the change of $w^{\star}$ between a current and a preceding round (gray dots) with the same treatment (at the current round's H_{tot}). The intersection of the regression lines of previous (gray line) and current (black line) discount factors reveals that the value of $w^{\star}$ increases from round to round if the group manages to establish a $w^{\star} > 8.5$ (corresponding to $H_{tot} = 323$) in the preceding round, or decreases if $w^{\star} < 8.5$. This intersection marks the sustainability threshold between positive and negative feedbacks in the system and sets the reference point for the users' expectations. **b,** H_{tot} in a preceding round determines the discount rate r, derived from the relation $r = (w^{\star} + 1)^{-1}$. The solid, sigmoidal line indicates a least-squares fit to a logistic equation (root mean square error $RMSE = 0.155 s^{-1}$).

Sustainable harvest strategies, characterized by high H_{tot} exceeding $400N$, are associated with rounds in which either communication is possible or users had experienced communication in previous rounds (Figure 3A). On the contrary, in the absence of communication or with prior experience of punishment, users harvest unsustainably throughout the simulation and, thus, realize poor harvests (below 300N). Rounds in which the positive and negative effects of previous communication and punishment are balanced represent a transition between clearly separated strategies of successful and unsuccessful harvests (black dots in Figure 3A).

The impact of previous experience of punishment on the maximum discount factor and therefore on the total harvest becomes clear when comparing the outcomes of NCP rounds that were preceded by differing treatments. While in rounds 4 and 5 of the C-NCP treatment users manage to sustain a $w^{\star} > 15$, the maximum discount factor drops considerably to values around 10 in corresponding rounds of the CP-NCP treatment (Figure 3A). In our model, this indirect effect on the harvest behavior is much stronger than any direct effect of punishment. As a tool to enhance the confidence of users into the future, punishment has, different from communication, no or even negative effects beyond the period of its availability [41]. Punishment alters the expectations of future returns, but communication is clearly more effective in raising the users' maximum discount factor and eventually in establishing cooperation [9,42,43]. In studies of real social-ecological systems, leadership significantly influences the successful management of the commons [12,13,44]. Supported by Janssen et al. 's observations [9], we argue that communication enables negotiation and promotes leadership in the artificial environment of this simple, computer-based CPR system.

Changes of the Maximum Discount Factor within One Treatment

A feedback between $w^{\star}$ and H_{tot} links the outcome of previous rounds with the same treatment to the current harvest strategy. Our results suggest that the group's total harvest will increase further in the following round with identical treatment if H_{tot} is above a certain threshold ($323 > N$, equivalent to $w^{\star} > 8.5$). Values below this reference lead to a further deterioration of the maximum discount factor and diminish the group's total harvest in most rounds.

The reference value hence marks a sustainability threshold for the system. This feedback between rounds is similar to the mechanism proposed by Fehr & Gächter [45], who explained the decay of cooperation in a public goods game as a feedback loop of disappointed expectations that leads to lower and lower endowments of the players of the experiment.

The discount rates r, which were derived from $w^{\star}$ assuming exponential discounting [17], are related to the H_{tot} of the previous round by a sigmoid function (Figure 3B). We suggest that the threshold determined in Figure 3A and the discount rate correspond respectively to the reference point of the value function and to the psychological value of an outcome in prospect theory [46]. Prospect theory, which is based on gains and losses rather than on absolute outcomes, explains the discrepancies between economic rationality and observed human behavior. According to prospect theory the perceived value of an outcome does not depend linearly on its economic value. Instead, it is an asymmetric, sigmoidal function of gains and losses with respect to a reference point, the value function, which can be influenced by expectations or the current status. Furthermore, the weight humans associate to an uncertain outcome is related, but not equal, to the corresponding probability, because the human ability to objectively estimate probabilities, in particular those of rare events, is limited [46,47].

We, thus, argue that users adjust the discount rates according to the psychological value of the total harvest realized in the previous round with identical treatment. Users are highly sensitive to losses, which are caused in this system by the lack of communication or by punishment. Losses with respect to the reference value in the previous round lead to a severely impaired perception of future certainty. In contrast, small gains can be sufficient for users to adopt lower discount rates in following rounds (Figure 3B). Note, however, that we excluded the first rounds with a new treatment from our analysis, because a drastic change of the institutional environment does clearly affect future expectations and obfuscates the relationship between harvest experience and discounting as shown in Figures 3A and B.

Conclusions

We have shown that it is possible to understand main features of a CPR and the harvest dynamics of a simplified SES by reducing the social environment to its impact on the perceived future certainty of the users. Our approach extends classic models of maximum sustainable [48,49] or maximum economic [34] yield by introducing a behavioral trait that accounts for the mutual dependency of current behavior and future expectations. The social environment including the first-tier variables "users" and "governance system" of Ostrom's framework for the analysis of SESs [23] is obviously more complex than assumed here and exhibits a dynamics of its own (indicated in Figure 3A and 3B). Despite this well-recognized complexity, the harvest behavior of CPR users can be analyzed, understood, and even roughly predicted with a simple model. Our model is able to describe the co-evolution of the renewable CPR and the adaptive harvest behavior of the users following a mechanistic trade-off, a disregarded feature in classic harvest models [16]. Furthermore, by showing the influence of user experience on the perception of future certainty, we presented an approach to understand the observed variability of user behavior in apparently similar or identical situations.

We conclude that unsustainable harvest leads to reduced discounted future pay-offs and low cooperation in two ways, first, as a consequence of reduced resource productivity and, second, as a consequence of a deteriorating discount factor. Once the temporal gradient of both terms has turned negative, it is difficult for users to escape from the downward vortex of decreasing expectations and diminishing pay-offs. This feedback works also in the opposite direction towards sustainable harvest strategies, high pay-offs, and sustained cooperation among resource users. Our findings illustrate the behavioral variability of users that act rationally according to their current opportunities and their perception of future returns. By this means, our approach opens up a perspective for predicting dynamics and identifying tipping points of coupled user-resource systems.

Methods

Our adaptive model consists of three ordinary differential equations and describes the combined dynamics of the resource R, the harvest H, and the harvest trait x for a constant number of users C.

The Renewable Resource

R is changing over time t according to the difference between new production and harvest

$$\frac{dR}{dt} = r_r R - r_h C, \tag{1}$$

where r_r is the productivity of R

$$r_r = \mu_R \left(1 - \frac{R}{K}\right), \tag{2}$$

with μ_R and K indicating the maximum specific growth rate and the carrying capacity of the resource system, respectively. Resource growth is hence logistic with highest growth rates at $R = 0.5K$ and declining rates towards $R = 0$ as well as towards $R = K$ [16].

Adaptive Harvesting

The current harvest rate r_h is based upon Monod kinetics

$$r_h = \mu_C \left(\frac{R}{R + K_S}\right), \tag{3}$$

with μ_C representing the maximum specific harvest rate and K_S the half-saturation constant. The harvest rate r_h is, thus, insensitive to changes in R for $R >> K_S$, but sensitive for $R < K_S$. In our model, K_S is variable and responds to changes in the harvest trait x.

$$K_S = \hat{K}_S + x^2 K_S^{\star} \tag{4}$$

where $\hat{K}_S$ and $K_S^{\star}$ denote the minimum and the variable part of K_S, respectively. Using an adaptive modeling approach [29–32], the temporal change of x is proportional to the fitness gradient F

$$\frac{dx}{dt} = \kappa \frac{\partial F}{\partial x}, \tag{5}$$

with $F = r_h + r_p + r_f$ and κ denoting the rate constant of the adaptive process [29]. κ hence parameterizes the speed of the adaption process, i.e. the speed of learning, of the group of users over the course of a round. While the punishment rate

$$r_p = \gamma e^{\frac{-(\hat{x} - x)^2}{2\sigma_p^2}} R K^{-1}, \tag{6}$$

is only available in some rounds, the discounted future productivity

$$r_f = w r_r R C^{-1}, \tag{7}$$

which stands for the future resource productivity expected by the users, is considered in all rounds. In Equation 6 γ is the specific punishment rate, $\hat{x}$ is the trait value of the punishment maximum, and σ_p is the punishment standard deviation. Consistent with the experimental results [9], $\hat{x}$ is set to intermediate values in the model so that punishment is applied mostly at intermediate levels of x. At high x users forgo a large fraction of their potential harvest, they cooperate, and punishment is therefore not necessary. By contrast, at low levels of x, which indicate egoistic harvest strategies and low importance of future pay-offs, users are not inclined to invest in a costly and uncertain measure that may support long-term sustainability. The discount factor w is a variable function of t and x

Table 1. Parameter values and variables (with initial conditions given in parenthesis).

Symbol	Name	Value	Unit
α	Shape parameter	0.055	$[s^{-1}]$
β	Shape parameter	137.5	[s]
C	Users	5	[N]
γ	Specific punishment rate	0.096	$[s^{-1}]$
H	Harvest	(0.0)	[N]
H_{tot}	Total harvest	(0.0)	[N]
μ_C	Max. specific harvest rate	1.37	$[s^{-1}]$
μ_R	Max. specific resource growth rate	0.0095	$[s^{-1}]$
K	Carrying capacity	841.0	[N]
K_S	Half-saturation const.	(320.0)	[N]
K_S^*	Variable half-saturation const.	35.0	[N]
$\hat{K}_S$	Min. half-saturation const.	50.0	[N]
κ	xVariance of	1.25	[]
ϑ	Cooperation	(0.6)	[]
r_f	Future productivity	$(-)$	$[s^{-1}]$
r_h	Current harvest rate	$(-)$	$[s^{-1}]$
r_p	Punishment rate	$(-)$	$[s^{-1}]$
r_r	Resource growth rate	$(-)$	$[s^{-1}]$
R	Resource	(210.0)	[N]
σ_p	Punishment standard deviation	0.5	[]
t	Time	$(-)$	[s]
w	Discount factor	$(-)$	[]
w^*	Max. discount factor	(*roundspecific*)	[]
x	Mean harvest trait	(3.0)	[]

$$w = w^*(1-e^{-x})\left(1-\frac{1}{1+e^{\alpha(-t+\beta)}}\right), \tag{8}$$

with w^*, the maximum discount factor, representing the only free parameter between rounds, and α and β, two shape parameters, determining the decay of w as the time in a round elapses. While the time dependence of w is similar to the discounted utility model [17], the parameters α and β allow for a modification of the timing and the speed of decay of w with time. w acts here as a weight on future productivities and is connected to the harvest behavior of the users via the harvest trait x. We assume here that users estimate the future productivity of the resource at the current resource level. Hence, the trade-off between r_h and r_f emerges, because an increase of x raises the current harvest r_h, but erodes r_f by reducing w. The functional dependence of r_f and r_h on x is determined by the highly non-linear shape of the trade-off in the data (cf. Figure 1C) and constrained by the requirement that r_f and r_h may not be negative for any x. Integrating the harvest over time while accounting for possible costs for punishment gives the temporal evolution of the harvest H

$$\frac{dH}{dt} = (r_h - r_p)C. \tag{9}$$

The total harvest realized over the 240s of an experimental round is then $H_{tot} = H(t=240)$.

Cooperation

Cooperation is a diagnostic variable in our model. It is defined by the non-realized current harvest normalized to the maximum possible harvest $r_h(x=0)$

$$\varphi = 1 - \frac{r_h(x)}{r_h(x=0)}. \tag{10}$$

In other words, not harvesting anything results in $\varphi = 1$, whereas maximizing the current harvest rate by adopting $x=0$ leads to $\varphi = 0$. Note that considering users as an adaptive entity implies that the properties x and also φ are mean properties of the group. Unlike in similar evolutionary dynamics models, the dynamics of the trait distribution is not determined by the reproductive fitness of individuals bearing a certain trait, because we assume that the change of a behavioral trait does not require sexual reproduction [50]. In our model, the fixed group of resource users is able to quickly adapt the harvest strategy according to the state of the resource and the maximum discount factor, an assumption that is corroborated by the observed variability of harvest rates in the laboratory experiments [9]. Consistent with the published results of the experimental study, the model is not spatially explicit, because the dynamics of the spatial averages of the resource and the group's harvest in the homogeneous system can be adequately described by zero-dimensional approach.

Simulations

The parameter-set of the model was manually calibrated to fit the temporal evolution of the resource and the total harvest observed in the 36 rounds of the laboratory experiment (cf. Table 1 and Figure S2). The data [9] are averages of five or six replicates for each round. All simulations were conducted with identical initial conditions and parameter values except for the maximum discount factor w^*. Changes of w^* account for all variability in the model results we show in the main text. The different treatments and the learning of the users are, thus reduced to their impact on the expectations of future pay-off, which is represented by the maximum discount factor w^* in the model. Additional results with three variable parameters are presented in the Supporting Information (Figures S4 and S5, Text S1, and Table S1).

Supporting Information

Figure S1 Total harvest for different treatments. Comparison of total harvest H_{tot} for combinations of different treatments, namely communication (C), costly punishment (P), communication and costly punishment (CP), neither communication nor punishment (NCP). Respectively, bars and dots with error bars denote mean values and standard deviations of experimental results obtained from the laboratory study of Janssen et al. [9].

Figure S2 Temporal dynamics of the resource. Times-series of resource levels for six experiments consisting of six rounds each. Treatments are communication (C), costly punishment (P), communication and costly punishment (CP), neither communication nor punishment (NCP) and change after three rounds. Solid lines indicate model results and dotted lines indicate the experimental results [9]. Only the maximum discount factor $w^\star$ was varied between rounds to fit the experimental results, all other parameter values are reported in Table 1.

Figure S3 Model-data comparison. Comparison of all experimental data from [9] shown in Figure S2 with corresponding model data. A linear regression yields $r^2 = 0.972$.

Figure S4 Temporal dynamics of the resource. Times-series of resource levels for six experiments consisting of six rounds each. Treatments are communication (C), costly punishment (P), communication and costly punishment (CP), neither communication nor punishment (NCP) and change after three rounds. Solid lines indicate model results and dotted lines indicate the experimental results obtained from the laboratory study of Janssen et al. [9]. Only the parameters $w^\star$, α, and β vary between rounds to fit the experimental results, all other parameter values are reported in Table 1.

Figure S5 Sensitivity analysis. Sensitivity of the root mean square (RMS) error between simulated and experimental data to changes in the three parameters $w^\star$ (A), α (B), and β (C), all other parameter values are reported in Table 1. The ranges of variation were $w^\star = [0 - 70]$, $\alpha = [0.00 - 0.04\mathrm{s}^{-1}]$, and $\beta = [10 - 200\mathrm{s}]$. Each of the 36 experimental resource time-series was compared to the results of 112000 model runs with unique combinations of the three variable parameters to find the optimal parameter values (cf. Figure S4 for the best results). The panels A–C show how the RMS error increases from the optimum when only one of the three parameters is varied while the other two are held constant at their optimum value.

Acknowledgments

We are grateful to two anonymous reviewers for their thoughtful and constructive comments, which helped to improve the manuscript considerably.

Author Contributions

Conceived and designed the experiments: GB AM. Performed the experiments: GB. Analyzed the data: GB AM. Contributed reagents/materials/analysis tools: GB. Wrote the paper: GB AM BV AS.

References

1. Pauly D, Christensen V, Dalsgaard J, Froese R, Torres Jr F (1998) Fishing down marine food webs. Science 279: 860–863.
2. Pauly D, Christensen V, Guenette S, Pitcher TJ, Sumaila UR, et al. (2002) Towards sustainability in world fisheries. Nature 418: 689–695.
3. Jackson JB, Kirby MX, Berger WH, Bjorndal KA, Botsford LW, et al. (2001) Historical overfishing and the recent collapse of coastal ecosystems. Science 293: 629–37.
4. Achard F, Eva H, Stibig H, Mayaux P, Gallego J (2002) Determination of deforestation rates of the world's humid tropical forests. Science 297: 999–1002.
5. Pandolfi J, Jackson J, Baron N, Bradbury R (2005) Are US coral reefs on the slippery slope to slime? Science 307: 1725–1726.
6. Wada Y, van Beek LPH, van Kempen CM, Reckman JWTM, Vasak S, et al. (2010) Global depletion of groundwater resources. Geophys Res Lett 37: 1–5.
7. Levin SA (2010) Crossing scales, crossing disciplines: collective motion and collective action in the Global Commons. Philos Trans R Soc London, Ser B 365: 13–8.
8. Dal Bó P, Fréchette G (2011) The evolution of cooperation in in_netely repeated games: experimental evidence. Amer Econ Rev 101: 411–429.
9. Janssen MA, Holahan R, Lee A, Ostrom E (2010) Lab experiments for the study of social-ecological systems. Science 328: 613–617.
10. Dal Bó P (2005) Experimental cooperation under the shadow of the future: evidence from infitely repeated games. Amer Econ Rev 95: 1591–1604.
11. Gintis H (2000) Beyond Homo economicus: evidence from experimental economics. Ecol Econ 35: 311–322.
12. Gutiérrez NL, Hilborn R, Defeo O (2011) Leadership, social capital and incentives promote successful fieries. Nature 470: 386–9.
13. Rustagi D, Engel S, Kosfeld M (2010) Conditional Cooperation and Costly Monitoring Explain Success in Forest Commons Management. Science 330: 961–965.
14. Dolsak N, Ostrom E, editors (2003) The Commons in the New Millennium: Challenges and Adaptations. Cambridge: The MIT Press.
15. Ostrom E (1990) Governing the Commons. Cambridge: Cambridge University Press.
16. Clark C (1976) Mathematical Bioeconomics. New York: Wiley.
17. Frederick S, Loewenstein G, O'Donoghue T (2002) Time discounting and time preference: a critical review. J Econ Lit XL: 351–401.
18. Holt Ca, Laury SK (2002) Risk aversion and incentive effects. Amer Econ Rev 92: 1644–1655.
19. Axelrod R (1984) The Evolution of Cooperation. New York: Basic Books.
20. Scheéér M, Carpenter S, Foley JA, Folke C, Walker B (2001) Catastrophic shifts in ecosystems. Nature 413: 591–6.
21. Kortenkamp KV, Moore CF (2006) Time, uncertainty, and individual differences in decisions to cooperate in resource dilemmas. Pers Soc Psychol B 32: 603–615.
22. Ostrom E (2009) A general framework for analyzing sustainability of social-ecological systems. Science 325: 419–422.
23. Ostrom E (2007) A diagnostic approach for going beyond panaceas. Proc Nat Acad Sci USA 104: 15181–15187.
24. Dietz T, Ostrom E, Stern PC (2003) The struggle to govern the commons. Science 302: 1907–12.
25. An L (2011) Modeling human decisions in coupled human and natural systems: Review of agentbased models. Ecol Modell 229: 25–36.
26. Liu J, Dietz T, Carpenter SR, Alberti M, Folke C, et al. (2007) Complexity of coupled human and natural systems. Science 317: 1513–6.
27. Pinsky ML, Jensen OP, Ricard D, Palumbi SR (2011) Unexpected patterns of fisheries collapse in the world's oceans. Proc Nat Acad Sci USA 108: 8317–8322.
28. Lees K, Pitois S, Scott C, Frid C, Mackinson S (2006) Characterizing regime shifts in the marine environment. Fish Fish 7: 104–127.
29. Abrams Pa, Matsuda H, Harada Y (1993) Evolutionarily unstable fitness maxima and stable fitness minima of continuous traits. Evol Ecol 7: 465–487.
30. Wirtz KW, Eckhardt B (1996) Effective variables in ecosystem models with an application to phytoplankton succession. Ecol Modell 92: 33–53.
31. Norberg J, Swaney DP, Dushoff J, Lin J, Casagrandi R, et al. (2001) Phenotypic diversity and ecosystem functioning in changing environments: a theoretical framework. Proc Nat Acad Sci USA 98: 11376–11381.
32. Merico A, Bruggeman J, Wirtz KW (2009) A trait-based approach for downscaling complexity in plankton ecosystem models. Ecol Modell 220: 3001–3010.
33. Janssen MA, Anderies JM (2007) Robustness trade-offs in social-ecological systems. Int J Comm 1: 43–66.
34. Grafton RQ, Kompas T, Hilborn RW (2007) Economics of overexploitation revisited. Science 318: 1601.
35. Lamba S, Mace R (2011) Demography and ecology drive variation in cooperation across human populations. Proc Nat Acad Sci USA 2011: 1–5.
36. Gächter S, Herrmann B (2009) Reciprocity, culture and human cooperation: previous insights and a new cross-cultural experiment. Philos Trans R Soc London, Ser B 364: 791–806.
37. Herrmann B, Thöni C, Gächter S (2008) Antisocial punishment across societies. Science 319: 1362–1367.
38. Henrich J (2004) Cultural group selection, coevolutionary processes and large-scale cooperation. J Econ Behav Organ 53: 3–35.
39. Henrich J, Boyd R, Bowles S, Camerer C, Fehr E, et al. (2001) In search of Homo economicus: behavioral experiments in 15 small-scale societies. Amer Econ Rev 91: 73–78.

40. Hayo B, Vollan B (2012) Group interaction, heterogeneity, rules, and co-operative behaviour: evidence from a common-pool resource experiment in South Africa and Namibia. J Econ Behav Organ 81: 9–28.
41. Gneezy URI, Rustichini A, Glaeser E, Levine D, Nyhus E, et al. (2000) A fine is a price. J Legal Stud XXIX: 1–17.
42. Rand DG, Dreber A, Ellingsen T, Fudenberg D, Nowak MA (2009) Positive interactions promote public cooperation. Science 325: 1272–5.
43. Dreber A, Rand DG, Fudenberg D, Nowak MA (2008) Winners don't punish. Nature 452: 348–51.
44. Kenward RE, Whittingham MJ, Arampatzis S, Manos BD, Hahn T, et al. (2011) Identifying governance strategies that effectively support ecosystem services, resource sustainability, and biodiversity. Proc Nat Acad Sci USA 108: 5308–12.
45. Fehr E, Fischbacher U (2003) The nature of human altruism. Nature 425: 785–791.
46. Kahneman D, Tversky A (1979) Prospect theory: an analysis of decision under risk. Econometrica 47: 263–292.
47. Tversky A, Kahneman D (1974) Judgment under uncertainty: heuristics and biases. Science 185: 1124–1131.
48. Lande R, Engen S, Saether B (1994) Optimal harvesting, economic discounting and extinction risk in uctuating populations. Nature 372: 88–90.
49. Clark C (1973) The economics of overexploitation. Science 181: 630–634.
50. Levin SA (2006) Learning to live in a global commons: socioeconomic challenges for a sustainable environment. Ecol Res 21: 328–333.

Genetic Control of Startle Behavior in Medaka Fish

Satomi Tsuboko[1], Tetsuaki Kimura[2,3], Minori Shinya[4,5], Yuji Suehiro[1,6], Teruhiro Okuyama[1,2], Atsuko Shimada[1], Hiroyuki Takeda[1], Kiyoshi Naruse[2,3], Takeo Kubo[1], Hideaki Takeuchi[1]*

1 Department of Biological Sciences, Graduate School of Science, The University of Tokyo, Bunkyo, Tokyo, Japan, **2** Laboratory of Bioresources, National Institute for Basic Biology, Okazaki, Aichi, Japan, **3** National Institute for Basic Biology Center of the Interuniversity Bio-Backup Project, National Institute for Basic Biology, Okazaki, Aichi, Japan, **4** Model Fish Genomics Resource, Genetic Strains Research Center, National Institute of Genetics, Mishima, Shizuoka, Japan, **5** Department of Biology, Keio University, Yokohama, Kanagawa, Japan, **6** Department of Physiology, Tokyo Women's Medical University School of Medicine, Shinjuku, Tokyo, Japan

Abstract

Genetic polymorphisms are thought to generate intraspecific behavioral diversities, both within and among populations. The mechanisms underlying genetic control of behavioral properties, however, remain unclear in wild-type vertebrates, including humans. To explore this issue, we used diverse inbred strains of medaka fish (*Oryzias latipes*) established from the same and different local populations. Medaka exhibit a startle response to a visual stimulus (extinction of illumination) by rapidly bending their bodies (C-start) 20-ms after the stimulus presentation. We measured the rates of the response to repeated stimuli (1-s interval, 40 times) among four inbred strains, HNI-I, HNI-II, HO5, and Hd-rR-II1, and quantified two properties of the startle response: sensitivity (response rate to the first stimulus) and attenuation of the response probability with repeated stimulus presentation. Among the four strains, the greatest differences in these properties were detected between HNI-II and Hd-rR-II1. HNI-II exhibited high sensitivity (approximately 80%) and no attenuation, while Hd-rR-II1 exhibited low sensitivity (approximately 50%) and almost complete attenuation after only five stimulus presentations. Our findings suggested behavioral diversity of the startle response within a local population as well as among different populations. Linkage analysis with F2 progeny between HNI-II and Hd-rR-II1 detected quantitative trait loci (QTL) highly related to attenuation, but not to sensitivity, with a maximum logarithm of odds score of 11.82 on linkage group 16. The three genotypes (homozygous for HNI-II and Hd-rR-II1 alleles, and heterozygous) at the marker nearest the QTL correlated with attenuation. Our findings are the first to suggest that a single genomic region might be sufficient to generate individual differences in startle behavior between wild-type strains. Further identification of genetic polymorphisms that define the behavioral trait will contribute to our understanding of the neural mechanisms underlying behavioral diversity, allowing us to investigate the adaptive significance of intraspecific behavioral polymorphisms of the startle response.

Editor: Harold A. Burgess, National Institutes of Health/NICHD, United States of America

Funding: This work was supported by the National Institute for Basic Biology Priority Collaborative Research Project (10-104, http://www.nibb.ac.jp/en/collabo/collabo.html), Grant-in-Aid for Challenging Exploratory Research (23650162, http://www.jsps.go.jp/english/e-grants/grants01.html), Grant-in-Aid for Scientific Research on Innovative Areas (Comprehensive Brain Science Network, https://www.hokatsu-nou.nips.ac.jp/), the Graduate Program for Leaders in Life Innovation (GPLLI) at the University of Tokyo Life Innovation Leading Graduate School from MEXT (to ST, http://square.umin.ac.jp/gplli/en/index.html), and Grant-in-Aid for a Japan Society for Promotion of Science (JSPS) fellowship (to ST, YS, and TO, http://www.jsps.go.jp/english/e-pd/index.html). The funders had no role in study design, data collection and analysis, decision to publish, or preparation of the manuscript.

Competing Interests: The authors have declared that no competing interests exist.

* Email: takeuchi@bs.s.u-tokyo.ac.jp

Introduction

Many animals, including humans [1], exhibit intraspecific behavioral diversities [2,3,4,5,6], which are generated not only by learning and development, but also by genetic factors [2,3]. Individual differences in behavioral traits are thought to be an important factor influencing fitness [7] and are suggested to be under natural selection [2]. In addition, intraspecific behavioral diversities in some animal groups may contribute to the emergence of social organization [8,9]. Thus, individual differences in behavioral characteristics have attracted the interest of many researchers in the field of evolutionary ecology [1,2,4,5]. Why individual differences are maintained in natural populations and how genetic polymorphisms generate behavioral diversities, however, remain unresolved.

To explore this issue, we used medaka fish (*Oryzias latipes*). Over 10 inbred strains of medaka generated from over 20 generations of sister-brother mating have been established from various origins, including wild and pet populations [10,11,12,13,14]. Importantly, the availability of a dense genetic linkage map [15,16,17,18] and large-scale genomic data [19,20,21,22] allows us to identify the quantitative trait loci (QTL) for complex traits and to compare genetic polymorphisms in candidate genomic regions among medaka inbred strains. Comparative analysis of the genomes of medaka inbred strains revealed a genome-wide single nucleotide polymorphism rate between the two inbred strains (HNI-II and Hd-rR-II1) of 3.42%, the highest single nucleotide polymorphism rate observed to date in any vertebrate species [19,20]. In the process of inbreeding, the genetic variation within the species led to differences among strains. Because each inbred strain is genetically homogeneous with its unique genotype, the genetic background itself is thought to define strain-specific traits [15,23]. Medaka are thus considered an ideal animal model for the search of genetic factors that underlie complex traits [15].

To date, there is only one report describing differences in behavioral traits among medaka inbred strains; some inbred strains tend to be attached to humans, while other inbred strains tend to escape from humans [13], responses that may be mediated by a visual stimulus. The difference in behavioral traits, however, has not been quantitatively analyzed based on paradigmatic experiments. In the present study, we quantitatively compared the visual response among medaka inbred strains and performed a genome-wide QTL analysis.

Results

Behavioral paradigm of visually-evoked startle response

We established a behavioral paradigm to quantify visually-evoked behavioral characteristics of medaka using a liquid crystal display (Figure 1A). Russel (1967) demonstrated that a shadow stimulus induces a startle response in guppies [24]. We found that extinction of illumination (color transition from white to black on the entire screen of the liquid crystal display) robustly elicited a startle response in medaka fish. At approximately 20-ms after the stimulus presentation, medaka exhibited rapid and transient body bending (Figure 1C and 1D). This startle response resembled the "C-start", which is broadly observed in the escape behaviors of teleost fish [25]. Medaka fish exhibited a response to the extinction of illumination (light-to-dark stimulus), but no apparent response to dark-to-light stimulus (Movie S1 and S2). We tested four inbred strains from the same and different populations (Table 1, [11,12,13,14]), and detected no prominent difference in either the latency of the response or the motion kinematics.

Differences in startle response properties among medaka inbred strains

To investigate changes in the startle response to repeated eliciting stimuli, we used blinking stimuli, where the entire screen flickered 40 times at 1-s intervals (Figure 1B). We examined whether fish exhibited the startle response to each eliciting stimulus based on recorded movies (Figure 1A and 1C). This behavioral test was performed in the same fish everyday for five consecutive days and the startle response rate of each fish was calculated for every stimulus (Figure 2A). Among the four strains, the most prominent differences were detected between HNI-II and Hd-rR-II1 in two features, although neither strain exhibited a significantly decreased maximum velocity during the response (i.e. response magnitude) and both continued to respond (one-way ANOVA; HNI-II: $F (39, 685) = 1.14$, $p>0.05$; Hd-rR-II1: $F (3, 28) = 2.11$, $p>0.05$; Figure S1A). HNI-II demonstrated the highest sensitivity and no attenuation of response probability, and Hd-rR-II1 demonstrated the lowest sensitivity and almost complete attenuation of response probability within five repeated stimulus presentations. To quantitatively compare the response properties, we defined two indexes. The "sensitivity index" was defined as the probability of a response to the first eliciting stimulus on each of the 5 days. The "response stability index" was defined as the number of eliciting stimuli needed before an integrated response reached half the total amount (see Materials and Methods), which negatively correlates with the degree of attenuation. No effect of day on either of the indices was detected (two-way repeated measures ANOVA; sensitivity: $F (4, 280) = 0.31$, $p>0.05$; response stability index: $F (4, 280) = 1.03$, $p>0.05$; Figure S2). The sensitivity index was significantly different between HNI-II and Hd-rR-II1 (Scheffe's F test, $p<0.01$; Figure 2B and S3A) among the four inbred strain, while the response stability index was significantly different among several combinations (Figure 2C and S3B). HNI-I and HNI-II both originated from the same local-wild population or a "deme" (Niigata stock), belonging to the Northern Japan population [13,14,20], while Hd-rR-II1 and HO5 originated from different stocks in the Southern Japan population [13,14,20].

Attenuation of the startle response probability with highly frequent stimuli in HNI-II

HNI-II exhibited no attenuation in response probability to repeated stimulus presentations, suggesting that some mutations in the HNI-II inbred strain resulted in a loss of the ability to exhibit

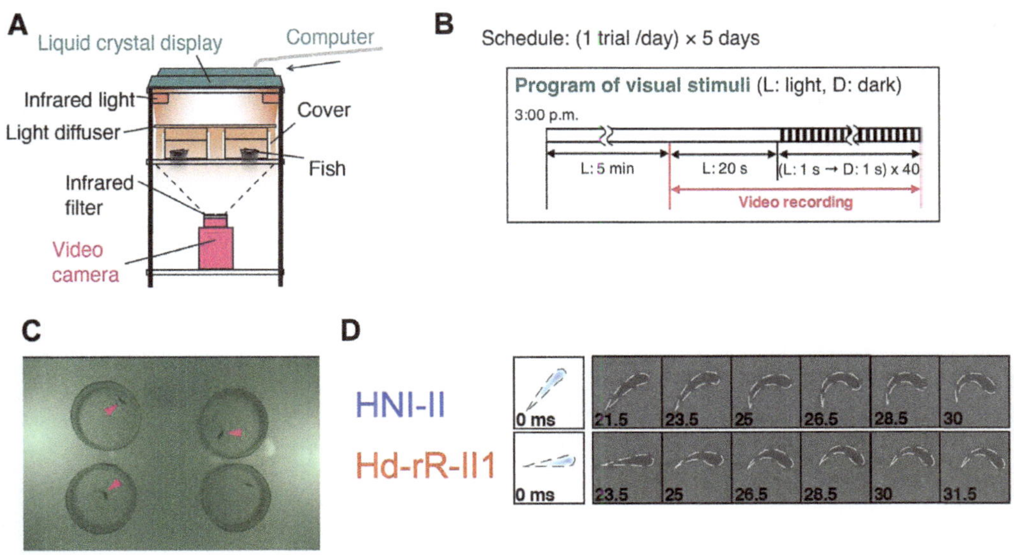

Figure 1. Experimental Design and Observed Startle Response. (A) Experimental apparatus with infrared imaging system. (B) Experimental program. (C) Images from infrared recording. Three individuals (top-left, top-right, and bottom-left, indicated by magenta arrowheads) exhibit C-start. (D) C-start of medaka. Fish bend their body in a "C" shape after a ~20-ms latency.

Table 1. Inbred Strains and Their Origins.

Strain	Local Population	Origin
HNI-I	Northern Japan	Wild population in Niigata city, Niigata Prefecture, Japan
HNI-II	Northern Japan	Same wild population as HNI-I
HO5	Southern Japan	Commercial stock purchased in Chiba prefecture, Japan
Hd-rR-II1	Southern Japan	Commercial stock purchased in Yatomi, Aichi prefecture, Japan

startle response attenuation. Therefore, to examine whether HNI-II could exhibit startle response attenuation, we presented visual stimuli with a shorter interval, i.e., the display flickered 40 times at 0.5-s intervals. Under these conditions, HNI-II exhibited an attenuated startle response and the response stability index was significantly lower than that when stimuli were presented at 1.0-s intervals (one-way ANOVA; $F(1, 24) = 83.32$, $p<0.001$; HNI-II, $n=6$; Hd-rR-II1, $n=8$; Figure 3C). This finding indicated that HNI-II exhibits attenuated startle response probability to repeated stimulus presentations. The index in Hd-rR-II1, however, did not differ significantly between the two stimulus presentation conditions (one-way ANOVA; $F(1, 26) = 1.68$, $p>0.05$; Figure 3C), suggesting that the response stability index of Hd-rR-II1 reached the lowest plateau with presentation at 1.0-s intervals. There was no significant difference in sensitivity in either HNI-II or Hd-rR-II1 (one-way ANOVA; HNI-II: $F(1, 24) = 0.02$ $p>0.05$; Hd-rR-II1: $F(1, 26) = 1.82$, $p>0.05$; Figure 3B). Thus, we demonstrated that HNI-II does not have defective plasticity of the visual response.

Identification of QTL highly related to startle response attenuation

To search for candidate genetic regions defining the behavioral diversity of startle response attenuation, we performed a QTL analysis using HNI-II and Hd-rR-II1. We selected the two strains for their most prominent behavioral difference (Figure 2B and 2C), abundance of polymorphic markers [15,17,26], and availability of genome sequence information [20] for further investigations. For QTL analysis, we crossed HNI-II and Hd-rR-II1 for two generations and obtained ~100 F2 individuals. The sensitivity and response stability indices were measured in each individual under the 1-s interval stimulus presentation condition. After the behavioral tests, each individual was genotyped with 147 polymorphic markers on the genome, indicated as black bars in Figure 4A and 4B, and black arrowheads in Figure 4C, with a marker interval length of ~10-cM.

Analysis of the response stability index detected revealed significant QTL with a maximum logarithm of odds (LOD) score of 11.82 located on linkage group 16, explaining 26.7% of the variance ($n=78$, $p<0.01$; Figure 4C, 4D and Table 2). The response stability index of F1 between HNI-II and Hd-rR-II1 was similar to that of Hd-rR-II1 (Figure S4C), suggesting that recessive genetic factors may have a major effect on the behavioral traits of HNI-II.

PVE, percent variance explained; HNI-II homo, homozygous for HNI-II alleles; Hetero, heterozygous, Hd-rR-II1 homo, homozygous for Hd-rR-II1 alleles.

Analysis of the sensitivity index with the entire F2 population revealed no significant QTL ($n=87$, $p>0.05$; Figure 4A). In the previous F1 generation, the sensitivity index was lower in progeny between Hd-rR-II1 females and HNI-II males (F1-a) than that of HNI-II, while the index of progeny obtained from the reciprocal cross (F1-b, from HNI-II females and Hd-rR-II1 males) was not significantly lower than that of HNI-II (Figure S4B). In the original four inbred strains we used, the sensitivity index of females was significantly higher than that of males (two-way ANOVA; sex: $F(1,52) = 9.61$, $p<0.01$; Figure S5A), while there was no significant difference in the response stability index between sexes (two-way ANOVA; sex: $F(1,52) = 2.63$, $p=0.11$; Figure S5B). Thus, the failure to detect a QTL for the sensitivity index might be due to a sexual dimorphism of this behavioral trait. Analysis discriminating F2 populations from different crosses suggested the presence of a QTL that was specifically detected with the offspring of F1-a ($n=29$, $p<0.05$; Figure S6A), not with that of F1-b ($n=58$, $p>0.05$; Figure S6B), which might support the hypothesis. This QTL was not on linkage group 1, corresponding to the sex chromosome of medaka [27].

Discussion

The findings of the present study demonstrated a significant difference in the startle response pattern among medaka inbred strains. The startle response is a reflexive reaction to potentially stressful environmental stimulation that is observed in most, if not all, animal species [28]. Extinction of illumination robustly induces the startle response (C-start) in medaka fish, suggesting that the visual stimulus mimics the sudden appearance of a natural predator above the water [24,29]. In rodents, rats exhibit a startle response to an acoustic stimulus, and there is a significant difference among rodent strains in both the mean amplitude of the startle response and startle response attenuation to repeated startle stimulus presentations (15-s interval) [30]. The genetic polymorphism that determines the difference in the startle response among rodent wild strains based on QTL analysis has not been clarified [31,32,33]. In the present medaka study, QTL analysis suggested that a single 30-cM genomic region (18 Mbp) mainly defines the difference in startle response attenuation. This genomic region contains approximately 500 predicted genes. Further generation of congenic lines will allow us to narrow candidate genomic regions to a restricted region containing ~100 predicted genes and to examine epistatic effects between other loci. The applicability of recently developed advanced molecular genetic techniques, such as transgenic and genomic editing [34,35,36,37], to medaka will allow for the identification of additional genetic polymorphisms [38]. The identification of genetic factors underlying the diversities of the startle response will allow us to investigate the ultimate causes of individual behavioral differences in medaka fish.

Medaka are thought to be a model organism for comparative population genetics [11,39,40,41]. The availability of various local-wild medaka populations further allows for investigations of the adaptive meaning of behavioral diversities defined by genetic polymorphisms among populations and estimations of the evolutionary trajectory of behavioral traits [11,39,40,41]. Medaka

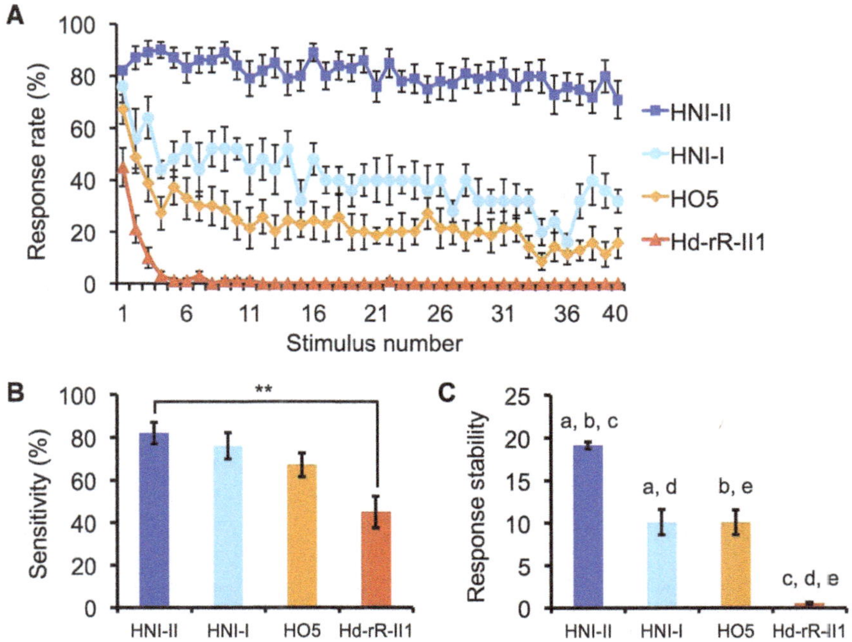

Figure 2. Comparison of Startle Response Properties among Inbred Strains. HNI-II (n = 20), HNI-I (n = 6), HO5 (n = 14), Hd-rR-II1 (n = 20). (A) Transition of response probability. Bars represent SEM. (B) Sensitivity of four inbred strains. Bars represent SEM. ** $p<0.01$ by Scheffe's F test. (C) Response stability index of four inbred strains. Bars represent SEM. a, b, c, d, and e indicate $p<0.01$ by Scheffe's F test.

fish exhibit various kinds of social behaviors, such as female mating preference [42], schooling [43,44], visually-guided group-foraging [45], and behavioral paradigms to quantify these social behaviors have been already established, which will allow us to examine how the behavioral diversities within animal groups are involved in the emergence of social organization [8] and collective decision-making [38]. Our results demonstrated differences in the startle response between HNI-I and HNI-II, which both originated from the same local-wild population. Further behavioral and genetic characterizations of individuals in wild populations will help us to

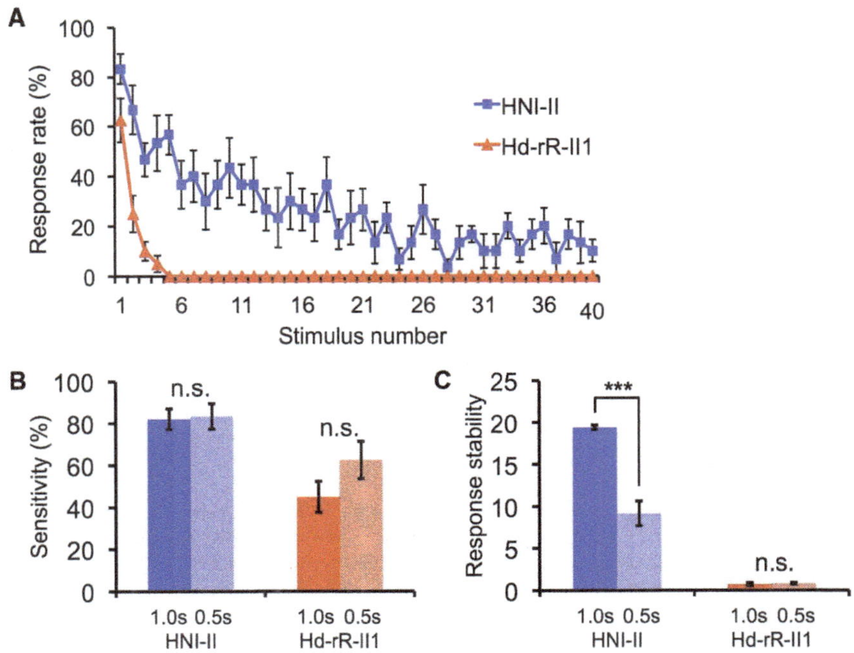

Figure 3. Comparison of Startle Response Properties between Different Stimulus Interval Conditions. HNI-II (n = 6), Hd-rR-II1 (n = 8). (A) Transition of response probability to frequent (0.5-s interval) stimulus presentation. (B) Sensitivity in different stimulus interval conditions. Bars represent SEM. In each inbred strain, no significant difference was detected between the 1.0-s and 0.5-s interval conditions, based on one-way ANOVA. (C) Response stability index of two inbred strains. Bars represent SEM. ***$p<0.001$ by one-way ANOVA.

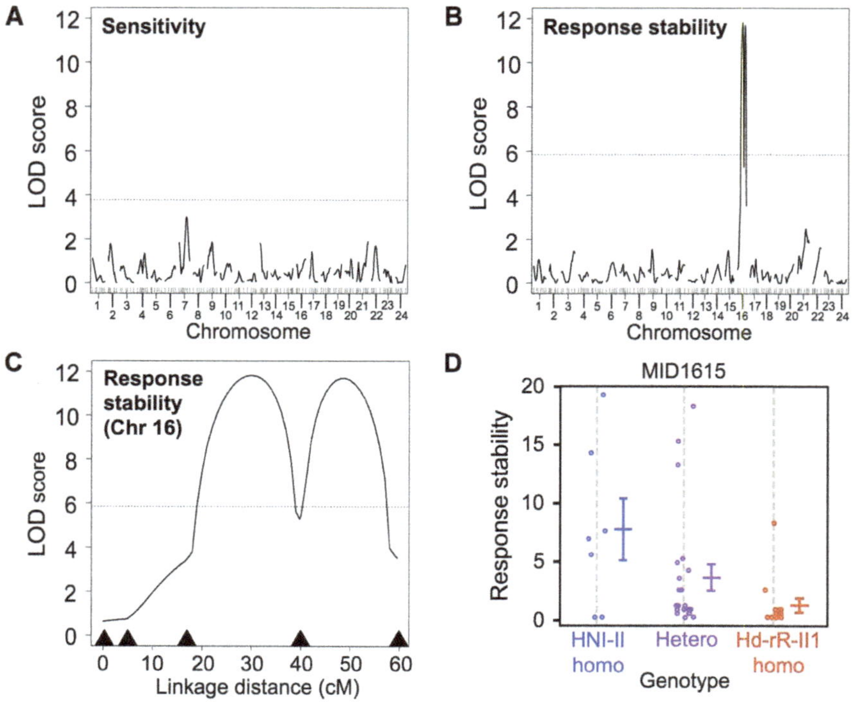

Figure 4. Quantitative Trait Analysis with F2 between HNI-II and Hd-rR-II1. In A, B, and C, dashed lines indicate the thresholds (p = 0.05). (A) QTL analysis on sensitivity. No significant QTL was detected. (n = 87) (B) QTL analysis on response stability index. Significant QTL was detected on chromosome 16. (n = 78) (C) QTL analysis on response stability index. The result of chromosome 16 is shown. Black arrowheads indicate marker positions. (D) Genotype-phenotype relationship at a polymorphic marker located at 39.75-cM on linkage group 16, corresponding to chromosome 16.

determine whether the behavioral differences reflect genetically-affected behavioral diversity within a local-wild population. Our study reveals the future potential of medaka as a new model animal for behavioral genetics as well as evolutionary behavioral ecology.

Materials and Methods

Ethics statement

The work in this paper was conducted using protocols approved by the Animal Care and Use Committee of the University of Tokyo (permit number: 12–07). All surgery was performed under cold anesthesia, and all efforts were made to minimize suffering.

Animals

Adult medaka (>3 months after hatch, 2.5–4 cm of length) were used for behavioral tests and genotyping. Four medaka inbred strains, HNI-I, HNI-II, Hd-rR-II1, and HO5 [11,12,13,14], were used for inter-strain behavioral comparison. These strains have been established and maintained by over 20 generations of full sister-brother mating. All of these strains were provided by the National BioResource Project Medaka. Two F1 hybrids, F1-a (female Hd-rR-II1 x male HNI-II) and F1-b (female HNI-II x male Hd-rR-II1), were used for the experiment shown in Figure S4. The two F2 hybrids used for the QTL analysis, F2-a and F2-b, were obtained from sibling mating within F1-a and F1-b, respectively. Water temperature was maintained at ~28°C. The light: dark ratio was 14 h:10 h, with the lights on from 08:00 to 22:00. Fish were maintained in plastic aquariums (12 cm × 13 cm × 19 cm). Each aquarium was separated by a transparent partition into two ID-tagged compartments, each of which contained two fish. Fish were fed brine shrimp (3 - 4 times daily). Rearing water was circulated among the aquariums, and the loss was replaced with dechlorinated tap water.

Behavioral tests

The behavioral tests were performed over 5 days. On each experimental day, fish were scooped from the aquariums at ~12:00, and placed into ID-tagged plastic cups with rearing water (one pair of fish per cup) for transportation to the observation room. Fish were kept in the observation room for ~3 hours for acclimation. For behavioral observation, a shelf with three transparent glass boards was used (Figure 1A). At ~15:00, each individual fish was placed in a tall Petri dish (8 cm in diameter, filled with water to a 2–3 cm depth, and covered on the side surface with white paper) on the middle shelf with a 4-cm-long plastic plate as a spatial scale. Up to four fish were tested at the same time, and the other fish were kept under an opaque cover during the trial. After the fish were set on the middle board, the light in the experiment room was turned off. Visual stimuli were presented with a liquid-crystal display (TSA4634JT, Iiyama, Tokyo, Japan. Response time: 72 ms) set on the upper board of the shelf. Fish were first presented with the light stimulus for 5 min for acclimation, and then for 20 s for start-up of the video recording. After the video recording was started, fish were shown a stimulus movie (Figure 1B). Video recording was stopped after the movie finished. Between trials, the light of the room was turned on to change fish. After all trials were complete, the fish were transported back to the rearing room.

Visual stimuli

In the trials, fish were presented a light-dark stimulus movie in which 40 light periods and 40 dark periods appeared alternately. In the experiments shown in Figure 2, 4, S1, S2, S3, S4, S5 and

Table 2. Location and Effect of the QTL Detected in the Analysis with the Whole F2 Population.

Trait	LG	Estimated QTL (cM)	LOD score	Nearest marker	Marker location (cM)	LOD score	p Value	PVE	HNI-II homo	Hetero	Hd-rR-II1 homo
Sensitivity	-	-	-	-	-	-	-	-	-	-	-
Response stability index	16	30	11.82	MID1615	39.75	5.26	p<0.001	26.7	8.9±1.1	3.1±0.7	1.4±1.0

PVE, percent variance explained; HNI-II homo, homozygous for HNI-II alleles; Hetero, heterozygous, Hd-rR-II1 homo, homozygous for Hd-rR-II1 alleles.

S6, the lengths of the light periods (interval of dark stimuli) and dark periods were each set to be 1.0-s long. In the experiment shown as "0.5 s" in Figure 3, the dark stimulus was presented every 0.5-s for a period of 1.0-s.

Video recording

The time-lapse image of C-start shown in Figure 1D was recorded at 600 frames per second with a digital camera (EXILIM EX-F1, CASIO, Japan). In the experiments shown in Figure 2, 3, 4, S1, S2, S3, S4, S5 and S6, the infrared imaging ("night-shot") mode of a video camera (HDR-HC3, SONY, Japan) was used to observe the behavior of the fish under consecutive light-to-dark transitions. Infrared LED lights (LB12-WP01, Ebisu-denshi, Osaka, Japan, with a peak wavelength of 850 nm that is out of the estimated visible wavelength range of medaka [46]), were used for infrared lighting. To illuminate the viewing field uniformly, an opalescent acrylic board (3 mm thick) was placed on tall Petri dishes as a light diffuser. The movement of the fish was recorded from underneath at 30 frames per second with the video camera. To cut off excess visible light from the recorded movies, an infrared filter (IR 78, Fuji film, Japan), which cuts off light with a wavelength <780 nm, was set on the lens of the video camera. A pinhole was created at the center of the filter, through which the light-dark changes of the display were recorded (Movie S2).

Quantification of response properties

The startle response was manually scored on the video recordings (30 frames per second), as a quick tail flip with C-shaped body bend starting within three recording frames after an onset of light-to-dark transition, with about>0.5 cm of body moving distance between two adjacent recording frames (corresponding to >15 cm/s in velocity). The "sensitivity" was calculated as response probability to the first dark period. The "response stability index" was calculated as the number of dark periods needed before an integrated amount of responses reached half the total amount of response. In the analysis of sensitivity, a mean value of 5 days was used as a characteristic index value for each individual. In the analysis of the response stability index, a mean value of 3 days, not including the highest and lowest values over the 5 days, was used for each individual.

Quantification of maximum velocity during a response

The maximum velocity during each response was calculated from fish velocities and time sequences of light-dark changes according to the following procedure using ImageJ (http://imagej.nih.gov/ij/). First, the recorded movies were imported to ImageJ (Ver.10.2) using QuickTime (Ver. 7.6.4) and each frame of the movies was converted to an 8-bit monochrome image. To calculate the fish velocities, the area inside of the tall Petri dish was defined as the regions of interests. The brightness threshold was set to distinguish the fish from the background, and a spatial scale was calibrated using a 4-cm-long plastic plate placed near the tall Petri dish. The positions of the fish were measured using the Particle Analyzer algorithm of ImageJ and the velocities of the fish were calculated at each frame. To track light-dark changes over time, the "Stacks" function of ImageJ was used. The small circular area at the center of the image, representing the light-dark changes recorded through the pinhole of the infrared filter, was selected as a region of interest (see Movie S2). A time sequence of brightness in the region was obtained by the "Plot Z-axis Profile" command, and the onset of each light-to-dark change was defined as <−1 of transition in brightness from a previous recording frame (in typical cases, mean brightness values in the region were ~85 in light states, and <80 in dark states). Based on the sequential values, the maximum velocity during each response was defined as the highest velocity in three recording frames after a light-to-dark transition.

Genotyping

For DNA extraction, fish were euthanized by submersion in ice water (4°C or less) until cessation of gill movement and heartbeat, and then decapitated with a sharp blade. Genomic DNA was extracted using a PI-50 automatic DNA isolation system (Kurabo Industries Ltd., Osaka, Japan) from the decapitated and gutted medaka bodies fixed in 100% ethanol. We used 147 polymerase chain reaction product length polymorphism primers, which were described previously [26]. Polymerase chain reactions and polymorphic analysis were performed as described previously [16,26].

QTL analysis

We used a total of 117 F2 individuals for 5 days of behavioral tests. Behavioral tests were performed in 10 divided batches. In each batch, 2 HNI-II (positive control), 2 Hd-rR-II1 (negative control), and 11 or 12 F2 individuals were used. In preliminary experiments, some batches exhibited a low response rate among all individuals, including HNI-II. Therefore, we used the response rate of HNI-II as a criterion for excluding batches with generally low response rates from a dataset for further analysis. We eliminated two batches (23 individuals) in which HNI-II exhibiting $<70\%$ of the mean response rates throughout the 40 stimulus presentations. We also eliminated four individuals that died after the behavioral tests, and three individuals for which the movie recordings were problematic. As a result, a total of 87 F2 individuals (offspring of F1-a: $n=29$, offspring of F1-b: $n=58$) were used for further QTL analysis. In the analysis of the response stability index, nine individuals with no response were additionally eliminated from the dataset, because we could not define the index in such cases. QTL analysis was performed using R/qtl [47]. Nonparametric interval mapping was performed with a "scanone" function, using the argument "model = np". The LOD threshold for interval mapping was determined by 1000 iterated permutations. The genome-wide significance thresholds were set at alpha $=0.05$. The bayesint function was used to calculate the Bayes credible interval (95%) for estimation of the QTL location. The fitqtl function was used to calculate the percentage variance explained by the marker nearest the estimated QTL.

Supporting Information

Figure S1 Transition of Fish Velocity. (A) Transition of maximum velocity during startle movement. HNI-II ($n=4$), Hd-rR-II1 ($n=6$), data from five days of trials. Bars represent SEM. In each strain, no significant effect of stimulus number was detected by one-way ANOVA ($p>0.05$). (B) Example of tracked velocity of one HNI-II individual through one trial.

Figure S2 Effect of Experimental Days on the Response Properties in the Four Inbred Strains. HNI-II ($n=20$), HNI-I ($n=6$), HO5 ($n=14$), Hd-rR-II1 ($n=20$). Bars represent SEM. *** $p<0.001$ by two-way repeated measures ANOVA. No significant effect of days, and no significant interaction between day and strain were detected.

Figure S3 Whole Distribution of Values of the Response Properties in the Four Inbred Strains. HNI-II ($n=20$), HNI-I ($n=6$), HO5 ($n=14$), Hd-rR-II1 ($n=20$). (A) Individual values of sensitivity in four inbred strains. (B) Individual values of response stability index in four inbred strains.

Figure S4 Startle Response Properties of F1 between HNI-II and Hd-rR-II1. F1-a are the progeny of Hd-rR-II1 females and HNI-II males. F1-b are the progeny obtained from an opposite cross, between HNI-II females and Hd-rR-II1 males. HNI-II ($n=18$), F1-a ($n=8$), F1-b ($n=7$) Hd-rR-II1 ($n=17$) (A) Transition of response probability. Bars represent SEM. (B) Sensitivity of F1-a, F1-b, and their parental strains. Bars represent SEM. *$p<0.05$ and ** $p<0.01$ by Scheffe's F test. (C) Response stability index of F1-a, F1-b, and their parental strains. Bars represent SEM. a, b, and c indicate $p<0.01$ by Scheffe's F test.

Figure S5 Sexual Difference in the Response Properties in the Four Inbred Strains. HNI-II (male, female = 11, 9), HNI-I (male, female = 3, 3), HO5 (male, female = 6, 8), Hd-rR-II1 (male, female = 10, 10). Bars represent SEM. ** $p<0.01$ and *** $p<0.001$ by two-way ANOVA. No significant effect of sex, and no significant interaction between sex and strain were detected.

Figure S6 Quantitative Trait Analysis for Sensitivity with F2 Obtained from Different Crosses. In A and B, dashed lines indicate the thresholds ($p=0.05$). (A) QTL analysis with F2 obtained from F1-a, the progeny of Hd-rR-II1 females and HNI-II males. QTL with a maximum LOD score of 4.42 located on linkage group 3 (explaining 20.9% of the variance) was detected ($n=29$). (B) QTL analysis with F2 obtained from F1-b, the progeny of HNI-II females and Hd-rR-II1 males. No significant QTL was detected ($n=58$).

Movie S1 Startle Response to Light-to-dark, not to Dark-to-light, Stimuli. HNI-II fish are in the top-left and bottom-left, and Hd-rR-II1 fish are in the top-right and bottom-right. This movie is from a preliminary experiment, in which only visible light, not infrared light, was video recorded.

Movie S2 Infrared Imaging under Entire Light-to-dark Background Changes. The blinking of the small circle at the center of the image corresponds to that of the liquid crystal display placed above the fish. When the circle is brighter than the surrounding area, the liquid crystal display is brightly lit. When the circle is the same level of brightness as the surrounding area, the display has turned from light to dark. Note that the display practically covers the entire field, but the light-dark changes were recorded through a pinhole at the center of an infrared filter on a camera lens, which is why the change in lightning is shown only in a small circle.

Acknowledgments

We thank the National BioResource Project Medaka, which is supported by the Ministry of Education, Culture, Sports, Science, and Technology (MEXT) of Japan, for supplying the medaka inbred strains. We thank Dr. Y. Takehana and Dr. T. Sasado for technical assistance.

Author Contributions

Conceived and designed the experiments: ST T. Kimura MS AS KN H. Takeuchi. Performed the experiments: ST T. Kimura. Analyzed the data: ST T. Kimura. Contributed reagents/materials/analysis tools: ST T.

Kimura YS. Wrote the paper: ST T. Kimura MS YS TO AS H. Takeda T. Kubo H. Takeuchi.

References

1. Gosling SD (2001) From mice to men: what can we learn about personality from animal research? Psychol Bull 127: 45–86.
2. Réale D, Reader SM, Sol D, McDougall PT, Dingemanse NJ (2007) Integrating animal temperament within ecology and evolution. Biol Rev 82: 291–318.
3. Bouchard TJ, Jr., Loehlin JC (2001) Genes, evolution, and personality. Behav Genet 31: 243–273.
4. Carter AJ, Feeney WE, Marshall HH, Cowlishaw G, Heinsohn R (2013) Animal personality: what are behavioural ecologists measuring? Biol Rev 88: 465–475.
5. Sih A, Bell A, Johnson JC (2004) Behavioral syndromes: an ecological and evolutionary overview. Trends Ecol Evol 19: 372–378.
6. Sloan Wilson D, Clark AB, Coleman K, Dearstyne T (1994) Shyness and boldness in humans and other animals. Trends Ecol Evol 9: 442–446.
7. Smith BR, Blumstein DT (2008) Fitness consequences of personality: a meta-analysis. Behav Ecol 19: 448–455.
8. Webster MM, Ward AJ (2011) Personality and social context. Biol Rev Camb Philos Soc 86: 759–773.
9. Bergmuller R, Taborsky M (2010) Animal personality due to social niche specialisation. Trends Ecol Evol 25: 504–511.
10. Hyodo-Taguchi Y(1996) Inbred strains of the medaka, *Oryzias talipes*: The Fish Biology Jounal MEDAKA 8: 11–14.
11. Spivakov M, Auer TO, Peravali R, Dunham I, Dolle D, et al. (2014) Genomic and phenotypic characterization of a wild medaka population: towards the establishment of an isogenic population genetic resource in fish. G3 (Bethesda) 4: 433–445.
12. Sasado T, Tanaka M, Kobayashi K, Sato T, Sakaizumi M, et al. (2010) The National BioResource Project Medaka (NBRP Medaka): an integrated bioresource for biological and biomedical sciences. Exp Anim 59: 13–23.
13. Hyodo-Taguchi Y (1990) Inbred strains of the medaka and their characters.(In Japanese); Egami N, Yamagami K, Shima A, editors. Tokyo: Tokyo University Press. pp. 129–142.
14. Hyodo-Taguchi Y (1983) Effects of UV irradiation on embryonic development of different inbred strains of the fish *Oryzias latipes*. J Radiat Res 24: 221–228.
15. Kimura T, Shimada A, Sakai N, Mitani H, Naruse K, et al. (2007) Genetic analysis of craniofacial traits in the medaka. Genetics 177: 2379–2388.
16. Kimura T, Naruse K (2010) M-marker 2009, a marker set for mapping medaka mutants using PCR length polymorphisms with an automated microchip gel electrophoresis system. Biotechniques 49: 582–583.
17. Kimura T, Yoshida K, Shimada A, Jindo T, Sakaizumi M, et al. (2005) Genetic linkage map of medaka with polymerase chain reaction length polymorphisms. Gene 363: 24–31.
18. Naruse K, Fukamachi S, Mitani H, Kondo M, Matsuoka T, et al. (2000) A detailed linkage map of medaka, *Oryzias latipes*: comparative genomics and genome evolution. Genetics 154: 1773–1784.
19. Ahsan B, Kobayashi D, Yamada T, Kasahara M, Sasaki S, et al. (2008) UTGB/medaka: genomic resource database for medaka biology. Nucleic Acids Res 36: D747–752.
20. Kasahara M, Naruse K, Sasaki S, Nakatani Y, Qu W, et al. (2007) The medaka draft genome and insights into vertebrate genome evolution. Nature 447: 714–719.
21. Kobayashi D, Takeda H (2008) Medaka genome project. Brief Funct Genomic Proteomic 7: 415–426.
22. Takeda H (2008) Draft genome of the medaka fish: a comprehensive resource for medaka developmental genetics and vertebrate evolutionary biology. Dev Growth Differ 50 Suppl 1: S157–166.
23. Ishikawa Y, Yoshimoto M, Yamamoto N, Ito H (1999) Different brain morphologies from different genotypes in a single teleost species, the medaka (*Oryzias latipes*). Brain Behav Evol 53: 2–9.
24. Russell EM (1967) Changes in the behaviour of Lebistes reticulatus upon a repeated shadow stimulus. Anim Behav 15: 574–585.
25. Eaton R, Lavender W, Wieland C (1981) Identification of Mauthner-initiated response patterns in goldfish: Evidence from simultaneous cinematography and electrophysiology. J Comp Physiol 144: 521–531.
26. Kimura T, Shinya M, Naruse K (2012) Genetic analysis of vertebral regionalization and number in medaka (*Oryzias latipes*) inbred lines. G3 (Bethesda) 2: 1317–1323.
27. Matsuda M, Shinomiya A, Kinoshita M, Suzuki A, Kobayashi T, et al. (2007) The *DMY* gene induces male development in genetically female (XX) medaka fish. Proc Natl Acad Sci U S A 104: 3865–3870.
28. Eaton RC, editor (1984) Neural Mechanisms of Startle Behavior: Springer US. 377 p.
29. Gwilliam GF (1963) The Mechanism Of The Shadow Reflex In Cirripedia. I. Electrical Activity In The Supraesophageal Ganglion And Ocellar Nerve. Biol Bull 125: 470–485.
30. Glowa J, Hansen C (1994) Differences in response to an acoustic startle stimulus among forty-six rat strains. Behav Genet 24: 79–84.
31. Flint J (2003) Analysis of quantitative trait loci that influence animal behavior. J Neurobiol 54: 46–77.
32. Flint J, Valdar W, Shifman S, Mott R (2005) Strategies for mapping and cloning quantitative trait genes in rodents. Nat Rev Genet 6: 271–286
33. Mackay TF, Stone EA, Ayroles JF (2009) The genetics of quantitative traits: challenges and prospects. Nat Rev Genet 10: 565–577.
34. Ansai S, Inohaya K, Yoshiura Y, Schartl M, Uemura N, et al. (2014) Design, evaluation, and screening methods for efficient targeted mutagenesis with transcription activator-like effector nucleases in medaka. Dev Growth Differ 56: 98–107.
35. Ansai S, Kinoshita M (2014) Targeted mutagenesis using CRISPR/Cas system in medaka. Biol Open 3: 362–371.
36. Ansai S, Sakuma T, Yamamoto T, Ariga H, Uemura N, et al. (2013) Efficient targeted mutagenesis in medaka using custom-designed transcription activator-like effector nucleases. Genetics 193: 739–749.
37. Okuyama T, Isoe Y, Hoki M, Suehiro Y, Yamagishi G, et al. (2013) Controlled Cre/loxP site-specific recombination in the developing brain in medaka fish, *Oryzias latipes*. PLoS One 8: e66597.
38. Michelena P, Jeanson R, Deneubourg J-L, Sibbald AM (2010) Personality and collective decision-making in foraging herbivores. Proc R Soc B Biol Sci 277: 1093–1099.
39. Katsumura T, Oda S, Mano S, Suguro N, Watanabe K, et al. (2009) Genetic differentiation among local populations of medaka fish (*Oryzias latipes*) evaluated through grid- and deme-based sampling. Gene 443: 170–177.
40. Hirayama M, Mukai T, Miya M, Murata Y, Sekiya Y, et al. (2010) Intraspecific variation in the mitochondrial genome among local populations of Medaka Oryzias latipes. Gene 457: 13–24.
41. Matsumoto Y, Oota H, Asaoka Y, Nishina H, Watanabe K, et al. (2009) Medaka: a promising model animal for comparative population genomics. BMC Res Notes 2: 88.
42. Okuyama T, Yokoi S, Abe H, Isoe Y, Suehiro Y, et al. (2014) A neural mechanism underlying mating preferences for familiar individuals in medaka fish. Science 343: 91–94.
43. Imada H, Hoki M, Suehiro Y, Okuyama T, Kurabayashi D, et al. (2010) Coordinated and cohesive movement of two small conspecific fish induced by eliciting a simultaneous optomotor response. PLoS One 5: e11248.
44. Perez-Escudero A, Vicente-Page J, Hinz RC, Arganda S, de Polavieja GG (2014) idTracker: tracking individuals in a group by automatic identification of unmarked animals. Nat Methods.
45. Ochiai T, Suehiro Y, Nishinari K, Kubo T, Takeuchi H (2013) A new data-mining method to search for behavioral properties that induce alignment and their involvement in social learning in medaka fish (*Oryzias latipes*). PLoS One 8: e71685.
46. Matsumoto Y, Fukamachi S, Mitani H, Kawamura S (2006) Functional characterization of visual opsin repertoire in Medaka (*Oryzias latipes*). Gene 371: 268–278
47. Broman KW, Wu H, Sen Ś, Churchill GA (2003) R/qtl: QTL mapping in experimental crosses. Bioinformatics 19: 889–890.

13

Coping with Daily Thermal Variability: Behavioural Performance of an Ectotherm Model in a Warming World

José M. Rojas[1,2], Simón B. Castillo[1], Guillermo Folguera[3], Sebastián Abades[1,4], Francisco Bozinovic[1]*

1 Departamento de Ecología and Center of Applied Ecology & Sustainability (CAPES), Facultad de Ciencias Biológicas, P. Universidad Católica de Chile, Santiago, Chile, **2** Centro de Investigación e Innovación para el Cambio Climático, Universidad Santo Tomás, Santiago, Chile, **3** Filosofía de la Biología Group, Facultad de Ciencias Exactas y Naturales & Facultad de Filosofía y Letras, Universidad de Buenos Aires, Buenos Aires, Argentina, **4** Instituto de Ecología y Biodiversidad, Santiago, Chile

Abstract

Global climate change poses one of the greatest threats to species persistence. Most analyses of the potential biological impacts have focused on changes in mean temperature, but changes in thermal variance will also impact organisms and populations. We assessed the effects of acclimation to daily variance of temperature on dispersal and exploratory behavior in the terrestrial isopod *Porcellio laevis* in an open field. Acclimation treatments were 24±0, 24±4 and 24±8°C. Because the performance of ectotherms relates nonlinearly to temperature, we predicted that animals acclimated to a higher daily thermal variation should minimize the time exposed in the centre of open field, – i.e. increase the linearity of displacements. Consistent with our prediction, isopods acclimated to a thermally variable environment reduce their exploratory behaviour, hypothetically to minimize their exposure to adverse environmental conditions. This scenario as well as the long latency of animals after releases acclimated to variable environments is consistent with this idea. We suggested that to develop more realistic predictions about the biological impacts of climate change, one must consider the interactions between the mean and variance of environmental temperature on animals' performance.

Editor: Bayden D. Russell, The University of Adelaide, Australia

Funding: This study was funded by Fondo Nacional de Ciencia y Tecnología grant #1130015 and Centro de Ecología Aplicada y Sustentabilidad, Line 3 to FB. The authors thank L. A. Ebensperger and F. M. Jaksic for helpful comments. Experimental protocols followed the rules of The Ethics and Biosafety Committee of the Faculty of Biological Sciences at the Catholic University of Chile, permit No. CBB-100/2012. The funders had no role in study design, data collection and analysis, decision to publish, or preparation of the manuscript.

Competing Interests: The authors have declared that no competing interests exist.

* Email: fbozinovic@bio.puc.cl

Introduction

Current environmental climate variability associated to global change poses one of the greatest threats to organismal functional diversity [1,2]. Anthropogenic impacts on the earth's climate and habitats will likely increase not only in mean temperature but in the frequency of extremely high temperatures and seasonal/daily variability in certain regions [3]. Although ecologists widely recognize the potential impacts of warming, less attention has been paid to changes in thermal variation on a scale that pertains directly to organisms [4,5,6]. Observations and experiments that quantify not only the effect of increases in mean environmental variables but also the effect of environmental variation on organismal traits are important for inferring ecological and evolutionary responses to climate change and the mechanisms by which organisms cope with this variation [7,8]. In this vein, how behavioural traits are affected by acclimation to temperature variability and their consequences for population persistence is poorly known [9]. Understanding the behavioural responses of animals, and their fitness consequences in variable environments, is important to predict the consequences of global warming on biodiversity. Indeed, as recently suggested by Sih [10] and Wingfield [11], the capability to explain and predict how organism respond behaviourally to human-induced rapid environmental changes has important implication for wildlife management as well as is becoming a focal point of basic and applied research within scenarios of global warming.

Terrestrial isopods are good models to test hypotheses in global change biology, since they exhibit a broad distribution under diverse abiotic conditions [12,13,14,15]. As suggested by Warburg et al. [16] and Hassall et al. [17] the main challenge for terrestrial isopod' fitness is to cope with desiccation stress and temperature fluctuation. In this way, some authors have evidenced the plasticity of diverse traits – behaviour included – that would help the persistence of woodlouse along different geographic as well as -climatic conditions [18,19,20,21,22,23,24]. It is common to observe seasonal changes in its daily activity pattern, together with the use of shelters and the formation of aggregations that hypothetically allows woodlice to avoid unfavourable environmental conditions [16,25,26]. Formally, the locomotory activity of terrestrial isopod is described as an uncorrelated discrete Random Walk [27,28,29,30]. This kind of displacement indicate that these animals moves straight ahead for a certain distance, then turn over a random angle (turning angle), and again moves straight ahead to turn without directional persistence [27,28]. The setting of this type of movement is mainly determined by the degree of autocorrelation of the angles of rotation. An increase in this variable determines a more linear movement of individuals. In laboratory studies, an increment in the autocorrelation level is associated mainly to an escape response, characterized by the temporal persistence of autocorrelation [28,30,31,32,33]. In the field, the search of suitable patches (e.g. soil moisture) or a dispersion behaviour is also characterized by a significant autocorrelation of turning angle but with a more unpredictable temporal memory, establishing more complex activity patterns such as correlated random walks or spiral displacement[29,30,34,35,36].

Here we experimentally tested the effect of acclimation to environmental thermal variability on dispersal behaviour in the terrestrial isopod *Porcellio laevis* in an open field. Traditionally, the inclusion of the behaviour in studies of thermal acclimation involves two kind of approaches, namely temperature selection [37,38,39] and locomotory performance [40]. In all cases, behaviour is used as descriptor of the flexibility or sensitivity of a whole-individual to a new thermal regime, contributing to understand process such as adaptation [37,41]. Nevertheless, behavioural responses behind ecological relationship of individuals or its implication on fitness still puzzling. Organismal studies in *P. laevis* evidence plasticity in physiological and life history traits in response to different geographic-climatic conditions [19,20,21,23,24]. Particularly, when the climatic conditions imply thermal variation, a decrease in locomotor and development performance is described in animals acclimated to an increment in the amplitude of ambient temperature variation [7,8,40]. Based on these data, and assuming that dispersal behaviour of *P. leavis* fits to a discrete random walk model, going from linear displacements under stressful condition to a more complex at normal condition [27,28,32,34], we predict that animals acclimated to a higher daily thermal variation should minimize the time exposed in the center of open field or in other words to increase the linearity of displacement. Consequently, Fig 1 shows that a dispersion trajectory is made by steps and the minimal length is determined by the interval sampling of each behavioural record, i.e. – the black line between open circles. Circles represent a sampling event. The absolute angle (α) is defined between the horizontal plane (dx) and the step. The relative angle (ρ) or turning, is the angle between successive steps. Distance (d), is the total distance during each record. The mean distance (R_n) is the net distance between the current location and the first relocation of the trajectory (Fig. 1). In our view this kind of approaches will help to understand if species physiological and behavioral flexibility may buffer (or not and how) the effect of global warming.

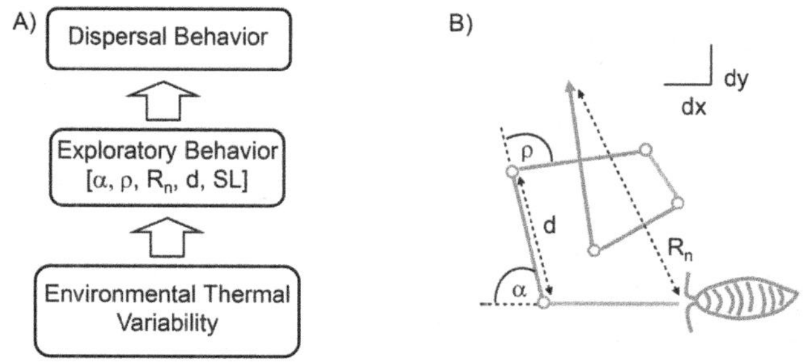

Figure 1. Dependence of behaviour of space use on the environmental thermal variability. (A). Dispersion trajectory is made by steps or SL (B).The minimal length is determined by the interval sampling of each behavioural record, i.e. gray line between open circles. Circles represent a sampling event. The absolute angle (α) is defined between the horizontal plane (dx) and the step. The relative angle (ρ) or turning, is the angle between successive steps. Distance (d), is the total distance during each record. The mean distance (R_n) is the net distance between the current location and the first relocation of the trajectory. Modified from [63].

Material and Methods

Ethics Statement

The study was performed with *P. laevis* specimens collected in the Mediterranean habitats of central Chile at San Carlos de Apoquindo, a Field Station of the Department of Ecology, Catholic University of Chile (33°23′ S, 70°31′ W at 1,230 m above sea level).

Experimental protocols followed the rules of The Ethics and Biosafety Committee of the Faculty of Biological Sciences at the Catholic University of Chile, permit No. CBB-100/2012. *Porcellio leavis* is not an endangered or protected species. We used animals of similar body size and both sexes.

Animals and experimental design

The mean minimum air temperature at San Carlos de Apoquindo is nearly 6°C below the mean and the maximum goes 7°C higher, on average [42]. Respect to yearly seasonality, the broadest range of temperatures is observed in Summer (16–17°C between mean maximum and mean minimum temperature), and the narrowest in Autumn and Winter (9–10°C).

After collection, animals were sorted by sex and maintained in plastic Petri dish (50 mm diameter; with a base layer of plaster-of-paris to maintain humidity). Food, in the form of dry spinach, and water was provided ad libitum [7]. Pregnant females were identified in the laboratory and placed under standard conditions of light (L:D = 12:12) and temperature (24°C) in culture boxes (2.2×2.2×2.4 cm) with a layer of damp sand one cm thick. Previous reports indicated that *P. laevis* exhibit a life cycle of about 12–18 months [43]. Thus, at first 10 days of development, F1 were maintained at same conditions as parents at 24°C to avoid higher rates of mortality [7]. Following Folguera et al. [7] methodology during 74 days, 60 individuals were randomly assigned to one of three thermal treatments in climatic chambers (20 replicates); namely constant temperature regime at 24°C ($\delta = 0$) and two treatments with variance in temperatures, reaching a maximum during daytime and a minimum at night with alternating temperature regimes of 28–20°C ($\delta = 4$) and 32–16°C ($\delta = 8$). These experimental temperatures were chosen because they are within the thermal tolerance range and daily variability of *P. laevis* [20,21,22,44].

Dispersion behaviour of woodlice was evaluated in an open-field setup. The open field consisted in a circular surface of 45 cm radius made of dry plaster and surrounded by a wall of opaque acrylic. Above the open-field we placed a webcam connected to a notebook which allow us to record animals behaviour. Studies were conducted in an acclimated room at 24°C. Displacement of animals was recorded during 15 minutes. Its location (X-Y coordinate) into the open field was determined using a video tracking system (Smart 2.0 – Panlab, Barcelona Spain) at 0.2 intervals seconds (event sampling). Previously to each record, each individual was placed in the centre of field within a dark chamber during 5 minutes to reduce handling stress. Upon the end of these observations, all experimental woodlice subjects were released back at their original sites of capture.

Movement analysis

To compare dispersal behaviour among treatments, we considered that displacement of woodlice follow basically a discrete random walk, without directional correlation. Therefore, a core point of analysis corresponded to checking randomness in the lineal and angular parameters of each paths (Fig. 1) Thus, each woodlouse was characterized by a trajectory. From the trajectory we estimated the parameters indicated in Fig. 1. Net displacement was quantified by the net square displacement (R^2_n) see [29,45]. Each movement variable included the elapsed time since each animal started the displacement until they reached the wall of the open field. As an additional descriptor of spatial behaviour we used the latency to first displacement (seconds). The exploration level or tortuosity of each path we estimated using the mean fractal dimension (D). This value varied from 1 (straight line) to 2 (a travel path which is so tortuous that it completely covers a two-dimensional plane, see [46,47]). The length step, R^2n, and turn angle (relative and absolute) for each path on each interval were estimated using the adehabitat package for R-Cran software [48]. The fractal dimension was estimated with the Fractal Mean estimator using a window range (or frame) of 0.25 to correct the border effect (see [49]) implemented in the Fractal software (Fractal ver. 5, V. Nams, Nova Scotia Agricultural College).

Statistical analysis

A generalized lineal model was used to test differences in the fractal dimension of displayed by isopods. Thermal treatment was used as categorical variable and body size as weight variable [50,51]. A similar statistical approach was used to evaluate the effect of acclimation temperature regime on the turning angle, absolute angle, step size and net distance. For each trajectory the temporal sequence of each parameter was evaluated as a time series. The median value of each time series was used as a statistical descriptor the each parameter, assuming that the median follows a gamma distribution with a log function as link [52]. The fractal dimension was also evaluated using a generalized lineal model, but in this case, using the mean as the central tendency descriptor. The significance of models was tested using loglikelihood ratios between the adjusted model and a null model (intercept). The significance of coefficients of models was estimated using an unpaired two tails t-test. Latency and velocity variables were rescaled to log(x). These variables were assessed using a one way ANOVA. Previously the assumptions of each test were evaluated. The temporal autocorrelation level of behavioural descriptors was evaluated using an autocorrelation function analysis (ACF). These analyses were conducted using the autocorrelation descriptors based in the square differences between successive steps to lineal parameters and chord distance between successive angles to test relative angle, estimated by Adehabitat package for the R-Cran software [47]. The significance level of these descriptors was assessed using the 95% confidence intervals estimated from permutation procedure implemented in Adehabitat package. Previously to each test, random and observed values were centred using the average of random values, and then transformed into absolute values [47,53]. We used correlograms to show mean values by thermal treatment for differences between ACF observed and the lower confidence interval estimated by individual. Finally, the central tendency of the dispersal behaviour was evaluated on frequency distribution of step length and turning angle using the Hartigan's dip test to unimodality implemented in the diptest package for R-Cran [48]. This analysis was pooled all individuals by treatment All parametric tests used were implemented using the R-Cran software (version 2.12.2).

Results

Experimental treatments did not affect body size of individuals (Table 1). Differences in dispersion behaviour as a result of acclimation to different thermal regimes were observed. Animals that experienced the largest thermal variation (δ8) showed a positive autocorrelation in the steps length to a delay of 3 lag. Similarly, the turning angle of these animals showed a lag autocorrelation to 4 (Fig. 2). Animals acclimated to δ0 and δ4 showed positive autocorrelation for the length of the steps to lag 3 only (Fig.2). No autocorrelation on turning behaviour was recorded (Fig. 2). These results were consistent with those observed in the frequency distribution of the relative angles. Animals exposed to a δ0 and δ4 thermal regime showed a symmetrical unimodal distribution centred at 0°. A different pattern is observed in animals acclimated to the larger thermal variation (δ8), that while it is centred at 0°, its distribution is nearly bimodal (Fig. 3).

In the case of the frequency distribution of the length of steps, a tendency to a bimodal distribution was observed, but mainly in animals acclimated to a daily variable temperature (Fig. 4). Particularly, isopods acclimated to highly variable environments (δ8) showed a distribution characterized by a predominance of short steps (Fig. 4). Animals acclimated to δ4 exhibited similar results described for δ8, but with a smaller representation of short steps. Finally, animals acclimated to a stable environment showed a more homogeneous distribution of length steps

A significant adjust of the generalized linear models was only detected for the descriptors of fractal dimension and absolute angle (Table 2). Thus, no differences were observed in the relative angle (turning) of paths among experimental groups (Table 2). Likewise, no significant differences in step length and net distance travelled (R^2n) were detected among experimental conditions (Table 2). Regarding to absolute angles, animals exposed to a constant thermal regime showed a significantly lower absolute angle in comparison to animals facing δ4 regimes (Tables 2, 3). Animals acclimated to a δ8 thermal regime did not change the magnitude of absolute angle (Tables 2, 3). A congruent for the absolute angle was observed for fractal dimension. That is, animals that did not

Table 1. Results for one-way analysis of variance for testing the homogeneity in individual body attributes and the general dispersal performance of woodlice among treatments.

	δ0	δ4	δ8	Statistics
Body length (mm)	10.9 (0.3)^a	11.5 (0.3)^a	10.8 (0.3)^a	$F_{(2,57)}=1.6$; SS = 5.12
Body mass (mg)	71.7 (3.2)^a	78.4 (3.2)^a	69.8 (3.2)^a	$F_{(2,57)}=1.9$; SS = 820.9
Speed (cm/s)	3.6 (0.4)^a	5.0 (0.6)^a	3.8 (0.4)^a	$F_{(2,57)}=1.0$; SS = 0.9
Latency (s)	81.1 (16.3)^a	163.2 (25.4)^b	100.8 (20.4)^{ab}	$F_{(2,57)}=4.5$; SS = 7.4

Different letters indicate significant differences among treatments (Tukey test at $P<0.05$). Values shown the mean estimated in each thermal variability treatments: δ0, δ4 and δ8 (i.e. 24±0, 24±4 and 24±8°C) for each parameter. In parentheses ±1 standard error.

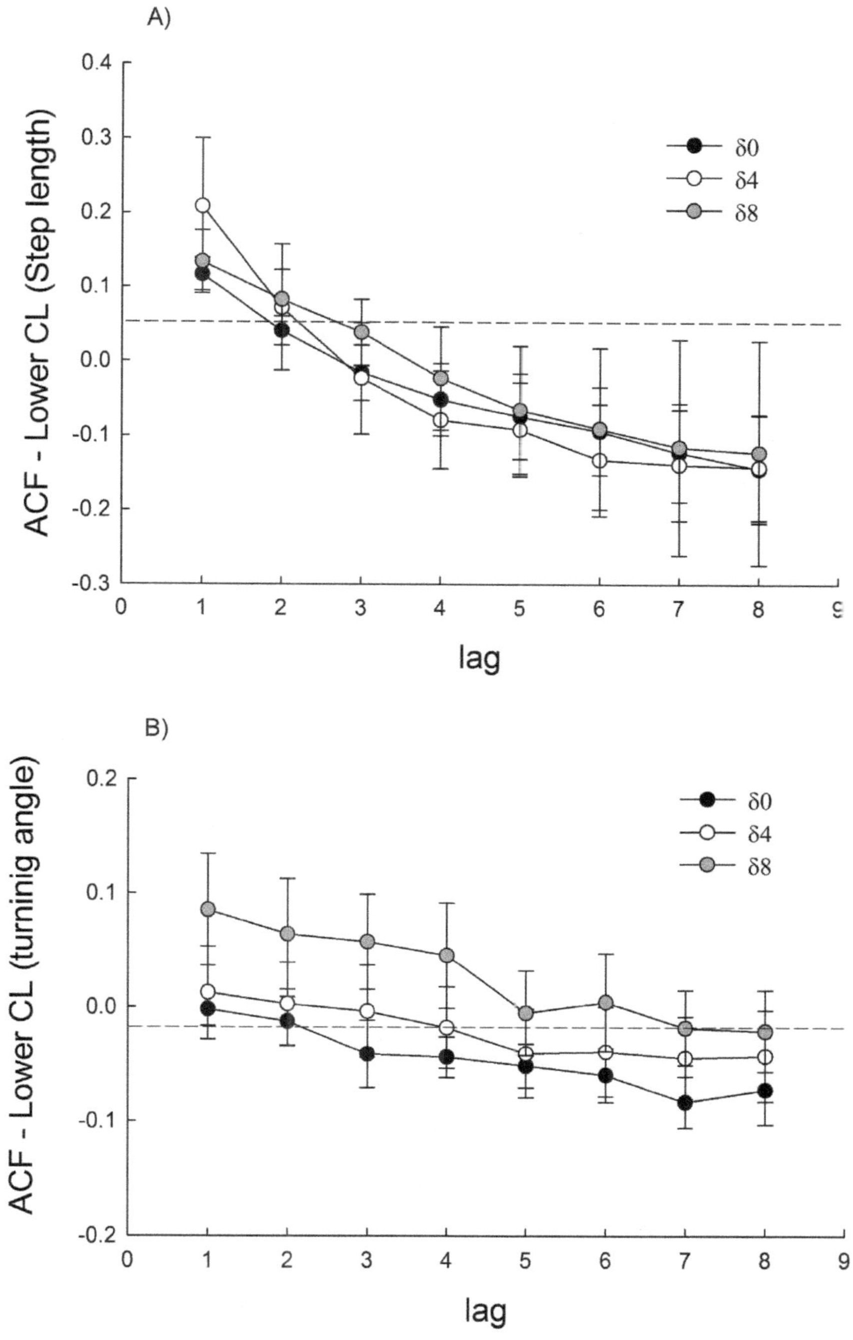

Figure 2. Autocorrelation mean (± standard error) estimated for three thermal treatments of acclimation for step length (A) and relative angle (B). Daily thermal variability treatments are: δ0 = 24±0°C, δ4 = 24±4°C and δ8 = 24±8°C. The ACF were calculated for lag = 0–8. Segmented line indicated the significance limits. Values correspond only to positive correlations. See text for details.

experienced thermal variation showed significant greater displacement (i.e. fractality) in the open field compared with animals that experienced a thermal variability of medium intensity (δ4) (Tables 2, 3, Fig. 5). No differences in the degree of fractality of movement between animals that were exposed to δ4 and δ8 were observed (Tables 2, 3, Fig. 5). When assessing the performance of animals, isopods acclimated to an environment without variability showed a latency to first displacement significantly lower than that observed in animals exposed to the δ4 treatment and similar to those acclimated to δ8 conditions (Table 1). Animals acclimated at δ4 and δ8 exhibited no difference in the latency time (Table 1). Finally, no statistical differences were recorded in speed among animals acclimated to different thermal regimes (Table 1).

Discussion

To predict responses to climate change, physiological and behavioural ecologists must understand the patterns of thermal variation and the mechanisms by which animals cope with this variation [6,10,11,54,55]. As pointed out before, impacts on the earth's climate will likely increase the frequency of extremely high

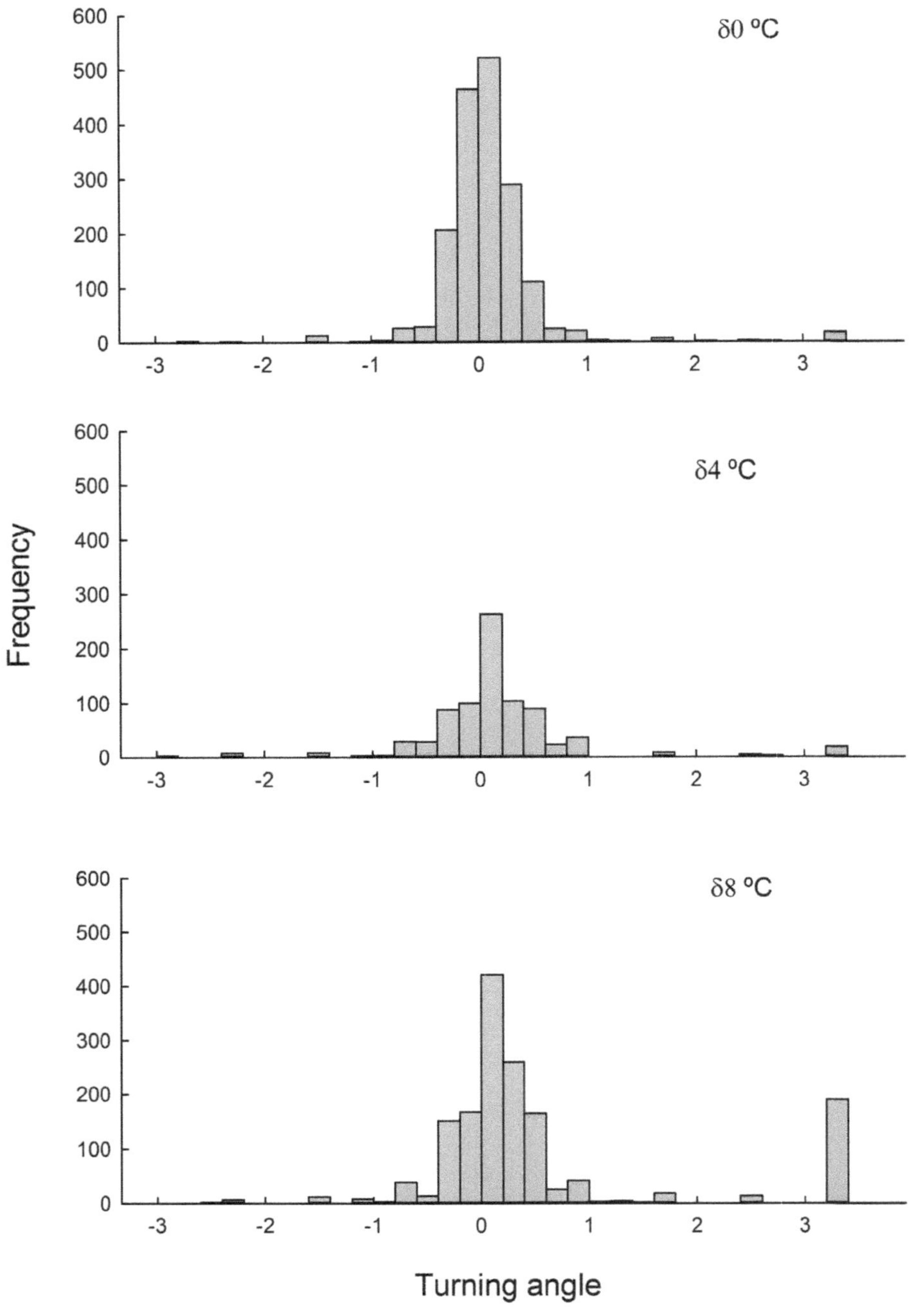

Figure 3. Frequency distribution of turning angle for different thermal treatments. Daily thermal variability treatments are: δ0 = 24±0°C, δ4 = 24±4°C and δ8 = 24±8°C. The Hartigan's dip statistical test revealed a significant bimodality pattern among all treatments. Namely, δ0: D = 0.0214, $P<0.05$; δ4: D = 0.0345, $P<0.05$ and δ8: D = 0.0396, $P<0.05$.

temperatures in certain regions [56]. Many organisms are expected to suffer a decrement in performance and fitness [57,58], but some may preserve their performance through behavioural responses [59]. For instance, we concur with Kearney et al [59] who enlighten the importance of behavioural traits to buffer the impact of global warming, arguing that this is an important and missing element from models of climatic change and predictions of impacts on biodiversity (see also [55]). Accordingly our results with terrestrial isopods represent an example of how changes in behavior allow animals to cope with variable habitats but also how environmental thermal variability may affect behavior.

Helmuth *et al.* [60] analyzed environmental variability at scales relevant to organisms to predict the responses of individuals and populations to thermal variability in the intertidal, where as previously suggested [61] some behavioral traits act as critical phase transition triggered by ambient temperatures. For instance, in hypervariable environments as the intertidal [60,62], the gastropod *Nerita atramentosa* behaviorally avoid the high temperatures through selection of thermal refuges as well as through huddling. Thus, as terrestrial isopods (this study), the

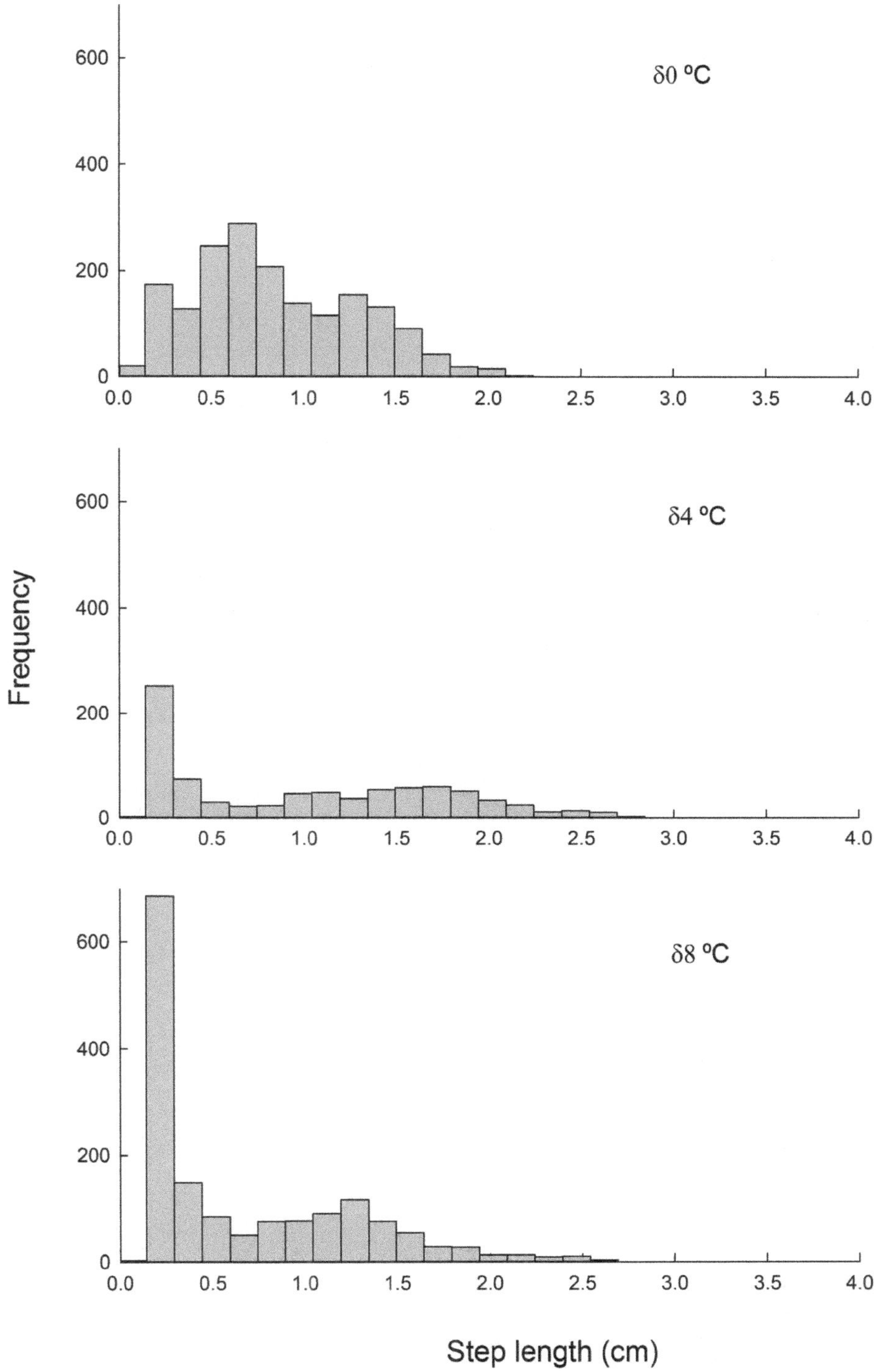

Figure 4. Frequency distribution of step length for different thermal acclimation conditions. Daily thermal variability treatments are: δ0 = 24±0°C, δ4 = 24±4°C and δ8 = 24±8°C. The Hartigan's dip statistical test revealed a significant bimodality pattern among all treatments. Namely, δ0: D = 0.013, $P>0.05$; δ4: D = 0.0123, $P>0.05$ and δ8: D = 0.0614, $P<0.05$.

behavior of marine gastropod during emersion seems to be related by local thermal stability or variability of the environment [63]. Theoretically, in variable habitats it is beneficial for ectotherms to select temperatures below thermal physiological optimum [64]. This hypothesis was tested by [65] in the intertidal snail *Chlorostoma funebralis* throughout experimental studies of behavioral thermoregulation in a thermal gradient. They found a "cold-biased" behavior which seems to guide snails to refuges in crevices. Indeed this snail performed a biased random walk along the gradient and huddle together at temperatures near their lower thermal limits. [66,67] also demonstrated that operative temperature in the intertidal may be one of the leading determinants of

Table 2. Results of generalized linear model for displacement descriptors.

Movement Descriptors	Null Model		Adjusted Model			*P*-value	AIC
	df	Loglik	Df	Loglik	L ratio		
Fractal dimension	5	49.3	7	7.9	15.91	0.001	−89.9
Absolute angle	5	−83.4	7	−79.5	7.92	0.02	172.9
Turning angle	5	48.1	7	48.7	1.16	0.55	−83.3
Step length	5	−51.9	7	−50.7	2.60	0.27	115.4
Net distance (R^2n)	5	−314.7	7	−.313.5	2.34	0.31	641.0

AIC corresponds to the Akaike information criterion; df: degree of freedom; Loglik: Loglikelihood estimator; L ratio: loglikelihood ratio Chi-square.

time/space variation of behavioral traits in the periwinkle *Echinolittorina peruviana*.

The constraints to behaviour of animals – though changes in ambient temperatures - are probably extremely important in determining when and how a habitat should be used. Climatic effects are of paramount importance since predicted increases in global thermal mean as well as in variance will change many behavioural displays including habitat use and time, among others. Therefore, thermal tolerance should depend on the variance of temperature as well as the mean. Consistent with this view, we found that woodlice exposed to highly variable thermal conditions showed a more conservative displacement. The long correlations of both the turning behaviour as well as the length of steps were greater than those observed in animals exposed to less variable or constant thermal conditions. Particularly for length step, which was not observed in the other test groups were noted. Thus, animals acclimated to a highly variable thermal environment display a more systematic or persistent displacement behaviour when exposed to a new environment, adjusting their behaviour to a more dispersive behaviour, namely less random and more linear [68]. Accordingly, the frequency distributions observed for the linear and angular parameters are consistent with a reduction of the random component in the movement. Among the parameters used to describe the movement of isopods, we consider the use of the fractal dimension, which allows describing the degree of homogeneity in which space was used or explored by the animals during their displacement [49]. Animals acclimated to stable thermal conditions showed a higher degree of scanning in comparison to the other experimental groups. The estimated magnitude for this group is similar to that reported for other terrestrial invertebrates [69], implying that the behaviour of isopods during our study resembles behaviour under natural conditions. This result is consistent with our previous suggestion that animals acclimated to variable conditions exhibit less exploratory behaviour. In this scenario interestingly, animals acclimated to an intermediate thermal variation are situated in an intermediate position. These animals showed a degree of fractalitiy, and bimodality in the frequency distribution of step length, as observed in animals acclimated to highly variable environments. Nevertheless, they also showed a temporal auto-correlation pattern for linear and angular parameters similar to those estimated for animals acclimated to stable environments.

Our results supported that thermal acclimation has significant effects on the behaviour of isopods, a finding not previously reported. Isopods exposed to a thermally variable environment reduce their exploratory behaviour, probably as a way to minimize their exposure to adverse environmental conditions [63,65]. This scenario as well as the long latency of animals acclimated to variable environments is consistent with this proposal. Clearly, further studies are needed to determine some of the ecological/ evolutionary consequences of these behavioural and physiological adjustments, yet our results highlight to the importance of the behavioural component when making predictions about population viability under a climate change scenario. Overall, changes in animals phenologies that result from global warming may produce changes in activity patterns, migration, and breeding timing, and ultimately all of these altered synchronization between trophic levels as well as changed species competitive ability [70]. The emerging picture is that some specific behaviours represent an integrated response to the abiotic environment and its variation and the behavioural ability of animals to respond to new climatic scenarios. According to Sih [10], the observed variation in the behavioural responses in our studied animals may be the result of a cue-response system or because unpredictable environmental

Table 3. Coefficients for significant models from Table 2.

Movement Descriptors		Thermal treatments		
		δ0	δ4	δ8
Fractal dimension	Estimate	1.07 (0.01)	−0.03 (0.01)	−0.02 (0.05)
	t-test; *P*-value	127.7; 0.001	−2,84; 0.006	−1.97; 0.05
Absolute angle	Estimate	−0.32 (0.15)	0.86 (0.30)	0.39 (0.26)
	t-test; *P*-value	−2.10; 0.04	2.79; 0.001	1.46; 0.14

In parentheses standard error of coefficient. Student t-test and *P*-values correspond to hypotheses test of the estimated coefficients (df = 57, 4).

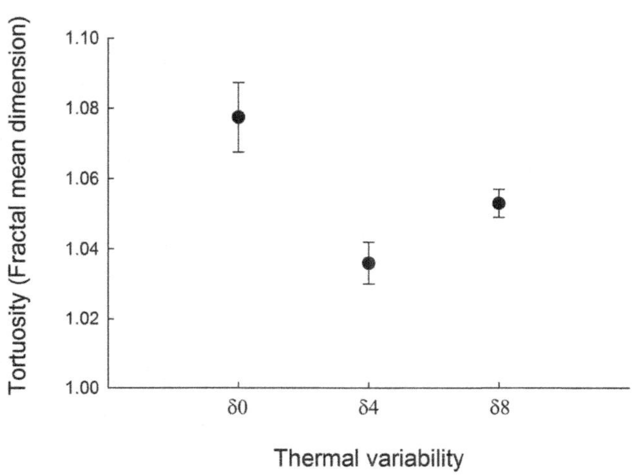

Figure 5. Fractal mean dimension recorded for animals acclimated to different variability thermal regimen. Daily thermal variability treatments are: δ0 = 24±0°C, δ4 = 24±4°C and δ8 = 24±8°C. Bar showed ±1 standard error.

information may affect woodlice responses to novel situations or because just behaviour plasticity may affect an animal's response. Probably a signal detection or a cost-benefit hypotheses on variation in learning may be the causes behind our results [5].

A main principle of physiological and behavioural ecology is that populations exposed to environmental change may crash when most individuals deteriorate, and that individuals decline when they reach a state that prevents them from maintaining homeostasis and behaviour –that is, the proper equilibrium through time with internal processes and foreign stimuli. It follows that the effects of climate change on species cannot be assessed from the type, magnitude or time scale of the perturbation, but from the physiological and behavioural states caused by it. Thus, a same set of environmental conditions may be deleterious for one species or population but harmless to another. The relatively narrow approaches used to the study of climate change on biodiversity are limited because they ignore the mechanisms that organisms use to cope with environmental changes [10,55].

Acknowledgments

We thank L.A. Ebensperger and F. M. Jaksic for helpful comments. Experimental protocols followed the rules of The Ethics and Biosafety Committee of the Faculty of Biological Sciences at the Catholic University of Chile, permit No. CBB-100/2012.

Author Contributions

Conceived and designed the experiments: JMR FB GF. Performed the experiments: SBC GF. Analyzed the data: JMR SA FB. Contributed reagents/materials/analysis tools: FB. Wrote the paper: JMR FB.

References

1. Hansen J, Kharecha P, Sato M, Masson-Delmotte V, Ackerman F, et al. (2013) Assessing "dangerous climate change": required reduction of carbon emissions to protect young people, future generations and nature. PLoS ONE 8: e81648.
2. Sunday JM, Bates AE, Kearney MR, Colwell RK, Dulvy NK, et al. (2014) Thermal-safety margins and the necessity of thermoregulatory behavior across latitude and elevation. Proc Natl Acad Sci USA 111: 5610–5615.
3. Rahmstorf S, Coumou D (2011) Increase of extreme events in a warming world. Proc Natl Acad Sci USA 108: 17905–17909.
4. Bozinovic F, Calosi P, Spicer JI (2011) Physiological correlates of geographic range in animals. Ann Rev Ecol Evol Syst 42: 155–179.
5. Sih A, Ferrari MCO, Harris DJ (2011) Evolution and behavioural responses to human-induced rapid environmental change. Evol Appl 4: 367–387.
6. Estay SA, Lima M, Bozinovic F (2014) The role of temperature variability on insect performance and population dynamics in a warming world. Oikos 123: 131–140.
7. Folguera G, Bastías DA, Caers J, Rojas JM, Piulachs MD, et al. (2011) An experimental test of the role of environmental temperature variability on ectotherm molecular, physiological and life-history traits: implications for global warming. Comp Biochem Physiol A Mol Integr Physiol 159: 242–246.
8. Folguera G, Mensch J, Munoz JL, Ceballos SG, Hasson E, et al. (2011) Ontogenetic stage-dependent effect of temperature on developmental and metabolic rates in a holometabolous insect. J Insect Physiol 56: 1679–1684.
9. Frost AJ, Thomson JS, Smith C, Burton HC, Davis B, et al. (2013) Environmental change alters personality in the rainbow trout, *Onchorhyncus mykiss*. Anim Behav 85: 1199–1207.
10. Sih A (2013) Understanding variation in behavioural responses to human-induced rapid environmental change: a conceptual overview. Anim Behav 85: 1077–1088.
11. Wingfield JC (2013) The comparative biology of environmental stress: behavioural endocrinology and variation in ability to cope with novel, changing environments. Anim Behav 85: 1127–1133.
12. Schultz GA (1972) Ecology and systematics of terrestrial isopod crustaceans from Bermuda (Oniscoidea). Crustaceana 3: 79–99.
13. Schultz GA (1961) Distribution and establishment of a land isopod in North America. Syst Zool 10: 193–196.
14. Miller RH, Cameron GN (1987) Effects of temperature and rainfall on populations of *Armadillidium vulgare* (Crustacea: Isopoda) in Texas. Am Midl Nat 117: 192–198.
15. Helden AJ, Hassall M (1998) Phenotypic plasticity in growth and development rates of *Armadillidium vulgare* (Isopoda: Oniscidea). Isr J Zool 44: 379–394.
16. Warburg MR, Linsenmair KE, Bercovitz K (1984) The effect of climate on the distribution and abundance of isopods. In: Sutton SL, Holdich DM, editors. The Biology of Terrestrial Isopods. Oxford: Clarendon Press. pp. 339–367.
17. Hassall M, Helden A, Goldson A, Grant A (2005) Ecotypic differentiation and phenotypic plasticity in reproductive traits of *Armadillidium vulgare* (Isopoda: Oniscidea). Oecologia 143: 51–60.
18. Greenaway P, Warburg MR (1998) Water fluxes in terrestrial isopods. Isr J Zool 44: 473–486.
19. Warburg MR, Calahorra Y, Amar K-O (2001) Non-seasonal breeding in a Porcellionid isopod. J Crust Biol 21: 375–383
20. Castañeda LE, Lardies MA, Bozinovic F (2004) Adaptive latitudinal shifts in the thermal physiology of a terrestrial isopod. Evol Ecol Res 6: 1–15.
21. Lardies MA, Bozinovic F (2006) Geographic covariation between metabolic rate and life history traits. Evol Ecol Res 8: 455–470.
22. Lardies MA, Bozinovic F (2008) Genetic variation for plasticity in physiological and life-history traits among populations of an invasive species, the terrestrial isopod *Porcellio laevis*. Evol Ecol Res 10: 747–762.
23. Lardies MA, Catalán TP, Bozinovic F (2004) Metabolism and life-history correlates in a lowland and highland population of a terrestrial isopod. Can J Zool 82: 677–687.
24. Bacigalupe LD, Araya NM, Carter MJ, Catalán TP, Lardies MA, et al. (2007) Maternal effects, maternal body size and offspring energetics: a study in the common woodlouse *Porcellio laevis*. Comp Biochem Physiol A Mol Integr Physiol 147: 349–354.
25. Hassall M, Tuck JM (2007) Sheltering behavior of terrestrial isopods in grasslands. Invertebrate Biology 126: 46–56.
26. Hassall M, Edwards DP, Carmenta R, Derh AM, Moss A (2010) Predicting the effect of climate change on aggregation behaviour in four species of terrestrial isopods. Behaviour 147: 151–164.
27. Hoffman G (1983) The search behavior of the desert isopod *Hemilepistus reaumuri* as compared with a systematic search. Behav Ecol Sociobiol 13: 93–106.
28. Hoffmann G (1983) The random elements in the systematic search behavior of the desert isopod *Hemilepistus reaumuri*. Behav Ecol Sociobiol 13: 81–92.
29. Turchin P (1998) Quantitative analysis of movements. Measuring and modeling population redistribution in animals and plants: Sinauer Associates.
30. Heinz SK, Conradt L, Wissel C, Karin F (2005) Dispersal behaviour in fragmented landscapes: Deriving a practical formula for patch accessibility. Landsc Ecol 20: 83–99.
31. Hughes RN (1967) Turn alternation in woodlice (*Porcellio scaber*). Anim Behav 15: 282–286.
32. Hughes RN (1978) Effects of blinding, antennectomy, food deprivation and simulated natural conditions on alternation in woodlice woodlice (*Porcellio scaber*). Journal Biological Psychology 20: 35–40.
33. Carbines GD, Dennis RM, Jackson RR (1992) Increased turn alternation by woodlice (*Porcellio scaber*) in response to a predatory spider, *Dysdera crocata*. Int J Comp Psychol 5: 138–144.

34. Paris O (1963) The ecology of *Armadillidium vulgare* (Isopoda: Oniscoidea) in California grassland: food, enemies and weather. Ecol Monog 33: 1–22.
35. Warburg MR (1968) Adaptation of terrestrial isopods. Am Zool 8: 545–559.
36. Baker M (2005) Experience influences settling behaviour in desert isopods, *Hemilepistus reaumuri*. Anim Behav 69: 1131–1138.
37. Hutchison VH, Maness JD (1979) Role of behavior in temperature acclimation and tolerance in ectotherms. Am Zool 19: 367–384.
38. Pörtner HO, Bennett AF, Bozinovic F, Clarke A, Lardies MA, et al. (2006) Trade-offs in thermal adaptation: the need for a molecular to ecological integration. Physiol Biochem Zool 79: 295–313.
39. Marais E, Chown SL (2008) Beneficial acclimation and the Bogert effect. Ecol lett 10: 1027–1036.
40. Folguera G, Bastías DA, Bozinovic F (2009) Impact of experimental thermal amplitude on ectotherm performance: Adaptation to climate change variability? Comp Biochem Physiol A Mol Integr Physiol 154: 389–393.
41. Angilletta Jr MJ (2009) Thermal adaptation: a theoretical and empirical synthesis. Oxford: Oxford University Press.
42. Jaksic FM (2001) Spatiotemporal variation patterns of plants and animals in San Carlos de Apoquindo, central Chile. Rev Ch Hist Nat 74: 477–502.
43. Nair GA (1976) Life cycle of *Porcellio laevis* (Latreille) (Isopoda, porcellionidae). Proc Indian Natl Sci Acad B Biol Sci 84: 165–172.
44. Husain MZ, Alikhan MA (1979) Physiological adaptations in crustacea to the environment: Oxygen consumption as a function of body weight and environmental temperature in the terrestrial isopod, *Porcellio laevis* Latreille (Isopoda, Oniscoidea). Crustaceana 36: 277–286.
45. Codling EA, Plank MJ, Benhamou S (2008) Random walk models in biology. J R Soc Interface 5: 813–834.
46. Nams VO (1996) The Vfractal: a new estimator for fractal dimension of animals movement paths. Landsc Ecol 11: 289–297.
47. Dray S, Royer-Carenzi M, Calenge C (2010) The exploratory analysis of autocorrelation in animal-movement studies. Ecological Research 25: 673–681.
48. Team RDC (2008) R: A language and environment for statistical computing. Version 2.12 ed: R Foundation for Statistical Computing.
49. Nams VO (2006) Improving accuracy and precision in estimating fractal dimension of animal movement path. Acta Biotheor 54: 1–11.
50. Bolker BM, Brooks ME, Clark CJ, Geange SW, Poulsen JR, et al. (2009) Generalized linear mixed models: a practical guide for ecology and evolution. Trends Ecol Evol 24: 127–135.
51. Zuur A, Ieno EN, Walker N, Saveliev AA, Smith GM (2009) Mixed effects models and extensions in ecology with R. New York: Springer Science+Business Media.
52. Westerberg L, Lindström T, Nilsson E, Wennerger U (2008) The effect on dispersal from complex correlations in small-scale movement. Ecol Modell 213: 263–272.
53. Calenge C (2006) The package adehabitat for the R software: A tool for the analysis of space and habitat use by animals. Ecol Modell 197: 516–519.
54. Bozinovic F, Bastías DA, Boher F, Clavijo-Baquet S, Estay SA, et al. (2011) The mean and variance of environmental temperature interact to determine physiological tolerance and fitness. Physiol Biochem Zool 84: 543–552.
55. Lankau R, Jorgensen PS, Sih A (2011) Incorporating evolutionary principles into environmental management and policy. Evol Appl 4: 315–325.
56. Solomon S, Qin D, Manning M, Chen Z, Marquis M, et al. (2008) Climate Change 2007: The Physical Science Basis. Cambridge: Cambridge University Press.
57. Deutsch CA, Tewksbury JJ, Huey RB, Sheldon KS, Ghalambor CK, et al. (2008) Impacts of climate warming on terrestrial ectotherms across latitude. Proc Natl Acad Sci USA 105: 6668–6672.
58. Bozinovic F, Catalán TP, Estay SA, Sabat P (2013) Acclimation to daily thermal variability drives the metabolic performance curve. Evol Ecol Res 15: 579–587.
59. Kearney MR, Shine R, Porter WP (2009) The potential for behavioral thermoregulation to buffer cold-blooded animals against climate change. Proc Natl Acad Sci USA 106: 3835–3840.
60. Helmuth B, Broitman BR, Yamane L, Gilman SE, Mach K, et al. (2010) Organismal climatology: analyzing environmental variability at scales relevant to physiological stress. J Exp Biol 213: 995–1003.
61. Canals M, Bozinovic F (2011) Huddling behavior as critical phase transition triggered by low temperatures. Complexity 17: 35–43.
62. Helmuth B (2009) From cells to coastlines: how can we use physiology to forecast the impacts of climate change? J Exp Biol 212: 753–760.
63. Chapperon C, Le Bris C, Seuront L (2013) Thermally mediated body temperature, water content and aggregation behaviour in the intertidal gastropod *Nerita atramentosa*. Ecological Research 28: 407–416.
64. Martin Tara L, Huey Raymond B (2008) Why "suboptimal" is optimal: Jensen's inequality and ectotherm thermal preferences. Am Nat 171: E102–E118.
65. Tepler S, Mach K, Denny MW (2011) Preference versus performance: body temperature of the intertidal snail *Chlorostoma funebralis*. Biol Bull 220: 107–117.
66. Muñoz JLP, Camus PA, Labra FA, Finke R, Bozinovic F (2008) Thermal constraints on daily patterns of aggregation and density along an intertidal gradient in the periwinkle *Echinolittorina peruviana*. Journal of Thermal Biology 33: 149–156.
67. Muñoz JLP, Finke R, Camus PA, Bozinovic F (2005) Thermoregulatory behavior, heat gain and thermal tolerance in the periwinkle *Echinolittorina peruviana* in central Chile. Comp Biochem Physiol A Mol Integr Physiol 142: 92–98.
68. Bartumeus F, Da Luz MGE, Viswanathan G.M., Catalan J (2005) Animal strategies: a quantitative random-walk analysis. Ecology 86: 3078–3087.
69. Weins JA, Crist TO, With KA, Milne B (1995) Fractal patterns of insect movement in microlandscape mosaic. Ecology 76: 663–666.
70. Root TL, Hughes L (2005) Present and future phenological changes in wild plants and animals. In: Lovejoy TE, Hanah L, editors. Climate change and biodiversity. London: Yale Press. pp. 61–74.

14

Continuous Variation Rather than Specialization in the Egg Phenotypes of Cuckoos (*Cuculus canorus*) Parasitizing Two Sympatric Reed Warbler Species

Szymon M. Drobniak[1], Andrzej Dyrcz[2]*, Joanna Sudyka[1], Mariusz Cichoń[1]

1 Institute of Environmental Sciences, Jagiellonian University, Kraków, Poland, **2** Department of Behavioural Ecology, Wroclaw University, Wroclaw, Poland

Abstract

The evolution of brood parasitism has long attracted considerable attention among behavioural ecologists, especially in the common cuckoo system. Common cuckoos (*Cuculus canorus*) are obligatory brood parasites, laying eggs in nests of passerines and specializing on specific host species. Specialized races of cuckoos are genetically distinct. Often in a given area, cuckoos encounter multiple hosts showing substantial variation in egg morphology. Exploiting different hosts should lead to egg-phenotype specialization in cuckoos to match egg phenotypes of the hosts. Here we test this assumption using a wild population of two sympatrically occurring host species: the great reed warbler (*Acrocephalus arundinaceus*) and reed warbler (*A. scirpaceus*). Using colour spectrophotometry, egg shell dynamometry and egg size measurements, we studied egg morphologies of cuckoos parasitizing these two hosts. In spite of observing clear differences between host egg phenotypes, we found no clear differences in cuckoo egg morphologies. Interestingly, although chromatically cuckoo eggs were more similar to reed warbler eggs, after taking into account achromatic differences, cuckoo eggs seemed to be equally similar to both host species. We hypothesize that such pattern may represent an initial stage of an averaging strategy of cuckoos, that – instead of specializing for specific hosts or exploiting only one host – adapt to multiple hosts.

Editor: Csaba Moskát, Hungarian Academy of Sciences, Hungary

Funding: This study was supported by the grant of the National Science Center no. N N304 030739 (to AD) and by funding from the Jagiellonian University within the SET project (to SMD; the project is co-financed by the European Union). The funders had no role in study design, data collection and analysis, decision to publish, or preparation of the manuscript.

Competing Interests: The authors have declared that no competing interests exist.

* Email: dyrcz@biol.uni.wroc.pl

Introduction

The common cuckoo (*Cuculus canorus*) is an obligate brood parasite utilizing a wide range of hosts that differ substantially in body size, egg morphology and reproductive behaviour. Brood parasites usually have strong impacts on host fitness, which induces strong selection pressures on the host to avoid parasites, leading to a co-evolutionary arms race [1]. The common cuckoo has particularly strong effects on its host as the parasite does not allow the host to rear any offspring. Thus, the naive hosts might quickly go extinct if they do not develop antiparasite adaptations. Egg recognition is one of the most prominent antiparasite adaptations. However, the parasite usually develops contradaptation in the form of egg mimicry. Cuckoo's potential hosts, however, have eggs that differ substantially in size, shape and colouration. In response to this diversity, cuckoos have evolved specialization to specific hosts. Within the cuckoo species in Europe, there are about 16 variants (called gentes) of egg types that match the most common host egg phenotypes [2]. Egg-phenotype matching between specific cuckoo gentes and their hosts is remarkable, and in some cases, almost perfect [2–5]. Gentes are genetically distinct and are inherited mainly through maternal genetic line [6,7].

Adaptation to parasitizing specific host species seems unquestionable in cuckoos and is apparent both on genetic and phenotypic levels. However, cuckoo females belonging to one gens may exploit a number of sympatric host species differing substantially in egg phenotypes [8–10]. In such a case, cuckoos should either 1) lay eggs that mimic the eggs of only one host species, thus laying dissimilar eggs in nests of other host species – which would be particularly expected in the early stages of a host-parasite arms race [11], or 2) show a generalist strategy by producing eggs dissimilar to eggs of any of the available host species [10]. Either strong specialization to the host species or a generalist strategy of averaging between potential hosts may be favoured by selection in such circumstances. Strong specialization should lead to the evolution of specific phenotypes matching specific host species even within a single gens, maximizing the breeding success of individual cuckoos. The evolution of such discrete egg morphs is supported by field data [12] and numerical simulations, which show that the equilibrium endpoint of an evolving brood parasite – host system is the formation of discrete egg morphs produced by the parasite [11,13]. Mathematical simulations prevent a continuous distribution of cuckoo egg-morphs from becoming a stable endpoint – discrete egg classes arise even if the initial distribution of egg phenotypes is continuous [11]. However, models allow for a continuous distribution of parasitic egg phenotypes in the initial stages of the evolutionary arms race [11,13]. Following the introduction of a new host species, producing continuously distributed egg morph, centred around an average phenotype, might be more advantageous in terms of the breeding success than immediate production of specialized eggs – particularly if such specialization requires slow genetic adaptation. Such averaging strategy should result in the disappearance of discrete mimicry to specific hosts, resulting in substantial variation in egg phenotypes observed locally (i.e. regardless of larger-scale patterns resulting from overlapping geographical variation in host egg phenotypes). In both cases, one possible scenario is the evolution of substantial variation in egg phenotype even within a single gens, but specialization should result in discontinuity of the distribution of cuckoo egg phenotypes with regard to the parasitized hosts. One has to also be aware that in the wild population, substantial imbalance in the numbers of available hosts of different species might constraint variation in cuckoo phenotypes: in the case of one prevalent host species and one or more other host species present at substantially lower numbers, cuckoos might evolve egg phenotypes that match the egg phenotype of the most common host [12,14]. Thus, the nature of observed intra- and inter-gentes variation in egg phenotypes, together with host species frequencies, are important when considering different evolutionary scenarios shaping cuckoo egg mimicry.

Here our goal was to test whether cuckoo eggs found in sympatrically occurring (Eurasian) reed warbler (*Acrocephalus scirpaceus*) and great reed warbler (*A. arundinaceus*) nests differ in size, colouration and the strength of the eggshell. Cuckoos parasitizing these two species are assigned to either the *Sylvia* egg morph or the *Acrocephalus* morph [8,15,16]. In the studied population, based on visual identification, both host species are parasitized with the same *Sylvia* morph [17,18]. Although exploited by one cuckoo egg morph, the eggs of these two host species are clearly distinct in terms of size and coloration. Reed warblers are also less vigilant towards cuckoos than great reed warbles and more readily accept cuckoo eggs [17]. In an egg discrimination experiment carried out in the study population, great reed warblers rejected 13.6% ($n = 44$ nests) of alien mimetic eggs (an egg of the host species taken from another nest) and reed warbler 3.6% ($n = 28$) of alien mimetic eggs. Similar statistics calculated for alien non-mimetic eggs (a conspecific egg painted plain brown or blue) showed 92.9% ($n = 70$) rejections in great reed warbler and 61.8% ($n = 204$) in reed warbler. Thus, clear differences in egg morphology and anti-parasite behaviour between these two hosts species provide a suitable case to study parasite egg mimicry strategies. In this paper, we try to disentangle whether the cuckoo (i) employs the averaging strategy to match the continuum of egg appearance between these two host species, (ii) shows a better match to one of the species (possibly more frequently utilized) or (iii) exhibits dichotomous specialization by laying eggs matching both host species. In our analysis, we employ morphological measurements, eggshell colour analysis [4,19], and eggshell strength measurements [20,21].

Materials and Methods

Field methods

The study was conducted from May to August in each of three consecutive years (2011–2013) in the population of two reed warbler species, located at the Stawy Milickie fishponds complex in southern Poland [17,22]. In this area reed warbler densities reach 57 breeding pairs/ha. We attempted to locate all reed warbler and great reed warbler nests in the area. Their locations were obtained through systematic search in reed-beds. Cuckoo eggs were identified during egg-laying and subsequently collected before hatching and frozen until further analysed. Shells of hatched eggs were also collected if possible. Additionally, in 2012 we collected single eggs from different nests of each of the two host species (15 of great reed warbler and 11 of reed warbler).

Egg measurements

All collected eggs were thawed and opened after the field season to collect cuckoos' genetic material. Prior to this, we photographed and colour-measured all collected eggs. Identical measurements were also done with respect to host-species eggs. Photographing was done using the Canon EOS 450D camera with a 17–70 mm lens. Photographs were taken at the focal length of approximately 25 mm with automatic shutter and iris settings. Photographs were used solely for measuring physical dimension of eggs. Colour measurements were performed using the Ocean optics JAZ portable spectrophotometer with xenon pulsed-light source (100 pulses per second) and a bifurcating optic fibre. With the probe held perpendicular to the shell surface, we lit the shell surface from a distance of 3 mm (which resulted in a lit area of approx. 3 mm^2). We measured each egg 12 times at equally-spaced spots on their surfaces. Both photographing and spectrophotometry were performed in different light conditions and at different locations each year, which might contribute to the naturally observed year differences – however all eggs were treated identically within each study year. Since all eggs were measured after the respective field seasons, their colour might change slightly due to chemical degradation of egg-shell compounds [19,23]. Unfortunately, we cannot assign precise collection rates to the eggs. We assume that the possible effects of prolonged storage (if any) would not be confounded with host-species identities, and thus would not affect subsequent comparisons. Importantly, all eggs were kept frozen in the darkness until assayed, which should further reduce any colour-related deterioration of eggshells.

Reflectance spectra obtained with JAZ were further analysed in the *pavo* package ver. 0.3–1 [24] in R ver. 2.15.0 [25]. After averaging the spectra within each measured egg and smoothing the resulting curves, we analysed resulting spectra using the average bird visual system, employing functions implemented in the *pavo* package. Briefly, individual spectra were transformed into cone receptor quantum catches (Q_i) defined as:

$$Q_i = \int_{\lambda} R_i(\lambda) S(\lambda) I(\lambda) d\lambda,$$

taking into account i-th receptor sensitivity, colour reflectance spectrum S and illuminant spectrum I. Quantum catches of each receptor where then transformed into Cartesian coordinates defining the position of each data point in a tetrahedral colour space spanned on the vertices of a tetrahedron (vertices represent full relative stimulations of red, blue, green and UV receptors). In this representation, the central point of the tetrahedron (equal to [0, 0] of the Cartesian coordinates system) is the achromatic point of black/white colour. Hue in this system is defined using two angles: theta θ defined on the RGB plane describes the human-visible part of the hue, whereas phi φ represent the deviation in the z-axis direction, measuring the UV component [26]. Chroma of the colour can be defined as the Euclidean distance r from this achromatic centre. In subsequent analyses, we used both θ and φ as hue measures, and r and achieved r_a (percent of maximum possible chroma for a given hue) as measures of chroma. Additionally, as tetrahedral-space colour model does not provide a measure of brilliance or brightness, we extracted mean brightness for each spectrum, defined as:

$$Br = \frac{\sum_{\lambda_{min}}^{\lambda_{max}} R_\lambda}{n},$$

which is the mean reflectance R over all n recorded wavelengths.

We also used generated quantum catches to analyse differences between eggs according to the receptor noise model, which returns colour distances between measured samples accounting for receptor-related noise in signal perception [27]. Colour contrasts (chromatic and achromatic) were generated for pairs of reed warbler vs. great reed warbler eggs, cuckoo eggs from reed warbler nests vs. cuckoo eggs from great reed warbler nests, and finally for reed warbler/great reed warbler vs. cuckoo eggs. Contrasts are expressed in the units of just noticeable differences (JND, see [24,27] and were generated assuming four colour receptors (long-, medium-, short- and very-short-wavelengths) with relative receptor densities characteristic for a blue tit retina [5,24].

Egg photographs were used to obtain their physical measurements (length and width in the widest place). Dimensions were measured to the nearest 0.1 mm in the ImageJ software [28]. These variables were then used in calculations of egg volumes (we used a commonly applied formula: $0.498 \times$ length $\times$ width^2 [29]; volume was then expressed in cm^3).

Additionally, all cuckoo eggs collected in 2011 were subject to shell-strength measurements. Briefly, eggshells were analysed using a pressure-sensitive specimeter that measures force and work required to break the shell. Shell resistance to breaking was analysed at a number of spots (sharp end, blunt end, side of the egg); both inward and outward breaking force and work were considered. Respective parameters were measured to the nearest 0.001 N (force) and 0.00001 J (work).

Statistical analyses

All measured variables for cuckoo eggs were analysed using a simple linear model with year and host species as fixed effects (*lm* procedure in R). Egg-thickness measures (available only for one year) were analysed using a two-tailed Student t test (*t.test* procedure in R). In all tests $\alpha = 0.05$ was used as the acceptable threshold type-I error. Prior to analyses, all variables were centered and standardized to have zero mean and unity variance. We ensured that all fitted models returned homoscedastic and approximately normal residuals.

All measured colour variables together with egg width, length, volume and shape, transformed for normality if necessary (see above) and standardized, were aggregated into principal components using the *prcomp* procedure in R. The first two components were used to represent each egg in a multivariate colour space to depict the presence or absence of clustering of eggs within the groups of two host species. This procedure was applied both to the cuckoo eggs and host species eggs for comparison. The resulting graph depicts the joint variation of reed warbler and cuckoo eggs in all measured variables. Additionally we provide a projection plot depicting positions of all eggs in the tetrahedral colour space projected on the circumscribed sphere of the tetrahedron. To visualize variation in hues/chromas and areas of overlapping colours, we generated in *pavo* convex hulls encapsulating respective groups of points in the tetrahedral colour space. We provide both numerical outputs from these analyses (proportions of overlapping hulls) and graphical depictions.

In total we collected and measured 15 cuckoo eggs from great reed warbler nests and 47 cuckoo eggs from reed warbler nests. Twelve eggs (great read warbler – 4, reed warbler – 8) were measured for shell-thickness.

Ethics statement

This study was carried out under the license granted by the division of the Polish Bioethical Committee at Wroclaw University (to AD). The license covered collection of the eggs, egg measurements and subsequent procedures. All procedures were performed so that they minimized the stress exhibited by birds during egg sampling and monitoring.

Results

We found no differences in shell strength in cuckoo eggs collected in nests of the two host species (Table S1). In particular, we found no significant differences in either the mean inward or outward breaking force (inward: $t_{df=6.81} = 0.13$, $P = 0.9$; outward: $t_{df=5.01} = 0.48$, $P = 0.65$). Eggs laid by cuckoos into nests of the two host species did not differ in their shape index (one means perfect sphere): $t_{df=59} = 0.01$, $P = 0.98$. Overall, size variables and colour measures did not indicate any differentiation (Table 1, Figure 1a).

Host species significantly differed with respect to several egg-morphology variables. We found significant differences in shell brightness ($t_{df=20.76} = 2.67$, $P = 0\ 01$), RGB hue θ ($t_{df=12.48} = 3.77$, $P = 0.002$), realized chroma r ($t_{df=17.68} = 5.65$, $P = 0.005$), and achieved chroma r_a ($t_{df=15.57} = 2.47$, $P = 0.03$); no significant differences in UV hue φ were found ($t_{df=15.44} = 0.34$, $P = 0.73$). Great reed warbler eggs are, on average, brighter, more saturated in the blue range, and generally lack the red hue (Figure 1b). Moreover, compared to reed warbler eggs, they are longer ($t = 9.97$, $df = 19.60$, $P < 0.0001$) and wider ($t = 11.38$, $df = 21.49$, $P < 0.0001$), which also translates into larger volume ($t = 14.29$, $df = 21.99$, $P < 0.0001$). Colour differences are also apparent in the tetrahedral colour space. The two host species clearly occupy separate clusters (Figure 2, Table 2), and the convex hulls surrounding colour measurements do not overlap.

Cuckoos' eggs occupy the space between the two host species, although they seem to group closer to the reed warbler eggs. This indicates that cuckoo eggs are closer to reed warbler eggs in terms of hue and chroma. This is further confirmed by a substantial overlap of convex hulls of cuckoo and reed warbler eggs (Table 2, Figure 2). In contrast, regions occupied by cuckoos' and great reed warblers' eggs do not overlap (Figure 2), however they are very close to each other. For clarity we did not present separate analogous analyses of overlap for cuckoos split by the host species – they occupy virtually the same region in the colour space and overlap extensively (over 90%).

Tetrahedral colour space representation does not account for differences in brightness. We have thus represented all measured eggs using principal components derived using all colour (i.e. hue, chroma and brightness) and size variables. Differences between host species are apparent after representing all measured host species eggs on the plane defined by the first two principal components (Figure 3). The two host species form two clearly separated clusters of points indicating they are clearly differentiated. An analogous plot representing all measured cuckoo eggs (colour and size variables) indicates that they form a uniform, non-differentiated population of points, with the variation among cuckoo eggs being substantially larger than within host species and clearly overlapping with variation in both host species (Figure 3). The first principal components represent mainly physical dimensions, chroma and RGB hue, whereas second components represent mainly brightness, RGB and UV hues (see Table S2 for loadings of all variables on the first two components). The first principal component explained 51% of total variance in eggs' morphology and the second component explained 34% of

Table 1. Comparison of differences in colour-related and size-related variables for cuckoo eggs laid in nests of reed warblers (rw) and great reed warblers (grw).

Variable name	Trait value ± SE (rw)	Trait value ± SE (grw)	t	P
Br	31.00±2.54	26.31±3.91	0.99	0.32
θ	2.80±0.04	2.81±0.06	0.20	0.84
φ	−0.48±0.06	−0.51±0.09	0.25	0.80
r	0.12±0.01	0.11±0.01	0.55	0.59
r_a	0.35±0.02	0.35±0.04	0.07	0.94
Length	22.10±0.16	22.20±0.16	0.28	0.78
Width	16.54±0.12	16.66±0.12	0.49	0.62
Volume	3.02±0.05	3.08±0.05	0.48	0.63

Estimates were obtained from linear models and are corrected for the effect of study years. All tests are performed with $df=59$.

variance. Correlations between original variables included in the PCA are provided in Table S3.

A receptor-noise model (Vorobyev & Osorio) further confirmed the above observations (Table 3). The average chromatic contrast between cuckoo eggs and great reed warbler was similar to the chromatic contrast between great reed warbler and reed warbler. On the other hand, the achromatic contrast between cuckoo eggs and either of the two host species appeared to be similar in value and slightly higher than the contrast between the host species. Both chromatic and achromatic contrasts between cuckoo eggs found in nests of reed warbler and great reed warbler were not significantly higher than contrasts calculated within the host-species groups (all $P>0.05$), supporting a lack of any differentiation among cuckoo eggs.

Discussion

Our study confirms clear differentiation in egg appearance between reed warbler and great reed warbler. The eggs differ in size and colour spectra forming two distinct clusters in the principal-components analysis. Despite these clear differences between hosts, we failed to show clear clustering among cuckoos parasitizing these two species with respect to their egg morphologies. A similar conclusion comes from a study performed on four sympatrically occurring warbler species in which no differences in egg colouration were found between cuckoo eggs from nests of those host species [8]. More importantly, our data show that cuckoo eggs exhibit substantial variability that – in specific morphology components – largely overlaps the variance observed in both species. Interestingly, this overlap is less apparent when analysing only hue and chroma. In terms of chromatic components of egg colouration, all sampled cuckoo eggs were more

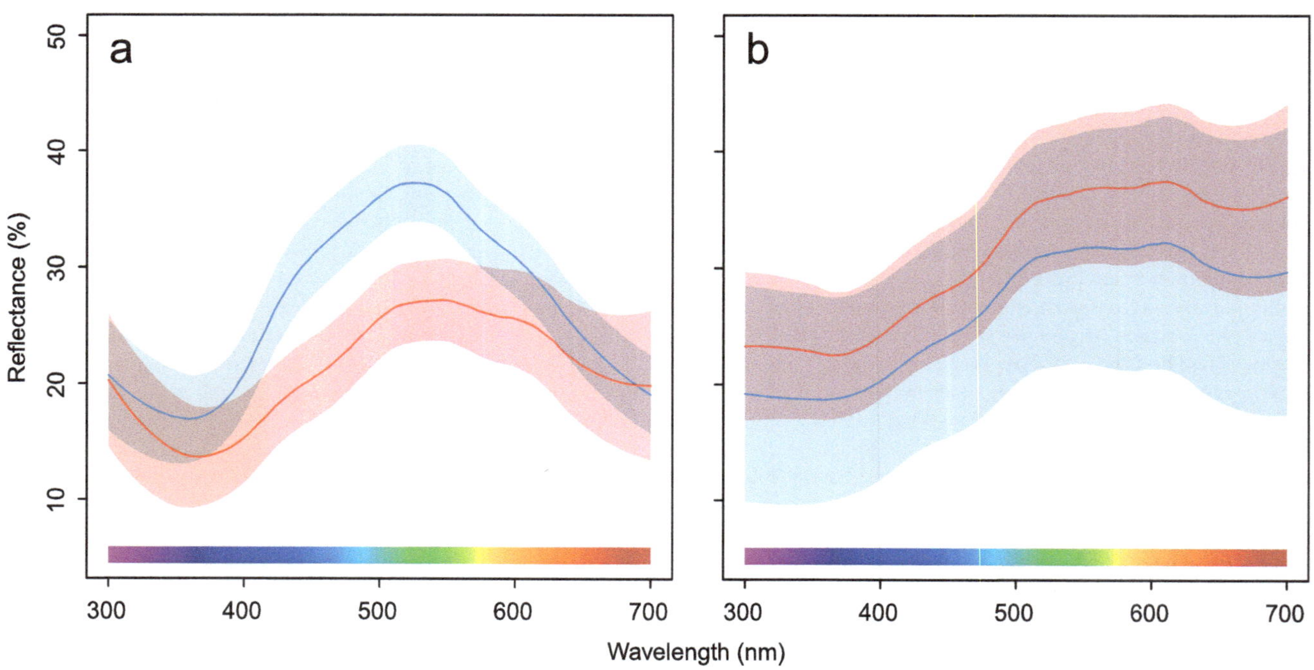

Figure 1. Average smoothed reflectance spectra of warbler eggs (a) and cuckoo eggs (b). Blue lines: eggs from great reed warbler nests; red lines: eggs from reed warbler nests. Bands represent 95% confidence regions.

Table 2. Summary of colour variables measured in the tetrahedral colour space for three analysed species.

Species	h_m	mean r_a	Convex hull overlap [%] with		
			A.arundinaceus	*A.scirpaceus*	*C.canorus*
A.arundinaceus	0.033	0.390	–	–	–
A.scirpaceus	0.047	0.325	0	–	–
Cuckoo	0.071	0.351	0	47	–

For each species we present: mean hue span (h_m), average saturation (mean r_a), and overlap of convex hulls calculated as: *volume_of_overlap/volume_of_smaller_hull* (thus the value of 1 would indicate perfect inclusion of the smaller hull in the bigger one; see [24] for more details).

similar to reed warbler host eggs than to great reed warbler eggs. However, achromatic colour components of cuckoo eggs not only exhibited larger variation, but they also overlapped both host species egg phenotypes. To our knowledge this is the first observation of such variation overlapping multiple hosts in the cuckoo brood parasitic system.

These results clearly indicate that sampled cuckoo eggs comprised phenotypes closely matching both host species as well as a range of intermediate phenotypes. This is very interesting since the proportion of parasitized nests clearly differs between these two host species, so one may expect cuckoo eggs to show a better match to the more frequently utilized host species. During the study, only 5.8% of nests of great reed warbler were parasitized

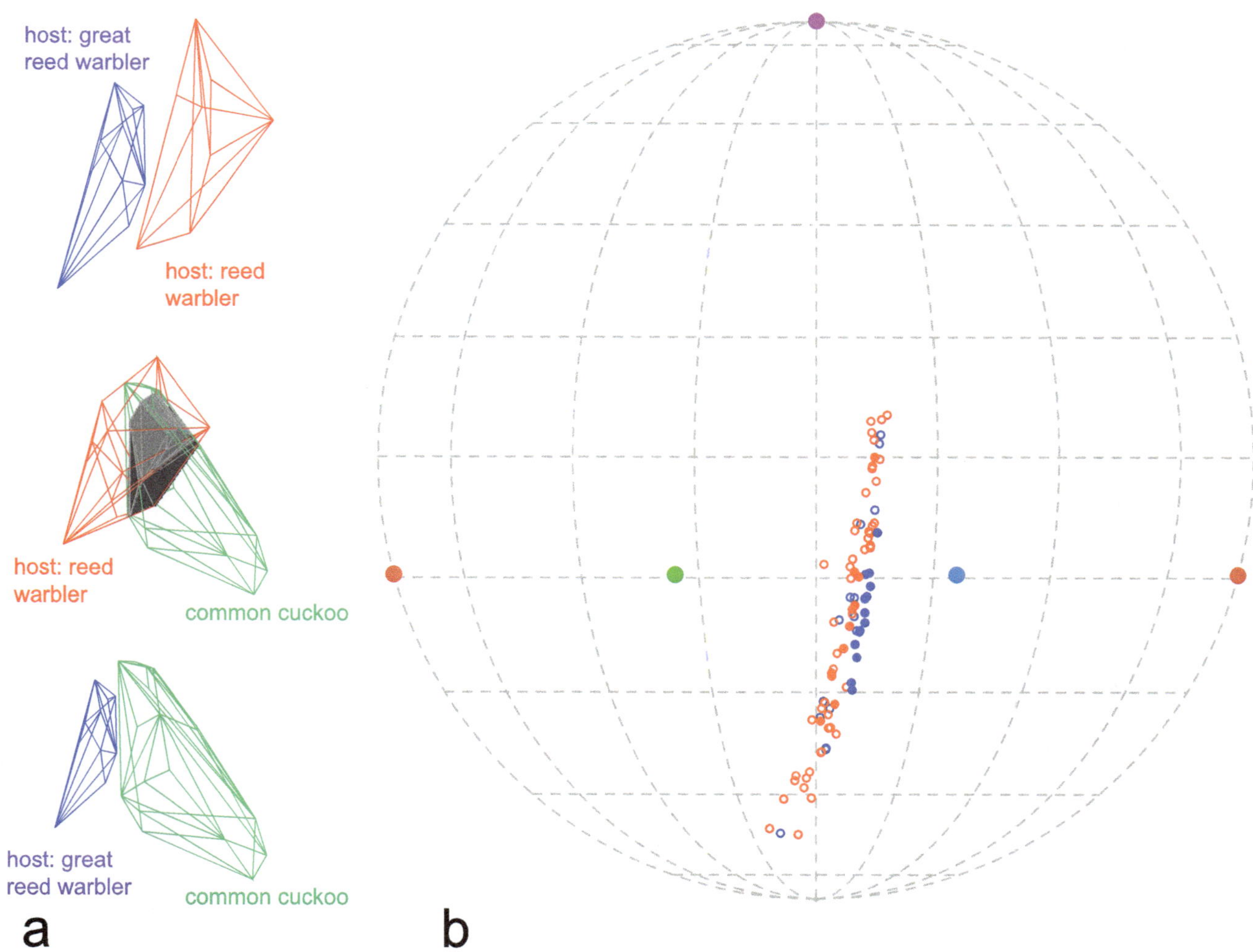

Figure 2. Cuckoo and host eggs in tetrahedral bird average visual colour space. (a) Overlap of convex hulls (grey polyhedron) encapsulating point measurements of host (red and blue) and cuckoo (green) eggs. Second and third plot are presented in identical orientations for ease of comparison and rotated to maximize the visibility of differences on the third plot (i.e. where there is no overlap). (b) Projection of colour measurement on a circumscribed sphere of the colour space tetrahedron; open symbols – cuckoos, filled symbols – hosts, red – reed warbler nests, blue – great reed warbler nests.

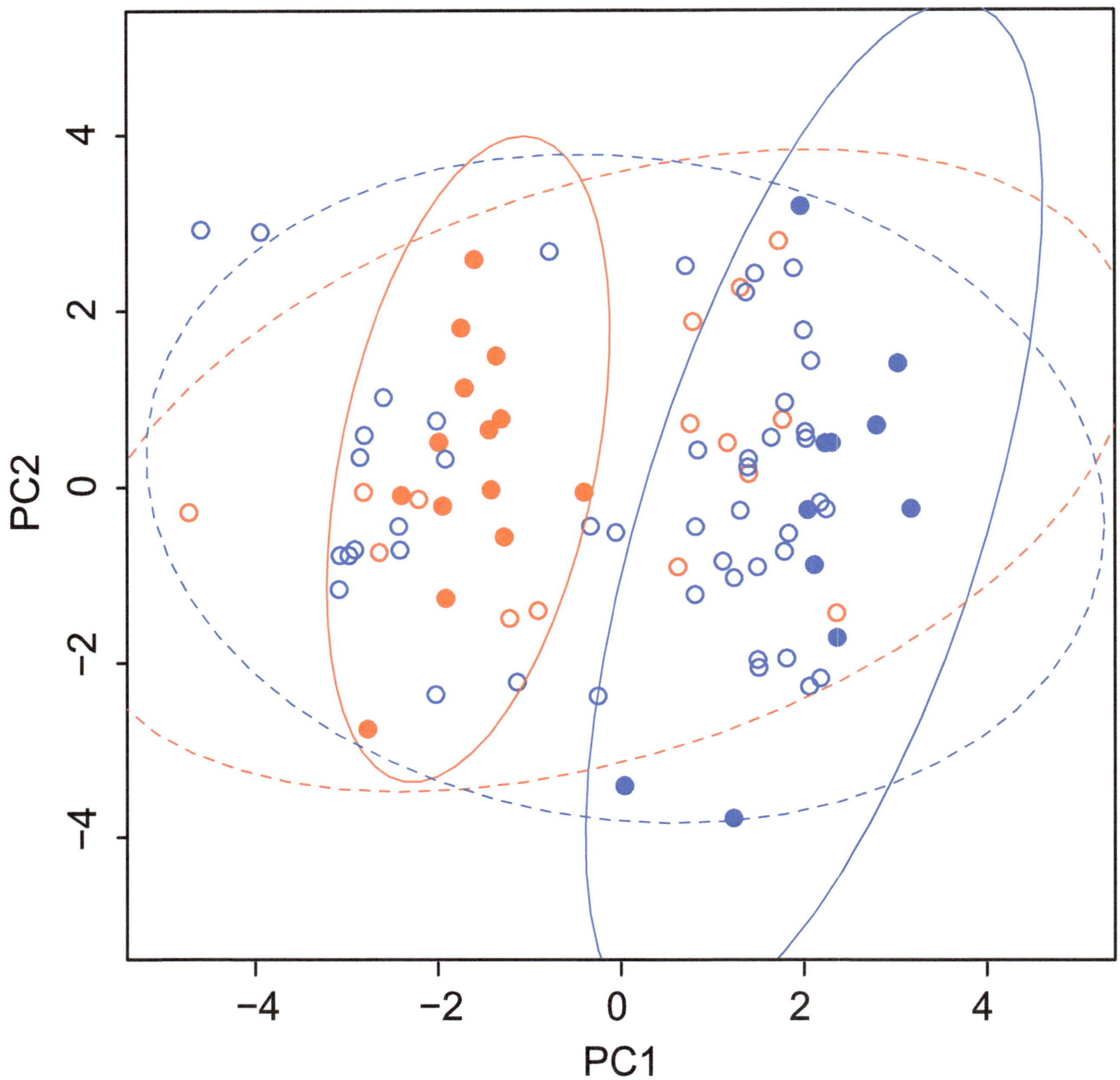

Figure 3. The first two principal components describing colour and size of cuckoo and warbler eggs. Blue symbols - great reed warbler eggs (filled) or cuckoo eggs found in great warbler nests (open); red symbols - reed warbler eggs (filled) or cuckoo eggs from reed warbler nests (open). Lines represent bivariate 95% confidence ellipses. Line colours correspond to circle colours. Solid lines - host species; dashed lines - cuckoos.

while 11.7% of reed warbler nests were parasitized. Moreover, our long-term observations suggest that great reed warbler has started to be utilized as a potential host only recently. Cuckoo eggs were found in only 2 out of over 700 nests of great reed warbler recorded in the years 1970–2006 [18]. In 2007 – out of 154 nests, 4 contained cuckoo eggs (2.6%); in 2009 – out of 136 nests, 6 cuckoo eggs were found (4.4%). In 2012–2013, 20 cuckoo eggs were found in 356 nests of the great reed warbler (5.6%). Although these proportions are similar to some other European populations (e.g. 5.3% in Moravia (Czech Republic); [30]), they are at the same time in contrast to figures from a number of other locations (Hungary and Bulgaria: 64% and 40% nests parasitized, respectively [7,31]). These figures indicate that host switching may be an underestimated but important phenomenon in brood-parasites occurring in areas where several suitable hosts are present. Cases of host-switching have rarely been reported (but see Jelinek et al. 2014) – clearly, more research is required in this matter as host switching may represent an important component of continuously an evolving host-parasite system of brood-parasitic species. Our colour measurements are consistent with the process of cuckoos slowly adapting to the newly acquired host species – interestingly, this adaptation seems to occur firstly in achromatic components of colouration. The picture of some components of egg morphology specializing and acquiring mimicry faster than the others is an important indication that egg phenotypes are multivariate entities, with their individual components undergoing

Table 3. Receptor-noise model output for chromatic and achromatic contrasts between cuckoos and host species.

Comparison	Chromatic contrast	Achromatic contrast
Host eggs		
GRW - RW	1.23±0.05	6.05±0.29
GRW - GRW	0.58±0.06	3.43±0.40
RW - RW	0.90±0.08	4.95±0.49
Cuckoo eggs		
CC_RW - CC_GRW	1.67±0.05	2.61±0.07
CC_RW - CC_RW	1.27±0.02	2.09±0.04
CC_GRW - CC_GRW	1.27±0.21	2.07±0.31
Host eggs vs. cuckoo eggs[1]		
CC - GRW	1.28±0.01	6.26±0.09
CC - RW	0.97±0.01	6.32±0.06

All values are in the units of just-noticeable differences (JND) and are presented with respective standard errors. Shortcuts: RW – reed warbler, GRW – great reed warbler, CC – common cuckoo (all eggs), CC_GRW – cuckoo eggs from great reed warbler nests, CC_RW – cuckoos eggs from reed warbler nests.
[1]Cuckoo eggs exhibited no significant differences in terms of respective chromatic and achromatic contrasts and thus all cuckoo eggs were compared with respective host species eggs.

partly independent evolutionary changes. Obviously our data lack the dynamic aspect as the egg morphologies have not been tracked in the past. However, the existence of partial specialization in some morphology components and its complete absence in the others suggests that our system may represent and early stage of acquiring a new host, which would be consistent with patterns of egg-morph variability predicted by numerical models [11,13]. Whether the observed patterns represent a genuine snapshot from a dynamic trajectory of egg-colouration traits evolving in response to host-switching remains to be confirmed. It is possible, that the observed variation is a part of a larger pattern of egg morphological variation. Numerical simulations indicate that host egg morphologies can fluctuate and follow a cyclic patterns changing relatively rapidly [32]. It is possible that similar dynamics occur in co-evolving brood parasites; long term tracking of egg morphologies in brood parasitic systems should allow these exciting questions to be answered. Importantly such tracking would not only put evolving parasitic eggs in the appropriate temporal scale – it would also reveal how parasites shape the evolution of host eggs. It has been recently shown, that egg mimicry can have profound effects on the evolution of host egg phenotype [33–35]. In our system a substantial difference exists between the two close cuckoo hosts – further, long-term studies are needed to see how these differences are maintained and evolve in the absence of cuckoo specialization.

The sole presence of substantial variance in cuckoo egg phenotypes suggests that this trait has large evolutionary potential. Rapidly evolving egg mimicry has been recently shown among cuckoos parasitizing reed warblers in Zealand (Denmark) [36]. The authors observed a significant increase in similarity of egg coloration of the cuckoo and their reed warbler host over a 24-year study period. However, the substantial variance in cuckoo egg phenotypes may constitute an adaptation on its own. It is possible that benefits related to the ability to exploit other, less typical host species may overcome any potential costs of failure due to imperfect egg mimicry.

Our analyses focus on size and background colouration of cuckoo eggs, ignoring possible differences in the spotting pattern of cuckoo eggs. However, studies published to date have usually demonstrated background colouration to be the most important trait differentiating gentes [4,37]. In contrast, studies considering spottiness of eggs failed to show any differences between gentes [36,38]. Similarly, Antonov et al. [39] found no significant differences in egg spotting between cuckoos parasitizing different hosts, in spite of clear differences in the analysed host species (great reed warbler, marsh warbler and corn bunting). We believe that ignoring egg spottiness in our study does not significantly bias our inference.

Lack of any differentiation in cuckoo egg phenotypes with respect to two parasitized host species was also apparent in the case of shell strength. It is well known that cuckoo eggs are equipped with strong egg shells; a number of hypotheses have been suggested to explain this observation [20,21,40]. One hypothesis is that stronger shells protect cuckoo eggs from being punctured and ejected from host nests. In our study, we predicted that cuckoo eggs laid in nests of great reed warblers would have stronger shells, which would be an adaptation to parasitizing hosts with larger and stronger beaks. We found no evidence of such differentiation. The sample size used for shell thickness measurements was relatively small, however given observed effect sizes, it is unlikely that lack of observed differentiation is a matter of power. Also, in this study we used dynamometry, whereas in a number of studies egg shell thickness was measured as a reliable proxy of egg shell strength [41,42]. In our opinion, the dynamometric approach is more appropriate as it more directly relates measured characteristics to the way the egg shell affects host anti-parasite responses and cuckoo hatching processes [20,21]. Lack of observed differences between cuckoos parasitizing hosts that differ greatly in their egg puncturing potential may on one hand suggest that the puncture-resistance hypothesis is not adequate in this system. On the other hand, it is possible that eggshell thickness is less evolutionarily labile than egg colouration and does not differentiate so easily with respect to parasitized hosts.

Another potentially confounding aspect of our study which cannot be properly accounted for is the identity of cuckoo females. Obviously, eggs laid by the same females should be expected to be more similar to each other than to eggs laid by other unrelated females. This issue would be particularly important in our study if specific females laid eggs in nests of different host species. Cuckoo females are believed to be specific with respect to chosen host-species although there are rare cases of females laying eggs in nests of multiple host species [1]. Unfortunately, genetic material gathered in this study does not allow for robust and confident identification of cuckoo eggs coming from specific females: the uncertainty of assigning parents deteriorates rapidly if neither of the parental genotypes is known. However, GPS coordinates collected for all sampled nests indicate that the average distance between sampled host nests in this study was 9 km. According to Davies (2011) [1] (see also [43]), the (mostly non-overlapping) breeding areas of female cuckoos range from 0.21 to 1.68 km^2, resulting in area radii ranging from 250 to 730 m. Even after assuming 730 m as a conservatively large territory size and excluding from the analysis all nests involved in pairs with nest-to-nest distances less than $2 \cdot 730 = 1460$ m (9 nests), we did not find evidence for the presence of significant differentiation in egg phenotypes of cuckoos laying eggs in nests of different hosts. Thus, we assume that our analyses are robust despite the lack of female identities.

Finally, potential bias in our analyses may also be related to egg collection procedure. It is well recognized that hosts tend to reject more dissimilar eggs at higher rates [44,45], and thus dissimilar eggs might be underrepresented in our data set if such cuckoo eggs disappeared prior to egg collection. This problem should be accounted for while interpreting the data, but it is extremely difficult to validate this source of bias. This would be particularly important if we find that the host species showing higher rejection rates matches cuckoo eggs better that the host species showing lower rejection rates. In fact, our data show a rather opposite pattern, so this potential bias is unlikely to alter our interpretation.

In conclusion, our data do not support the hypothesis of specialization of the common cuckoo in parasitizing two species of reed warblers. In the studied population, cuckoos parasitizing two studied hosts seem to lay phenotypically indistinguishable eggs. Our results directly contrasts with other published studies looking at cuckoos exploiting sets of sympatric hosts [5,36]. We failed to find significant differences in egg morphology between cuckoos parasitizing these two species despite clear differentiation between host species. Instead, observed colouration patterns strongly depend on the colour components studied. Chromatic components clearly cluster cuckoo eggs close to the reed warbler eggs, supporting the expectation that cuckoos should match most efficiently the most prevalent host species in a given area. However, achromatically, cuckoo eggs exhibit much more variation that spreads continuously between achromatic components of egg morphology of both host species. Finally, as indicated by the subspaces occupied by cuckoo egg phenotypes in the tetrachromatic colour space, ample amount of chromatic variation in cuckoo eggs spreads in a direction opposite to either of the two available host species. The cuckoo from this gens may exploit other host species present in the study area, and this may explain this additional variation in cuckoo egg morphology. Further studies should focus on analysing cuckoo egg phenotypes found in other sympatrically occurring host species. It may be particularly interesting to explore selection regimes that may lead to homogenising of cuckoo egg phenotypes even in the presence of clearly distinct host species egg phenotypes. The studied population is a dynamic system, where host species frequencies undergo continuous changes. Our study demonstrates that in colouration studies, a multivariate approach is the one that yields more realistic results. Ignoring size and brightness in our study would clearly favour greater similarity of cuckoo eggs to those of the reed warbler. However, after including brightness and physical dimensions, the variation in egg phenotypes spreads much wider and may overlap phenotypic values of the other host species. Future studies should look more closely at such patterns. In particular, it seems important to establish the relative importance of different aspects of egg morphology (size, brightness, saturation, hue) in the evolution of host egg discrimination and parasite egg mimicry. The amount of observed differentiation may depend on the choice and assembly of characteristics used in such studies. Different aspects of egg morphology may evolve at different rates in brood-parasitic species, substantially biasing simple univariate comparisons of specific traits.

Acknowledgments

We thank two anonymous reviewers for their valuable comments that helped to improve the manuscript. We are also thankful to Kimberly S. Sheldon for useful comments and English editing. Finally, we would like to thank two filed assistants – Monika Czuchra and Wanda Zdunek – for help in searching for new reed warblers' nests.

Author Contributions

Conceived and designed the experiments: AD MC SMD JS. Performed the experiments: AD JS. Analyzed the data: SMD MC JS. Contributed reagents/materials/analysis tools: AD MC SMD. Wrote the paper: AD MC JS SMD.

References

1. Davies N (2011) Cuckoos, cowbirds and other cheats. Poyser Monographs. New York: Princeton University Press. 328p.
2. De L. Brooke M, Davies NB (1988) Egg mimicry by cuckoos *Cuculus canorus* in relation to discrimination by hosts. Nature 335: 630–632. doi:10.1038/335630a0.
3. Moksnes A, Røskaft E (1995) Egg-morphs and host preference in the common cuckoo (*Cuculus canorus*): an analysis of cuckoo and host eggs from European museum collections. J Zool 236: 625–648. doi:10.1111/j.1469-7998.1995.tb02736.x.
4. Stoddard MC, Stevens M (2011) Avian vision and the evolution of egg color mimicry in the common cuckoo. Evol Int J Org Evol 65: 2004–2013. doi:10.1111/j.1558-5646.2011.01262.x.
5. Honza M, Šulc M, Jelínek V, Požgayová M, Procházka P (2014) Brood parasites lay eggs matching the appearance of host clutches. Proc Biol Sci 281: 20132665. doi:10.1098/rspb.2013.2665.
6. Gibbs HL, Sorenson MD, Marchetti K, Brooke MD, Davies NB, et al. (2000) Genetic evidence for female host-specific races of the common cuckoo. Nature 407: 183–186. doi:10.1038/35025058.
7. Fossøy F, Antonov A, Moksnes A, Røskaft E, Vikan JR, et al. (2011) Genetic differentiation among sympatric cuckoo host races: males matter. Proc Biol Sci 278: 1639–1645. doi:10.1098/rspb.2010.2090.
8. Edvardsen E, Moksnes A, Røskaft E, Øien IJ, Honza M (2001) Egg Mimicry in Cuckoos Parasitizing Four Sympatric Species of *Acrocephalus Warblers*. The Condor 103: 829. doi:10.1650/0010-5422(2001)103[0829:EMICPF]2.0.CO;2.
9. Honza M, Taborsky B, Taborsky M, Teuschl Y, Vogl W, et al. (2002) Behaviour of female common cuckoos, *Cuculus canorus*, in the vicinity of host nests before and during egg laying: a radiotelemetry study. Anim Behav 64: 861–868. doi:10.1006/anbe.2002.1969.
10. Takasu F, Moskát C, Muñoz AR, Imanishi S, Nakamura H (2009) Adaptations in the common cuckoo (*Cuculus canorus*) to host eggs in a multiple-hosts system

of brood parasitism. Biol J Linn Soc 98: 291–300. doi:10.1111/j.1095-8312.2009.01288.x.

11. Takasu F (2003) Co-evolutionary dynamics of egg appearance in avian brood parasitism. Evol Ecol Res 5: 345–362.
12. Yang C, Liang W, Cai Y, Shi S, Takasu F, et al. (2010) Coevolution in Action: Disruptive Selection on Egg Colour in an Avian Brood Parasite and Its Host. PLoS ONE 5: e10816. doi:10.1371/journal.pone.0010816.
13. Takasu F (2005) A theoretical consideration on co-evolutionary interactions between avian brood parasites and their hosts. Ornithol Sci 4: 65–72. doi:10.2326/osj.4.65.
14. Kleven O, Moksnes A, Røskaft E, Rudolfsen G, Stokke BG, et al. (2004) Breeding success of common cuckoos *Cuculus canorus* parasitising four sympatric species of *Acrocephalus* warblers. J Avian Biol 35: 394–398. doi:10.1111/j.0908-8857.2004.03359.x.
15. Moskat C, Takasu F, Munoz AR, Nakamura H, Ban M, et al. (2012) Cuckoo parasitism on two closely-related *Acrocephalus* warblers in distant areas: a case of parallel coevolution. Chin Birds 3: 320–329.
16. Honza M, Moksnes A, Røskaft E, Stokke BG (2001) How are different Common Cuckoo Cuculus canorus egg morphs maintained? An evaluation of different hypotheses. Ardea 89: 341–352.
17. Dyrcz A, Hałupka L (2006) Great reed warbler *Acrocephalus arundinaceus* and reed warbler *Acrocephalus scirpaceus* respond differently to cuckoo dummy at the nest. J Ornithol 147: 649–652. doi:10.1007/s10336-006-0097-x.
18. Dyrcz A, Hałupka K (2007) Why Does the Frequency of Nest Parasitism by the Cuckoo Differ Considerably Between Two Populations of Warblers Living in the Same Habitat? Ethology 113: 209–213. doi:10.1111/j.1439-0310.2006.01308.x.
19. Cassey P, Thomas GH, Portugal SJ, Maurer G, Hauber ME, et al. (2012) Why are birds' eggs colourful? Eggshell pigments co-vary with life-history and nesting ecology among British breeding non-passerine birds. Biol J Linn Soc 106: 657–672. doi:10.1111/j.1095-8312.2012.01877.x.
20. Antonov A, Stokke B rd G, Moksnes A, Kleven O, Honza M, et al. (2006) Eggshell strength of an obligate brood parasite: a test of the puncture resistance hypothesis. Behav Ecol Sociobiol 60: 11–18. doi:10.1007/s00265-005-0132-6.
21. Honza M, Picman J, Grim T, Novak V, Capek Jr M, et al. (2001) How to hatch from an egg of great structural strength. A study of the Common Cuckoo. J Avian Biol 32: 249–255. doi:10.1111/j.0908-8857.2001.320307.x.
22. Dyrcz A (1981) Breeding ecology of Great Reed Warbler *Acrocephalus arundinaceus* and Reed Warbler *Acrocephalus scirpaceus* at fish-ponds in SW Poland and lakes in NW Switzerland. Acta Ornithol 18: 307–334.
23. Cassey P, Ewen JG, Blackburn TM, Hauber ME, Vorobyev M, et al. (2008) Eggshell colour does not predict measures of maternal investment in eggs of *Turdus* thrushes. Naturwissenschaften 95: 713–721. doi:10.1007/s00114-008-0376-x.
24. Maia R, Eliason CM, Bitton P-P, Doucet SM, Shawkey MD (2013) pavo: an R package for the analysis, visualization and organization of spectral data. Methods Ecol Evol 4: 906–913. doi:10.1111/2041-210X.12069.
25. R Development Core Team (2012) R: A Language and Environment for Statistical Computing.
26. Stoddard MC, Prum RO (2008) Evolution of avian plumage color in a tetrahedral color space: a phylogenetic analysis of new world buntings. Am Nat 171: 755–776. doi:10.1086/587526.
27. Vorobyev M, Osorio D (1998) Receptor noise as a determinant of colour thresholds. Proc Biol Sci 265: 351–358. doi:10.1098/rspb.1998.0302.
28. Rasband WS (2012) ImageJ: image processing and analysis in Java. Bethesda: National Institutes of Health.
29. Hoyt D (1979) Practical methods of estimating volume and fresh weight of bird eggs. The Auk 96: 73–77. Available: http://www.jstor.org/stable/4085401.
30. Moksnes A, Røskaft E, Bičík V, Honza M, Øien IJ (1993) Cuckoo *Cuclus canorus* parasitism on *Acrocephalus* Warblers in Southern Moravia in The Czech Republic. J Ornithol 134: 425–434. doi:10.1007/BF01639833.
31. Moskát C, Honza M (2002) European Cuckoo *Cuculus canorus* parasitism and host's rejection behaviour in a heavily parasitized Great Reed Warbler *Acrocephalus arundinaceus* population. Ibis 144: 614–622. doi:10.1046/j.1474-919X.2002.00085.x.
32. Liang W, Yang C, Stokke BG, Antonov A, Fossøy F, et al. (2012) Modelling the maintenance of egg polymorphism in avian brood parasites and their hosts. J Evol Biol 25: 916–929. doi:10.1111/j.1420-9101.2012.02484.x.
33. Stoddard MC, Kilner RM, Town C (2014) Pattern recognition algorithm reveals how birds evolve individual egg pattern signatures. Nat Commun 5: 4117. Available: http://www.nature.com/ncomms/2014/140618/ncomms5117/full/ncomms5117.html. Accessed 2014 Aug 2.
34. Spottiswoode CN, Stevens M (2011) How to evade a coevolving brood parasite: egg discrimination versus egg variability as host defences. Proc R Soc B Biol Sci 278: 3566–3573. doi:10.1098/rspb.2011.0401.
35. Lahti DC (2005) Evolution of bird eggs in the absence of cuckoo parasitism. Proc Natl Acad Sci U S A 102: 18057–18062. doi:10.1073/pnas.0508930102.
36. Avilés JM, Stokke BG, Moksnes A, Røskaft E, Asmul M, et al. (2006) Rapid increase in cuckoo egg matching in a recently parasitized reed warbler population. J Evol Biol 19: 1901–1910. doi:10.1111/j.1420-9101.2006.01166.x.
37. Spottiswoode CN, Stevens M (2010) Visual modeling shows that avian host parents use multiple visual cues in rejecting parasitic eggs. Proc Natl Acad Sci U S A 107: 8672–8676. doi:10.1073/pnas.0910486107.
38. Avilés JM, Stokke BG, Moksnes A, Røskaft E, Møller AP (2006) Environmental conditions influence egg color of reed warblers *Acrocephalus scirpaceus* and their parasite, the common cuckoo *Cuculus canorus*. Behav Ecol Sociobiol 61: 475–485. doi:10.1007/s00265-006-0275-0.
39. Antonov A, Stokke BG, Vikan JR, Fossøy F, Ranke PS, et al. (2010) Egg phenotype differentiation in sympatric cuckoo *Cuculus canorus* gentes. J Evol Biol 23: 1170–1182. doi:10.1111/j.1420-9101.2010.01982.x.
40. Rothstein SI (1975) An Experimental and Teleonomic Investigation of Avian Brood Parasitism. The Condor 77: 250–271.
41. Igic B, Hauber ME, Galbraith JA, Grim T, Dearborn DC, et al. (2010) Comparison of micrometer- and scanning electron microscope-based measurements of avian eggshell thickness. J Field Ornithol 81: 402–410. doi:10.1111/j.1557-9263.2010.00296.x.
42. Hargitai R, Moskat C, Ban M, Gil D, Lopez-Rull I, et al. (2010) Eggshell characteristics and yolk composition in the common cuckoo *Cuculus canorus*: are they adapted to brood parasitism? J Avian Biol 41: 177–185.
43. Nakamura H, Miyazawa Y, Kashiwagi K (2005) Behavior of radio-tracked Common Cuckoo females during the breeding season in Japan. Ornithol Sci 4: 31–41.
44. Cherry MI, Bennett ATD, Moskát C (2007) Host intra-clutch variation, cuckoo egg matching and egg rejection by great reed warblers. Naturwissenschaften 94: 441–447. doi:10.1007/s00114-007-0216-4.
45. Bártol I, Karcza Z, Moskát C, Røskaft E, Kisbenedek T (2002) Responses of great reed warblers Acrocephalus arundinaceus to experimental brood parasitism: the effects of a cuckoo Cuculus canorus dummy and egg mimicry. J Avian Biol 33: 420–425. doi:10.1034/j.1600-048X.2002.02945.x.

15

Group Selection as Behavioral Adaptation to Systematic Risk

Ruixun Zhang[1], Thomas J. Brennan[2], Andrew W. Lo[3,4]*

1 MIT Department of Mathematics, Cambridge, Massachusetts, United States of America, **2** Northwestern University School of Law, Chicago, Illinois, United States of America, **3** MIT Sloan School of Management, CSAIL, and EECS, Cambridge, Massachusetts, United States of America, **4** AlphaSimplex Group, LLC, Cambridge, Massachusetts, United States of America

Abstract

Despite many compelling applications in economics, sociobiology, and evolutionary psychology, group selection is still one of the most hotly contested ideas in evolutionary biology. Here we propose a simple evolutionary model of behavior and show that what appears to be group selection may, in fact, simply be the consequence of natural selection occurring in stochastic environments with reproductive risks that are correlated across individuals. Those individuals with highly correlated risks will appear to form "groups", even if their actions are, in fact, totally autonomous, mindless, and, prior to selection, uniformly randomly distributed in the population. This framework implies that a separate theory of group selection is not strictly necessary to explain observed phenomena such as altruism and cooperation. At the same time, it shows that the notion of group selection does captures a unique aspect of evolution—selection with correlated reproductive risk–that may be sufficiently widespread to warrant a separate term for the phenomenon.

Editor: Claude Wicker-Thomas, CNRS, France

Funding: Research support from the MIT Laboratory for Financial Engineering is gratefully acknowledged. The funders had no role in study design, data collection and analysis, decision to publish, or preparation of the manuscript.

Competing Interests: The authors have read the journal's policy and the authors of this manuscript have the following competing interests: Andrew W. Lo has the following affiliations and research grants: AlphaSimplex Group, LLC, Chairman and Chief Investment Strategist; U.S. Treasury (Office of Financial Research), Consultant; National Bureau of Economic Research, Research Associate; NY Fed Financial Advisory Roundtable, Member; FINRA Economic Advisory Committee, Member; Consortium for Systemic Risk Analysis, Academic Advisory Committee, Member; Beth Israel Deaconess Medical Center Board of Overseers, Member; research grant from the National Science Foundation on systemic risk; research grant from Citigroup; patent pending on cryptographic methods for computing systemic risk; patent awarded (U.S. Patent No. 7,599,876, "Electronic Market-Maker"); patent awarded (U.S. Patent No. 7,562,042, "Data Processor for Implementing Forecasting Algorithms"). According to PLOS ONE, a competing interest is "anything that interferes with, or could reasonably be perceived as interfering with, the full and objective presentation, peer review, editorial decision making, or publication of research or non-research articles submitted to one of the journals." None of the above affiliations are remotely involved with the subject of the submission to PLOS ONE, nor do they or the authors stand to gain from the publication of this manuscript in PLOS ONE. Moreover, these affiliations do not alter the authors' adherence to all the PLOS ONE policies on sharing data and materials.

* Email: alo@mit.edu

Introduction

Since the publication of the path-breaking work of Wynne-Edwards [1] and Hamilton [2,3] in the 1960s, the theory of evolution has been applied to much broader contexts than self-replicating genes. By appealing to notions of inclusive fitness and kin and group selection, compelling explanations for previously inexplicable behaviors such as altruism and cooperation have been developed. This approach has generated a number of additional insights such as reciprocity [4], the Price equation [5], sociobiology [6,7], and the theory of multi-level selection [8]. Moreover, empirical studies have lent further support to the theory of group and kin selection, including social behavior in bacteria [9–11], sterility in social insects [3,12], and the avoidance of cannibalism in salamanders [13].

The behavioral implications of group selection have also received considerable attention from economists, who have used evolutionary models to explain apparent conflicts between individual rationality and human behavior [14,15], including attitudes toward risk and utility functions [16,17], time preference [18], and financial markets [19,20]. As an alternative to the traditional view that "market prices fully reflect all available information" [21,22], the Adaptive Markets Hypothesis [23] provides an evolutionary interpretation of financial market dynamics.

Despite these applications, group selection is still one of the most hotly contested issues in evolutionary biology. An enormous body of research has been dedicated to understanding the relationship between genetic, organismic, and group selection [24–26], and the relationship between group and kin selection [27–32]. Critics over the last few decades have argued forcefully against group selection as a major mechanism of evolution, and recent attempts to revive it [30,33,34] have been met with swift and broad rebuke [35,36].

Here we propose a reconciliation of the two opposing perspectives by arguing that what appears to be group selection may, in fact, simply be the consequence of natural selection occurring in stochastic environments with reproductive risks that are correlated across individuals. Those individuals with highly correlated risks will appear to form "groups", even if their actions are, in fact, totally autonomous, mindless, and, prior to natural selection, uniformly randomly distributed in the population.

We illustrate our approach with the following simple example. Consider a population of individuals, each facing a binary choice between one of two possible actions, a and b, and suppose the environment consists of two possible states of nature, rain or sunshine, with probability 20% and 80%, respectively. If it rains, action a leads to 0 offspring for any given individual and action b leads to 3 offspring; if it shines, the reverse occurs and action a leads to 3 offspring while action b leads to 0 offspring. From an individual's perspective, choosing a will lead to more reproductive success on average given the higher likelihood of sunshine. However, if all individuals in the population behaved in this manner, and rain or sunshine occurred for all individuals at the same time, the first time that a negative environment appears, the entire population of individuals that always choose a will become extinct. If we assume that offspring behave identically to their parents (perfect transmission of traits across generations), the behavior "always choose a" cannot survive over time. In fact, we show below that the behavior with the highest reproductive success over time in this very specialized example is to randomize between a and b using the same probability as the probability of sunshine, 80%; the group of individuals exhibiting this probability-matching behavior achieves the maximum possible growth rate. As a result, it appears as though selection operates at the group level and that this group—all individuals i who randomize their actions with probability $p_i = 80\%$—is the fittest.

The key to this outcome is the fact that the reproductive risk facing all individuals in the population, rain or sunshine, is perfectly correlated, which we refer to as systematic risk. If we had assumed, instead, that reproductive risk was idiosyncratic–that the state of nature is independently and identically distributed (IID) for each individual–then the evolutionarily dominant strategy is, in fact, the purely "selfish" one in which a is chosen all the time.

This framework demonstrates that a separate theory of group selection is not strictly necessary to explain observed phenomena such as altruism and cooperation. But our results also show that the notion of group selection does capture a unique aspect of evolution—selection with correlated reproductive risk–that may be sufficiently widespread and distinct to warrant a separate term for the phenomenon. There is no controversy around the fact that selection occurs at the genetic level because of the basic biology of reproduction. However, we show that selection can also appear to operate at coarser levels if environmental forces affect a particular subset of individuals in similar fashion, i.e., if their reproductive risks are highly correlated. We use the term "appear" intentionally because in our framework, selection does not occur at the level of the group, but the behavior that is evolutionarily dominant is consistent with some of the empirical implications of group selection.

By studying the impact of selection on behavior rather than on genes, we are able to derive evolutionary implications that cut across species, physiology, and genetic origins. In the same way that different magnifications of a microscope reveal different details of a specimen, applying evolutionary principles to behavioral variations leads to different insights that may be more relevant for economics, psychology, and behavioral ecology. Because evolution is essentially a passive "process of elimination" [37], selection and adaptation operate at all levels, from genes to populations, depending on the nature of the corresponding environmental challenges. Our focus on behavior as the object of selection is a different lens through which the effects of evolution may be studied.

In the remainder of this paper, we provide a review of the literature, followed by an introduction to the binary choice model with two systematic environmental factors. In this framework, we derive a behavioral adaptation to stochastic environments with systematic risk in which groups seem to be the unit of selection, but which is purely the result of natural selection operating on individuals in the population. We conclude with a brief discussion of the results. A generalization of the binary choice model to the multinomial case with multiple factors is provided in the Supplementary Information, as well as all proofs and derivations.

Literature Review

The literature on evolution and behavior can be overwhelming, spanning the disciplines of evolutionary biology, ecology, evolutionary and social psychology, and economics. While a comprehensive survey is well beyond the scope of this paper, we attempt to provide a representative sampling of the many strands that are most relevant to our focus.

The role of stochastic environments in evolution has been investigated extensively by biologists and ecologists. Stochastic environments cause high genetic variation and, in the extreme cases, extinction [38–44]. Environmental uncertainty that is associated with stochasticity over time [45–47] or heterogeneity across space [48] can cause natural selection to favor a gene that randomizes its phenotypic expression. Gillespie and Guess [49] describe a heuristic method to study selection in random environments. Frank [50–52] analyzes how variability in reproductive success affects fitness and relates it to the geometric mean principle. Geometric-mean fitness has also appeared in the financial context as the "Kelly criterion" for maximizing the geometric growth rate of a portfolio [53–56]. However, the motivation for geometric-mean fitness in population biology is considerably more compelling than in financial investments because maximizing the geometric-mean return of a portfolio is optimal only for individuals with a very specific risk preference, i.e., those with logarithmic utility functions [54].

In the evolutionary biology literature, Maynard Smith [57] has developed the concept of an "evolutionarily stable strategy", specific behaviors that survive over time by conferring reproductive advantages or "fitness", typically measured by the rate of population growth. Using this notion of fitness, Fretwell [58], Cooper and Kaplan [59], and Frank and Slatkin [60] observe that randomizing behavior can be advantageous in the face of stochastic environmental conditions. The impact of variability in reproductive success among individuals in a population has been shown to yield a kind of risk aversion (which increases average reproductive success) and "bet-hedging" (which reduces the variance of reproductive success) [47,61–67]. Frank and Slatkin [60] propose a framework that highlights the importance of correlations among individual reproductive success in determining the path of evolution.

Similar results have been derived in the behavioral ecology literature, in which the maximization of fitness via dynamic programming has been shown to yield several observed behaviors, including risk-sensitive foraging in non-human animal species [68–79] and seed dispersal strategies in plants [80,81]. Recently, the neural basis of risk aversion has also received much attention as researchers discovered that the activity of specific neural substrates correlates with risk-taking and risk-averse behaviors [82–85].

The relationship between risk-spreading behavior and kin selection has also been considered. Yoshimura and Clark [86] show that the risk-spreading polymorphism, in which a given genotype consists of a mixture of two or more forms each employing different behavioral strategies [48,59], makes sense only for groups. Yoshimura and Jansen [87] argue that risk-spreading adaptation is a form of kin selection [59], and the strategies of kin can be very important in stochastic environments even if there are no interactions at all. Mcnamara [88] introduces the profile of a strategy and relates the geometric mean fitness to a deterministic game.

In the economics literature, evolutionary principles were first introduced to understand cooperation and altruism [89–91], and evolutionary game theory is considered a foundation for altruistic behavior [17,92]. Evolutionary models of behavior are especially important for economists in resolving conflicts between individual rationality and human behavior [14,18], including attitudes toward risk and utility functions [16,17,93–95], time preference [96–100], financial markets and firm selection [19,20,101,102], and the economic analysis of social institutions [103–105].

Evolutionary models of behavior have also been used to justify the existence of utility functions and to derive implications for their functional form [16,95,106] (see Robson [18] and Robson and Samuelson [99] for comprehensive reviews of this literature). For example, Robson [16] investigates expected and non-expected utility behaviors, and finds that randomized behavior may be optimal from a population perspective even though it is sub-optimal from an individual perspective [107,108]. Robson [95] argues that the kind of predictable behavior capable of being captured by a utility function emerged naturally as an adaptive mechanism for individuals faced with repeated choices in a nonstationary environment. Robson and Samuelson [97] find that exponential discounting in utility functions is consistent with evolutionarily optimal growth of a population.

Binary Choice Model with Systematic Risk

Consider a population of individuals that live for one period, produce a random number of offspring asexually, and then die. During their lives, individuals make only one decision: they choose from two actions, a and b, and this results in one of two corresponding random numbers of offspring, x_a and x_b. Now suppose that each individual chooses action a with some probability $p\in[0,1]$ and action b with probability $1-p$, denoted by the Bernoulli variable I^p, hence the number of offspring of an individual is given by the random variable:

$$x^p = I^p x_a + (1-I^p)x_b,$$

where

$$I^p = \begin{cases} 1 & \text{with probability } p \\ 0 & \text{with probability } 1-p. \end{cases}$$

We shall henceforth refer to p as the individual's *behavior* since it completely determines how the individual chooses between action a and b. Note that p can be 0 or 1, hence we are not requiring individuals to randomize—this will be derived as a consequence of natural selection under certain conditions.

Now suppose that there are two independent environmental factors, λ_1 and λ_2, that determine reproductive success, and that x_a and x_b are both linear combinations of these two factors:

$$\begin{cases} x_a = \beta_1\lambda_1 + (1-\beta_1)\lambda_2 \\ x_b = \beta_2\lambda_1 + (1-\beta_2)\lambda_2 \end{cases}.$$

where $\lambda_i \geq 0$, $0 \leq \beta_i \leq 1$, $i=1,2$. Examples of such factors are weather conditions, the availability of food, or the number of predators in the environment. Because these factors affect the fecundity of all individuals in the population, we refer to them as *systematic*, and we assume that:

(A1) λ_1 and λ_2 are independent random variables with some well-behaved distribution functions, such that (x_a, x_b) and $\log(px_a+(1-p)x_b)$ have finite moments up to order 2 for all $p\in[0,1]$, $\beta_1\in[0,1]$, $\beta_2\in[0,1]$, and

(A2) (λ_1, λ_2) is IID over time and identical for all individuals in a given generation.

We shall henceforth refer to (β_1, β_2) as an individual's *characteristics*. For each action, individuals are faced with a tradeoff between two positive environmental factors because of limited resources.

Under this framework, an individual is completely determined by his behavior p and characteristics (β_1, β_2). We shall henceforth refer to $f \equiv (p, \beta_1, \beta_2)$ as an individual's *type*. We assume that offspring behave in a manner identical to their parents, i.e., they have the same characteristics (β_1, β_2), and choose between a and b according to the same p, hence the population may be viewed as being comprised of different types indexed by the triplet f. In other words, offspring from a type-f individual are also of the same type f, hence we are assuming perfect genetic transmission from one generation to the next (once a type f, always a type f). We also assume that the initial population contains an equal number of all types, which we normalize to be 1 each without loss of generality. In summary, an individual i of type $f=(p, \beta_1, \beta_2)$ produces a random number of offspring:

$$x_i^{p,\beta_1,\beta_2} = I_i^p x_{a,i}^{\beta_1} + (1-I_i^p)x_{b,i}^{\beta_2} \tag{1}$$

where

$$\begin{cases} x_{a,i}^{\beta_1} = \beta_1\lambda_1 + (1-\beta_1)\lambda_2 \\ x_{b,i}^{\beta_2} = \beta_2\lambda_1 + (1-\beta_2)\lambda_2. \end{cases} \tag{2}$$

Now consider an initial population of individuals with different types. Suppose the total number of type $f=(p, \beta_1, \beta_2)$ individuals in generation T is n_T^f. Under assumptions (A1)–(A2), it is easy to show that $T^{-1}\log n_T^f$ converges in probability to the log-geometric-average growth rate:

$$\mu(p, \beta_1, \beta_2) = \mathbb{E}\left[\log\left(px_a^{\beta_1} + (1-p)x_b^{\beta_2}\right)\right], \tag{3}$$

as the number of generations and the number of individuals in each generation increases without bound (we provide the proof for a more general case in the Supplementary Information). Define

$$\begin{cases} \alpha_1 = p\beta_1 + (1-p)\beta_2 \\ \alpha_2 = p(1-\beta_1) + (1-p)(1-\beta_2), \end{cases}$$

then $\alpha_1 + \alpha_2 = 1$, and (3) can be equivalently written as

$$\mu(p, \beta_1, \beta_2) = \mathbb{E}[\log(\alpha_1\lambda_1 + (1-\alpha_1)\lambda_2)]. \tag{4}$$

We shall henceforth refer to (α_1, α_2) as the *factor loadings* of type-f individuals, and (3) and (4) characterize the log-geometric-average growth rate of individuals as a function of their type f in terms of both behavior p and characteristics (β_1, β_2).

Over time, individuals with the largest growth rate will dominate the population geometrically fast [14]. The optimal factor loading α_1^* that maximizes (4) is given by:

$$\alpha_1^* = \begin{cases} 1 & \text{if } \mathbb{E}[\lambda_1/\lambda_2] > 1 \text{ and } \mathbb{E}[\lambda_2/\lambda_1] < 1 \\ \text{solution to (6)} & \text{if } \mathbb{E}[\lambda_1/\lambda_2] \geq 1 \text{ and } \mathbb{E}[\lambda_2/\lambda_1] \geq 1 \\ 0 & \text{if } \mathbb{E}[\lambda_1/\lambda_2] < 1 \text{ and } \mathbb{E}[\lambda_2/\lambda_1] > 1 \end{cases}. \quad (5)$$

where α_1^* is defined implicitly in the second case of (5) by

$$\mathbb{E}\left[\frac{\lambda_1}{\alpha_1^*\lambda_1 + (1-\alpha_1^*)\lambda_2}\right] = \mathbb{E}\left[\frac{\lambda_2}{\alpha_1^*\lambda_1 + (1-\alpha_1^*)\lambda_2}\right]. \quad (6)$$

As a result, the growth-optimal type is $f^* = (p^*, \beta_1^*, \beta_2^*)$, which is given explicitly in Table 1.

The optimal characteristics and associated optimal behaviors in Table 1 show that, when α_1^* is 1 or 0, one of the factors, λ_1 or λ_2, is significantly more important than the other, and the optimal strategy places all the weight on the more important factor. However, when α_1^* is strictly between 0 and 1, a combination of factors λ_1 and λ_2 is necessary to achieve the maximum growth rate. Individual characteristics (β_1^*, β_2^*) need to be distributed in such a way that one of the two choices of action puts more weight on one factor, while the other choice puts more weight on the other factor. Eventually, the behavior p^* randomizes between the two choices and therefore achieves the optimal combination of factors. This is a generalization of the "adaptive coin-flipping" strategies of Cooper and Kaplan [59], who interpret this behavior as a form of altruism because individuals seem to be acting in the interest of the population at the expense of their own fitness. However, Grafen [107] provides a different interpretation by proposing an alternate measure of fitness, one that reflects the growth rate of survivors.

This result may be viewed as a primitive form of herding behavior—where all individuals in the population choose to act in the same manner—especially if the relative environmental factors, $\mathbb{E}[\lambda_1/\lambda_2]$ and $\mathbb{E}[\lambda_2/\lambda_1]$, shift suddenly due to rapid environmental changes. To an outside observer, behaviors among individuals in this population may seem heterogenous before the shift, but will become increasingly similar after the shift, creating the appearance (but not the reality) of intentional coordination, communication, and synchronization. If the reproductive cycle is sufficiently short, this change in population-wide behavior may seem highly responsive to environmental changes, giving the impression that individuals are learning about their environment. This is indeed a form of learning, but it occurs at the population level, not at the individual level, and not within an individual's lifespan.

Individually Optimal versus Group Optimal Behavior

It is instructive to compare the optimal characteristics and behavior in Table 1 that maximize growth with the behavior that maximizes an individual's reproductive success. According to (1) and (2), the individually optimal behavior maximizes

$$\mathbb{E}\left[x^{p,\beta_1,\beta_2}\right] = \mathbb{E}\left[I^p x_a^{\beta_1} + (1-I^p)x_b^{\beta_2}\right]$$

$$= \alpha_1\mathbb{E}[\lambda_1] + \alpha_2\mathbb{E}[\lambda_2]$$

over p, β_1, β_2. Therefore, the individually optimal factor loading, denoted by $\hat{\alpha}_1$, is simply:

$$\hat{\alpha}_1 = \begin{cases} 1 & \text{if } \mathbb{E}[\lambda_1] > \mathbb{E}[\lambda_2] \\ 0 & \text{if } \mathbb{E}[\lambda_1] < \mathbb{E}[\lambda_2] \\ \text{arbitrary} & \text{if } \mathbb{E}[\lambda_1] = \mathbb{E}[\lambda_2]. \end{cases}$$

Given a particular environment (λ_1, λ_2), the individually optimal behavior depends only on the expectation of two factors, and this selfish behavior is generally sub-optimal for the group.

In contrast, individuals of type $f^* = (p^*, \beta_1^*, \beta_2^*)$ described in Table 1 are optimal in the group sense, attaining the maximum growth rate as a group by behaving differently than the individually optimal behavior. We shall refer to f^* henceforth as the *growth-optimal* behavior to underscore the fact that it is optimal from the population perspective, not necessarily from the individual's perspective. This provides a prototype of group selection as a consequence of stochastic environments with systematic risk. We define ***groups*** to be individuals with the same characteristics. More precisely, in our model, individuals with the same (β_1, β_2) are considered a group. Nature selects the groups with optimal characteristics (β_1^*, β_2^*), and p^* is a reflection of different behaviors for each group.

Table 1. Optimal type $f^* = (p^*, \beta_1^*, \beta_2^*)$ for the binary choice model

	Optimal characteristics	Optimal behavior
If $\alpha_1^* = 1$	$\{(\beta_1, \beta_2) : \beta_1 = 1 \text{ or } \beta_2 = 1\}$	$p^* = \begin{cases} \frac{\alpha_1^* - \beta_2^*}{\beta_1^* - \beta_2^*} = 1 & \text{if } \beta_1^* = 1, \beta_2^* \neq 1 \\ \frac{\alpha_1^* - \beta_2^*}{\beta_1^* - \beta_2^*} = 0 & \text{if } \beta_1^* \neq 1, \beta_2^* = 1 \\ \text{arbitrary} & \text{if } \beta_1^* = \beta_2^* = 1 \end{cases}.$
If $\alpha_1^* = 0$	$\{(\beta_1, \beta_2) : \beta_1 = 0 \text{ or } \beta_2 = 0\}$	$p^* = \begin{cases} \frac{\alpha_1^* - \beta_2^*}{\beta_1^* - \beta_2^*} = 1 & \text{if } \beta_1^* = 0, \beta_2^* \neq 0 \\ \frac{\alpha_1^* - \beta_2^*}{\beta_1^* - \beta_2^*} = 0 & \text{if } \beta_1^* \neq 0, \beta_2^* = 0 \\ \text{arbitrary} & \text{if } \beta_1^* = \beta_2^* = 0 \end{cases}$
If $0 < \alpha_1^* < 1$	$\{(\beta_1, \beta_2) : (\beta_1 - \alpha_1^*)(\beta_2 - \alpha_1^*) \leq 0\}$	$p^* = \begin{cases} \frac{\alpha_1^* - \beta_2^*}{\beta_1^* - \beta_2^*} & \text{if } \beta_1^* \neq \beta_2^* \\ \text{arbitrary} & \text{if } \beta_1^* = \beta_2^* \end{cases}.$

Like altruism, cooperation, trust, and other behaviors that do not immediately benefit the individual, the growth-optimal characteristics and behaviors derived in our framework flourish because they allow these individuals to pass through the filter of natural selection. However, unlike theories of group selection that are based on sexual reproduction and genetic distance, our version of group selection is based on behavior itself. Those individuals with types other than f^* will not reproduce as quickly, hence from an evolutionary biologist's perspective, group selection is operating at the level of those individuals with characteristics (β_1^*, β_2^*) and behaving according to p^*. Of course, we cannot measure all forms of characteristics and behavior as readily as we can measure genetic make-up, but in the stark case of the binary choice model, it is clear that selection can and does occur according to groups defined by characteristics and behavior.

The Supplementary Information contains a generalization of the binary choice model to multinomial choices with multiple environmental factors. In general, it is possible that the optimal growth rate μ^* corresponds to multiple groups, and each group corresponds to multiple optimal behaviors. In terms of group selection, this fact means that several different groups—each defined by a specific combination of characteristics—could simultaneously be optimal from an evolutionary perspective. Within each group, natural selection will determine the behavior that achieves the optimal growth rate. The optimal behavior is not necessarily unique for each group. Also, the optimal behaviors for different groups might overlap.

A Numerical Example

Consider an island that is isolated from the rest of the world, and suppose λ_1 is a measure of weather conditions of the local environment, and λ_2 is a measure of the local environment's topography where, without loss of generality, we assume that larger values of each factor are more conducive to reproductive success. Moreover, λ_1 and λ_2 are independent random variables and described by:

$$\lambda_1 = \begin{cases} 1 & \text{with probability } \frac{1}{2} \\ 2 & \text{with probability } \frac{1}{2}, \end{cases}$$

$$\lambda_2 = \begin{cases} 1 & \text{with probability } \frac{1}{2} \\ 2 & \text{with probability } \frac{1}{2}. \end{cases}$$

An individual on this island lives for one period, has one opportunity to choose one of two actions—farming (action a) or mining (action b)—which determines its reproductive success, and then dies immediately after reproduction. The number of offspring is given by $x_a^{\beta_1}$ if action a is chosen and $x_b^{\beta_2}$ if b is chosen, where $x_a^{\beta_1}$ and $x_b^{\beta_2}$ are given by:

$$\begin{cases} x_a^{\beta_1} = \beta_1\lambda_1 + (1-\beta_1)\lambda_2 \\ x_b^{\beta_2} = \beta_2\lambda_1 + (1-\beta_2)\lambda_2. \end{cases}$$

Here, β_1 captures an individual's farming ability as determined by the two factors, weather and topography; β_2 captures an individual's mining ability as determined by the same two factors. According to Table 1, the optimal factor loadings are ..., which indicates that individuals should have a balanced exposure to both weather and topography. The optimal characteristics are

$$\left\{(\beta_1^*, \beta_2^*) | (\beta_1^* - \frac{1}{2})(\beta_2^* - \frac{1}{2}) \leq 0\right\}, \quad (7)$$

and each group is associated with the optimal behavior:

$$p^* = \begin{cases} \frac{\frac{1}{2} - \beta_2^*}{\beta_1^* - \beta_2^*} & \text{if } \beta_1^* \neq \beta_2^* \\ \text{arbitrary} & \text{if } \beta_1^* = \beta_2^* = \frac{1}{2}. \end{cases} \quad (8)$$

For example, $\{(\beta_1^*, \beta_2^*) | \beta_1^* = \frac{1}{2}\}$ is an optimal group associated with $p^* = 1$. These are individuals who can perfectly balance the output of farming with respect to weather and topography. Therefore, they choose farming with probability 1 and appear as a "group" of farmers.

On the other hand, $\{(\beta_1^*, \beta_2^*) | \beta_2^* = \frac{1}{2}\}$ is another optimal group, but associated with the optimal behavior $p^* = 0$. These individuals can perfectly balance the output of mining with respect to weather and topography. Therefore, they choose mining with probability 1 and appear as a "group" of miners.

Finally, there are also other optimal groups, described by (7) in general, in which individuals randomize their choices between farming and mining according to (8), to achieve the optimal exposure to weather and topography.

Figure 1 shows the optimal behavior for each group. The optimal groups described by (7) correspond to the upper-left and lower-right blocks. Randomized behaviors are optimal for these groups. Interestingly, all the sub-optimal groups (upper-right and lower-left blocks) correspond to deterministic behaviors ($p^* = 0$ or 1) except when $\beta_1 = \beta_2$. Figure 2 shows the optimal log-geometric-average growth rate for each group. It is clear that all groups described by (7) have the largest growth rate.

We can see that multiple optimal groups co-exist through natural selection, and within each group, individuals share the same characteristics. A particular behavior must be paired with a particular set of characteristics to achieve the optimal growth rate. Note that the individuals in (7) are optimal only in the group sense. As a single entity, a group possesses survival benefits above and beyond an individual, and in our framework, these benefits arise purely from stochastic environments with systematic risk.

The usual notion of group selection in the evolutionary biology literature is that natural selection acts at the level of the group instead of at the more conventional level of the individual, and interaction among members within each group is much more frequent than interaction among individuals across groups. In this case, similar individuals are usually clustered geographically. However, in our model, individuals do not interact at all, nevertheless, the fact that individuals with the same behavior generate offspring with like behavior makes them more likely to cluster geographically and appear as a "group". In addition, imagine that the environment (λ_1, λ_2) experiences a sudden shift. To an outside observer, behaviors among individuals in this population will become increasingly similar after the shift, creating the appearance—but not the reality—of intentional coordination, communication, and synchronization.

Here we use the phrase "appear as a group" because our derived behavior is not strictly the same as group selection as defined in the evolutionary biology literature. Instead, we show that behavior which is evolutionarily dominant through the traditional mechanism of natural selection is consistent with the

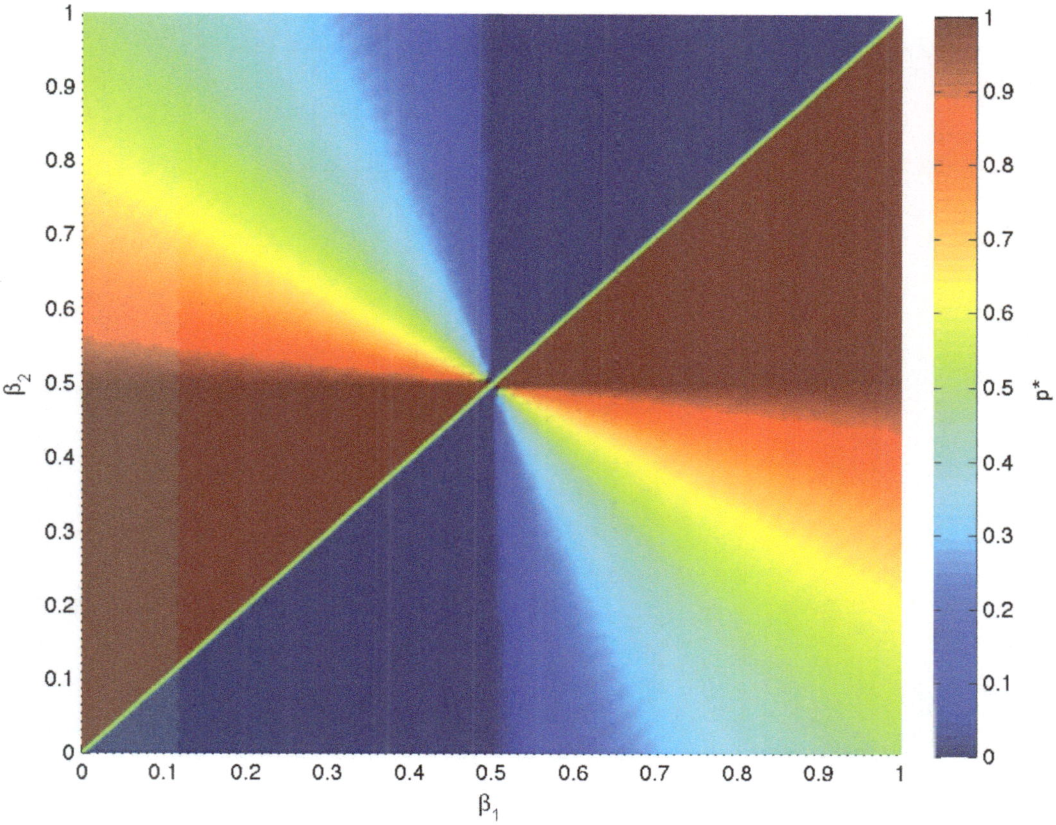

Figure 1. The optimal behavior for each group in the numerical example of group selection.

implications of group selection. We purposefully model individuals in our population as mindless creatures engaging in random choices to demonstrate that group-like selection can arise purely through common factors in reproductive success. If we include more complex features such as sexual reproduction, limited resources, and competitive/cooperative interactions among individuals, even more sophisticated group dynamics can be generated.

Discussion

Many species have a social structure in which individuals form groups and the aggregation of individuals promotes the fitness of group members. When selection for a biological trait in such populations depends on the difference between groups rather than individual differences within a group, it is described as "group selection" in evolutionary biology.

The debate surrounding genetic, kin, and group selection began over four decades ago, but has recently become more animated thanks to Nowak et al. [30], who challenge inclusive fitness theory in the study of social evolution by arguing that it is not a constructive theory that allows a useful mathematical analysis of evolutionary processes. Moreover, they conclude that inclusive fitness is neither useful nor necessary to explain the evolution of eusociality or other phenomena. However, this view was sharply criticized by a flurry of responses by many leading evolutionary biologists [35,109–112], who observe that the more general inclusive fitness theory has stimulated the extensive empirical literature over the past 40 years in the fields of behavioral and evolutionary ecology [35], and that kin selection is a strong, vibrant theory that forms the basis for our understanding of how social behavior has evolved [110].

On the other hand, a significant amount of research suggests that group selection and kin selection (inclusive fitness) are essentially one process [28, 29, 31, 113–115], both seeking to characterize the genetic structure of a population but in different ways. These authors argue that it is now time to step back from the details of the specific arguments and consider the more general question of how evolution works in structured populations. This line of inquiry has the potential to generate insights beyond areas to which it has traditionally been applied [115].

Instead of entering into this debate, we propose to reconcile these opposing perspectives by studying the impact of selection on behavior and deriving evolutionary implications that cut across species, physiology, and genetic origins. As a direct consequence of this behavioral approach, we have shown that what appears to be group selection may, in fact, simply be the consequence of natural selection occurring in stochastic environments with reproductive risks that are correlated across individuals. In particular, we provide an evolutionary model with population dynamics for the simplest form of behavior, a binary choice, and derive the implications of selection on the behavior of individuals that share certain characteristics. Not surprisingly, individuals with similar characteristics experience similar selective pressures, hence evolution in stochastic environments with systematic risk can generate empirical phenomena that are consistent with group selection. In fact, Nature does select for "groups" of individuals with optimal characteristics and the optimal behavior within selected groups is simply a reflection of optimality of the characteristics of that group with respect to the given environment. Moreover, it is possible that

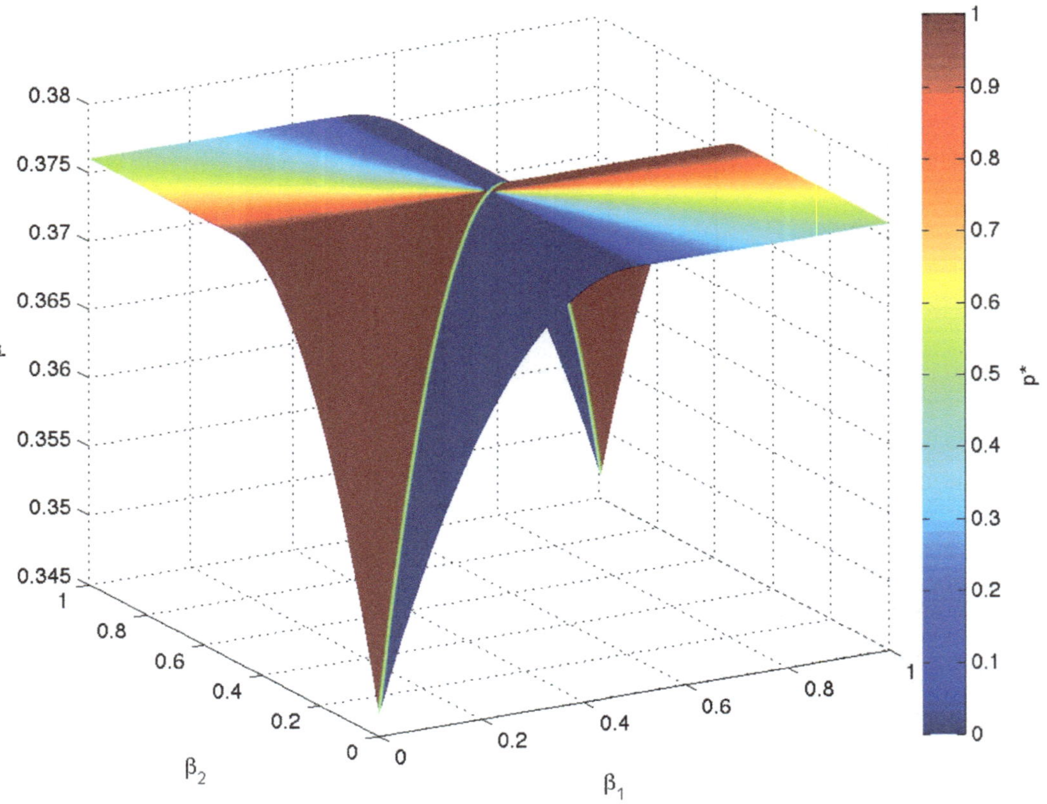

Figure 2. The optimal log-geometric-average growth rate for each group in the numerical example of group selection.

multiple groups achieve the same optimal growth rate but through different means, i.e., many combinations of behavior and characteristics can be optimal, leading to considerable variation in the types of behavior and characteristics in the population.

Hamilton's great insight was that individual fitness is not maximized by social evolution; inclusive fitness is [2, 3]. The idea that something other than the individual organism could be the fitness-maximizing unit was completely revolutionary at the time and opened new research areas that are still being explored [111]. We have shown that, in addition to Hamilton's insight, individuals with highly correlated risks will appear to form "groups" in evolution, even if their actions are, in fact, totally autonomous, mindless, and, prior to natural selection, uniformly randomly distributed in the population. Although this result seems to eliminate the need for a separate theory of group selection, the unique and important evolutionary implications of multiple sources of correlated systematic risk suggest that a separate term for this phenomenon may be worthwhile.

Acknowledgments

Research support from the MIT Laboratory for Financial Engineering is gratefully acknowledged. We thank the editor and the anonymous reviewers for helpful comments and suggestions, and Jayna Cummings for editorial assistance.

Author Contributions

Conceived and designed the experiments: RZ TB AL. Performed the experiments: RZ TB AL. Analyzed the data: RZ TB AL. Contributed reagents/materials/analysis tools: RZ TB AL. Wrote the paper: RZ TB AL.

References

1. Wynne-Edwards VC (1963) Intergroup selection in the evolution of social systems. Nature 200: 623–626.
2. Hamilton WD (1963) The evolution of altruistic behavior. American Naturalist 97: 354–356.
3. Hamilton WD (1964) The genetical evolution of social behavior. i and ii. Journal of Theoretical Biology 7: 1–52.
4. Trivers RL (1971) The evolution of reciprocal altruism. The Quarterly Review of Biology 46: 35–57.
5. Price GR (1970) Selection and covariance. Nature 227: 520–521.
6. Wilson EO (1975) Sociobiology: The New Synthesis. Cambridge, MA: Harvard University Press.
7. Wilson EO (2005) Kin selection as the key to altruism: its rise and fall. Social Research 72: 159166.
8. Wilson DS, Sober E (1994) Reintroducing group selection to the human behavioral sciences. Behavioral and Brain Sciences 17: 585–607.
9. Ben-Jacob E, Schochet O, Tenenbaum A, Cohen I, Czirok A, et al. (1994) Generic modelling of cooperative growth patterns in bacterial colonies. Nature 368: 46–49.
10. Ben-Jacob E, Becker I, Shapira Y, Levine H (2004) Bacterial linguistic communication and social intelligence. TRENDS in Microbiology 12: 366–372.
11. Ingham C, Ben-Jacob E (2008) Swarming and complex pattern formation in *paenibacillus vortex* studied by imaging and tracking cells. BMC Microbiology 8: 36.
12. Queller DC, Strassmann JE (1998) Kin selection and social insects. Bioscience 48: 165–175.

13. Pfennig DW, Collins JP, Ziemba RE (1999) A test of alternative hypotheses for kin recognition in cannibalistic tiger salamanders. Behavioral Ecology 10: 436–443.
14. Brennan TJ, Lo AW (2011) The origin of behavior. Quarterly Journal of Finance 1: 55–108.
15. Brennan TJ, Lo AW (2012) An evolutionary model of bounded rationality and intelligence. PloS one 7: e50310.
16. Robson AJ (1996a) A biological basis for expected and non-expected utility. Journal of Economic Theory 68: 397–424.
17. Samuelson L (2001) Introduction to the evolution of preferences. Journal of Economic Theory 97: 225–230.
18. Robson AJ (2001a) The biological basis of economic behavior. Journal of Economic Literature 39: 11–33.
19. Blume L, Easley D (1992) Evolution and market behavior. Journal of Economic Theory 58: 9–40.
20. Kogan L, Ross SA, Wang J, Westerfield MM (2006) The price impact and survival of irrational traders. Journal of Finance 61: 195–229.
21. Samuelson PA (1965) Proof that properly anticipated prices fluctuate randomly. Industrial Management Review 6: 41–49.
22. Fama EF (1970) Efficient capital markets: A review of theory and empirical work. Journal of Finance 25: 383–417.
23. Lo AW (2004) The adaptive markets hypothesis: Market efficiency from an evolutionary perspective. Journal of Portfolio Management 30: 15–29.
24. Wilson DS (1975) A theory of group selection. Proceedings of the National Academy of Sciences 72: 143–146.
25. Wright S (1980) Genic and organismic selection. Evolution 34: 825–843.
26. Williams GC (1996) Adaptation and natural selection: a critique of some current evolutionary thought. Princeton, NJ: Princeton University Press.
27. Maynard Smith J (1964) Group selection and kin selection. Nature 201: 1145–1147.
28. Queller DC (1992) Quantitative genetics, inclusive fitness, and group selection. American Naturalist 139: 540–558.
29. Wild G, Gardner A, West SA (2009) Adaptation and the evolution of parasite virulence in a connected world. Nature 459: 983–986.
30. Nowak MA, Tarnita CE, Wilson EO (2010) The evolution of eusociality. Nature 466: 1057–1062.
31. Wade MJ, Wilson DS, Goodnight C, Taylor D, Bar-Yam Y, et al. (2010) Multilevel and kin selection in a connected world. Nature 463: E8–E9.
32. Frank SA (2013) Natural selection. vii. history and interpretation of kin selection theory. Journal of Evolutionary Biology 26: 1151–1184.
33. Wilson DS, Wilson EO (2008) Evolution "for the good of the group". American Scientist 96: 380–389.
34. Wilson EO (2013) The Social Conquest of the Earth. New York: Liveright.
35. Abbot P, Abe J, Alcock J, Alizon S, Alpedrinha JAC, et al. (2011) Inclusive fitness theory and eusociality. Nature 471: E1–E4.
36. Pinker S (2012) The false allure of group selection. Technical report, Edge.com. Available: http://edge.org/conversation/the-false-allure-of-group-selection.
37. Mayr E (2001) What Evolution Is. New York: Basic Books.
38. Lynch M, Lande R (1993) Evolution and extinction in response to environmental change. In: Karieva PM, Kingsolver JG, Huey RB, editors, Biotic Interactions and Global Change, Sunderland, MA: Sinauer Associates. pp. 235–250.
39. Burger R, Lynch M (1995) Evolution and extinction in a changing environment: A quantitative-genetic analysis. Evolution 49: 151–163.
40. Burger R, Gimelfarb A (2002) Fluctuating environments and the role of mutation in maintaining quantitative genetic variation. Genetical Research 80: 31–46.
41. Pekalski A (1998) A model of population dynamics. Physica A: Statistical Mechanics and its Applications 252: 325–335.
42. Pekalski A (1999) Mutations and changes of the environment in a model of biological evolution. Physica A: Statistical Mechanics and its Applications 265: 255–263.
43. Pekalski A (2002) Evolution of population in changing conditions. Physica A: Statistical Mechanics and its Applications 314: 114–119.
44. De Blasio FV (1999) Diversity and extinction in a lattice model of a population with fluctuating environment. Physical Review 60: 5912–5917.
45. Cohen D (1966) Optimizing reproduction in a randomly varying environment. Journal of theoretical biology 12: 119–129.
46. Oster GF, Wilson EO (1979) Caste and Ecology in the Social Insects. Princeton, NJ: Princeton University Press.
47. Bergstrom TC (2014) On the evolution of hoarding, risk-taking, and wealth distribution in nonhuman and human populations. Proceedings of the National Academy of Sciences 111: 10860–10867.
48. Levins R (1968) Evolution in Changing Environments: Some Theoretical Explorations.(MPB-2) (Vol. 2). Princeton, NJ: Princeton University Press.
49. Gillespie JH, Guess HA (1978) The effects of environmental autocorrelations on the progress of selection in a random environment. American Naturalist 112: 897–909.
50. Frank SA (2011a) Natural selection. i. variable environments and uncertain returns on investment. Journal of Evolutionary Biology 24: 2299–2309.
51. Frank SA (2011b) Natural selection. ii. developmental variability and evolutionary rate. Journal of Evolutionary Biology 24: 2310–2320.
52. Frank SA (2012a) Natural selection. iii. selection versus transmission and the levels of selection. Journal of Evolutionary Biology 25: 227–243.
53. Kelly JL Jr (1956) A new interpretation of information rate. Bell System Technical Journal 35: 917–926.
54. Samuelson PA (1971) The fallacy of maximizing the geometric mean in long sequences of investing or gambling. Proceedings of the National Academy of Sciences of the United States of America 68: 2493–2496.
55. Thorp EO (1971) Portfolio choice and the kelly criterion. Proceedings of the Business and Economics Section of the American Statistical Association: 215–224.
56. Thorp EO (2006) The kelly criterion in blackjack, sports betting and the stock market. In: Zenios SA, Ziemba WT, editors, Handbook of Asset and Liability Management Oxford, UK: North-Holland. pp. 385–428.
57. Maynard Smith J (1982) Evolution and the Theory of Games. Cambridge university press.
58. Fretwell SD (1972) Populations in a seasonal environment. 5. Princeton University Press.
59. Cooper WS, Kaplan RH (1982) Adaptive "coin-flipping": A decision-theoretic examination of natural selection for random individual variation. Journal of Theoretical Biology 94: 135–151.
60. Frank SA, Slatkin M (1990) Evolution in a variable environment. American Naturalist 136: 244–260.
61. Slatkin M (1974) Hedging one's evolutionary bets. Nature 250: 704–705.
62. Caraco T (1980) On foraging time allocation in a stochastic environment. Ecology: 119–128.
63. Real LA (1980) Fitness, uncertainty, and the role of diversification in evolution and behavior. American Naturalist: 623–633.
64. Rubenstein DI (1982) Risk, uncertainty and evolutionary strategies. Current problems in sociobiology Cambridge University Press, Cambridge: 91–111.
65. Seger J, Brockmann HJ (1987) What is bet-hedging? Oxford surveys in evolutionary biology 4: 182–211.
66. Beaumont HJ, Gallie J, Kost C, Ferguson GC, Rainey PB (2009) Experimental evolution of bet hedging. Nature 462: 90–93.
67. Childs DZ, Metcalf C, Rees M (2010) Evolutionary bet-hedging in the real world: empirical evidence and challenges revealed by plants. Proceedings of the Royal Society B: Biological Sciences: rspb20100707.
68. Real L, Caraco T (1986) Risk and foraging in stochastic environments. Annual Review of Ecology and Systematics: 371–390.
69. Stephens DW, Krebs JR (1986) Foraging theory. Princeton University Press.
70. Deneubourg JL, Aron S, Goss S, Pasteels JM (1987) Error, communication and learning in ant societies. European Journal of Operational Research 30: 168–172.
71. Harder LD, Real LA (1987) Why are bumble bees risk averse? Ecology 68: 1104–1108.
72. Pasteels JM, Deneubourg JL, Goss S (1987) Self-organization mechanisms in ant societies. i: Trail recruitment to newly discovered food sources. Experientia Supplementum.
73. Mangel M, Clark CW (1988) Dynamic modeling in behavioral ecology. Princeton University Press.
74. Hölldobler B, Wilson EO (1990) The Ants. Cambirdge, MA: Belknap Press.
75. Kirman A (1993) Ants, rationality, and recruitment. Quarterly Journal of Economics 108: 137–156.
76. Thuijsman F, Peleg B, Amitai M, Shmida A (1995) Automata, matching and foraging behavior of bees. J Theor Biol 175: 305–316.
77. Smallwood P (1996) An introduction to risk sensitivity: The use of Jensen's Inequality to clarify evolutionary arguments of adaptation and constraint. American Zoologist 36: 392–401.
78. Keasar T, Rashkovich E, Cohen D, Shmida A (2002) Bees in two-armed bandit situations: foraging choices and possible decision mechanisms. Behavioral Ecology 13: 757–765.
79. Ben-Jacob E (2008) Social behavior of bacteria: From physics to complex organizations. European Physics Journal B: 1–8.
80. Levin SA, Cohen D, Hastings A (1984) Dispersal strategies in patchy environments. Theoretical population biology 26: 165–191.
81. Levin SA, Muller-Landau HC, Nathan R, Chave J (2003) The ecology and evolution of seed dispersal: a theoretical perspective. Annual Review of Ecology, Evolution, and Systematics: 575–604.
82. Knoch D, Gianotti LR, Pascual-Leone A, Treyer V, Regard M, et al. (2006) Disruption of right prefrontal cortex by low-frequency repetitive transcranial magnetic stimulation induces risk-taking behavior. The Journal of neuroscience 26: 6469–6472.
83. Fecteau S, Pascual-Leone A, Zald DH, Liguori P, Thoret H, et al. (2007) Activation of prefrontal cortex by transcranial direct current stimulation reduces appetite for risk during ambiguous decision making. The Journal of neuroscience 27: 6212–6218.
84. Tom SM, Fox CR, Trepel C, Poldrack RA (2007) The neural basis of loss aversion in decision-making under risk. Science 315: 515–518.
85. Christopoulos GI, Tobler PN, Bossaerts P, Dolan RJ, Schultz W (2009) Neural correlates of value, risk, and risk aversion contributing to decision making under risk. The Journal of Neuroscience 29: 12574–12583.
86. Yoshimura J, Clark CW (1991) Individual adaptations in stochastic environments. Evolutionary Ecology 5: 173–192.
87. Yoshimura J, Jansen VA (1996) Evolution and population dynamics in stochastic environments. Researches on Population Ecology 38: 165–182.
88. McNamara JM (1995) Implicit frequency dependence and kin selection in fluctuating environments. Evolutionary Ecology 9: 185–203.

89. Alexander RD (1974) The evolution of social behavior. Annual Review of Ecology and Systematics 5: 325–383.
90. Hirshleifer J (1977) Economics from a biological viewpoint. Journal of Law and Economics 20: 1–52.
91. Hirshleifer J (1978) Natural economy versus political economy. Journal of Social and Biological Structures 1: 319–337.
92. Bergstrom TC (2002) Evolution of social behavior: Individual and group selection. Journal of Economic Perspectives 16: 67–88.
93. Waldman M (1994) Systematic errors and the theory of natural selection. American Economic Review 84: 482–497.
94. Robson AJ (1996b) The evolution of attitudes to risk: Lottery tickets and relative wealth. Games and Economic Behavior 14: 190–207.
95. Robson AJ (2001b) Why would nature give individuals utility functions? Journal of Political Economy 109: 900–914.
96. Rogers AR (1994) Evolution of time preference by natural selection. American Economic Review 84: 460–481.
97. Robson AJ, Samuelson L (2007) The evolution of intertemporal preferences. American Economic Review 97: 496–500.
98. Robson AJ, Szentes B (2008) Evolution of time preferences by natural selection: Comment. American Economic Review 98: 1178–1188.
99. Robson AJ, Samuelson L (2009) The evolution of time preferences with aggregate uncertainty. American Economic Review 99: 1925–1953.
100. Burnham TC (2013) Toward a neo-darwinian synthesis of neoclassical and behavioral economics. Journal of Economic Behavior and Organization 90: S113–S127.
101. Luo GY (1995) Evolution and market competition. Journal of Economic Theory 67: 223–250.
102. Luo GY (1998) Market efficiency and natural selection in a commodity futures market. Review of Financial Studies 11: 647–674.
103. Herrmann-Pillath C (1991) A darwinian framework for the economic analysis of institutional change in history. Journal of Social and Biological Structures 14: 127–148.
104. van den Bergh JC, Gowdy JM (2009) A group selection perspective on economic behavior, institutions and organizations. Journal of Economic Behavior and Organization 72: 1–20.
105. Safarzyńska K, van den Bergh JC (2010) Evolving power and environmental policy: Explaining institutional change with group selection. Ecological Economics 69: 743–752.
106. Hansson I, Stuart C (1990) Malthusian selection of preferences. The American Economic Review: 529–544.
107. Grafen A (1999) Formal darwinism, the individual–as–maximizing–agent analogy and bet–hedging. Proceedings of the Royal Society of London Series B: Biological Sciences 266: 799–803.
108. Curry PA (2001) Decision making under uncertainty and the evolution of interdependent preferences. Journal of Economic Theory 98: 357–369.
109. Boomsma JJ, Beekman M, Cornwallis CK, Griffin AS, Holman L, et al. (2011) Only full-sibling families evolved eusociality. Nature 471: E4–E5.
110. Strassmann JE, Page RE Jr, Robinson GE, Seeley TD (2011) Kin selection and eusociality. Nature 471: E5–E6.
111. Ferriere R, Michod RE (2011) Inclusive fitness in evolution. Nature 471: E6–E8.
112. Herre EA, Wcislo WT (2011) In defence of inclusive fitness theory. Nature 471: E8–E9.
113. Lehmann L, Kekker L, West S, Roze D (2007) Group selection and kin selection: Two concepts but one process. Proceedings of the National Academy of Sciences 104: 6736–6739.
114. Marshall JAR (2011) Group selection and kin selection: Formally equivalent approaches. Trends in Ecology and Evolution 26: 325–332.
115. Lion S, Jansen VAA, Day T (2011) Evolution in structured populations: Beyond the kin versus group debate. Trends in Ecology and Evolution 26: 193–201.

16

Usual Populations, Unusual Individuals: Insights into the Behavior and Management of Asian Elephants in Fragmented Landscapes

Nishant M. Srinivasaiah[1]*, Vijay D. Anand[2†], Srinivas Vaidyanathan[3], Anindya Sinha[4,5]

1 Postgraduate Program in Wildlife Biology and Conservation, National Centre for Biological Sciences, Tata Institute of Fundamental Research, GKVK Campus, Bangalore, India, **2** A Rocha India, Bangalore, India, **3** Foundation for Ecological Research, Advocacy and Learning, Appavounagar, Vazhakulam, Pondicherry, India, **4** School of Natural Sciences and Engineering, National Institute of Advanced Studies, Indian Institute of Science Campus, Bangalore, India, **5** Nature Conservation Foundation, Mysore, India

Abstract

Background: A dearth in understanding the behavior of Asian elephants (*Elephas maximus*) at the scale of populations and individuals has left important management issues, particularly related to human-elephant conflict (HEC), unresolved. Evaluation of differences in behavior and decision-making among individual elephants across groups in response to changing local ecological settings is essential to fill this gap in knowledge and to improve our approaches towards the management and conservation of elephants.

Methodology/Principal Findings: We hypothesized certain behavioral decisions that would be made by Asian elephants as reflected in their residence time and movement rates, time-activity budgets, social interactions and group dynamics in response to resource availability and human disturbance in their habitat. This study is based on 200 h of behavioral observations on 60 individually identified elephants and a 184-km^2 grid-based survey of their natural and anthropogenic habitats within and outside the Bannerghatta National Park, southern India during the dry season. At a general population level, the behavioral decisions appeared to be guided by the gender, age and group-type of the elephants. At the individual level, the observed variation could be explained only by the idiosyncratic behaviors of individuals and that of their associating conspecific individuals. Recursive partitioning classification trees for residence time of individual elephants indicated that the primary decisions were taken by individuals, independently of their above-mentioned biological and ecological attributes.

Conclusions/Significance: Decision-making by Asian elephants thus appears to be determined at two levels, that of the population and, more importantly, the individual. Models based on decision-making by individual elephants have the potential to predict conflict in fragmented landscapes that, in turn, could aid in mitigating HEC. Thus, we must target individuals, in addition to populations, in our efforts to manage and conserve this threatened species, particularly in human-dominated landscapes.

Editor: Wayne M. Getz, University of California, Berkeley, United States of America

Funding: This work was supported by the Department of Science and Technology, Government of India. The funders had no role in study design, data collection and analysis, decision to publish, or preparation of the manuscript.

Competing Interests: The authors have declared that no competing interests exist.

* E-mail: msnishant@gmail.com

† Deceased.

Introduction

An important, often-neglected, aspect of behavioral ecology concerns the ability of animal populations and individuals to respond to changes in their immediate environment, both in the long- and the short- term. Systematic changes in behavioral patterns in response to predictable variation in the environment such as resource availability, including those that may occur seasonally, could be innate and selected for [1]. For example, behavioral responses encompassed by varying time-activity budgets, spatial and temporal preferences for certain habitats or the avoidance of areas in response to certain ecological parameters have been documented extensively [1,2]. Such responses may often be determined by relatively more biologically determined factors such as the gender, age or the typical social organization of the species concerned. What might be more problematic for animals, however, are the demands placed on them by short-term, unpredictable ecological changes in their environment, a classic example being of animal populations that largely occur in increasingly human-dominated landscapes. This might be more challenging for large-bodied mammalian species such as the elephant that lives in complex societies but which may be able to cope with rapid environmental changes with their learning capacities and sophisticated cognitive decision-making abilities [3].

The availability and distribution of resources seem to primarily influence the occurrence of elephants in a particular region [3–7]. Although this species does not have any natural predator [3], humans and anthropogenic disturbances have, over time, emerged as major threats to these mammals, occasionally threatening their very survival. Demographic changes brought about by illegal hunting may demand that elephants rapidly adapt behaviorally to such drastic ecological changes [8–10], but which have rarely been documented. Moreover, what remain virtually unknown are the behavioral differences displayed by groups and individuals in response to these change. Such a dearth in our understanding of the behavioral decisions made by elephants, influenced both by their biology and their ecology, has left important management issues, particularly related to human-elephant conflict (HEC), unresolved and rendered ineffective several mitigation measures that have been adopted to reduce conflict in anthropogenic habitats.

Our current knowledge of elephants show that they are highly polygynous and sexually dimorphic animals [11]. They are also long-lived with well-defined growth phases, punctuated importantly by puberty. In males, body size dictates dominance and reproductive success [11–14]. Older and larger male elephants have been observed to come into *musth* more often and be in that state for longer durations of time than do younger males [3,14,15]. These males are also known to control the *musth* and regulate the social behavior of younger bulls [15]. Bulls in *musth* are known to have higher reproductive success than those not in *musth* [14]. In the African elephants, females, are known to form a fission-fusion social structure with a strong social hierarchy among herds and clans, which are herds that associate with one another [16–17]. Recent pioneering studies on Asian elephants by de Silva *et al* show that, unlike their African counterparts, female societies in Asian elephants do not maintain coherent core groups and exhibit significantly less social bonding at the population level. They suggest, however, that strong social networking does occur at different levels of association and that this unique multilevel social organization, which does not follow the classical hierarchical structure, must be recognized among Asian elephants [18]. Social dominance and rank hierarchy among herds may, in turn, determine their distribution in an area and the use of certain habitats [17,19,20]. In females, dominance has been suggested to be an offshoot of the age of the individual. The oldest female acts as the matriarch and influences the behavior of other individuals in the herd [3,20]. Studies in Africa have shown the existence of exclusive bull areas and the differential seasonal use of different areas by males and females depending on their resource requirements and the need to avoid human-related threats [14].

Behavioral patterns displayed by elephants are a result of their decision-making processes and could be influenced by their innate biology as well as the prevailing ecology. For example, a number of studies, both in Asia and Africa, have attributed crop-raiding behavior mostly to adult males [3,21,22]. In the Asian elephant, males, are typically born in a herd, reach puberty at about the age of 10 to 15 years, form loose associations with other males or live solitarily, and associate with herds thereafter only for mating [3]. These decisions seem to be primarily governed by the requirements of a certain age or a growth phase. A detailed assessment of the biological and ecological factors that are pivotal to decision-making in elephants have, however been rarely explored.

A notable exception is the recent study by Chiyo *et al.* that shows how life-history characteristics and inter-individual interactions influence decision-making by the male African elephant [22]. They argue that simplistic models of exposure of an animal to crops do not fully explain its crop-raiding behavior and establish the influence of life-history traits such as age and energy requirements, as well as cultural traits such as social learning, on their raiding behavior. This study, however, explores the behavior of individual male elephants in all-male groups alone and does not take into account the presumably important influence of mixed age-sex herds on such decision-making processes. More importantly, this study does not consider the often-idiosyncratic crop-raiding decisions made by individual elephants, independently of their biological attributes and socioecological environments, but which, we think, are crucial determinants of behavioral decisions, including crop-raiding, made by Asian elephants. It is, therefore, crucial that we comprehensively identify the factors driving behavioral decisions by the Asian elephant not only to understand its biology better but also to develop more effective management strategies for it; this is especially true in the light of its increasing conflict with humans, which has emerged as possibly its most important conservation threat in the 21st century. It is perhaps essential that we distinguish between decision-making processes that are relatively more influenced by innate biological features of the species such as gender or age and those that are more idiosyncratic, being determined at the level of the individual. Management strategies would then have to be designed to address problems posed by particular individuals rather than for the population at large.

In this study, we assessed the variation in decision-making by individuals of an Asian elephant population, residing in a fragmented human-dominated landscape, in response to varying levels of resource availability (forage, water and shade) as well as human disturbance. Residence time and movement rates, time-activity budgets, social interactions and group dynamics were used as quantitative measures of such decision-making processes. We first explored the influence of innate biological variables such as gender, age and group type on these behavioral responses. We further investigated the possibility of individual idiosyncrasies impacting these decisions and manifest as variability in residence time of individually identified elephants in a particular habitat and area. This was considered important as elephants could conceivably have strategies to acquire resources and avoid threats that are unique to them as individuals.

We hypothesized the following behavioral manifestations of the decisions that could possibly be made by our study elephants in their highly fragmented habitat.

At a general population level, Elephants would prefer resource-rich areas and avoid highly disturbed areas; The time-activity budget of individual elephants would be influenced by human disturbance and could lead to an increase in vigilance-related behaviors with a concomitant decrease in the time spent in foraging-related activities; Individuals would associate in smaller groups, reduce social interactions and increase movement rates in highly disturbed areas; Elephants would differ in their levels of occurrence in disturbed areas and in their utilization of resources in relation to their gender, age and group type; At an individual level, The residence time in different kinds of habitats, characterized by varying levels of resource availability and human disturbance, could vary across individuals idiosyncratically, independent of their gender, age and group type.

Results

General population-level behavioral pattern

Elephants Occur Mostly in Resource-Rich and Undisturbed Areas. We quantified the probability of elephants to occur with different residence times in strata that were independently classified as high, medium or low with reference to their resource (forage, water or shade) availability and human disturbance (Figure 1). We used Pearson's correlation to assess the multicollinearity amongst the measured variables. The variables forage and shade alone were found to be significantly correlated ($r = 0.48$, $p<0.001$) while the other variables were not. Our field observations on elephants, however, indicated that vegetation cover during the day and the cover of darkness at night played a key role in influencing decision-making in elephants. We, therefore, treated shade as a surrogate for cover availability and have, accordingly, presented the results of the G-tests of goodness of fit and independence for shade. We hypothesized that elephants would occur more in strata where forage, water and shade were relatively more abundant and less in strata with high human imprint. Our results show that indeed elephants did have significantly higher Observed Residence Times in strata that had relatively greater availability of forage (G-test of goodness of fit, $G = 270.33$, $df = 2$, $p<0.001$), water ($G = 403.50$, $p<0.001$) and shade ($G = 168.13$, $p<0.001$), and those with relatively low human disturbance ($G = 231.99$, $p<0.001$) than what would be expected if Observed Residence Times depended on the area available in each kind of stratum independent of resource availability and human disturbance in these strata (Figure 1).

Increasing Human Disturbance Reduces Effective Feeding Time, Alters Movement and Increases Time Spent Standing Alert. An examination of the overall time-activity budget of the study elephants showed that they spent a majority of their time in foraging-related behaviors, including feeding (45.63%) and moving (30.48%). These behaviors and resting, measured as proportion time spent in these behaviors, increased in habitats with increasing forage and shade, but decreased in those with high human disturbance (Figure 2). These behaviors, however, did not differ significantly with water availability. The proportion of time spent feeding increased with increasing forage (G-test of goodness of fit, $G = 40.57$, $df = 6$, $p<0.001$) and decreased with increasing disturbance ($G = 54.20$, $p<0.001$). It should be noted that the proportion of time spent in feeding was highest during morning (57.14%) than during the other times of the day including the night. Moving, decreased with increasing forage ($G = 40.57$, $p<0.001$) and shade ($G = 13.15$, $p = 0.041$). It, however, increased with increasing disturbance ($G = 54.20$, $p<0.001$). Moving occurred most significantly during the night (49.12%). The elephants increased the proportion of time standing alert with increase in human disturbance ($G = 40.57$, $p<0.001$) but also with increasing forage ($G = 54.20$, $p<0.001$) and water ($G = 13.58$, $p = 0.035$). Resting was virtually absent in highly disturbed areas ($G = 40.57$, $p<0.001$) and increased significantly with increasing shade ($G = 13.15$, $p = 0.041$).

Elephants Associate in Smaller Groups, Reduce Social Interactions and Increase Movement Rates in Highly Disturbed Areas. The group size of elephants in the study area varied from a minimum of 1 to a maximum of 12, with a mean (± SE) of 6.50 (±0.06) individuals (Figure 3 and Figure S1). The mean (± SE, range) group size in high-disturbance – low-resource, medium-disturbance – medium-resource and low-disturbance – high-resource areas were 5.67 (±0.13, 1–8), 4.89 (±0.07, 1–8) and 7.09 (±0.08, 1–12) respectively. Group sizes greater than seven were not observed in high-disturbance – low-resource areas. Elephant group size significantly differed across varying levels (high, medium and low) of forage (G-test of

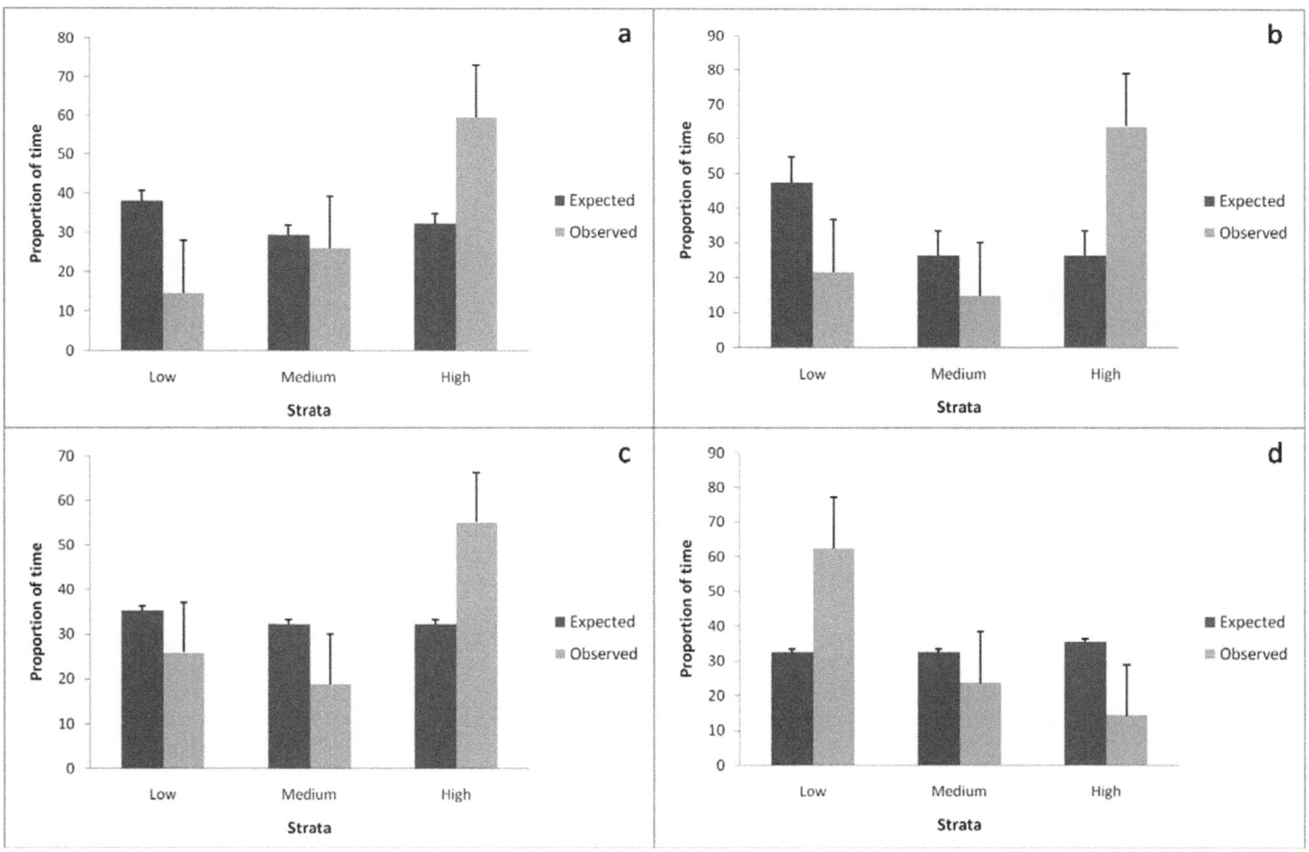

Figure 1. Expected and Observed Residence Time of elephants in the different strata. Low-, medium- and high levels of (a) forage-, (b) water-, (c) shade availability and (d) human disturbance.

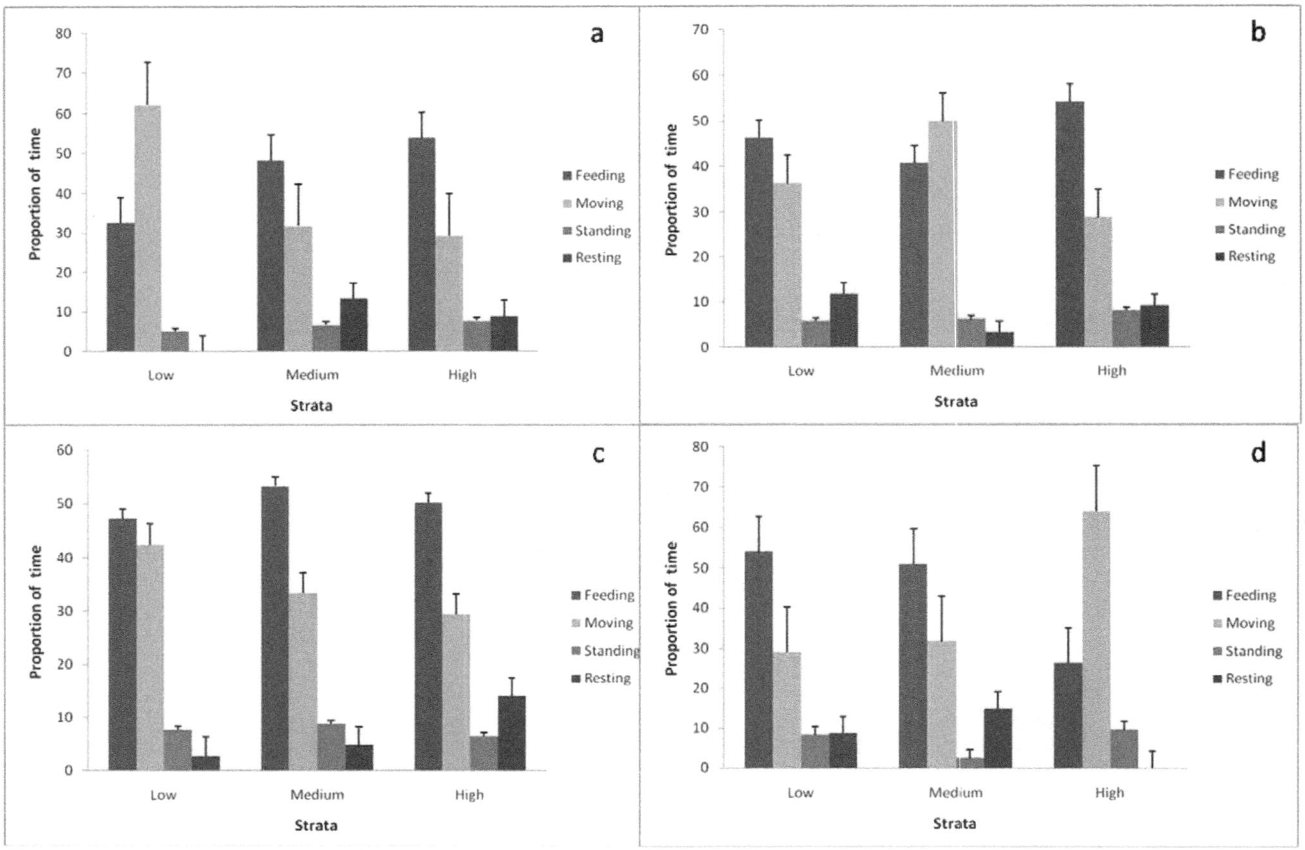

Figure 2. Proportion of time spent in feeding, moving, standing and resting in different strata. Low-, medium- and high levels of (a) forage-, (b) water- and (c) shade availability and (d) human disturbance.

independence, G = 710.56, df = 18, p<0.001), water (G = 620.88, p<0.001), shade (G = 439.44, p<0.001) and human disturbance (G = 844.88, p<0.001) than what would be expected of these habitats by chance alone.

The study elephants displayed increasing levels of bunching behavior, with lowered Inter-Individual Distance values, with increasing levels of human disturbance (Wilcoxon rank sum test with continuity correction, W = 465.50, N = 113, p<0.01). They vocalized audibly more frequently when in less-disturbed areas than they did in those with high human disturbance (G test of goodness of fit, G = 16.11, df = 2, p<0.001).

We hypothesized that elephants would move rapidly across highly disturbed areas in order to minimize their exposure to threat by humans. Our study elephants moved at a rate four times higher across high-disturbance areas as compared to that in low-disturbance areas (Figure 4). The movement rate was least in high-forage (8.97 m/min) and high-shade (8.90 m/min) areas and highest in areas with high human disturbance (31.65 m/min) and medium forage (37.06 m/min). The rate of movement also varied significantly in strata with different levels of forage (G-test of goodness of fit = 24.30, df = 2, p<0.001), shade (G = 10.02, p = 0.006) and disturbance (G = 11.09, p = 0.003). It did not, however, vary with differing water availability (G = 0.03, p = 0.98).

Elephants Vary in Their Residence Time in Different Habitats Depending on Their Gender, Age and Group-type. Both male and female elephants occurred primarily in high-resource – low-disturbance areas (Figure 5 and Figure S2). The genders, however, differed in their residence time across different strata with varying levels of forage (G-test of independence, G = 112.18, df = 2, p<0.001), water (G = 35.45, p<0.001) and human disturbance (G = 87.25, p<0.001). No difference, however, could be detected with varying shade (G = 6.48, p>0.05). Female elephants thus occurred in medium-forage – medium-disturbance areas at levels higher than that of males (Figure 5; G = 112.18, p<0.001). Male elephants, in contrast, occurred at significantly higher levels in high-forage areas (G = 112.18, p<0.001) and at significantly lower levels in low-disturbance areas (G = 87.25, p<0.001).

We observed no difference between the different age classes of elephants across the sexes (adult, subadult, juvenile and calf) in their residence time in habitats with differing levels of forage, water and human disturbance. Their occurrence, however, differed with varying levels of shade (G = 6.43, p = 0.04). When only males were considered, adult elephants occurred in high-forage – low-disturbance areas at levels higher than did subadults. The latter age class, on the other hand, occurred in medium-forage – medium-disturbance areas at levels higher than the former. Adult and subadult males differed in their residence time in strata with different forage (Figure 6 and Figure S3; G = 85.69, p<0.001), shade (G = 54.89, p<0.001) and disturbance (G = 44.25, p<0.001). Adult and subadult female elephants also showed differences in their occurrence in habitats with varying levels of forage (Figure S4; G = 16.94, p<0.001) and disturbance (G = 17.90, p<0.001).

When individual adult and subadult male elephants associated to form All Male Groups (AMG), they showed a higher propensity

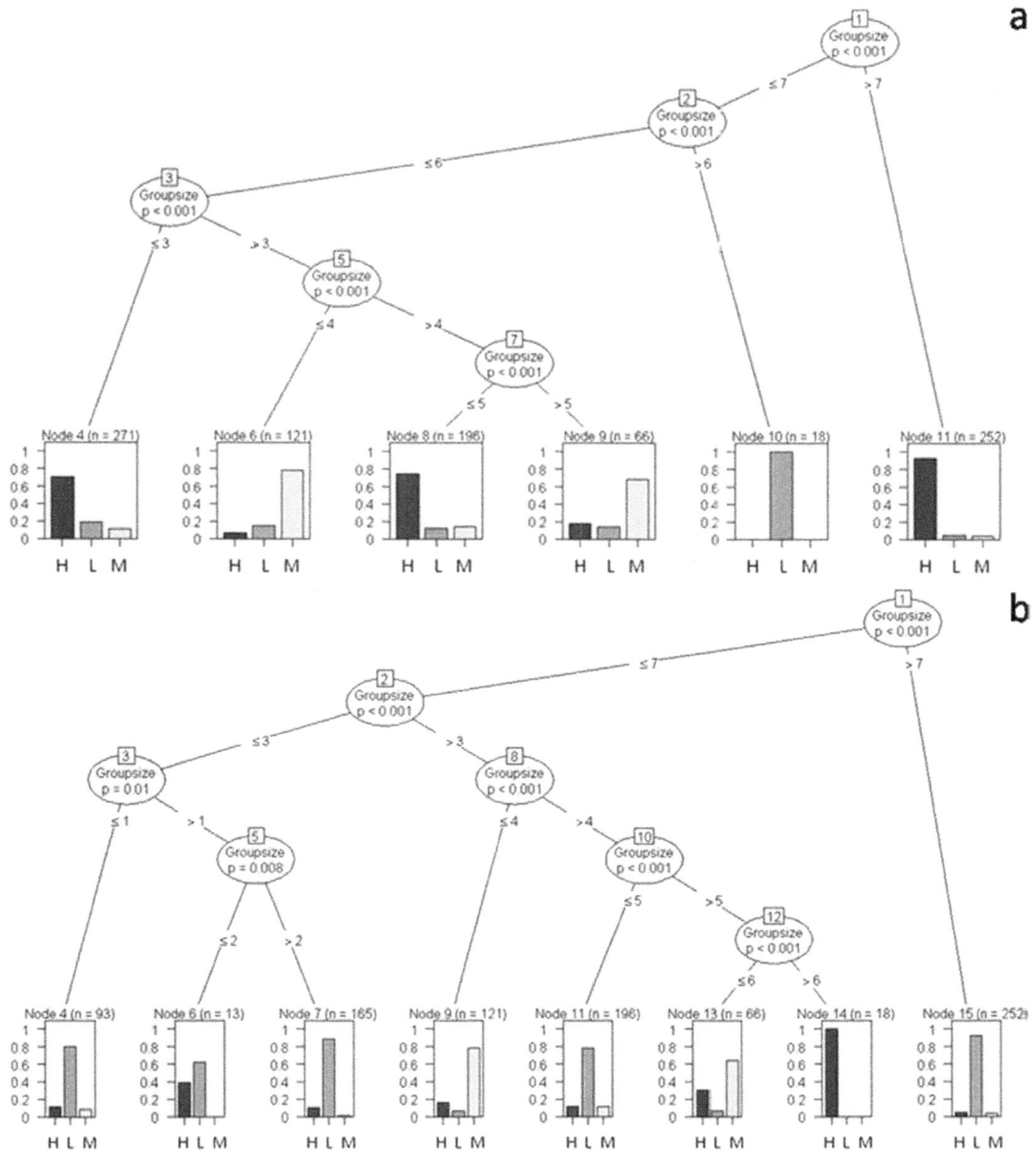

Figure 3. Classification trees showing the partitioning of elephant group size in the different strata. L: low-; M: medium- and H: high (a) forage availability and (b) human disturbance. The y-axis of each graph indicates the proportion of groups observed in the different strata. For the trees depicting the influence of water- and shade availability, see Figure S1.

of occurrence in high-disturbance areas. Solitary elephants, in contrast, exhibited the least propensity to occur in such areas, followed by Herds (Figure 7). There were significant differences in the occurrence of these group types in different habitats as a function of forage availability ($G = 539.36$, $df = 4$, $p < 0.001$) and human disturbance ($G = 400.85$, $p < 0.001$). Solitary individuals and AMG differed from Herds in their residence time with varying levels of water ($G = 90.81$, $p < 0.001$) and shade ($G = 41.754$, $p < 0.001$). No difference was, however, observed between Solitary elephants and AMG across differing availability of water (Node 2, $p > 0.05$) and shade (Node 2, $p > 0.05$).

Specific individual-level behavioral patterns

Recursive partitioning classification trees constructed for residence time of individual elephants indicated that the primary decisions made by our study elephants occurred at the level of the individual independent of their biological and ecological attributes. This variability in behavioral decisions could, at best, only be attributed to individual idiosyncrasies. Some of these decisions made by identified individuals, however, changed when they associated with certain conspecific individuals.

The total number of scans on identified individual adult and subadult males (>10 years) ranged from 5 to 119 with a mean (± SE) of 32.60 (±11.83) and that on identified herds from 4 to 560 with a mean (± SE) of 238 (±61.10). We have, however, chosen individual adult and subadult males with a minimum number of 35 scans and 8.75 h of observation on each individual and herds with 419 scans and 104.75 h of observation on each herd for further analysis. As an example, let us consider a particular classification tree shown in Figure 8. This tree depicts decision-

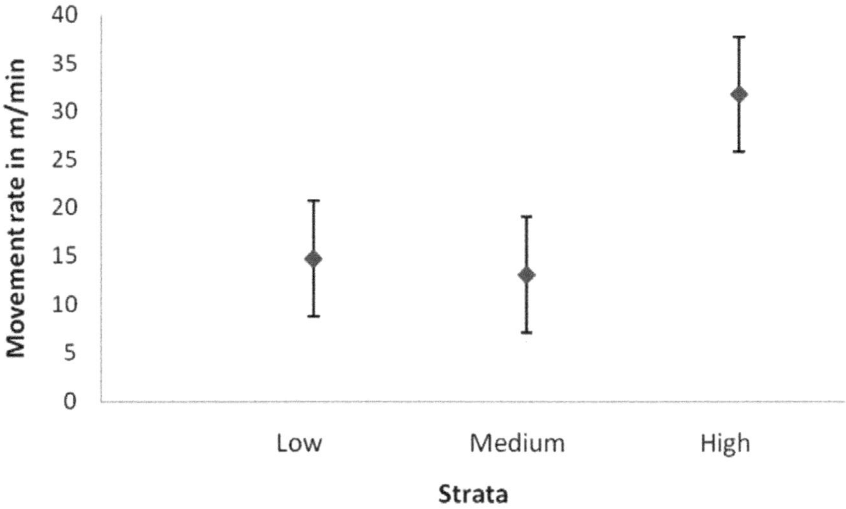

Figure 4. Movement rate (mean ± SE) of elephants in different strata of human disturbance.

making among our study adult and subadult male elephants, either when Solitary or in AMG, in response to resource availability and human disturbance measured on ordinal scales. The adult male VKT displayed a high propensity to occur in high-disturbance areas (Node 1) as compared to the other males and his behavior did not change even when he associated independently with other conspecific individuals. Two other males, SID and Stalker, in contrast, clearly differed in their residence time in different habitats as a function of the group in which they occurred (Node 2, $p = 0.004$). For example, when these males were solitary, they preferentially occurred in low-disturbance areas (Node 3); when in AMG, however, they increased their propensity to occur in high-disturbance areas (G-test of independence, $G = 17.11$, $df = 2$, $p < 0.001$). When decision-making by SID and Stalker was considered in relation to a different ecological variable, forage availability, however, they showed an increased propensity to spend significantly higher residence time in high-forage areas when compared to VKT (Node1). When in AMG, however, SID showed a tendency to spend more time in low-forage areas as compared to Stalker ($G = 14.30$, $df = 2$, $p < 0.001$). Figure S5 lists the classification trees that reveal variability in residence time displayed by individual adult and subadult male elephants under water- and shade availability.

Individuality was also observed across Herds (Figure 9). Herd QH, for example, displayed the highest propensity of occurrence in low-forage – high-disturbance areas (Nodes 3 and 2 respectively) as compared to the Herds FFH and MH, which had highest residence times in high-forage – low-disturbance areas (Node 5 and 3 respectively). Figure S6 shows the classification trees that reveal variability in residence time displayed by individual Herds under water- and shade availability.

Discussion

The habitat of the Asian elephant is increasingly becoming fragmented due to anthropogenic influences across the distribution range of the species [23]. Many protected areas in the Indian subcontinent that harbor a high density of elephants nevertheless record intense human-elephant conflict in their fringe villages if not within the protected area itself [24]. The Bannerghatta National Park in southern India is one such elephant landscape that faces high levels of conflict, particularly related to crop raiding by elephants. Any attempt to manage and conserve this species is thus crucially dependent on our understanding of the behavioral decisions made by elephants in such disturbed ecological settings. The current study is possibly the first to extensively investigate the variation in decision-making by Asian elephants as reflected in their residence time, time-activity budgets, social interactions and grouping patterns under varying levels of resource availablity and human disturbance. It appears that such decisions are critically influenced, on the one hand, by biological and socioecological factors at the population level, such as gender, age and grouping patterns, and, on the other, by individual idiosyncracies.

The elephants in the study area appear to selectively respond to locally dynamic ecological situations by preferring to reside in resource-rich, but low human-disturbance areas, a pattern that has been observed earlier in both Asian and African elephants [4–6,25–32]. Such risk-avoidance behavior may have aided in the long-term survival of elephants in fragmented habitats by way of their spatial and temporal separation from humans thus keeping conflict levels low [8,19,20]. Increasing human disturbance, by way of degradation of the quality and quantity of vegetation within the protected area and disrupting free access to and the use of crucial resources by the elephants within the Park, however, may lead to escalated conflicts in the near future and may, in turn, affect stress-related mortality patterns in such populations.

The rapidly-changing land-use patterns that characterize the neighborhood of the Bannerghatta National Park and the seasonal variability in resource availability within the Park, especially during the dry season, not only alters the time-activity budgets of elephants but also leads to significant changes in their use of the habitat, depending on their gender, age and group-types.

A reduction in foraging-related activities with a corresponding increase in vigilance behaviors was, thus, observed among our study elephants in areas with high human disturbance. The proportion of time spent feeding, for example, reduced from 54.08% in low human-disturbance areas to 26.44% in high human-disturbance areas. Resting ceased in highly human-disturbed areas, with the proportion of time spent moving increasing to 64.07% in these areas from 29.07% in areas relatively less disturbed by people, a finding that reflects in Asian elephants suggestions of similar patterns in their African counter-

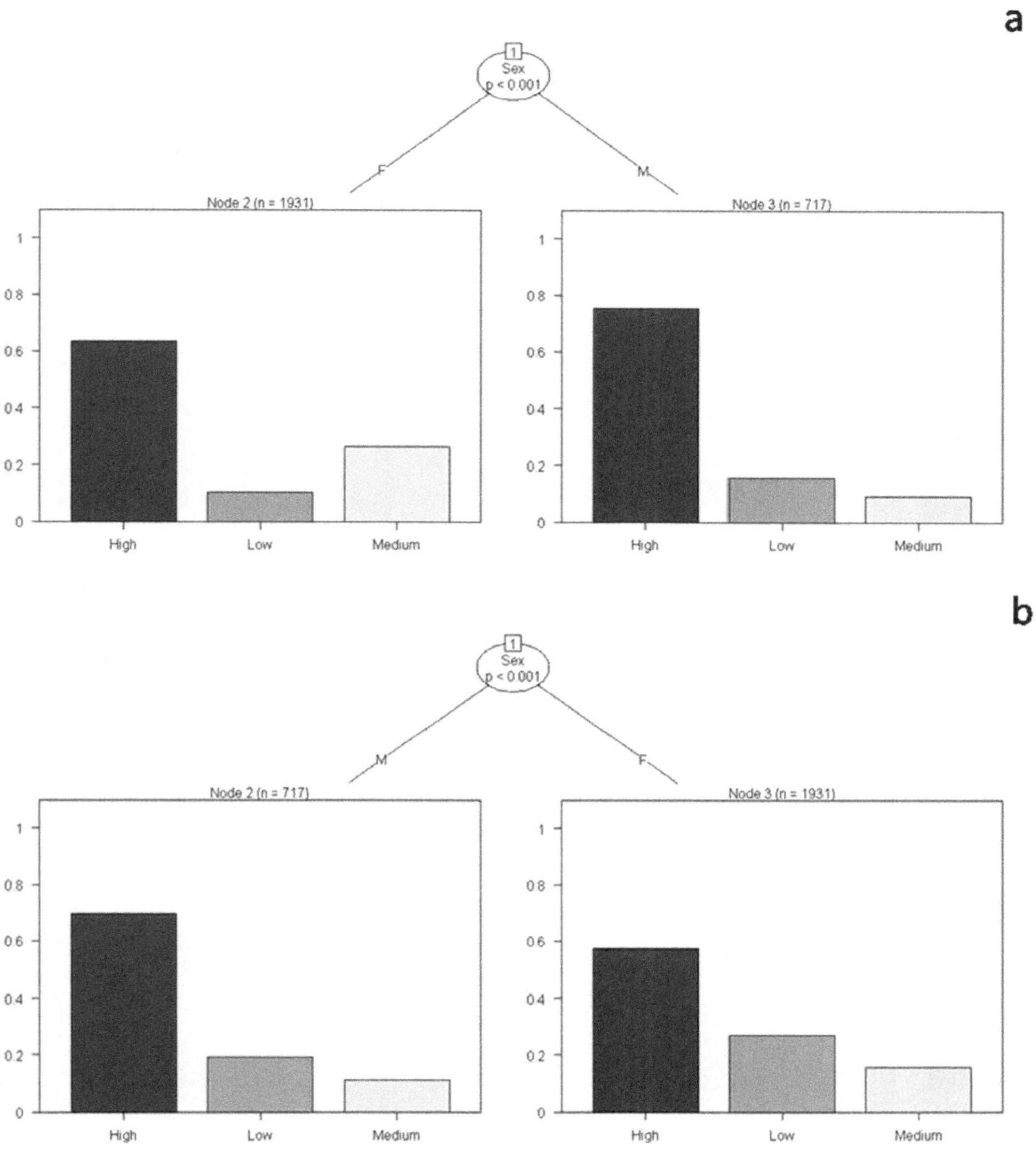

Figure 5. Classification trees for residence time of female and male elephants in the different strata. (a) Forage- and (b) water availability. The y-axis of each graph indicates the proportion of groups observed in the different strata. For the trees depicting the influence of shade availability and human disturbance, see Figure S2.

parts [19,33]. Elephants in and around the study area are constantly persecuted by local communities who depend on these habitats for their resources, especially farmers whose cropfields are raided by the elephants. The time spent standing alert also increased significantly in disturbed areas; the elephants usually froze as a first reaction to the close proximity of humans while they waited for the perceived threat to pass. Increased movement away from people was observed only when targeted efforts were made to drive them away, either from cropfields or from areas where livestock had been taken to graze.

We used recursive partitioning classification trees, as an exploratory analysis tool, to statistically test and visualize the differences observed in behavioral decision-making by elephants in relation to low-, medium- and high levels of resource availability and human disturbance. The observed variation in residence time of elephants, as a function of more general level attributes such as group size, gender, age and group-type, as well as at the individual level were all assessed in these strata using classification trees. Such an analysis has possibly never been used before to unravel behavioral decision-making in elephants and can potentially be a powerful tool to explain elephant occurrence and habitat use in different environments.

Group size in elephants is known to increase with a concomitant increase in resource availability and reduce when resources become limiting [3]. The group size of Asian elephants in dry habitats has been found to usually range between 5 and 10 individuals [3]. The elephants in the study area, all identified individuals, actively reduced their group size not only in relation to reduction in resource availability but also to increasing human disturbance, all observed differences in group size being statisti-

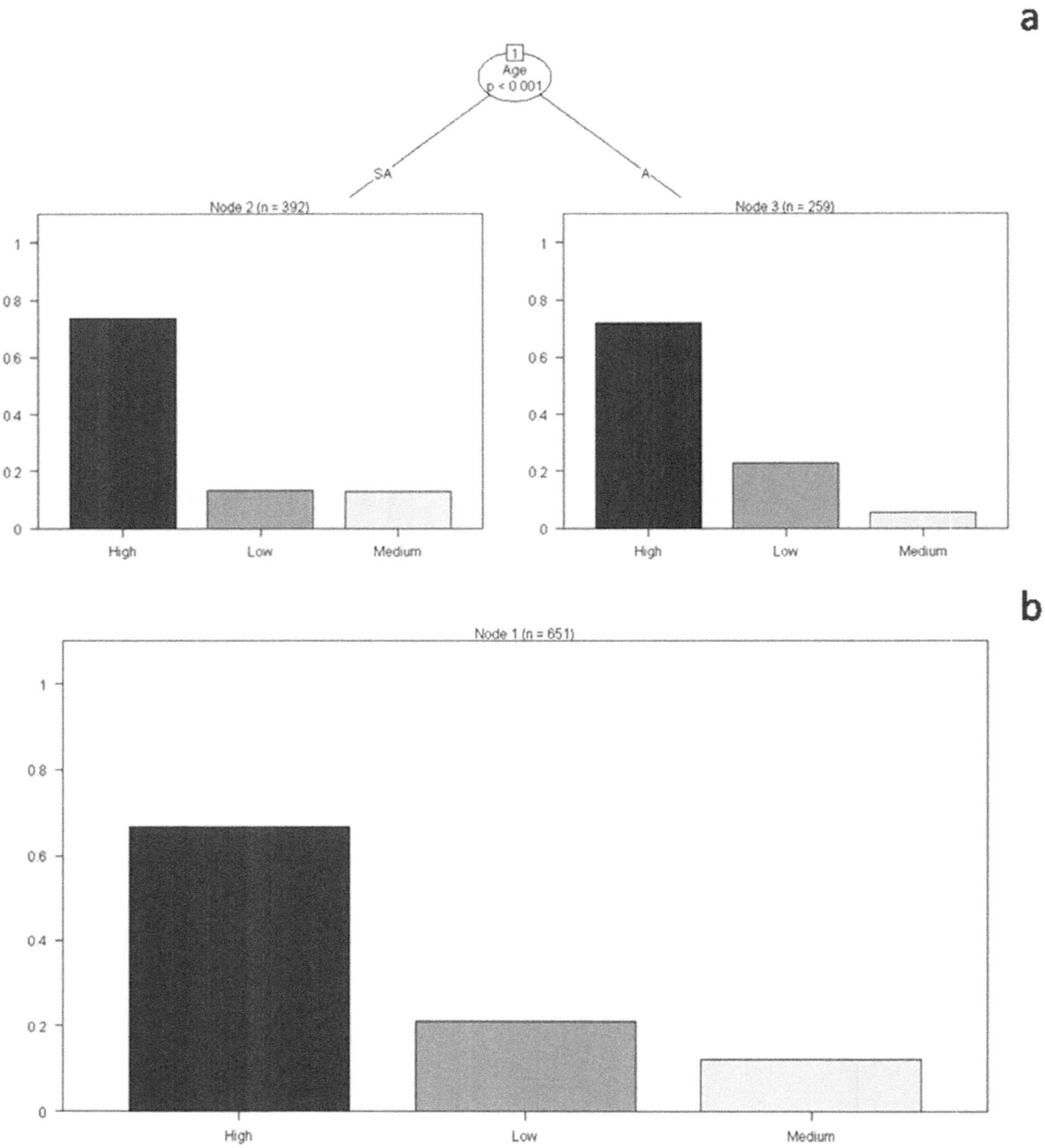

Figure 6. Classification trees for residence time of adult and subadult male elephants in the different strata. (a) Forage- and (b) water availability. The y-axis of each graph indicates the proportion of groups observed in the different strata. For the trees depicting the influence of water- and shade availability, see Figure S3. For the trees depicting the influence of forage-, water-, shade availability and human disturbance on residence time of adult and subadult female elephants, see Figure S4.

cally significant. Associating in relatively larger numbers in highly human-disturbed areas could lead to greater detectability and hence, crop-raiding in smaller units could be a more prudent strategy for these elephants. Support for such a hypothesis comes from our own personal observations when we were able to detect larger groups of elephants with relatively greater ease, even in darkness. A second possibility could be that elephants prefer to raid cropfields in family units alone, especially during times of resource limitation, as direct benefits would then only accrue to highly related kin groups.

Our observations on the bunching behavior of the study elephants, associated with a reduction in inter-individual distance in highly human-disturbed areas has been observed in other studies too, both in the Asian and African elephant [8]. The demonstrated increase in stress in elephants, possibly associated with a heightened sensitivity to human disturbance and a lack of resources, could have other significant effects on their behavior and physiology [33]. Our study elephants, for example, significantly reduced their audible vocalizations when in highly human-disturbed areas. This was particularly true when they raided cropfields or moved through areas with rampant cattle grazing or firewood collection. Actively reducing loud vocalizations could be a learnt behavior, as we observed that such vocalizations, even though rare, elicited frequent antagonistic responses, usually first from dogs and then from people; such learning could well be reinforced in an area such as the Bannerghatta National Park which has had a long history of human-elephant conflict [34]. This study, however, did not attempt to establish whether these elephants shifted from audible to infrasound communication during these times. Another behavioral adaptation that we

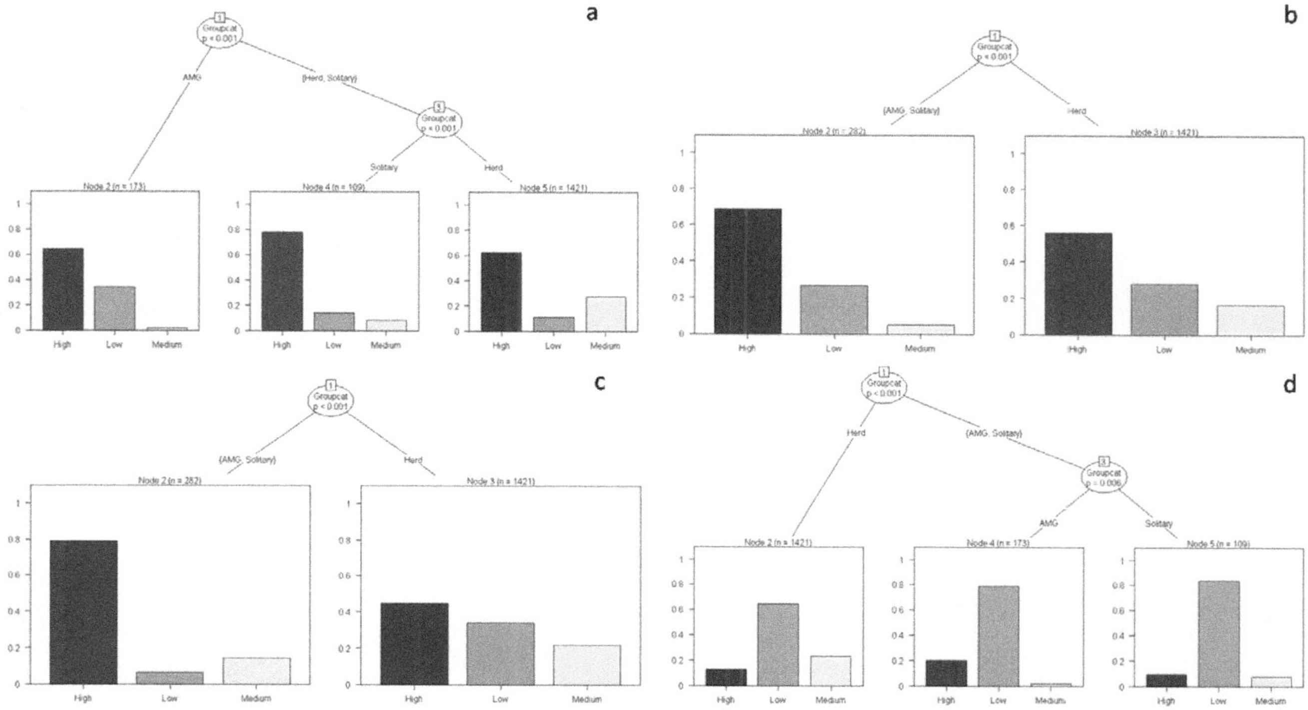

Figure 7. Classification trees for residence time of adult and subadult male elephants in the different strata as a function of group types. (a) Forage-, (b) water- and (c) shade availability and (d) human disturbance. The y-axis of each graph indicates the proportion of groups observed in the different strata.

observed was a significant increase in the rate of movement displayed by the elephants especially when travelling through areas of high human disturbance. Such a potential adaptation to anthropogenic threats has been documented earlier when elephants moved faster across forested tracts that were highly human-disturbed but connected resource-rich patches [4,19,33].

Gender and age are two important attributes that influence decision-making by elephants. Earlier studies on both Asian and African elephants have described differential patterns of habitat use by the two sexes although there have been very few investigations on the influence of age on such patterns [3]. What needs to be explored further, however, are the underlying motivations that drive individuals of different age-sex categories to selectively utilize areas of differing resource availability and human disturbance.

The high observed propensity of herds to occur in medium-forage – medium-disturbance areas and actively avoid high-forage – low-disturbance areas, which are usually frequented by adult males, for example, could be a behavioral response of females to avoid such males, possibly due to an inter-sexual dominance hierarchy that we believe exists in this species just as it does in the African elephant [20,35,36]. Such a dominance hierarchy usually manifests itself when adult males seek out herds with receptive females and display assertive behaviors towards its members, especially receptive females and young males [3]. Thus, the use of medium-forage – medium-disturbance areas by adult females could be a strategy [19] to ensure resources not only for themselves but for the rest of the herd, especially calves and juveniles, while reducing risk of encountering humans as well as certain male elephants. Female elephants, accompanied by calves and juveniles, also showed a high propensity to occur in areas with high water availability.

In polygynous species, the physical and physiological condition of males determines their positions in the dominance hierarchy and usually influences the outcome of inter-specific competition [12,37,38]. Male elephants need to build body mass, come into the energetically demanding state of *musth* and seize opportunities to mate. Male elephants, thus, as is typical of most polygynous species, track both forage and females [13,14]. The occurrence of adult male elephants in high-forage – low-disturbance areas could therefore be explained as a need to maximize their energy gain and simultaneously minimize anthropogenic threats in order to improve their fitness. Subadult males, moving out of their natal herds and attempting to acquire social status [3,20,39], may also prefer high-forage – low-disturbance strata areas in order to build up body mass and to associate with adult males. Their presence in medium-forage – medium-disturbance areas at levels higher than that of adult males could possibly be expected as subadult individuals occasionally continue to associate with their natal herds.

The tendency of individual elephants to associate with one another to form transient or stable same-sex or mixed-sex groups, or to remain solitary may be an important component of their life-history strategies. Although the influence of all-male associations on crop raiding by African elephants has only very recently been explored [22], the factors motivating individual Asian elephants to form specific associations in different socioecological environments needs to be thoroughly investigated. All-male groups (AMG), formed by the association of adult and subadult males and which have been noted earlier in both Asian and African elephants [3,12], could act as transient cooperative groups in which individuals could learn survival skills from one another [12,22]. These associations could also represent genetically related kin groups, as has been demonstrated in African elephants [37]. These

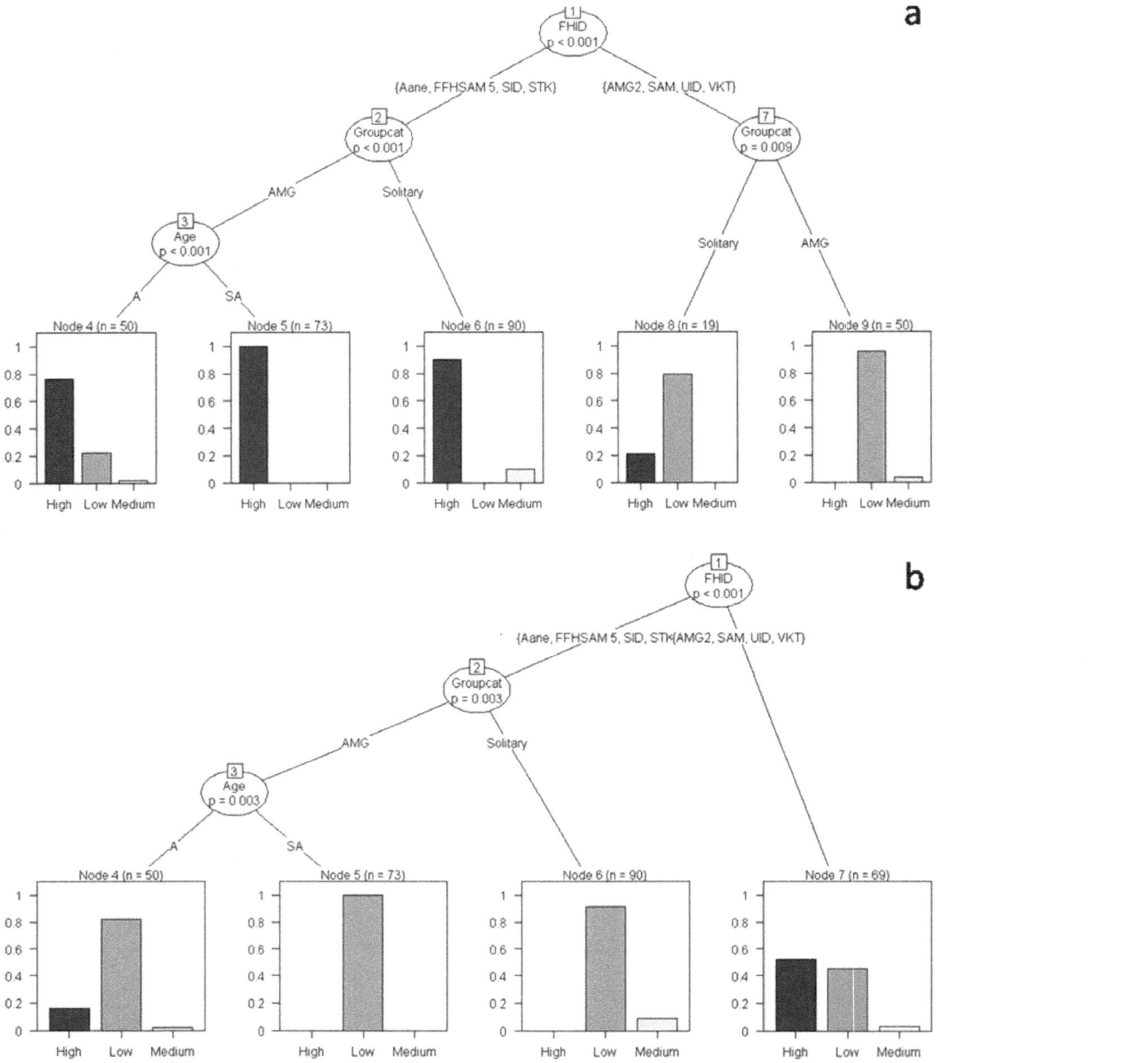

Figure 8. Classification trees of the probability of individual adult and subadult male elephants occurring in the different strata. (a) Forage availability and (b) human disturbance. The y-axis of each graph indicates the proportion of groups observed in the different strata. For the trees depicting the influence of water- and shade availability, see Figure S5.

groups most often occurred in the high-disturbance – low-forage strata than in any other kind of forage- or disturbance regime during our study. Although such a choice of habitat could appear counter-intuitive, individual males in AMG tended to be more exploratory and often raided cropfields that required them to traverse high-disturbance tracts. They were perhaps able to withstand impending threats due to their close association with one another without the encumbrance of dependent young, as faced by herds.

Herds, in contrast to AMG or solitary males, typically occurred in medium-forage – medium-disturbance regimes, as has been discussed above. There were, however, a few notable exceptions. Herd QH differed from the others by showing a high tendency to frequent highly disturbed areas. Although a single study each on both Asian and African elephants have suggested that herds lower in the social hierarchy within a population are forced to reside in areas with relatively higher human disturbance [17,19], such unusual herd behavior could also depend on two other factors, the first, as discussed above, being group size. A second, hitherto unexplored factor, could be the behavioral propensities of individual elephants comprising a particular herd. Understanding the behavioral profiles and social relationships of individual elephants within herds could serve to explain certain patterns of herd distribution and behavior that remain inexplicable, especially in fragmented human-dominated landscapes.

One of the most significant findings of this study, we believe, is that the occurrence of an individual elephant in a particular stratum could be a consequence of its age or grouping pattern,

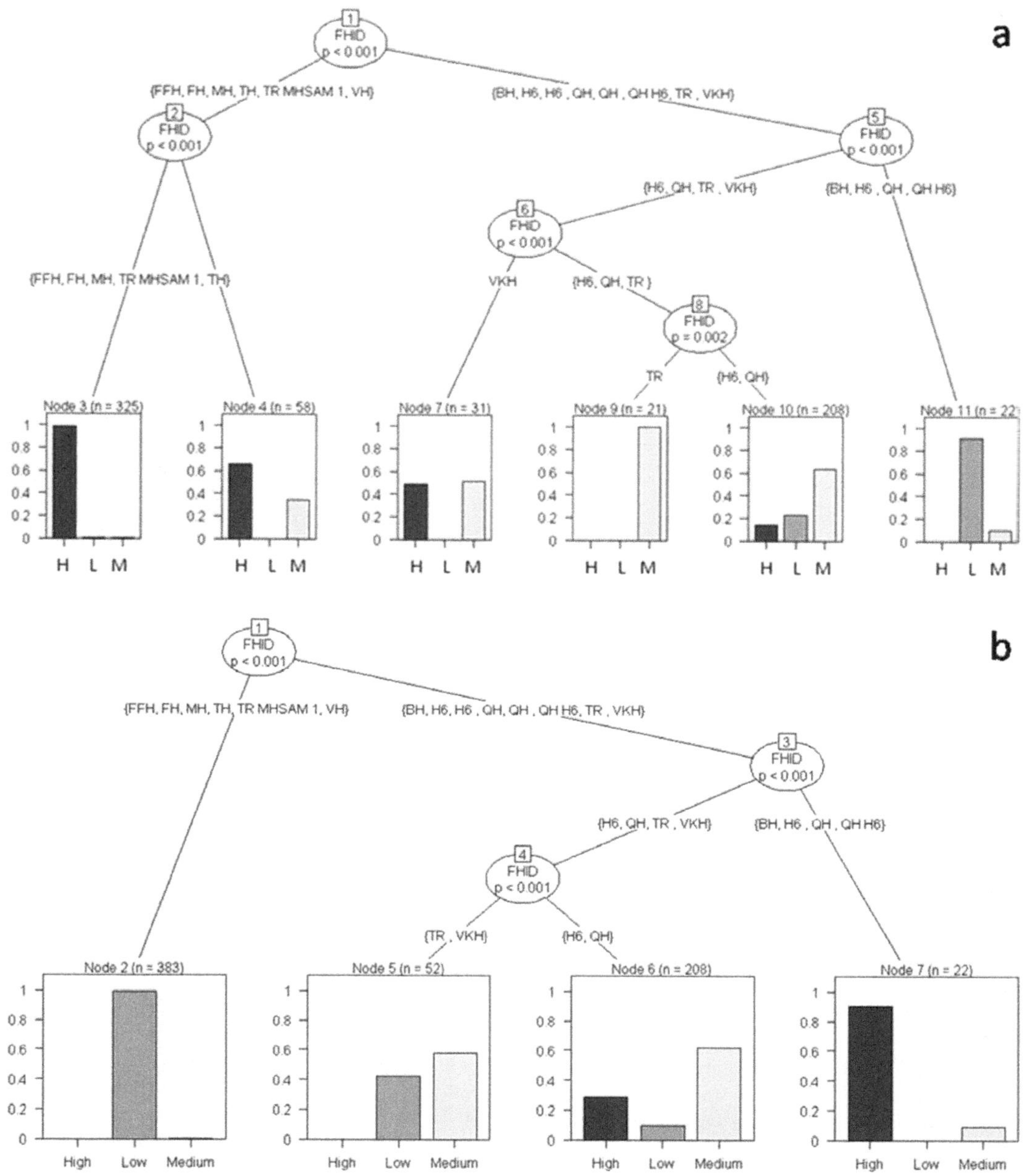

Figure 9. Classification trees of the probability of an individual herd occurring in the different strata. L: low-; M: medium- and H: high (a) forage availability and (b) human disturbance. The y-axis of each graph indicates the proportion of groups observed in the different strata. For the trees depicting the influence of water- and shade availability, see Figure S6.

neither of which is static over time. We have thus demonstrated the ability of individual elephants to change behavioral strategies depending on whether they were solitary or in different group-types. What must be recognized, therefore, is that the adoption of particular strategies usually reflect the different demands placed on individuals depending on their life-history stage. Understanding this behavioral variability of individual elephants within a particular population to the extent that their behavioral strategies can be predicted is a challenging exercise but one that must be attempted if we are to develop viable management approaches to mitigate human-elephant conflict (HEC) or, more specifically, engage with conflict animals that occur in high-disturbance areas, comprising mostly of crop fields and human habitations. This becomes particularly crucial if we plan to implement any management strategy such as capture, translocation or any other action that additionally interferes with and disrupts the social fabric of individual elephants.

In conclusion, this study clearly shows that elephants behaviorally select high-resource areas and avoid anthropogenically disturbed areas. Several behavioral activities of elephants, especially foraging, movement and social interactions, are affected by increasing disturbance levels. Thus, the management of an area for elephants must aim at reducing these disturbances. At a population level, elephants differ in their residence times within particular areas, depending on their gender, age and grouping patterns. This clearly indicates differential needs and strategies among elephants within a population. Individual level studies are thus vital in order to identify the needs of particular animals and correctly predict their behavior. This also suggests that the management of an area should be dynamic and that we need to develop predictive models that would allow us to cater to the needs of a particular elephant population and also to the demands of each individual within it. Such an approach can immensely aid managers to take on-gound well-informed decisions in order to manage conflict, especially when dealing with highly endangered temperamental animals such as elephants.

The Asian elephant is clearly a species that displays extensive social behavior, complex cognitive abilities and sophisticated decision-making processes. Given these capabilities, both at the population- and individual level, a thorough understanding of elephant behaviour is absolutely crucial for any science-based management action that aims to conserve this increasingly endangered species.

Materials and Methods

Study area

This study was conducted in the Bannerghatta National Park and its surrounding human-dominated landscape (Figure 10), which forms a part of the Nilgiris – Eastern Ghats Elephant Reserve [34,40]. It is one of the largest Elephant Reserves in India, with an area of approximately 11,000 km^2. The Eastern Ghats Mountains also remains the last largest remaining scrub forest for elephants among its range countries and, along with another mountain chain, the Western Ghats, harbors the single largest contiguous Asian elephant population in the world, with an estimated number of about 9000 individuals [41].

The Park measures approximately 26 km in length and 0.30 to 5 km in width [34]. It has an area of about 104.27 km^2 and a perimeter of 137 km. The terrain is highly undulating, with a mean altitude of 865 m above msl, and receives an average annual rainfall of 937 mm. The vegetation in the Park is predominantly deciduous to scrub woodland with riparian patches along the streams. Geographically, the Park is contiguous with larger patches of forests on its southeast and southwest.

The Park is surrounded on three sides by well-irrigated croplands and human habitations. There are five human settlements within the Park and around 117 settlements located within five km around the Park [34]. The high-density human population in and around the Park is largely comprised of subsistence farmers, livestock-grazers and manual laborers engaged in sand-mining and granite-quarrying. These communities depend on the forests for their non-timber forest produce, firewood and livestock-grazing. Amongst the farming communities, the majority are marginal farmers practicing subsistence agriculture with a handful of progressive farmers who grow commercial crops and have plantations. The major crops grown in this region are rice (*Oryza sativa*), finger millet (*Eleusine coracana*), castor (*Ricinus communis*), maize (*Zea mays*), banana (*Musa paradisiaca*), mango (*Mangifera indica*), sapota (*Achras zapota*), coconut (*Cocos nucifera*) and jackfruit (*Artocarpus heterophyllus*).

The number of incidents of crop depredation by elephants registered every year within and around the Park ranges from 470 to 1477 with a mean (± SE) of 900 (±151.33). The Forest Department records indicate that the conflict situation was at its peak in the years 2005–2006, recording 1477 cases and resulting in the Department spending INR 18,48,250 (USD 42,982) as compensation towards life and property damages alone. On an average, two people have been killed and two injured by wild elephants, and two elephants killed or poached in retaliation, each year since 1997 [40].

This study was conducted during the dry season [34], from December 2009 to May 2010, both within the Park and in the surrounding human-modified landscape. The study area, as a whole, could be classified into at least seven different land cover types, namely, dry deciduous forest (32.44%), scrub forest (11.44%), rocky outcrops and hills (5.22%), plantations (5.11%), cropfields (29.11%), sand and granite mines (8%) and human habitations (8.67%; see also Table 2).

Study species and population

The Asian elephant has today been estimated to number about 35,000–50,000, spread across its range over thirteen Asian countries [23]. India has approximately 50% of the total population of wild Asian elephants (20,000 to 25,000), with southern India supporting around 10,000 elephants in the wild [42]. Owing largely to the pressures of hunting as well as habitat loss, fragmentation and degradation, the geographic range of the Asian elephant has, however, declined by more than 70% since the 1960's. The surviving populations are highly fragmented and the species is listed as Endangered in the *2011 IUCN Red List of Threatened Species* [43]; it is also included in the Appendix I of the Convention on International Trade in Endangered Species of Wild Fauna and Flora (CITES, 43) and in the Schedule 1 of the Wildlife Protection Act of India 1972 [44].

The study population, a sub-set of the larger Nilgiris – Eastern Ghats elephant population, is free ranging in the study area and occurs at a fairly high density of 0.90 elephants km^{-2} [45]. Elephants regularly move in and out of the study area, however, which could be expected given their migratory nature and the contiguity of the Park with neighboring elephant habitats to its southeast and southwest. The study population of 60 individuals (in 10 herds, 3 all-male groups and as 8 solitary males), which has been monitored since 2007 [34] for its demographic parameters, consists of 25 adult and 17 subadult elephants, their juveniles and calves.

Habitat survey

In order to assess the effect of varying levels of resource availability and human-induced disturbance on the behavior of the study elephants, a habitat survey was carried out as a pre-requisite to quantify available resources and the intensity of human-induced disturbance. For this purpose, a total of 46 grids of 4 km^2 each were overlaid on the study area using a topographic sheet (1:250,000). The maximum average values of Normalized Difference Vegetation Index (NDVI) and Leaf Area Index (LAI) were extracted using MODIS data products for each of the 2×2 grids using the GIS software Quantum GIS [46] and Geographic Resources Analysis Support System [47]. The values of NDVI and LAI were used as surrogate measures of forage [48–51] and shade availability respectively. A sampling effort of four km (with eight straight line segments, each of 500 m) was invested in each grid with a total sampling effort of 184 km for habitat surveys to measure the number of water sources and human-induced disturbance, and evaluate the major land-use types. We stratified the study area into zones of low-, medium- and high levels of forage-, water- and shade availability, and human-induced disturbance, all the four variables being considered independently (Table 1). The values of NDVI, LAI, number of water holes per grid and the Human Disturbance Index (defined here as the encounter rate of human or human-associated disturbances×proportion of segments in which these disturbances were found) were respectively used as surrogate measures of these four variables and a quantile classification algorithm (Quantum GIS, version 1.4.0)

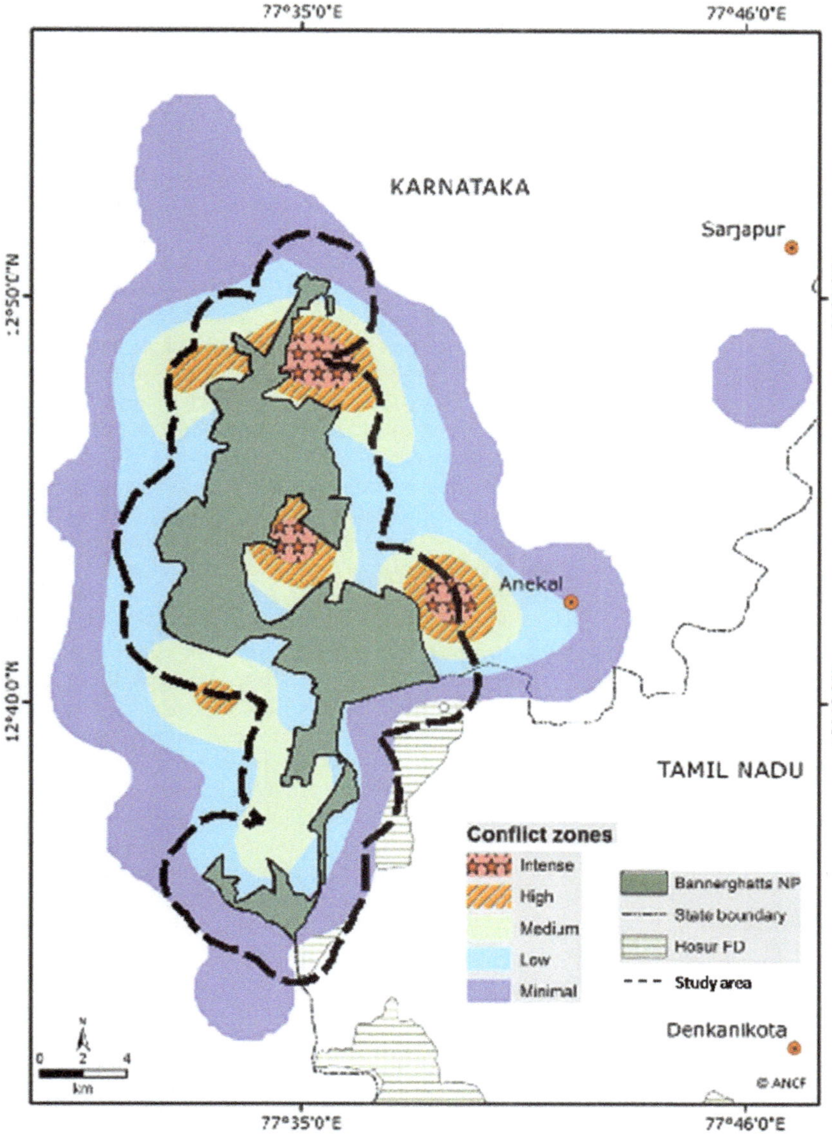

Figure 10. Map of the study area. The level of human-elephant conflict in the surrounding human-dominated landscape is shown.

was used to determine the cut-off ranges for each stratum. The distribution of the seven land-use patterns, mentioned above, across these resource availability and human disturbance strata was evaluated in order to examine the habitat characteristics of these strata (Table 2).

Table 1. Stratification of the study area into zones of low-, medium- and high levels of forage-, water- and shade availability, and human disturbance.

Forage	Range value of NDVI	Water	Range value of number of water holes per grid	Shade	Range value of LAI	Human disturbance	Range value of HDI
Low	4.29 to 5.11	**Low**	1 to 3	**Low**	5.30 to 7.10	**Low**	0 to 18
Medium	5.12 to 5.66	**Medium**	3 to 5	**Medium**	7.10 to 8.35	**Medium**	18 to 60.50
High	5.67 to 6.08	**High**	5 to 10	**High**	8.35 to 17.15	**High**	60.50 to 112

The stratification is on the basis of Normalized Difference Vegetation Index (NDVI), number of water holes per grid, Leaf Area Index (LAI) and Human Disturbance Index (HDI) respectively; The range value of each variable used for the classification has been shown.

Table 2. Distribution of land-use types observed in the study area across the zones of low-, medium- and high levels of forage-, water- and shade availability, and human disturbance.

	Percentage of zone under land-use type						
	DD	**SC**	**RO**	**PL**	**CF**	**MI**	**HH**
Low forage	14	14	6	4	41	7	13
Medium forage	45	12	6	7	17	6	7
High forage	44	7	3	5	25	11	5
Low water	32	5	6	7	29	8	15
Medium water	63	0	15	3	15	0	5
High water	54	0	4	14	19	0	9
Low shade	24	12	6	4	39	8	8
Medium shade	37	11	8	6	23	9	6
High shade	37	12	2	5	24	8	12
Low human disturbance	58	11	6	6	9	4	6
Medium human disturbance	18	12	6	3	38	15	8
High human disturbance	25	11	4	8	38	2	13

DD: Dry deciduous forest, SC: Scrub forest, RO: Rocky outcrops and hills, PL: Plantations, CF: Cropfields, MI: Sand and granite mines, HH: Human habitations.

Demographic and behavioral sampling

This study is based on approximately 200 field-hours of demographic and behavioral observations (amounting to 663.25 elephant-hours) conducted on 60 individually identified elephants, when they were in herds, all-male groups or solitary. The difficult terrain and habitat, combined with tracking on foot, however, limited the hours of observations per day to a mean (± SE, range) of 3.03 (±0.45, 0.75–13.75) h. A photographic file with sighting number, location, gender, approximate age and distinguishing morphological features was thus created for each individual elephant. The age of each individual was estimated from its physical characteristics and comparative measures of shoulder height [34,52]. The study elephants were classified into three group types, namely, solitary (single male elephant), all-male group (AMG, a coalition of male elephants comprising at least one adult and one or more subadult males, mean ± SE group size of 2.50±0.50, range 2 to 3) and herd (one or more family units with mean ± SE group size of 6.60±2.10, range 3 to 12) on the basis of their group size and structure. They were further grouped into four age classes, namely, adult (>15 years, mean ± SE of 34.22±1.52 years, range 20 to 50 years), subadult (range of 5 to 15 years, mean ± SE of 11.79±0.53 years), juvenile (1 to 5 years, mean ± SE of 3.75±0.28 years, range 2 to 5 years) and calf (<1 year). The adult male to female ratio was 1:2.57 and the adult female to calf ratio 1: 0.28.

The location of each identified elephant, whether solitary or in a herd or all-male group, was obtained once every 15 min using an e-trex Global Positioning System (GPS; Garmin Corp, Kansas, USA). Each recorded point was then assigned to the grid in which the elephant was observed. This data was further processed to class the grid into one of the 12 strata described above on the basis of resource availability and human disturbance (Table 2). The probability of individual elephants occurring in one of the 12 strata as a function of their gender, age and group type was then assessed. Observed Residence Time (ORT) has been defined as the total time spent by elephants in grids assigned to a particular stratum of resource availability or human disturbance. Expected Residence Time (ERT) was calculated by multiplying the sample size obtained from ORT with the area under each stratum.

Behavioral states including feeding, resting, moving, bathing/drinking and social interactions were recorded by instantaneous scan sampling of all visible individuals, whether solitary or in an association, at 15-min intervals [53]. Sampling was conducted during three sessions of three-hour duration each (Session 1: 0700 to 1000 h; Session 2: 1100 to 1400 h; Session 3: 1500 to 1800 h) within each observation day. The scan data were used to calculate the time-activity budget of the study elephants and also to examine the effect of resource availability and human disturbance on their behavioral decision-making. Extensive *ad libitum* observations were carried out during a fourth nocturnal session, which extended from 1900 to 0600 h. These data were largely used to estimate elephant occurrence and residence time in the different strata.

The study elephants were classified into three group types described above, on the basis of their group size and structure. These association data were recorded at first- and last sighting for each of the four sessions of observations (Session 1: 0700 to 1000 h; Session 2: 1100 to 1400 h; Session 3: 1500 to 1800 h and Session 4: 1900 to 0600 h) within each observation day. The frequency with which these group types were sighted in the 12 different strata during each session was then analyzed.

Inter-Individual Distance and the frequency of audible vocalizations were observed across all four sampling sessions to assess the effect of human disturbance on the social interactions of the study elephants including their auditory communication. For Inter-Individual Distance, the distance between all visible individuals of the focal herd was measured once every 60 min using scan sampling. The frequency of auditory vocalizations displayed by the elephants was recorded either during focal herd or group sampling or focal animal sampling (for solitary individuals), conducted for a period of 10 min each, following each instantaneous scan described above. Each focal group or animal sampling was followed by a 5-min rest period. Each event of vocalization was assigned to the grid in which the vocalization was heard and later classed into one of the three human disturbance zones. The movement rate of elephants was measured in terms of the total distance travelled by them within each stratum of human disturbance. The GPS locations of the elephants were also used to obtain the movement rate of solitary individuals, herds or all-male groups, measured in terms of the Euclidean distance covered in unit time using the package Adehabitat in R, version 2.9.2 [54].

Statistical analysis

In order to examine decision-making by individual elephants, when solitary or in association with conspecific individuals, we constructed recursive partitioning classification trees. This method estimates a regression relationship by binary recursive partitioning in a conditional inference framework [55] in the following manner. The global null hypothesis of independence between any of the input variables (the gender, age, group type and individual identity of an elephant) and the response variable (resource availability or human disturbance) was first tested and the hypothesis accepted if it could not be rejected. If the null hypothesis could be rejected, the input variable with the strongest association to the response variable was selected and their association measured by a p-value corresponding to a test for the partial null hypothesis of a single input variable and the response variable. A binary split was then implemented in the selected input variable. The above steps were repeated recursively until a statistically significant binary partitioning of the input variable could not be derived further. The implementation utilized a unified framework for conditional inference or permutation tests, developed by Strasser and Weber (1999). The criterion for the first binary spilt in the input variable was based on multiplicity-adjusted Monte-Carlo-simulated p-values. A split was implemented at the node when the simulated p-value was smaller than 0.05. This statistical approach ensured that the right-sized tree was grown and no form of pruning or cross-validation was required.

We assessed differences in the behavioral responses of elephants (probabilities of occurrence in particular strata, time-activity budgets, social interactions and demographic structure) to varying levels of resource availability and human-induced disturbance as a function of their gender, age and group type using G-tests of independence and goodness of fit. The former test was also used, where appropriate, as a post-hoc procedure to assess the statistical significance of the recursive partitioning classification trees obtained by exploratory analysis. Data on the display of IID and AV as a function of human-induced disturbance were analyzed using the Wilcoxon rank sum test with continuity correction and G-test of goodness of fit respectively [53].

Supporting Information

Figure S1 Classification trees showing the partitioning of elephant group size in the different strata. L: low-; M: medium- and H: high (a) water- and (b) shade availability. The y-axis of each graph indicates the proportion of groups observed in the different strata.

Figure S2 Classification trees for residence time of female and male elephants in the different strata. The trees depict the influence of (a) shade availability and (b) human disturbance. The y-axis of each graph indicates the proportion of groups observed in the different strata.

Figure S3 Classification trees for residence time of adult and subadult male elephants in the different strata. The trees depict the influence of (a) shade availability and (b) human disturbance. The y-axis of each graph indicates the proportion of groups observed in the different strata.

Figure S4 Classification trees for residence time of adult and subadult female elephants in the different strata. (a) Forage-, (b) water- and (c) shade availability and (d) human disturbance. The y-axis of each graph indicates the proportion of groups observed in the different strata.

Figure S5 Classification trees of the probability of individual adult and subadult male elephants occurring in the different strata. (a) water- and (b) shade availability. The y-axis of each graph indicates the proportion of groups observed in the different strata.

Figure S6 Classification trees of the probability of an individual herd occurring in the different strata. L: low-; M: medium- and H: high (a) water- and (b) shade availability. The y-axis of each graph indicates the proportion of groups observed in the different strata.

Acknowledgments

We thank the two reviewers for their valuable comments and suggestions on the manuscript that has immensely helped us improve its quality. We thank the Department of Science and Technology, Government of India for a research grant, the Karnataka Forest Department for research permits, the Deputy Conservator of Forests, Assistant Conservator of Forests and all the Range Forest Officers of the Bannerghatta National Park for their enthusiastic support in the field, all the field staff of the Park and Kaggalipura Range for their assistance in fieldwork, and A Rocha India and the Nature Conservation Foundation for logistic support. We express our gratitude to Ajith Kumar and Surendra Varma for their constant advice and encouragement, and all individuals from the local communities and conservation organizations who provided valuable food and shelter during fieldwork and without whose cooperation this study could not have been accomplished. We also owe a special debt of gratitude to Girish Arjun Punjabi for giving this paper a wonderful title!

We dedicate this paper to our close friend and colleague, Vijay D. Anand, who passed away recently but whose passion, we hope, will continue to inspire research and conservation efforts on the Asian elephant in the future.

Author Contributions

Performed the experiments: NMS. Analyzed the data: NMS SV AS. Contributed reagents/materials/analysis tools: NMS VDA SV AS. Wrote the paper: NMS SV AS.

References

1. Morrison ML, Marcot BG, Mannan RW (1998) Wildlife-Habitat Relationships: Concepts and Applications – 3rd ed. Washington DC: Island Press. 458 p.
2. Manning A, Dawkins MS (2002) An Introduction to Animal Behaviour – 5th ed. Cambridge: Cambridge University Press. 450 p.
3. Sukumar R (2003) The Living Elephants: Evolutionary Ecology, Behaviour and Conservation. New York: Oxford University Press. 478 p.
4. Graham MD, Douglas-Hamilton I, Adams WM, Lee PC (2009) The movement of African elephants in a human-dominated land-use mosaic. Anim Conserv 5: 445–455.
5. Buij R, Mcshea W, Campbell P (2007) Patch-occupancy models indicate human activity as major determinant of forest elephant *Loxodonta cyclotis* seasonal distribution in an industrial corridor in Gabon. Biol Conserv 135: 189–201. Available: http://linkinghub.elsevier.com/retrieve/pii/S0006320706004289. Accessed 2011 Jun 29.
6. Martin J, Chamaillé-Jammes S, Nichols JD (2010) Simultaneous modeling of habitat suitability, occupancy, and relative abundance: African elephants in Zimbabwe. Ecol Appl 20: 1173–82.
7. Pretorius Y, Boer FW de, Waal C van der (2011) Soil nutrient status determines how elephant utilize trees and shape environments. J Anim Ecol 80: 875–883. Available: http://www.ncbi.nlm.nih.gov/pubmed/21418208. Accessed 2011 Jun 14.
8. Mavatur AK, Singh M (2011) Behavior of Asian elephant (*Elephas maximus*) in a land-use mosaic: Conservation implications for human-elephant coexistence in the Anamalai hills, India. Wildlife Biology in Practice 6: 69–80. Available: http://socpvs.org/journals/index.php/wbp/article/view/10.2461-wbp.2010.6.6. Accessed 2011 Jun 10.
9. Whyte IJ (1993) The movement patterns of elephant in the Kruger National Park in response to culling and environmental stimuli. Pachyderm 16: 72–80.
10. Gobush KS, Mutayoba BM, Wasser SK (2008) Long-term impacts of poaching on relatedness, stress physiology, and reproductive output of adult female African elephants. Conserv Biol 22: 1590–1599.
11. Poole JH (1994) Sex differences in the behavior of African elephants. In: Short R, Balaban E editors. The Differences Between the Sexes. Cambridge University Press: pp. 331–346.
12. Poole JH (1987) Rutting behavior in African elephants: The phenomenon of musth. Behavior 102: 283–316.
13. Poole JH (1982) Musth and male-male competition in the African elephant. Ph.D. Thesis. University of Cambridge, Cambridge, UK.
14. Slotow R, Dyk GV (2004) Ranging of older male elephants introduced to an existing small population without older males: Pilanesberg National Park. Analysis 2: 91–104.
15. Slotow R, Dyk G V, Poole JH, Page B, Klocke A (2000) Older bull elephants control young males. Nature 408: 425–426.
16. Archie EA, Moss CJ, Alberts SC (2006) The ties that bind: Genetic relatedness predicts the fission and fusion of social groups in wild African elephants. P Roy Soc B-Biol Sci 273: 513–522.

17. Wittemyer G, Getz WM, Vollrath F, Douglas-Hamilton I (2007) Social dominance, seasonal movements, and spatial segregation in African elephants: A contribution to conservation behavior. Behav Ecol Sociobiol 61: 1919–1931.
18. de Silva S, Ranjeewa A, Kryazhimskiy S (2011) The dynamics of social networks among female Asian elephants. BMC Ecology 11: 17. Available: http://www.biomedcentral.com/1472-6785/11/17. Accessed 2012 Jun 21.
19. Bhaskaran N (1998) Ranging and foraging behavior of Asian elephants in Mudumalai Wildlife Sanctuary. Ph.D. Thesis. Bharathidasan University, Tiruchirappalli, India.
20. Douglas-Hamilton I (1972) The Ecology and Behaviour of the African Elephant. Ph.D. Thesis. University of Oxford, Oxford, UK.
21. Chiyo PI, Lee PC, Moss CJ (2011) No risk, no gain: Effects of crop raiding and genetic diversity on body size in male elephants. Behav Ecol 22: 552–558. Available: http://www.beheco.oxfordjournals.org/cgi/doi/10.1093/beheco/arr016. Accessed 2011 May 6.
22. Chiyo PI, Moss CJ, Alberts SC (2012) The influence of life history milestones and association networks on crop-raiding behavior in male African elephants. PLoS ONE 7: e31382. doi:10.1371/journal.pone.0031382. Available: http://www.plosone.org/article/info%3Adoi%2F10.1371%2Fjournal.pone.0031382. Accessed 2012 Mar 22.
23. Blake S, Hedges S (2004) Sinking the flagship: The case of forest elephants in Asia and Africa. Conserv Biol 18: 1191–1202.
24. Gubbi S (2012) Patterns and correlates of human-elephant conflict around a south Indian reserve. Biol Conserv 148: 88–95.
25. Wittemyer G, Polansky L, Douglas-Hamilton I, Getz WM (2008) Disentangling the effects of forage, social rank, and risk on movement autocorrelation of elephants using Fourier and wavelet analyses. P Natl Acad Sci USA 105: 19108–19113.
26. Hibert F, Calenge C, Fritz H (2010) Spatial avoidance of invading pastoral cattle by wild ungulates: Insights from using point process statistics. Biodivers Conserv 19: 2003–2024. Available: http://www.springerlink.com/index/10.1007/s10531-010-9822-0. Accessed 2011 Jun 16.
27. Loarie SR, Aarde RJV, Pimm SL (2009) Fences and artificial water affect African savannah elephant movement patterns. Biol Conserv 142: 3086–3098. Available: http://linkinghub.elsevier.com/retrieve/pii/S0006320709003759. Accessed 2010 July 1.
28. Shannon G, Matthews WS, Page BR, Parker GE, Smith RJ (2009) The affects of artificial water availability on large herbivore ranging patterns in savanna habitats: A new approach based on modelling elephant path distributions. Divers Distrib 15: 776–783. Available: http://doi.wiley.com/10.1111/j.1472-4642.2009.00581.txt. Accessed 2011 Feb 27.
29. Fernando P, Wikramanayake E, Janaka H (2008) Ranging behavior of the Asian elephant in Sri Lanka. Mamm Biol 73: 2–13. Available: http://linkinghub.elsevier.com/retrieve/pii/S1616504707001425. Accessed 2011 Jul 1.
30. Douglas-Hamilton I, Krink T, Vollrath F (2005) Movements and corridors of African elephants in relation to protected areas. Naturwissenschaften 92: 158–163.
31. Loarie SR, Aarde RJ van, Pimm SL (2009) Elephant seasonal vegetation preferences across dry and wet savannas. Biol Conserv 142: 3099–3107. Available: http://linkinghub.elsevier.com/retrieve/pii/S0006320709003760. Accessed 2011 Apr 21.
32. Sukumar R (1989) The Asian Elephant: Ecology and Management. Cambridge: Cambridge University Press. 272 p.
33. Burke T, Page B, Dyk GV, Millspaugh J, Slotow R (2008) Risk and ethical concerns of hunting male elephant: Behavioural and physiological assays of the remaining elephants. PloS ONE 3: e2417. doi:10.1371/journal.pone.0002417. Available: http://www.plosone.org/article/info:doi%2F10.1371%2Fjournal.pone.0002417. Accessed 2012 Mar 22.
34. Anand VD, Varma S, Gopalakrishna SP (2009) Conservation of Asian elephant through mitigation of human-elephant conflict in Bannerghatta National Park, southern India Technical Report II: 2009. Bangalore: A Rocha India.
35. Clutton-Brock TH, Guinness FE, Albon SD (1982) Red Deer: Behavior and Ecology of Two Sexes. Chicago: University of Chicago Press. 378 p.
36. Archie EA, Hollister-Smith JA, Poole JH (2007) Behavioural inbreeding avoidance in wild African elephants. Mol Ecol 16: 38–48.
37. Chiyo PI, Archie EA, Hollister-Smith JA, Lee PC, Poole JH et al. (2011) Association patterns of African elephants in all-male groups: The role of age and genetic relatedness. Anim Behav 81: 1093–1099. Available: http://linkinghub.elsevier.com/retrieve/pii/S0003347211000662. Accessed 2011 Jul 5.
38. Rasmussen HB, Okello JB, Wittemyer G (2007) Age- and tactic-related paternity success in male African elephants. Behav Ecol 19: 9–15. Available: http://www.beheco.oxfordjournals.org/cgi/doi/10.1093/beheco/arm093. Accessed 2010 Sep 23.
39. Evans KE, Harris S (2008) Adolescence in male African elephants, *Loxodonta africana*, and the importance of sociality. Anim Behav 76: 779–787. Available: http://linkinghub.elsevier.com/retrieve/pii/S0003347208002133. Accessed 2011 Jun 16.
40. Gopalakrishna SP, Somashekar RK, Anand VD, Varma S (2010) Asian Elephant and Bannerghatta National Park in Eastern Ghats. Gajah 33: 47–52.
41. Vidya TNC, Sukumar R (2005) Social organization of the Asian elephant (*Elephas maximus*) in southern India inferred from microsatellite DNA. J Ethol 23: 205–210. Available: http://www.springerlink.com/index/10.1007/s10164-005-0144-8. Accessed 2011 Jun 16.
42. Directorate of Project Elephant (2008), Ministry of Environment and Forests, Government of India. Available: http://envfor.nic.in/pe/population_we.pdf. Accessed 2007 Jun 10.
43. Choudhury A, Lahiri Choudhury DK, Desai A, Duckworth JW, Easa PS, et al. (2008) *Elephas maximus*. In: IUCN 2011. IUCN Red List of Threatened Species. Version 2011.2. Available: www.iucnredlist.org. Accessed 2012 Mar 22.
44. The Indian Wildlife (Protection) Act 1972. Ministry of Law and Justice, Government of India (2011). Available: http://www.envfor.nic.in/legis/wildlife/wildlife1.html. Accessed 2012 Mar 22.
45. Bhaskaran N, Sukumar R (2010) Karnataka Elephant Census 2010. Technical Report to Karnataka Forest Department. Bangalore: Asian Nature Conservation Foundation. Available: http://www.asiannature.org/sites/default/files/2010%20Synchronized%20Elephant%20Census%20Karntaka%20Final%20Report%20-%20June%202011.pdf. Accessed 2012 Mar 22.
46. Quantum GIS Development Team (2009) Quantum GIS 1.4.0: Geographic Information System User Guide. Open Source Geospatial Foundation Project. Available: http://download.osgeo.org/qgis/doc/manual. Accessed 2009 Jun 12.
47. GRASS Development Team (2007) GRASS 6.2 Users Manual. ITC-irst, Trento, Italy. Available: http://grass.osgeo.org/grass62/manuals/html62_user. Accessed 2009 Jun 12.
48. Okello MM, D'Amour DE (2008) Agricultural expansion within Kimana electric fences and implications for natural resource conservation around Amboseli National Park, Kenya. J Arid Environ 72: 2179–2192. Available: http://linkinghub.elsevier.com/retrieve/pii/S0140196308001936. Accessed 2010 Sep 27.
49. Wittemyer G, Rasmussen HB, Douglas-Hamilton I (2007) Breeding phenology in relation to NDVI variability in free-ranging African elephant. Ecography 30: 42–50. Available: http://doi.wiley.com/10.1111/j.2006.0906-7590.04900.txt. Accessed 2011 Feb 1.
50. Young KD, Ferreira SM, Aarde RJ Van (2009) The influence of increasing population size and vegetation productivity on elephant distribution in the Kruger National Park. Austral Ecol 34: 329–342. Available: http://doi.wiley.com/10.1111/j.1442-9993.2009.01934.txt. Accessed 2011 Jun 16.
51. Druce H, Pretorius K, Slotow R (2008) The response of an elephant population to conservation area expansion: Phinda Private Game Reserve, South Africa. Biol Conserv 141: 3127–3138. Available: http://linkinghub.elsevier.com/retrieve/pii/S0006320708003674. Accessed 2011 Apr 4.
52. Arivazhagan C, Sukumar R (2008) Constructing age structures of Asian elephant populations: A comparison of two field methods of age estimation. Gajah 29: 11–16.
53. Sokal RR, Rohlf FJ (1995) Biometry: The Principles and Practice of Statistics in Biological Research – 3rd ed. New York: WH Freeman and Co. 887 p.
54. R Development Core Team (2009) R: A language and environment for statistical computing. R Foundation for Statistical Computing, Vienna, Austria. Available: http://www.R-project.org/. Accessed 2009 Jun 12.
55. Hothorn T, Hornik K, Zeileis A (2006) Unbiased recursive partitioning: A conditional inference framework. J Comput Graph Stat 15: 651–674.

When Parasitoid Males Make Decisions: Information Used when Foraging for Females

Claire M-S Dufour*¤a, Philippe Louâpre, Joan van Baaren⁹, Véronique Martel⁹¤b

Université de Rennes I, Unité Mixte de Recherche (UMR) Centre National de la Recherche Scientifique (CNRS) 6553 EcoBio, Rennes, France

Abstract

Optimal foraging models predict how an organism allocates its time and energy while foraging for aggregated resources. These models have been successfully applied to organisms such as predators looking for prey, female parasitoids looking for hosts, or herbivorous searching for food. In this study, information use and patch time allocation were investigated using male parasitoids looking for mates. The influence of the former presence of females in absence of mates and the occurrence of mating and other reproductive behaviours on the patch leaving tendency was investigated for the larval parasitoid *Asobara tabida*. Although males do not modify their patch residence time based on the number of females that visited the patch, they do show an increase in the patch residence time after mating a virgin female and performing courtship behaviour such as opening their wings. These results are in concordance with an incremental mechanism, as it has been described for females of the same species while foraging for hosts. The similarities between males and females of the same species, and the conditions under which such a patch-leaving decision rule is fitted are discussed. This is the first study describing an incremental effect of mating on patch residence time in males, thus suggesting that similar information use are probably driving different organisms foraging for resource, regardless of its nature.

Editor: Sean A. Rands, University of Bristol, United Kingdom

Funding: This research is part of the CLIMEVOL programme funded by the Agence Nationale de la Recherche. The funders had no role in study design, data collection and analysis, decision to publish, or preparation of the manuscript.

Competing Interests: The authors have declared that no competing interests exist.

* E-mail: claire.dufour@univ-montp2.fr

¤a Current address: Institut des Sciences de l'Évolution de Montpellier, Montpellier, France
¤b Current address: Natural Resources Canada, Canadian Forest Service, Laurentian Forestry Centre, Quebec City, Canada

⁹ These authors contributed equally to this work.

Introduction

Optimal foraging models predict how organisms should behave in order to maximise their lifetime fitness gain when foraging for resources (e.g. food, hosts or sexual partners) [1,2]. When facing a resource of variable quality aggregated in patches of different sizes, organisms are expected to adjust their residence time to the patch quality to maximise their resource intake rate as predicted by Charnov's Marginal Value Theorem [3]: the higher the patch quality, the longer the residence time. This adaptive foraging behaviour was observed in a wide range of taxa such as insects, birds, rodents, cattle or humans [4–8].

Several authors attempted to explain patch time allocation of parasitoids or predators using different patch-departure decision rules (the so-called proximate mechanisms) such as giving-up time, giving-up density, motivation, time expectation rule, etc. [9–15]. McNamara and Houston [16] pointed out recently the importance of studying the evolution of these mechanisms in relation with the complexity of the environment and state that behavioural ecologists have to consider the internal state of the forager in order to understand how evolution and environment shaped the behaviour. Some of these patch-departure decision rules are based on the motivation (or responsiveness) to stay in the patch. Motivation is defined as a variable energizing, directing and selecting a given action [17–20]. Even if it is a hidden and thus not directly observable variable [21], a variety of animals (e.g. parasitoids, bumblebees, human) with different cognitive abilities were shown to make their decision to leave a patch in manners consistent with motivation models (e.g. [7,22]).

Models of patch departure assume that sensorial cues must be perceived during foraging [12,23,24]. The tendency to stay at arrival in the patch thus depends on the estimated patch quality. In the absence of any resource encountering event (see explanation below), the tendency to stay is assumed to decrease steadily with time until a threshold is reached; the patch is then abandoned [12] (Figure 1). As a result, as long as the decrease in the tendency to stay (or motivation) is independent of patch quality, patch residence time depends on the estimated patch quality: in many predatory mites, chemicals emitted by the prey inform predators about the patch quality and they increase their residence time accordingly [25,26]. The tendency to stay during foraging also depends on different intra-patch factors such as the resource encountering. Waage [12] suggested that the motivation rises suddenly when the ichneumon *Venturia canescens* lays an egg in a healthy host, thus defining an incremental motivation model of patch departure (Figure 1). This incremental model was suggested in several parasitoid species (reviewed by [7]), bumblebees [27] and humans [22]. In some cases, encountering the resource decreases instead of increasing the tendency to stay: such a decremental mechanism (Figure 1) has been suspected in a

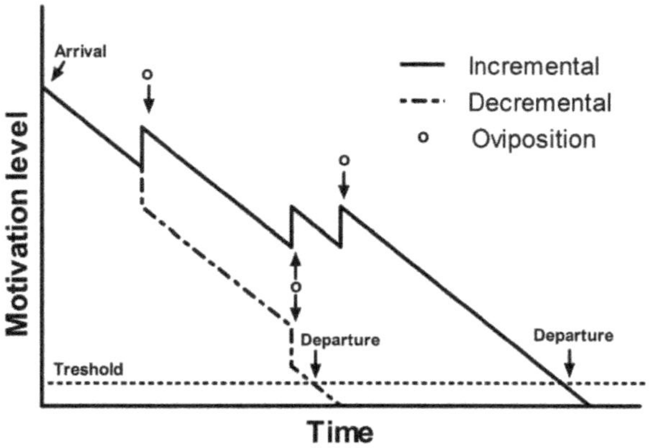

Figure 1. The main motivation-based mechanisms of decision-making described for female parasitoids. In the incremental mechanism, motivation level increases with an oviposition while in the decremental mechanism, motivation level decreases with an oviposition, leading to a shorter patch residence time with the latter. In both cases, the organism leaves the patch when its motivation level reaches a threshold.

number of insect parasitoids (reviewed by [7,28]) and bumblebees [29].

The use of an incremental or a decremental mechanism depends on several factors such as the quality of the resource (increment/decrement if the resource is of high/poor quality) [24], the resource distribution among patches (increment/decrement when clumped/uniform) [30], and the cues reliability of the patch quality at arrival (increment/decrement when the forager has unreliable/reliable information) [31]. A switch between both mechanisms has even been observed depending on the experience of the forager in *Aphidius rhopalosiphi* [32].

Both the patch-leaving decision rules and the information use by foragers have been extensively studied in insect parasitoids, mainly in females searching for hosts in which to lay their eggs (but see [33,34]). However, in many species such as gregarious or quasi-gregarious insect parasitoid (e.g. [34]), males, foraging for females to mate with, just like females foraging for hosts to parasitize, exploit patches of resources, and this foraging behaviour should thus have important effect on the male's fitness. As males' fitness is maximized while maximizing the number of mates [35], males should optimize their time allocation to maximize their number of mates [34] if (i) they are time-limited (and not sperm-limited), (ii) females occurs in patches, which is likely to happen on either emergence or foraging patches in many parasitoid species, especially the gregarious or quasi-gregarious ones with synchronized emergence and local mating, and (iii) the fitness gain on a patch decreases with time. Fitness gain is expected to decrease with patch exploitation as in insects, one mating is normally sufficient to fill the female's spermatheca [36–37]. Little is known however about the information used by males underlying the decision to leave a patch of females. Martel *et al.* [34] recently showed that the number of virgin females present as well as the number of encounters with parasitized hosts influence the patch residence time of males of a minute egg parasitoid (*Trichogramma euproctidis*); however, in that study the number of females on the patch was not controlled during the observation and the decision process is still unknown.

The main objective of this study is to understand which factors or information, in male parasitoids, influence the decision to leave a patch of females. We hypothesize that (i) males adjust their patch residence time (and their behaviours) with the perceived patch quality at arrival, staying longer in patches of higher perceived quality, and that (ii) courtship and mating events influence the decision of males to leave the patch. The studied species is the *Drosophila* larval parasitoid, *Asobara tabida* (Hymenoptera, Braconidae).

Materials and Methods

Insects

This study does not include any field work. The *Asobara tabida* strain used in this study originated from Avignon, France and was collected using banana baited traps in 2009; the insects were collected at a research station in Avignon and permits were obtained when necessary. The rearing was started using approximately 50 females and 20 males. The parasitoid was reared on *Drosophila subobscura* (collected in the Netherlands at the end of the 1980 decade) in glass jars (Ø 4.5 cm) containing a substrate of agar-nipagine (the agar serves as a substrate and the nipagine is an anti-fungi) topped with a thin layer of yeast *Saccharomyces cerevisiae* (Saccharomycedacae) (produced by Fermipan ®); the yeast serves as food for the *Drosophila* larvae. The cultures of both hosts and parasitoids were maintained in a climate room (20°C, 60% RH, and LD 16:8 h).

General Biology

Asobara tabida is a larval parasitoid of *Drosophila* species: females lay their eggs preferentially in late-first to early-second instar larvae and one adult emerges from the host pupa [38]. *Drosophila* parasitoids are protandrous, i.e. males emerge before females [39]. Soon after emergence, mating is expected to occur mainly locally, leading to frequent inbreeding [39]. The typical male mating behaviour in *Drosophila* parasitoids, as for parasitic wasps in general, is as follow: attraction, recognition, orientation, wing vibration (or wing fanning), antennation, head movement, leg tapping, copulation and post-copulatory grooming [39–40]. Wing fanning is the most conspicuous element of mating behaviour in these insects [40].

Once mated, females will search for hosts for oviposition. Their habitat is mainly fermented fruits or sap fluxes containing different numbers of host larvae aggregately distributed [41]. Females respond to chemical cues related to hosts [42–43] and to vibrations [38], and adjust their patch residence time depending on oviposition events [43], perceived patch quality [44] and experience [44–45]. Females are likely to be aggregated like their hosts – even if the former are more mobile than hosts – either in their emergence patch or on their foraging patches [43], although males are not attracted to oviposition sites [46]. Males have mature sperm at emergence (V. Martel, personal observation), but the possible production during their adult life is not known.

General method

Two days before emergence, parasitised pupae from the culture were isolated in gelatin capsules (0.68 mL) to obtain virgin individuals of known age. Males used in all experiments were 24 h old (except when otherwise mentioned), virgin, fed with a diluted honey solution and placed in a clean Petri dish (Ø 8.5 cm) containing a thin agar-nipagine layer the day of their emergence. All females used were virgin, 3 to 6 days old, fed and kept in pots (Ø 4.5 cm) containing a substrate of agar-nipagine.

In every experiment, a light table was used to maintain the individuals on the bottom of the Petri dish by phototaxis. All experiments were video-taped and analysed using the computer software package The Observer v3.0 [47].

1. Preliminary experiment: Male mate choice. In the experiments evaluating patch residence time (see below), only virgin females have been used. However, once the male mated one female, he was in presence of a mated female. Because of that, mate choice tests were performed first to evaluate if males would preferentially mate females depending on their mating status, suggesting either discriminative capacities or rejection by the female. One 1–4 d old virgin male was introduced in a Petri dish (Ø 8.5 cm) containing two 3–5 d old females (one virgin and one mated). Mated females were obtained 30 minutes before the beginning of the experiment by putting one virgin female with one virgin male until copulation was observed. Observations started as soon as the male was introduced in the Petri dish and ended after the first mating, or after 30 minutes if no mating occurred (only one out of 16 observations). The following events were recorded: first courtship and first mating.

2. Influence of the former presence of females on patch residence time. To determine if the chemicals deposited by females that previously visited the patch (a potential proxy of patch quality) influences male patch residence time, patches that have been visited by females were created. One or 10 virgin females were placed in the centre of a Petri dish (Ø 8.5 cm) containing a thin layer of agar-nipagine, under the lid of a smaller Petri dish (Ø 4 cm) delimiting the patch. After 30 minutes, the females and the lid of the small Petri dish were removed and a male was immediately introduced into the large Petri dish (Ø 8.5 cm), but outside the patch (n = 15 males for each treatment). A similar experiment was conducted, but on clean patches on which no female has walked (n = 24), as a control to determine if males perceive the former presence of females. Observations started as soon as the male walked inside the patch and ended when he left it for at least 60 s or when he landed on the Petri dish cover or sides. The following events were recorded: entering the patch, leaving the patch, wing opening and wing fanning (typical courtship behaviours). Males open their wings quite frequently when in presence of virgin females, but not so frequently in front of a mated one (V. Martel, personal observations).

3. Influence of mating on patch residence time. To evaluate the impact of mating and courtships events on male patch residence time, a patch of females (*i.e.* an area where the male could freely enter, mate and leave without the female escaping) has been created (Figure 2). A Petri dish (Ø 8.5 cm) containing a thin layer of agar-nipagine in the centre of which two alcohol-rinsed hair were inserted before the substrate solidified was used. The other end of the hair was glued (glue SADER max) to the fore right leg of an anesthetised virgin female (Video S1), delimiting a patch of two tethered females whose radius is the hair length (Ø 4 cm). Hence, females can move freely within the patch area. The use of tethered females allows keeping the number of females present on the patch constant during the whole observation (in opposition to the previous study on male patch time allocation [34]). Each patch was previously visited by five virgin females (same procedure as in experiment 2) in order to increase the attractiveness of the patch for the tested males, and consequently decreases the frequency of males not visiting the patch. Although the female behaviour might be altered by the experimental procedure [48–49], preliminary observations comparing males in a Petri dish with a tethered female, or with a free female showed similar males' reproductive behaviours (C. Dufour, personal observations).

Observations started as soon as the male entered the patch and ended when the male left the patch for more than 60 s or landed on the cover or sides of the Petri dish (n = 16). The following events were recorded: entering the patch, leaving the patch, wing opening, wing fanning, mating and female escape.

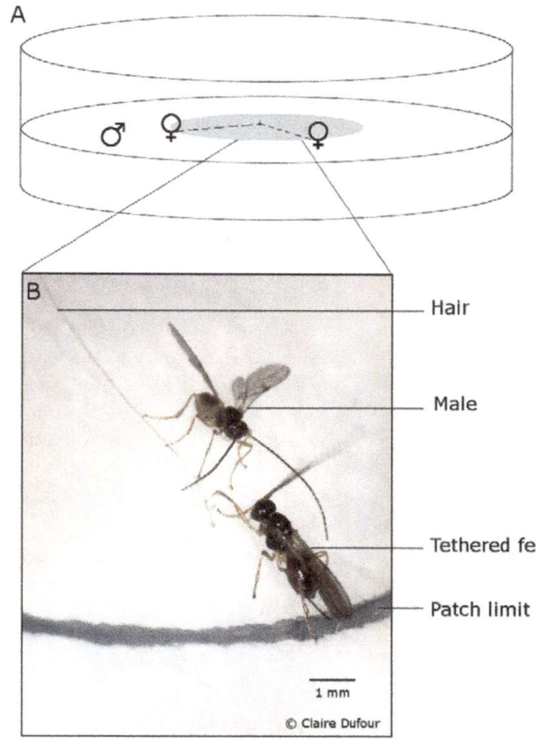

Figure 2. The experimental set-up used to determine the impact of mating on males' patch residence time. A. Two virgin females are attached by the leg to alcohol-washed hair (dashed line) on a delimited patch (grey area). **B.** A picture of a free male approaching a virgin tethered female in a patch.

Statistical analyses

All statistical analyses were performed using R 2.12.0 software [50].

In the male mate choice experiment, the ratio of virgin females and already mated females courted or mated were compared to a 50:50 probability using χ^2 tests.

The proportion of males entering the patch previously visited by females was compared to the proportion of males entering the clean patch using a contingency table followed by a χ^2 test. To estimate the effect of both the number of females that visited the patch and mating on the patch residence time, Cox proportional hazards models were used [51,52]. The following covariates were used when applicable: number of females that previously visited the patch, mating with a virgin female, re-mating a female, female escape, wing opening, wing fanning and temporary exit. The parameter "female encounter" was not tenable, as in such a small arena, both individuals are almost always in proximity. In addition, the female to which the behaviours "wing opening" or "wing fanning" was addressed was not discriminated as it was not always clear. A thorough description of this model can be found in the literature dealing with patch-leaving decision rules [53]. The proportional hazards model expresses the data in terms of patch-leaving tendency (*i.e.* hazard rate), which is the probability per unit of time that a male leaves the patch, given that he is still on it. A hazard ratio above one indicates an increase in the male's patch-leaving tendency, while a hazard ratio below one shows a decrease in the patch-leaving tendency.

Results

1. Preliminary experiment: Male mate choice

When given the choice, males prefer to court ($\chi^2 = 7.14$, d.f. = 1, P = 0.008) and to mate ($\chi^2 = 15$, d.f. = 1, P<0.001) virgin females.

2. Influence of the former presence of females on patch residence time

Fourteen out of 24 males did not enter the clean patch, which is significantly higher than in the patch previously visited by females, where only 6 out of 36 males did not visit the patch ($\chi^2 = 11.25$, d.f. = 1, P = 0.002).

Neither the number of females that previously visited the patch, nor any of the tested events significantly influenced the patch-leaving tendency for males (Table 1); the time spent on the patch (mean patch residence time ± SD: patches visited by one female, 127.95±185.83 s; patches visited by 10 females, 157.09±212.19 s) is thus independent of the number of females that previously visited the patch and the performed behaviour in absence of females.

3. Influence of mating on patch residence time

While mating with a virgin female significantly decreased the patch leaving tendency by a factor of 3.959, re-mating an already mated female had no effect on the patch leaving tendency (mean patch residence time ± SD: 1095.54±937.88 s) (Table 2). Opening their wings also decreased the patch leaving tendency, while temporary exits from the patch increased it. No other covariates had a significant effect.

Discussion

Our results show that male *A. tabida* use different information sources in their exploitation of a patch of resource, just like females do. Their residence time in a patch of females depends on the number of mating with virgin females and courtship behaviour such as wings opening, but not on the number of mating with already mated ones, nor on the number of females that previously visited the patch. The type of information used by male parasitoids for patch departure decision (i.e. the resource, mates) and its impact on patch residence time is thus similar to the one used by females of this species (i.e. the resource, hosts) even though some differences will be discussed.

Male *A. tabida* visit patches that have been previously visited by females significantly more frequently than clean patches, thus indicating that females lay a trail of a substance, which could be a sex pheromone, perceived by males. Sex pheromones are commonly used by insect males to locate and choose females [54]. After perceiving such infochemicals emitted by virgin females, males of parasitoid species such as *Trichogramma brassicae* and *Aphelinus asychis* tend to stay on the substrate, even in the absence of females [55].

Although male *A. tabida* do respond to the former presence of females by entering the patch, the decision of when to leave a patch previously visited by females did not depend on the number of females which visited it. In female parasitoids, infochemicals such as kairomones emitted by hosts were shown to be a highly detectable and reliable proxy of patch quality [28]. For example, patch residence of female *A. tabida* in patches containing host kairomones (but no hosts) depends on their amount: the higher the kairomone level, the longer the residence time [42,44]. In the present study, males' lack of response to the number of females that previously visited the patch could be explained by the saturation of the patch medium absorption capacity independently of the number of virgin females present, or on the incapacity of males to evaluate the number of females that previously visited the patch. As females *A. tabida* are more mobile than hosts (i.e. *Drosophila* larvae), using chemicals emitted by females, potentially a sex pheromone deposited on the substrate might be less reliable for males than hosts' kairomones: females might have already flown away when males arrive on the patch. Other cues such as mating or encounters might then be more reliable. In addition, males'

Table 1. Estimated regression coefficients (β) and hazard ratios (exp (β)) for covariates of the Cox model on the patch-leaving tendency of males *A. tabida* in a patch previously visited by females.

	β	exp (β)	z	P-value
Number of females that visited the patch	−0.0284	0.972	−0.598	0.550
Wings opening	0.526	1.692	0.807	0.420
Wing fanning	−0.125	0.883	−1.485	0.137
Temporary exit	0.395	1.484	1.934	0.053

Table 2. Estimated regression coefficients (β) and hazard ratios (exp (β)) for covariates of the Cox model on the patch-leaving tendency of males *A. tabida* in a patch of females.

	β	exp (β)	z	P-value
Mating with a virgin female	−3.959	0.019	−2.837	0.005
Re-mating a female	0.266	1.305	1.255	0.209
Wings opening	−0.434	0.648	−2.657	0.008
Wing fanning	0.080	1.083	1.293	0.196
Female escape	−0.146	0.864	−1.525	0.127
Temporary exit	1.656	5.237	3.172	0.002

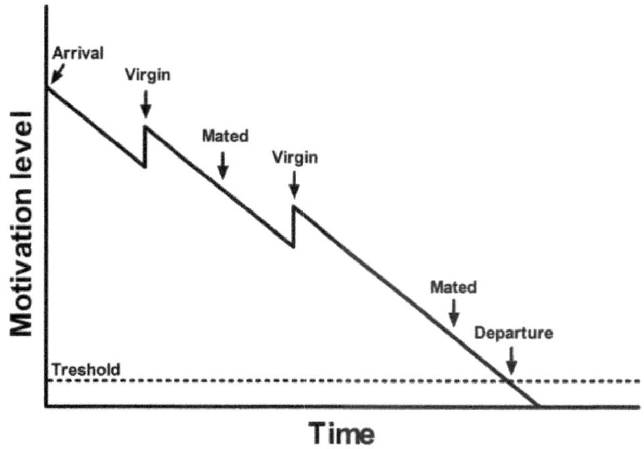

Figure 3. The suggested proximate mechanism underlying decision-making in male *A. tabida*. While mating a virgin female increases the tendency to stay on the patch, consequently increasing patch residence time, re-mating an already mated female has no impact on the residence time.

tendency to stay in the patch, even if independent of the number of females that previously visited it, could depend on other physiological factors (such as the male's age, nutritional status, or mating status) or on inter-patch travel time as observed in females (reviewed by [7]). Finally, the use of visual cues, tactile cues, or cues related to hosts in combination to cues related to females – as it has been shown to be attractive for males in other parasitoids species [56] – would be interesting to test, even though males were not shown to be attracted to oviposition sites [46].

In male *A. tabida*, each mating with a virgin female decreases the patch-leaving tendency, consequently increasing patch residence time on the patch of females. Both male and female *A. tabida* are thus expected to similarly use information in their patch-leaving decision, in concordance with an incremental mechanism (see Figure 1), as females seem to be distributed like hosts. The tendency of the male to stay on the patch of females thus seems to depend on the number of virgin females inseminated: the initial tendency to stay probably decreases with time and suddenly increases each time the male mates with a virgin female (Figure 3). Such an incremental mechanism was found in females of many parasitoid species [7]. However, mating with already mated females did not modify the male's decision to leave the patch, in concordance with our choice tests where male *A. tabida* discriminate females according to their mating status (even if in the choice test, the mated female was inseminated by another male). As in insects, one mating is normally sufficient to fill the female's spermatheca [36,37], fitness is thus maximized when a male inseminate a virgin rather than an already mated female in which sperm competition might occur. This difference in benefits (through the progeny fathered) probably explains the lack of increment of mating with an already mated female, especially as the female had been already mated by the same male; if the female was mated by another male, he could have gain by inseminating her and transferring at least some sperm, but the male does not gain any additional progeny by re-inseminating the same female. In males, opening their wings also increases patch residence time, although by a smaller factor than mating a virgin female (0.434 vs. 3.959). This is an expected result, since wings opening should equal the male perceiving the presence of a female.

Males discriminate the female quality and their tendency to stay on the patch increases only when mating with the high quality ones, i.e. the virgin ones, or performing behaviour linked to their presence such as wing opening. Similarly, females increase their patch time allocation when encountering healthy hosts, while encounters with parasitized hosts have no effect [43]; males and females *A. tabida* thus manage the information on resource quality the same way. In the parasitoid *Trichogramma euproctidis*, males also discriminate and prefer virgin females [57], but adjust their patch residence time based not on the number of mating with virgin females, but rather on the number of virgin females present on a patch [34]. In *A. tabida*, temporary exits are the only event decreasing the patch residence time. With the experimental set-up used in this study, exiting the patch also meant a drop in the perception of the former presence of females as no females were ever present outside the patch. It could then give the male an indication on the patch desertion, or on the patch surface. Alternatively, it could also mean that males that are about to leave the patch make more numerous temporary exits.

The efficiency of a given patch leaving rule of any organism depends on the ability to estimate patch quality, which in turns depends both on the ability to perceive patch quality [31] and the distribution of resource among patches [30]. The distribution of females among patches may explain the information use in decision mechanisms suggested in this study. Indeed, an incremental mechanism is adaptive when the resource is heterogeneously distributed and aggregated in a few patches in the environment, and no external cue is available to estimate patch quality [30,31,58], which is likely to be the case for both females in *A. tabida* [41,43].

Conclusion

This study extends the scope of the information used in the decisional mechanisms leading to patch departure. Until now, this type of decisional mechanism was suspected in a few numbers of organisms such as female parasitoids, bumblebees and human foraging for food and/or hosts [12,22,29]. This is the first study suggesting that males, as females, are able to manage the informative value of resource exploitation in accordance to an incremental proximal mechanism of patch departure. However, some studies are needed for a better understanding of the patch-leaving decision; testing the number of females, their quality (virgin or mated, by the same male or by another male, of different sizes and ages), the experience and physiological status of the male, and more would complete the information-use and patch-leaving decision rules understanding. Natural selection may favour emergence and evolution of mechanisms leading to efficient decision-making, such as patch departure, regardless of the biological basis of the mechanisms [16,59,60]. This adaptiveness of decision-making mechanisms may thus explain why organisms exploiting different types of resources (e.g. food, hosts or mates) with similar distribution and cues use the same proximal mechanism when facing the same environmental constraints.

Supporting Information

Video S1 A female *Asobara tabida* being glued on a hair during the patch preparation. The female is anesthetised before the procedure, and the hair is unglued from the agar substrate once she awakens.

Acknowledgments

We would like to thank Maxime Guenier, Benjamin Hornoy, Barbara Martineau, Joffrey Moiroux, Christoph Nicolai and Aymeric Mauroy for their technical help and advices, and to Youna Le Rouzo for her hair. All experiments conducted in this study comply with the current laws of France.

Author Contributions

Conceived and designed the experiments: CD PL JVB VM. Performed the experiments: CD. Analyzed the data: CD FL VM. Wrote the paper: CD PL VM JVB.

References

1. Krebs JR, Davies NB (1997) Behavioural ecology: an evolutionary approach, 4th ed. Oxford: Blackwell. 456 p.
2. Danchin E, Giraldeau L-A, Cézilly F (2008) Behavioural Ecology: An Evolutionary Perspective on Behaviour. Oxford: Oxford University Press. 912 p.
3. Charnov EL (1976) Optimal foraging: The marginal value theorem. Theor Pop Biol 9: 129–136.
4. Morrison ML, Ralph CJ, Verner J, Jehl JR, Jr (eds) (1990) Avian Foraging: Theory, Methodology, and Applications. Stud Avian Biol 13: 1–515.
5. Podolsky RH, Price MV (1990) Patch use by *Dipodomys deserti* (Rodentia: Heteromyidae): profitability, preference, and depletion dynamics. Oecologia 83: 83–90.

6. Laca EA, Demment MW (1996) Foraging strategies of grazing animals. In: Illius AW, Hodgson J, editors. The ecology and management of grazing systems. Wallingford: CAB International. 137–158.
7. Wajnberg E (2006) Time allocation strategies in insect parasitoids: from ultimate predictions to proximate behavioural mechanisms. Behav Ecol Sociobiol 60: 589–611.
8. Hutchinson JMC, Wilke A, Todd PM (2008) Patch leaving in humans: can a generalist adapt its rules to dispersal of items across patches? Anim Behav 75: 1331–1349.
9. Krebs JR (1973) Behavioral aspects of predation. In: Bateson PPG, Hopfer PH, editors. Perspectives in ethology. New York: Plenum. 73–111.
10. Krebs J, Ryan J, Charnov E (1974) Hunting by expectation or optimal foraging: A study of patch use by chickadees. Anim Behav 22: 953–964.
11. Gibb JA (1962) Tinbergen's hypothesis of the role of specific search images. Ibis 104: 106–111.
12. Waage JK (1979) Foraging for patchily-distributed hosts by the parasitoid, *Nemeritis canescens*. J Anim Ecol 48: 353–371.
13. Wajnberg E, Gonsard P-A, Tabone E, Curty C, Lezcano N, Colazza S (2003). A comparative analysis of patch-leaving decision rules in a parasitoid family. J Anim Ecol 72: 618–626.
14. Ydenberg RC (1984) Great tits and giving-up times: decision rules for leaving patches. Behaviour 90: 1–24.
15. Brown JS, Kotler BP (2004) Hazardous duty pay and the foraging cost of predation. Ecology Letters 7: 999–1014.
16. McNamara JM, Houston AI (2009) Integrating function and mechanism. TREE 24: 670–675.
17. Lorenz K (1970) Essais sur le comportement animal et humain: les leçons de l'évolution de la théorie du comportement. Paris: Seuil. 483 p.
18. McNamara JM (1985) An optimal sequential policy for controlling a markov renewal process. J Appl Prob 22: 324–335.
19. McClelland DC, Koestner R, Weinberger J (1989) How do self-attributed and implicit motives differ? Psychol Rev 96: 690–702.
20. Schultheiss OC, Wirth MM, Waugh CE, Stanton SJ, Meier EA et al. (2008) Exploring the motivational brain: effects of implicit power motivation on brain activation in response to facial expressions of emotion. Soc Cogn Affect Neurosci 3: 333–343.
21. Hinde RA (1956) Ethological models and the concept of drive. Brit J Phil Sci 6: 321–331.
22. Louâpre P, van Alphen JJM, Pierre JS (2010) Human and insects decide in similar ways. PLoS ONE 5:e14251.
23. Driessen G, Bernstein C, van Alphen JJM, Kacelnik A (1995) A count-down mechanism for host search in the parasitoid *Venturia canescens*. J Anim Ecol 64: 117–125.
24. van Alphen JJM (1993) Patch residence time and encounters with parasitized hosts: a reaction. Neth J Zool 43: 340–349.
25. Mayland H, Margolies DC, Charlton RE (2000) Local and distant prey-related cues influence when an acarine predator leaves a prey patch. Entomol Exp Appl 96: 245–252.
26. Maeda T, Takabayashi J (2001) Patch-leaving decision of the predatory mite *Amblyseius womersleyi* (Acari: Phytoseiidae) based on multiple signals from both inside and outside a prey patch. J Insect Behav 14: 829–839.
27. Taneyhill DE (2010) Patch departure behavior of bumble bees: rules and mechanisms. Psyche 2010: 1–9.
28. van Alphen JMJ, Bernstein C, Driessen G (2003) Information acquisition and time allocation in insect parasitoids. Trend in Ecology and Evolution 18: 81–87.
29. Lefebvre D, Pierre J, Outreman Y, Pierre JS (2007) Patch departure rules in bumblebees: evidence of a decremental motivational mechanism. Behav Ecol Sociobiol 61: 1707–1715.
30. Iwasa Y, Higashi M, Yamamura N (1981) Prey distribution as a factor determining the choice of optimal foraging strategy. Am Nat 117: 710–723.
31. Driessen G, Bernstein C (1999) Patch departure mechanisms and optimal host exploitation in an insect parasitoid. J Anim Ecol 68: 445–459.
32. Outreman Y, Le Ralec A, Wajnberg E, Pierre JS (2005) Effects of within- and among-patch experiences on the patch-leaving decision rules in an insect parasitoid. Behav Ecol Sociobiol 58: 208–217.
33. Parker GA, Stuart RA (1976) Animal behaviour as a strategy optimiser: evolution of resource assessment strategies and optimal emigration thresholds. Am Nat 110: 1055–1076.
34. Martel V, Wajnberg E, Boivin G (2008b) Patch time allocation in male parasitoids. Ecol Entomol 10: 1–6.
35. Bateman AJ (1948) Intra-sexual selection in *Drosophila*. Heredity 2: 349–368.
36. Walker WF (1980) Sperm utilization strategies in nonsocial insects. Am Nat 115: 780–799.
37. Knowlton N, Greenwell SR (1984) Male sperm competition avoidance mechanisms: the influence of female interests. In: Smith RL, editor. Sperm competition and the evolution of mating systems. Orlando: Academic Press. 61–84.
38. van Alphen JJM, Drijver RA (1982) Host selection by *Asobara tabida* Nees (Braconidae, Alysiinae), a larval parasitoid of fruit inhabiting *Drosophila* species. I. Host stage selection with *Drosophila melanogaster* as host species. Neth J Zool 32: 194–214.
39. Carton Y, Boulétreau M, van Alphen JJM, van Lenteren JC (1986) The *Drosophila* Parasitic Wasps. In: Ashburner M, Carson H, Thompson JN, eds. The Genetics and Biology of *Drosophila* Vol. 3e. London: Academic Press. 347–394.
40. Matthews RW (1975) Courtship in parasitic wasps. In: Price PW, ed. Evolutionary strategies of parasitic insects and mites. New York: Plenum. 66–82.
41. Ellers J, van Alphen JJM (1997) Life history evolution in *Asobara tabida*: plasticity in allocation of fat reserves to survival and reproduction. J Evol Biol 10: 771–785.
42. Galis F, van Alphen JJM (1981) Patch time allocation and search intensity of *Asobara tabida* Nees (Braconidae), a larval parasitoid of Drosophila. Neth J Zool 31: 596–611.
43. van Alphen JJM, Galis F (1983) Patch time allocation and parasitization efficiency of *Asobara tabida*, a larva parasitoid of Drosophila. J Anim Ecol. 52: 937–952.
44. Louâpre P, van Baaren J, Pierre JS, van Alphen JJM (2011) Information gleaned and former patch quality determine foraging behaviour of parasitic wasps. Behav Eco 90: 1064–1069.
45. Thiel A, Hoffmeister TS (2006) Selective information use in parasitoid wasps. Anim Biol 56: 233–245.
46. Hardy ICW, Godfray HCJ (1990) Estimating the frequency of constrained sex allocation in field populations of Hymenoptera. Behaviour 114: 137–147.
47. Noldus LPJJ (1995) The Observer: a software system for collection and analysis of observational data. Behav Res Meth Instr Comp 23: 415–429.
48. Mason AC, Oshinsky ML, Hoy RR (2001) Hyperacute directional hearing in a microscale auditory system. Nature 410: 686–690.
49. Pelletier Y, Giguère MA (2009) Effect of manipulations on the host selection behavior of *Sitobion avenae* (Homoptera: Aphididae). J Insect Behav 22: 165–171.
50. R development core team (2010) R: A language and environment for statistical computing. R foundation for statistical computing, Vienna, Austria. URL http://cran.r-project.org/manuals.html.
51. Cox DR (1972) Regression models and life tables. Biometrics 3: 67–77.
52. Collett D (1994) Modelling survival data in medical research. London: Chapman & Hall. 369 p.
53. Wajnberg E, Rosi MC, Colazza S (1999) Genetic variation in patch time allocation in a parasitic wasp. J Anim Ecol 68: 121–133.
54. Bonduriansky R (2001) The evolution of male mate choice in insects: a synthesis of ideas and evidence. Biol Rev 76: 305–339.
55. Fauvergue X, Fouillet P, Mesquita ALM, Boulétreau M (1998) Male orientation to trail sex pheromones in parasitoid wasps: does the spatial distribution of virgin females matter? J Insect Physiol 44: 667–675.
56. Metzger M, Fischbein D, Auguste A, Fauvergue X, Bernstein C, Desouhant E (2010) Synergy in information use for mate finding: demonstration in a parasitoid wasp. Anim Behav 79: 1307–1315.
57. Martel V, Damiens D, Boivin G (2008) Male Mate Choice in *Trichogramma turkestanica*. J Insect Behav 21: 63–71.
58. Pierre JS (2011) Neuroeconomics in parasitoids: computing accurately with a minute brain. Oikos 120: 77–83.
59. McNamara JM, Houston AI (1980) The application of statistical decision theory to animal behaviour. J Theor Biol 85: 673–690.
60. Louâpre P (2011) De la plante à l'homme via les guêpes parasitoïdes: comment décider sans calculer? Rennes: Université de Rennes I. 203 p.

18

A New Way of Assessing Foraging Behaviour at the Individual Level Using Faeces Marking and Satellite Telemetry

Marie-Andrée Giroux[1]*, Christian Dussault[1,2], Nicolas Lecomte[3,4], Jean-Pierre Tremblay[1], Steeve D. Côté[1]*

1 NSERC-Produits Forestiers Anticosti Industrial Research Chair, Département de Biologie and Centre d'Études Nordiques, Université Laval, Québec, Québec, Canada, **2** Direction Générale de l'Expertise sur la Faune et ses Habitats, Ministère des Ressources Naturelles et de la Faune, Québec, Québec, Canada, **3** Département de Biologie, Université du Québec à Rimouski, Rimouski, Québec, Canada, **4** Department of Environment, Government of Nunavut, Igloolik, Nunavut, Canada

Abstract

Heterogeneity in foraging behaviour can profoundly influence ecological processes shaping populations. To scale-up from individual foraging behaviour to processes occurring at the population scale, one needs to sample foraging behaviour at the individual level, and over large temporal scales or during critical seasons known to influence life-history traits. We developed an innovative technique to monitor foraging behaviour at the individual level in secretive species, a technique that can be ultimately used to investigate the links between foraging behaviour and life-history traits. First, the technique used a novel approach, namely the combination of telemetry tracking and biomarking of faeces with food dyes to locate fresh signs of presence left by individuals equipped with GPS collars. Second, the technique is based on the simultaneous or successive sampling of life-history traits and individual foraging behaviour, using tracks with high probabilities of recovery of dyed faeces. We first describe our methodological approach, using a case study of a large herbivore, and then provide recommendations and guidelines for its use. Sampling single snow tracks of individuals equipped with a GPS collar was a reliable way to assess individual winter foraging behaviour in a white-tailed deer (*Odocoileus virginianus* Zimmermann) population. During that period, the probability of recovery of dyed faeces within the range of the collar precision was very high for single snow tracks of equipped deer (97%). Our approach is well suited to study individual foraging behaviour, and could ultimately be used to investigate the interplay between intra-population heterogeneity in foraging behaviour, life-history traits, and demographic processes.

Editor: Kentaro Q. Sakamoto, Hokkaido University, Japan

Funding: This research was funded by the Natural Sciences and Engineering Research Council of Canada (NSERC)-Produits Forestiers Anticosti Industrial Research Chair, Université Laval, and the Ministère des Ressources naturelles et de la Faune du Québec. MAG received scholarships from NSERC (Alexander Graham Bell, Michael Smith, and EnviroNorth scholarships) and the Norwegian Research Council. The funders had no role in study design, data collection and analysis, decision to publish, or preparation of the manuscript.

Competing Interests: The authors have declared that no competing interests exist.

* E-mail: marie-andree.giroux.1@ulaval.ca (MAG); steeve.cote@bio.ulaval.ca (SDC)

Introduction

Individual variation in foraging behaviour can profoundly influence ecological processes shaping populations [1]. Inter-individual variation in foraging behaviour within populations may, for instance, reduce intraspecific competition [2,3] and modulate the risk of predation and parasitism [4]. Such heterogeneity can be a strong determinant of life-history traits [5,6,7], thus shaping demographic processes underlying population dynamics [8], and recent years have seen a surge in studies focusing on the importance of inter-individual variations at various temporal scales (e.g. [9,10,11]).

To scale-up from individual foraging behaviour to ecological processes occurring at the population scale, one first needs to link individual foraging behaviour with life-history traits. This requires the sampling of foraging behaviour at the individual level, but also over a large temporal scale or during critical seasons known to influence life-history traits. Sampling techniques have been developed to investigate foraging behaviour at the individual level, yet several constraints can limit their applicability in specific study systems, for instance in the study of secretive species or for investigating processes over large temporal scales [12,13]. We developed a technique to monitor individual foraging behaviour over variable temporal scales in secretive species, a technique that could be ultimately used to investigate the influence of foraging behaviour on life-history traits. First, our technique relies on a novel approach, namely the combination of telemetry tracking and biomarking of faeces with food dyes to sample fresh signs of presence left by individuals equipped with GPS collars. Second, the technique is also based on sampling life-history traits simultaneously or successively to individual foraging behaviour, using tracks with high probabilities of recovery of dyed faeces. We here describe the methodological approach we developed, using a case study of a large herbivore population, and then provide guidelines for determining whether it could be used in a given study system and which type of questions could be investigated.

Methods

Ethics Statement

All animals handling protocols were approved by the Université Laval Animal Care Committee of the Canadian Council on Animal Care (Permit number: UL 2008017-2). To conduct field work on Anticosti Island, we also obtained an authorisation for scientific studies from the Ministère des Ressources naturelles et de la Faune du Québec (Permit number: 09052701409SF).

Methodological Approach

To verify the reliability of GPS telemetry to sample foraging behaviour of marked individuals throughout winter, we administred food dyes orally to mark individuals' faeces. We subsequently visited in the field recent GPS locations of collared animals received via a data transmission system (satellite, GSM or radio transmission system) to estimate the probability of recovery of dyed faeces near different types of locations or tracks. Our proximate goal was to be able to use this technique beyond the effective marking period, i.e. when the dye is no longer effective, to sample foraging behaviour near the types of locations or tracks with high probabilities of recovery of dyed faeces (hereafter "the validated types of locations or tracks"). Our ultimate goal was to assess life-history traits of the collared individuals simultaneously or successively to individual foraging behaviour in order to establish the links between those two concepts. This technique used the four steps illustrated in Figure 1. We describe below these four steps, using the case study of a white-tailed deer (*Odocoileus virginianus* Zimmermann) population on Anticosti Island (Quebec, Canada, 49.0–49.9°N, 61.6–64.5°W; 7,943 km^2) to illustrate how the technique can be used in winter. Deer on Anticosti Island do not gather in deer yards during winter and single snow tracks are frequently encountered [14].

First step – Capturing and marking individuals. We captured 15 and 12 female deer during winters 2008–2009 and 2009–2010, respectively, using Stephenson box traps, cannon nets and net guns fired from a helicopter. We anesthetized deer by injecting a mixture of Telazol (Fort Dodge Animal Health, Fort Dodge, USA) and Xylazine (Bimeda-MTC Animal Health Inc., Cambridge, Canada) at doses of 3.2 mg/kg and 1.6 mg/kg, respectively [15], administered with a syringe following physical contention of the animal. After chemical immobilization, we fitted deer with GPS collars equipped with an Iridium satellite transmission device allowing the frequent reception of locations (Vectronic-Aerospace, Berlin, Germany).

To mark their faeces, each deer was given 5 to 6 gelatine capsules (5 cc) each containing 2 g of blue food dyes in powder (FD&C Blue No. 1, also called Brilliant Blue FCF; International Foodcraft Corporation, Linden, USA). Capsules were administered orally into the oesophagus with a dosing-gun (Cotran Corporation, Portsmouth, USA). The choice of FD&C Blue No. 1 was based on a captivity experiment aiming to test the effectiveness of the three following dyes at staining faeces: Beet Juice Powder, FD&C Blue No. 1 and FD&C Blue No. 2 (International Foodcraft Corporation, Linden, USA). Among those three dyes, FD&C Blue No. 1 was the only one providing an easily detectable stain of the faeces on the forest floor or snow. To avoid marking deer with overlapping home ranges, we captured deer at >2 km from one another (the maximal length of winter home range in this study population was 1.8 km; Massé and Côté, unpublished results). Prior to release, we administered Yohimbine (Lloyd Laboratories, Shenandoah, USA), an antagonist to Xylazine, at doses of 0.2 mg/kg [16].

Second step – Validating the telemetry technique with dyed faeces. We quantified the probability of recovery of dyed faeces in the snow tracks found in the vicinity of deer locations transmitted by the Iridium-GPS satellite collars. For each collared deer, we recorded one telemetry location every five minutes between 4:00 and 20:00 daily; we chose this schedule because it encompassed day and dusk, i.e. the two periods of greatest deer activity during winter in this population [17]. We transferred all telemetry locations received by email to a handheld GPS (Garmin eTrex, Southampton, UK) and visited locations shortly after their satellite transmission (mean time interval between data recordings and field visits: 22 hours, range: 5 hours to 36 hours). Trails sampled in the field corresponded to the tracks formed by linking successive locations recorded every five minutes (Figure 2) over a period varying between 1 and 12 hours (mean = 6 hours, 95% CI = [5,7]) up to 6 days after dye administration.

At each visited GPS-location along a trail, we classified deer tracks found within a 5 m-radius around the location in the following, mutually exclusive categories: single track (ST), track network (N, i.e. two or more single tracks found within the 5 m-radius), and path (P, i.e. tracks used more than once). The 5 m-radius corresponds to the average accuracy of collar locations (4.3 m, 95% CI = [1.7, 6.9]). We estimated the accuracy of collars using field trials, measuring the average distance between locations recorded by stationary paired GPS collars and handheld GPSs in both open and forest habitats [18]. Accuracy estimates were similar in these two habitats types (see Methods S1 in the Supporting Information for details). We followed snow tracks linking two successive locations and classified tracks as ST, N or P. Each trail was subdivided into segments and each segment was characterised by a single type of tracks (including a 5-m buffer on each side of the track; Figure 2). For track networks, a segment was rather defined as the 5-m buffer around the starting point of the network, as tracks forming a network ran in different directions that did not lead directly to subsequent locations on the trail.

To estimate the probability of recovery of dyed faeces during the effective marking period for all track types, we counted the number of dyed and undyed faeces on each trail segment. For each trail sampled, we noted the occurrence of snowfall between the time of GPS locations and sampling in the field to test for an effect of snowfall on the probability of detection. We only sampled trails following light snowfalls (approximately<3–4 cm) because we could not detect trails after heavier snowfalls (>4 cm). We performed the validation from December 11th to January 12th in 2008–2009, and from February 10th to 16th in 2009–2010.

To determine the probability of recovery of dyed faeces for all types of snow tracks, we fitted mixed logistic regression models with the number of dyed and undyed pellet groups as a binomial response (logit link). All models included the type of tracks and trail ID as fixed and random effects, respectively. To take into account variability in the detection rate of dyed faeces vs. undyed ones, we also modelled the effects of the occurrence of snowfall between the recording of locations and their sampling, the time of sampling during the day, as well as the interactions between track types and these two variables. To test whether the probability of detecting dyed faeces decreased with time elapsed since capture (i.e. to determine if we sampled beyond the effective marking period), we used the time interval between the capture and the location as a covariate. In this analysis, we excluded all sampled segments without faeces, i.e. 115 segments excluded over a total of 211 sampled (a large majority of segments were shorter than 100 m: average length of segments = 79 m, 95% CI = [55,103]). We then built 19 biologically relevant candidate models that could explain variations in the probability of recovering only dyed faeces (see

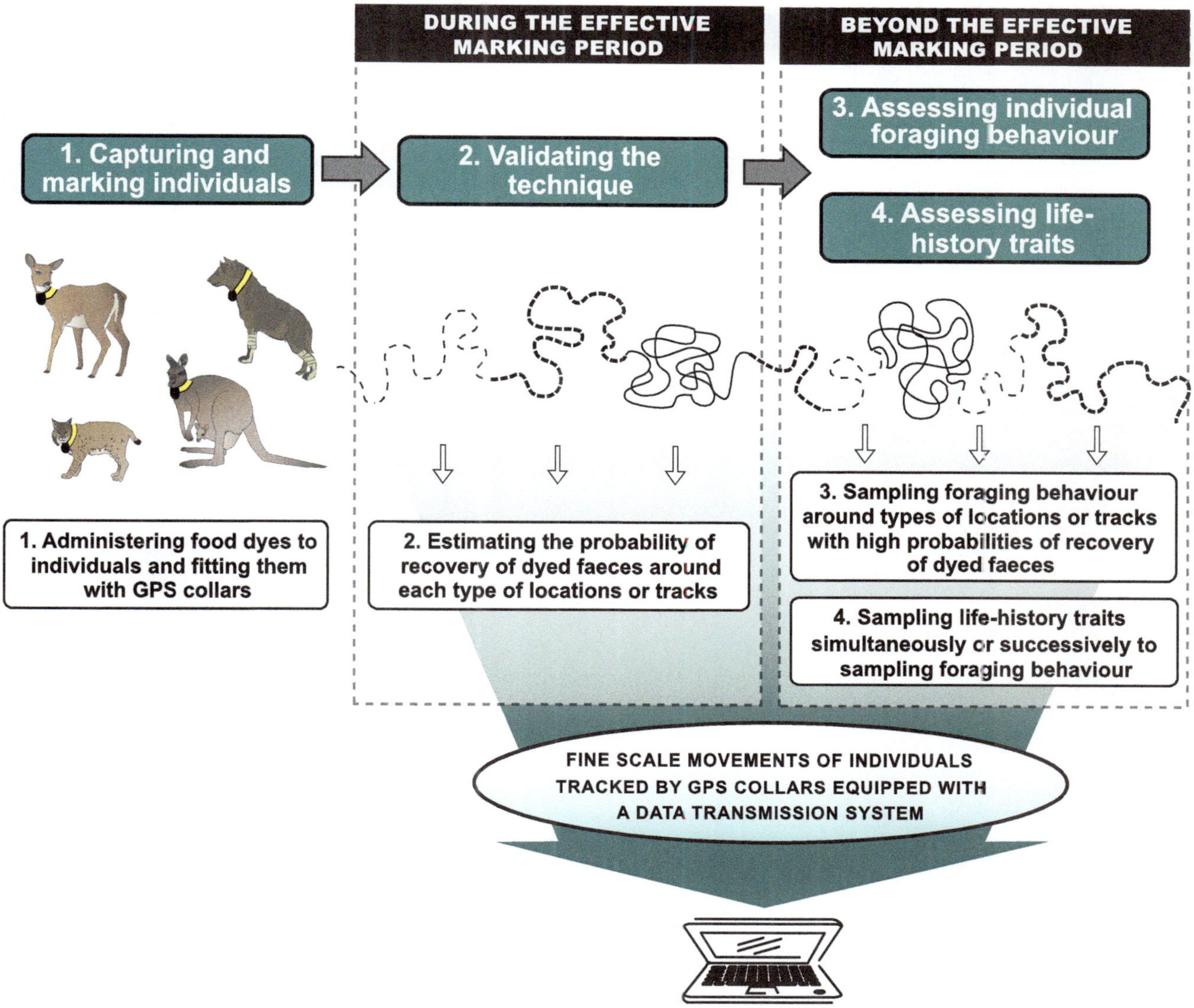

Figure 1. Schematic representation of the four-step approach to sample foraging behaviour at the individual level using faeces marking with food dyes and GPS telemetry.

Table S1 in Supporting Information for the full list of models). We used the Akaike information criterion corrected for small sample sizes (AICc) to select the mixed logistic regression model that received the highest empirical support. Models were fitted using maximum likelihood estimation, and final parameters were estimated using the restricted maximum likelihood [20]. We used the model with the smallest AICc to estimate parameter values [21], and we assessed statistical significance using their 95% confidence intervals [22]. We assessed goodness-of-fit with the Pearson's Chi-Square test. We carried out all statistical analyses with R 2.11.1 [23], using the function glmer in package lme4 [24].

Third step – Assessing individual foraging behaviour beyond the effective marking period. We here briefly described the third step of the technique, as well as the fourth step below, to allow a better understanding of the full potential of our technique. Yet, in this study, we will not report results obtained from these two steps. To describe individual diet, we repeatedly visited trails (as defined in the "*Second step*" section above) of individuals equipped with GPS collars throughout winter, and we collected faeces on validated types of tracks (see Results) found along those trails. Second, we assessed fine-scale resource selection by comparing the abundance of food resources on both validated types of tracks used by a given individual and paired transects of available habitat running parallel to those tracks (located within 15–30 m from those tracks). Third, we determined how movement rate of individuals was influenced by the abundance of food resources on the validated types of tracks, using GPS locations to estimate movement rates.

Fourth step – Assessing life-history traits simultaneously or successively to sampling foraging behaviour. We sampled critical life-history traits of large temperate herbivores such as winter survival, and female and foetus body mass at the onset of spring [25,26,27]. We detected winter mortality simultaneously to sampling foraging behaviour by 1- checking movement patterns of individuals equipped with GPS collars weekly, and 2- locating carcasses of individuals that seemed stationary for more than a few days in a row. We sampled female and foetus body mass at the onset of spring by culling females equipped with collars at the end of the study period, i.e. after having sampled their respective winter foraging behaviour.

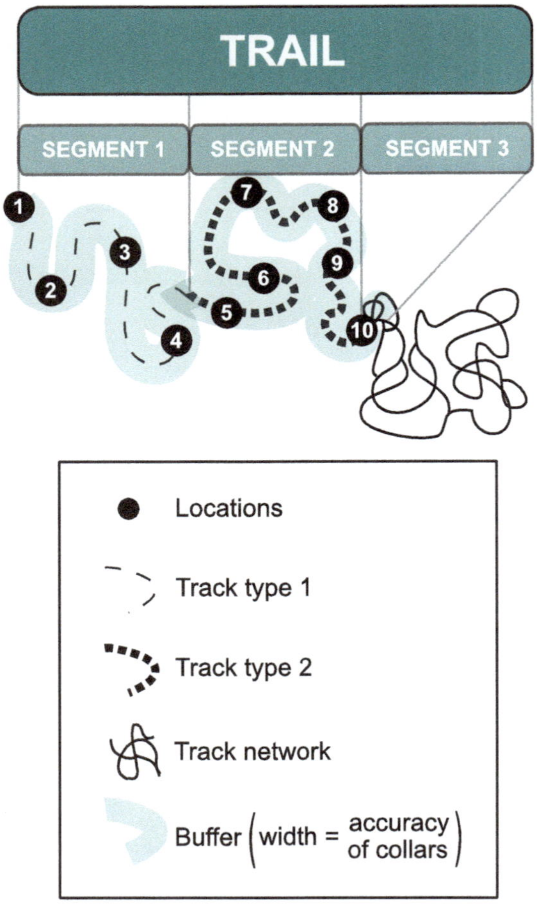

Figure 2. Schematic representation of a trail as defined in our methodological approach. The trail is formed by tracks linking successive GPS locations recorded frequently (e.g. every five minutes in our study of a white-tailed deer population). The trail is divided into different segments. Each segment is characterised by a single type of tracks (including a buffer on each side of the track, with buffer's width corresponding to the average accuracy of the GPS collar). In the case of a network of tracks, a segment corresponds to the buffer around the beginning of the network.

Results

Blue dyes appeared in the faeces of female white-tailed deer as soon as 20 h after ingestion. Across all track types, the probability of recovery of dyed faeces (number of dyed faeces/total number of faeces found on a given segment with faeces) averaged 0.44 (95% CI = [0.35,0.59], n = 98). Track type best explained variations in the probability of recovery of dyed faeces (best-fitting model: track type as a single fixed effect, see Table S1 for model selection). According to the best-fitting model, the predicted probability of recovery of dyed faeces for single tracks was very high (0.97, 95% CI = [0.90,1.00], n = 13, Figure S1) compared to other track types (networks: 0.33, 95% CI = [0.19,0.48], n = 73; paths: 0.15, 95% CI = [0.04,0.47], n = 11). Our results also confirmed that we sampled all trails within the effective marking period because the probability of recovery of dyed faeces did not vary with time elapsed since animal capture (slope = 6% per day, 95% CI = [-36%, 74%]; according to the model including the additive effect of time elapsed since animal capture and track types, see Table S1). In our study, the maximal time interval between any capture and trail sampling was 6 days, so the effective marking period was at least 6 days. Adding occurrence of snowfall and time of sampling during the day to the model with track types did not improve model fit (see Table S1).

Discussion

We developed and described an innovative technique for sampling foraging behaviour at the individual level in secretive species over variable temporal scales. This technique is based on a combination of telemetry tracking of animal movements and biomarking. The novelty in our approach is that we fitted individuals with GPS collars equipped with a data transmission system and marked their faeces with food dyes effective during a short-term period; this allowed the identification of tracks exclusively used by equipped individuals within the range of the GPS collar accuracy. This technique allows to sample individual foraging behaviour with high certainty on validated tracks, and could ultimately allow the determination of how foraging behaviour influences life-history traits.

To discuss the implications of this technique, we will address the three following points: 1- Linking foraging behaviour with life-history traits, 2- add-ons to existing methods of sampling foraging behaviour, and 3- guidelines to determine whether this technique can be used in a given study system and which type of questions could be investigated.

Linking Foraging Behaviour with Life-history Traits

Quantification of individual foraging behaviour is an essential step to link foraging behaviour with life-history traits, provided that: 1- such quantification is performed over a temporal scale long enough to capture its influence on life-history traits, and 2- those traits can be estimated simultaneously or succesively to the sampling of foraging behaviour. The novel component of our technique was designed to assess individual foraging behaviour beyond the effective marking period (Figure 1) over seasonal or larger temporal scales in secretive species. Indeed, one can confidently assess the foraging behaviour of any particular individual equipped with a GPS collar as long as the validated type of tracks or locations can be found, and GPS collars have enough battery power to record and transmit data. In our study, description of individual foraging behaviour was possible throughout the winter or as long as we could find single snow tracks. From the validated type of tracks or locations, we described individual diet by repeatedly collecting faeces, quantified individual food resource selection by assessing the availability and use of different food resources, and estimated animal movement rate in relation with the use of different resources.

A possible outcome of our technique, i.e. determining how foraging behaviour could influence life-history traits (Figure 1), may be achieved by using existing methods for sampling individual foraging behaviour (as discussed later) and life-history traits. This component is critical to scale-up from individual foraging behaviour to processes occurring at the population scale. This can be performed by monitoring, for instance, survival rates, reproductive success, number and mass of offspring, individual mass or growth rate during or at the end of the foraging behaviour study period. In our study, we sampled critical life-history traits of large temperate herbivores [25,26,27] both simultaneously (winter survival) and successively (female and foetus body mass at the onset of spring) to the foraging behaviour study period.

The probability of recovery of dyed faeces along networks and paths was low in our study (0.33 and 0.15, respectively) because conspecifics used the same tracks. Sampling individual foraging behaviour with our technique was therefore only possible along single tracks in this study population. This conclusion might be limited by the small sample size of single track segments sampled in our study (n = 13 segments). We remain confident, however, that with a larger sample size, we would have found similar results because GPS collars had a very high accuracy (mean = 4.3 m, 95% CI [1.7, 6.9]), thus reducing the probability that single tracks located within that buffer may have been used by conspecifics.

Our technique was only reliable for single tracks in our study. This could generate a bias if foraging behaviour differs between track types. For example, preferred resources might attract higher densities of individuals compared to less preferred ones. Thus, consumption of preferred resources could be more frequent on networks than on single tracks, where foraging would be biased towards less preferred resources. In our study system, the main facet of foraging behaviour studied was individual diet derived from faeces collected on single tracks. Diet estimates derived from those faeces are not biased towards resources consumed only on single tracks, because those faeces contained resources consumed over the range of track types used during the past days [28,29]. While interpreting results from the sampling of fine-scale selection or movement rates, we will take into account this potential bias by restricting our conclusions to foraging behaviour as described on single tracks.

Add-ons to Existing Methods for Sampling Foraging Behaviour

Many sampling techniques exist to assess foraging behaviour, yet most techniques have constraints that can limit their applicability for studying processes at the individual level over specific spatio-temporal scales. Behavioural observations are probably the most direct and exhaustive technique to assess foraging behaviour because they can provide an assessment of the entire foraging process, i.e. how individuals search for, acquire and consume food resources. This technique, however, has limited applicability in secretive species, remote areas or in species living in habitats where observations from a distance are impaired by the lack of visibility (e.g. forests; [13]). To overcome these constraints of working with forested secretive species, researchers have conducted detailed behavioural observations on hand-raised cervids [30,31]. Our technique overcomes these constraints by remotely monitoring free-ranging individuals, using their tracks to identify precise foraging locations, and using biomarking to rule-out the possibility that single tracks found in the vicinity of GPS-locations could have been used by conspecifics.

Genetic identification of individuals from faecal samples [32,33] could also be used to rule-out the possibility that signs of presence associated to a collared individual could have been left by conspecifics. This method would allow the validation of individual signs of presence during the entire study period and would thus extend the restricted window of validation of our technique (i.e. the effective marking period). In addition, faecal samples found around signs of presence at sites potentially used by conspecifics (e.g. networks of tracks and paths in our study) could be genotyped, providing additional samples to describe individual diet. Despite increasing efficiency and decreasing costs of genetic identification of individuals from faecal samples, such technique remains more costly than our approach in studies when telemetry is required anyway. For instance, telemetry monitoring is required in most studies that aim at linking foraging behaviour to life-history traits as it allows the assessment of life-history traits during and after the foraging study period. In addition, the degrading effect of weather on DNA decreases genotyping power [34,35]. A robust protocol to detect genotyping errors also needs to be developed to avoid spurious results [32,33,35]. Finally, in the case of cervids, the high number of pellets found on tracks requires a sampling protocol designed to reduce the number of pellets analysed without introducing any sampling bias [32].

Stable isotope and fatty acid analyses are two relatively recent techniques designed to trace the contribution of food sources to the diet of individuals over a range of temporal windows, using chemical analyses of animal tissues [12,36]. Despite all their advantages, these techniques hinge on a key requirement that is not met in many studies including ours, namely differences in isotopic or fatty acids composition among food sources. To characterise diet over a temporal scale similar to those assessed by stable isotope and fatty acid analyses, on can repeat faeces collection over the temporal scale of interest on the validated type of tracks or locations. The combined use of telemetry tracking and biomarking allowed us to overcome a main limitation of faecal analysis, i.e. that the collection of faeces from a known individual is not always possible in free-ranging populations [13], while avoiding the costs and time associated with developing and using an efficient genetic identification protocol.

Guidelines on the Use of this Technique in Other Study Systems

We here provide guidelines for determining whether this technique could be used in a given study system and which type of questions could be investigated (see also the decision tree in Figure 3).

A) Determine the possible use of GPS data transmission systems on the study animals. Our technique requires to monitor the movements of animals remotely and to receive locations rapidly to avoid the loss of individual signs of animal presence through time. Here, we used GPS collars equipped with an Iridium satellite transmission device receiveing a high number of locations by email shortly after their recording (within 1–24 hours in our study). GPS collars equipped with other data transmission systems can also be used.

B) Determine the possible use of food dyes to mark animal faeces. We recommend verifying the effectiveness of different food dyes at different doses with the study species in captivity before using it in the wild. We also recommend using the captivity experiment to estimate the length of the effective marking period. In addition to FD&C Blue No. 1, dyes of other colours have been shown to stained ruminant faeces [37], or urine and faeces of small mammals and lagomorphs [38]. The fluorescent marker Rhodamine B was used to mark faeces and urine in lagomorphs [39], while glass beads marked faeces of canids and collared peccaries [40].

C) Determine the likelihood of recovering faeces or tracks leading to faeces near telemetry locations, within the accuracy radius of the GPS collar and during the effective marking period. In our study, we used snow tracks but faeces could be found near locations or on tracks left in other soft substrates such as mud, sand, or soft soil.

D) Determine the likelihood of sampling a sufficient number of tracks or locations with faeces during the effective marking period to statistically estimate the probability of recovering only dyed faeces. This would vary with the variability in the recovery of dyed faeces per type of locations or tracks, the length of the effective marking period, the frequency of occurrence of faeces near locations or on tracks, the number of individuals marked, and the speed of the GPS data transmission system. If such a sufficient sampling is not possible, our technique could still be used to describe individual diet during the effective marking period by analysing content of dyed faeces.

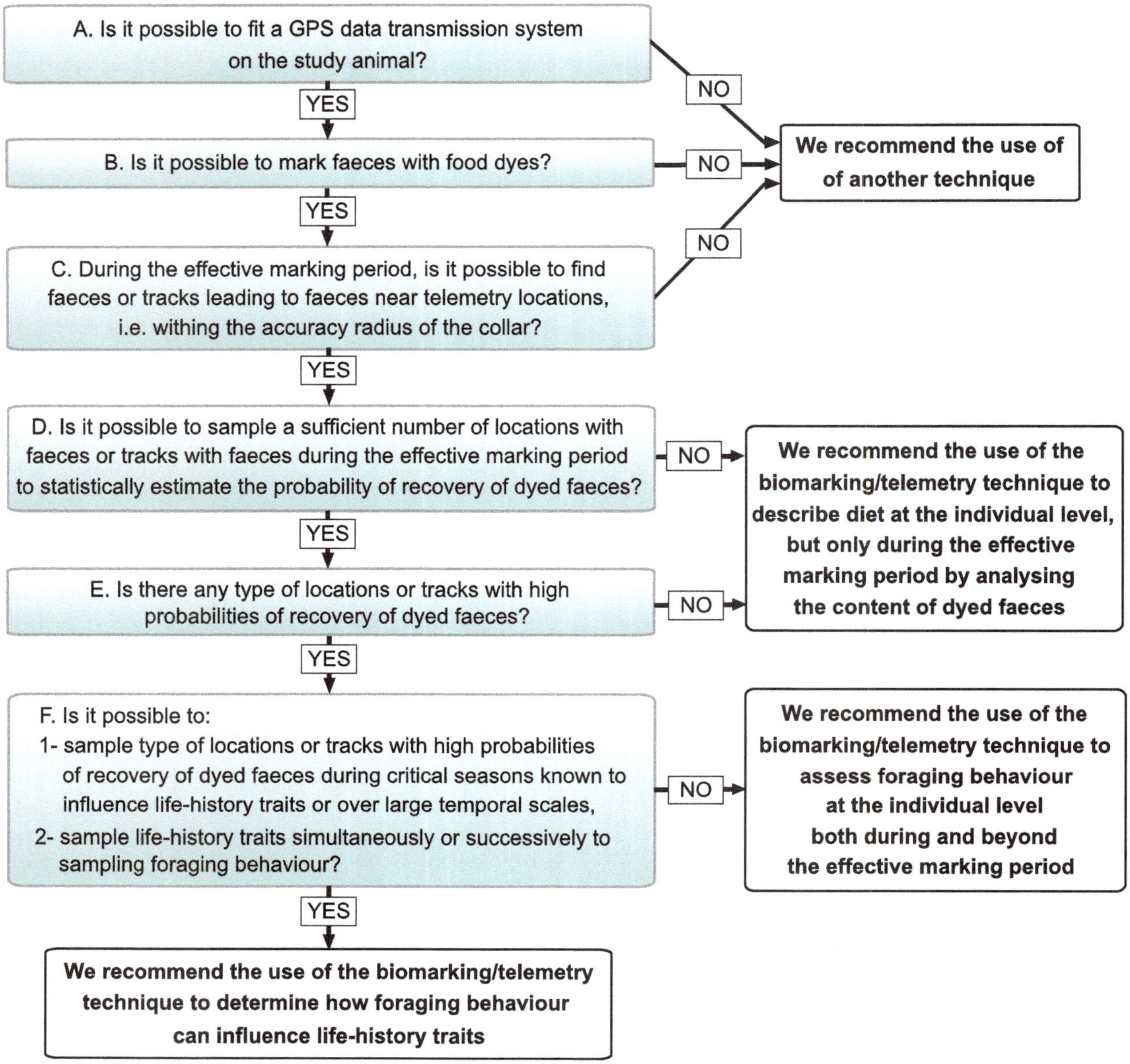

Figure 3. Decision tree for determining whether the biomarking/telemetry technique could be used in a given study system and which type of questions could be investigated.

E) Determine whether there is any type of locations or tracks with a high probability of recovery of dyed faeces. If there are such locations or tracks, then sampling them could allow assessing foraging behaviour following our approach, both during and outside the effective marking period. To maximize the probability of validating locations or tracks, it is critical to use collars that are accurate enough to isolate individual signs of presence in a given study system. Higher accuracy will be required in study systems with higher spatial overlap between individuals. As accuracy of GPS collars in ecological studies generally ranges from 3 to 30 m and can vary between brands and models [19], we recommend chosing a collar brand and model matching the accuracy-related constraints of the study system (i.e. canopy closure and topography [19], as well as the degree of spatial overlap between individuals).

F) Determine the likelihood of sampling foraging behaviour over large temporal scales or during critical seasons known to influence life-history traits, and of sampling those traits during or at the end of sampling periods.

In conclusion, our sampling technique can be added in the toolbox of ecologists aiming at assessing individual foraging behaviour in secretive species, and moreover at ultimately investigating the links between foraging behaviour and life-history traits. This technique will contribute to a better understanding of how inter-individual variation within animal populations can shape demographic processes, and ultimately provide insights on how individual heterogeneity may drive more complex ecological processes.

Supporting Information

Figure S1 Probability of recovery of dyed faeces by track type. Open circles represent raw data and closed squares represent predicted values with their 95% confidence intervals.

Table S1 Detailed results of model selection for mixed logistic regressions fitted to estimate the probability of sampling only dyed faeces for the different types of tracks (n = 96). We took into account the non-independence of data from the same trail (random variable: trail ID). We modelled the effects of track type (Type), occurrence of snowfall (Snow), time of sampling during the day (Time), time elapsed since the capture (Capture-Trail), as well as biologically relevant interactions (i.e. Type × Snow, and Type × Time). We report the number of parameters (*k*), the Akaike information criterion for small sample sizes (AICc) relative to the model with the lowest AICc (ΔAICc), as well as the AICc weight (ωAICc). Models are ranked by their AICc values and the best model is shown in bold.

Acknowledgments

We thank S. Rivard, B. Baillargeon and Héli-Boréal for help capturing deer, and S. De Bellefeuille and C. Hins for logistical support. We are grateful to D. Duteau, J.-F. Adam, B. Chénard, J.-P. Harnois, A. Hemery, J. Hénault-Richard, K. Ladhiri, G. Laprise, L. L'Italien, J. Malouin, D. Morin, G. Reuchet, L. Richard, M. O'Connor, M. Sicilia, R. Théoret-Gosselin, and particularly A. Veillette, for their valuable assistance in the field. We thank R. Freckleton, K.Q. Sakamoto and 2 anonymous reviewers for insightful comments on earlier versions of the manuscript. MAG is grateful to N.G. Yoccoz, R.A. Ims, as well as P. Hale for hosting her during her research stays in Tromsø (Norway) and Igloolik (Canada), respectively.

Author Contributions

Conceived and designed the experiments: MAG SC CD JPT. Performed the experiments: MAG NL JPT. Analyzed the data: MAG. Contributed reagents/materials/analysis tools: CD SC. Wrote the paper: MAG SC CD NL JPT.

References

1. Bolnick DI, Svanback R, Fordyce JA, Yang LH, Davis JM, et al. (2003) The ecology of individuals: incidence and implications of individual specialization. American Naturalist 161: 1–28.
2. Swanson BO, Gibb AC, Marks JC, Hendrickson DA (2003) Trophic polymorphism and behavioral differences decrease intraspecific competition in a cichlid, Herichthys minckleyi. Ecology 84: 1441–1446.
3. Beauchamp G, Giraldeau L-A, Ennis N (1997) Experimental evidence for the maintenance of foraging specializations by frequency-dependent choice in flocks of spice finches. Ethology Ecology & Evolution 9: 105–117.
4. Knudsen R, Curtis MA, Kristoffersen R (2004) Aggregation of helminths: the role of feeding behavior of fish hosts. Journal of Parasitology 90: 1–7.
5. Cucherousset J, Acou A, Blanchet S, Britton J, Beaumont W, et al. (2011) Fitness consequences of individual specialisation in resource use and trophic morphology in European eels. Oecologia 167: 75–84.
6. Annett CA, Pierotti R (1999) Long-term reproductive output in western gulls: consequences of alternative tactics in diet choice. Ecology 80: 288–297.
7. Parker KL, Barboza PS, Gillingham MP (2009) Nutrition integrates environmental responses of ungulates. Functional Ecology 23: 57–69.
8. Lomnicki A (1988) Population ecology of individuals; Simon A. Levin and Henry S. Horn SE, editor. Princeton, New Jersey: Princeton University Press. 223 p.
9. Estes JA, Riedman ML, Staedler MM, Tinker MT, Lyon BE (2003) Individual variation in prey selection by sea otters: patterns, causes and implications. Journal of Animal Ecology 72: 144–155.
10. Svanbäck R, Persson L (2004) Individual diet specialization, niche width and population dynamics: implications for trophic polymorphisms. Journal of Animal Ecology 73: 973–982.
11. Votier SC, Bearhop S, Witt MJ, Inger R, Thompson D, et al. (2010) Individual responses of seabirds to commercial fisheries revealed using GPS tracking, stable isotopes and vessel monitoring systems. Journal of Applied Ecology 47: 487–497.
12. Budge SM, Iverson SJ, Koopman HN (2006) Studying trophic ecology in marine ecosystems using fatty acids: a primer on analysis and interpretation. Marine Mammal Science 22: 759–801.
13. Litvaitis JA (2000) Investigating food habits of terrestrial vertebrates. In: Boitani L, Fuller TK, editors. Research techniques in animal ecology: controversies and consequences. New York: Columbia Universtiy Press.
14. Lefort S, Tremblay J-P, Fournier F, Potvin F, Huot J (2007) Importance of balsam fir as winter forage for white-tailed deer at the northeastern limit of their distribution range. Écoscience 14: 109–116.
15. Miller B, Muller L, Storms T, Ramsay E, Osborn D, et al. (2003) A comparison of carfentanil/xylazine and Telazol/xylazine for immobilization of white-tailed deer. Journal of Wildlife Diseases 39: 851–858.
16. Murray S, Monfort S, Ware L, McShea W, Bush M (2000) Anesthesia in female white-tailed deer using Telazol and xylazine. Journal of Wildlife Diseases 36: 670–675.
17. Massé A (2011) Comportement d'approvisionnement et sélection de l'habitat d'un grand herbivore à haute densité et en absence de prédation - Le cerf de Virginie à l'île d'Anticosti. [PhD thesis]: Université Laval. 250 p.
18. Hansen MC, Riggs RA (2008) Accuracy, precision, and observation rates of global positioning system telemetry collars. Journal of Wildlife Management 72: 518–526.
19. Frair JL, Fieberg J, Hebblewhite M, Cagnacci F, DeCesare NJ, et al. (2010) Resolving issues of imprecise and habitat-biased locations in ecological analyses using GPS telemetry data. Philosophical Transactions of the Royal Society B: Biological Sciences 365: 2187–2200.
20. McCulloch C, Searle S (2001) Generalized, linear, and mixed models. New York: Wiley-Interscience.
21. Burnham KP, Anderson DR (2002) Model selection and multimodel inference. New-York: Springer-Verlag. 488 p.
22. Yoccoz NG (1991) Use, overuse, and misuse of significance tests in evolutionary biology and ecology. Bulletin of the Ecological Society of America 72: 106–111.
23. R-Development-Core-Team (2010) R: A language and environment for statistical computing 2.11.1 ed: R Foundation for Statistical Computing, Vienna.
24. Bates D, Maechler M, Bolker B (2011) lme4: Linear mixed-effects models using S4 classes. R package version 0.999375–38. 0.999375–38 ed.
25. Côté SD, Festa-Bianchet M (2001) Birthdate, mass and survival in mountain goat kids: effects of maternal characteristics and forage quality. Oecologia 127: 230–238.
26. Fairbanks WS (1993) Birthdate, birthweigth and survival in pronghorn fawns. Journal of Mammalogy 74: 129–135.
27. Gaillard JM, Festa-Bianchet M, Yoccoz NG (1998) Population dynamics of large herbivores: variable recruitment with constant adult survival. Trends in Ecology & Evolution 13: 58–63.
28. Mautz WW, Petrides GA (1971) Food passage rate in the white-tailed deer. Journal of Wildlife Management 35: 723–731.
29. Baker DL, Hobbs NT (1987) Strategies of digestion: digestive efficiency and retention time of forage diets in montane ungulates. Canadian Journal of Zoology 65: 1978–1984.
30. Parker KL, Gillingham MP, Hanley TA, Robbins CT (1999) Energy and protein balance of free-ranging black-tailed deer in a natural forest environment. Wildlife Monographs 143: 5–48.
31. Gill RB, Carpenter LH, Bartmann RM, Baker DL, Schoonveld GG (1983) Fecal analysis to estimate mule deer diets. Journal of Wildlife Management 47: 902–915.
32. Brinkman TJ, Person DK, Chapin FS, Smith W, Hundertmark KJ (2011) Estimating abundance of Sitka black-tailed deer using DNA from fecal pellets. The Journal of Wildlife Management 75: 232–242.
33. Valière N, Bonenfant C, Toïgo C, Luikart G, Gaillard J-M, et al. (2007) Importance of a pilot study for non-invasive genetic sampling: genotyping errors and population size estimation in red deer. Conservation Genetics 8: 69–78.
34. Brinkman TJ, Schwartz MK, Person DK, Pilgrim KL, Hundertmark K (2010) Effects of Time and Rainfall on PCR Success Using DNA Extracted from Deer Fecal Pellets. Conservation Genetics 11: 1547–1552.
35. Maudet C, Luikart G, Dubray D, Von Hardenberg A, Taberlet P (2004) Low genotyping error rates in wild ungulate faeces sampled in winter. Molecular Ecology Notes 4: 772–775.
36. Kelly JF (2000) Stable isotopes of carbon and nitrogen in the study of avian and mammalian trophic ecology. Canadian Journal of Zoology 78: 1–27.
37. Kindel F (1960) Use of dyes to mark ruminant feces. Journal of Wildlife Management 24: 429.
38. Brown LN (1961) Excreted dyes used to determine movements of cottontail rabbits. Journal of Wildlife Management 25: 199–202.
39. Evans J, Griffith RE Jr. (1973) A fluorescent tracer and marker for animal studies. Journal of Wildlife Management 37: 73–81.
40. Sowls LK, Minnamon PS (1963) Glass beads for marking home ranges of mammals. Journal of Wildlife Management 27: 299–302.

19

Foraging Ecology Predicts Learning Performance in Insectivorous Bats

Theresa M. A. Clarin[1]*, Ireneusz Ruczyński[1,2], Rachel A. Page[3], Björn M. Siemers[1†]

1 Sensory Ecology Group, Max Planck Institute for Ornithology, Seewiesen, Germany, 2 Mammal Research Institute PAS, Białowieża, Poland, 3 Smithonian Tropical Research Institute, Panama, Panama

Abstract

Bats are unusual among mammals in showing great ecological diversity even among closely related species and are thus well suited for studies of adaptation to the ecological background. Here we investigate whether behavioral flexibility and simple- and complex-rule learning performance can be predicted by foraging ecology. We predict faster learning and higher flexibility in animals hunting in more complex, variable environments than in animals hunting in more simple, stable environments. To test this hypothesis, we studied three closely related insectivorous European bat species of the genus *Myotis* that belong to three different functional groups based on foraging habitats: *M. capaccinii*, an open water forager, *M. myotis*, a passive listening gleaner, and *M. emarginatus*, a clutter specialist. We predicted that *M. capaccinii* would show the least flexibility and slowest learning reflecting its relatively unstructured foraging habitat and the stereotypy of its natural foraging behavior, while the other two species would show greater flexibility and more rapid learning reflecting the complexity of their natural foraging tasks. We used a purposefully unnatural and thus species-fair crawling maze to test simple- and complex-rule learning, flexibility and re-learning performance. We found that *M. capaccinii* learned a simple rule as fast as the other species, but was slower in complex rule learning and was less flexible in response to changes in reward location. We found no differences in re-learning ability among species. Our results corroborate the hypothesis that animals' cognitive skills reflect the demands of their ecological niche.

Editor: Danilo Russo, Università degli Studi di Napoli Federico II via Università, Italy

Funding: The study was undertaken with the budget of the Max Planck Society (www.mpg.de). IR was supported by the Polish Ministry of Science and Higher Education (http://www.nauka.gov.pl). Program: "Support for International Mobility of Scientists". The funders had no role in study design, data collection and analysis, decision to publish, or preparation of the manuscript.

Competing Interests: The authors have declared that no competing interests exist.

* E-mail: tclarin@orn.mpg.de

† Deceased.

Introduction

Ecological demands have been postulated as a driving factor in the evolution of cognitive complexity and intelligence [1,2]. The cognitive abilities of animals are often well adapted to the requirements of their ecological niche (e.g. [3,4]). Migratory bird species, for example, have much longer long-term memory than non-migratory species, and food-storing bird species show better spatial memory than non-storing species [5,6]. The ecological context of feeding affects learning abilities in crabs: mobile species show experience-dependent modifications of foraging behavior while sedentary species do not [7]. In lizards, congeneric species with different foraging strategies display the same learning abilities but actively foraging species performed better in a reversal visual discrimination task than sit-and-wait predators [8]. The spatial learning abilities of voles reflect the complexity of their foraging habitats and their dietary specializations, with slower learning and decreased flexibility in more specialized species that forage in less complex habitats [9]. Game theory modeling also indicates that the unpredictability of food resources increases social foraging as well as generalism in diet; factors which can shape the evolution of cognition [1]. According to the "Environmental Complexity Thesis" [10], the heterogeneity of an environment in space and time is thought to be one of the key factors that determine the rate of the evolution of cognitive skills [1,10,11]. In our study, we compare learning and flexibility in insectivorous bats to investigate the influence of an animal's ecological niche – specifically the complexity of its foraging habitat and the degree of stereotypy in its foraging behavior – on its cognitive abilities.

Bats are especially well-suited for investigations of ecological adaptations. Following rodents they are the second most species rich mammalian order [12] and show great ecological diversity. Their diet ranges from nectar, pollen, and fruit to insects, small vertebrates, and blood [13]. We find high ecological diversity even among closely related species, and similar ecologies have developed convergently in many distantly related groups. Despite phylogenetic distance, similar wing morphology [14] and echolocation patterns (e.g. [15–18]) have emerged in bats foraging in similar habitat types. *Macrophyllum macrophyllum*, for instance, is the only phyllostomid bat that forages exclusively over water [19]. It uses distinct terminal groups of echolocation calls prior to catching its prey, an echolocation behavior that is unique among phyllostomid bats but similar to distantly related trawling bat species [20,21] and is thus clearly shaped by the species foraging behavior rather than its phylogeny [18].

The similarities in morphology and behavior among distantly related but ecologically similar bats make it reasonable to expect that the cognitive abilities of bats are also shaped by the demands of their respective niche [22] and that closely related but ecologically divergent species will differ in their abilities to solve cognitive tasks. In bats, it has been shown that wing size, which reflects foraging habitat density, is correlated to larger hippocampi, which are known to relate to better spatial memory in a wide range of animal taxa [23]. We investigated whether learning performance and behavioral flexibility vary with foraging ecology by comparing closely related species that differ in their foraging behavior. Can one predict the learning performance of a species from the complexity of the habitat in which it forages? We hypothesize that species hunting in structurally complex habitats that fluctuate over time should learn faster and should be more flexible than species hunting in less complex, more stable habitats.

Insectivorous bats can be categorized into different functional groups according to habitat complexity, temporal habitat stability, and the sensory basis of prey detection. In a simplified overview, three groups can be distinguished: bats that use echolocation to forage in the open, either in open air or over water; bats that use prey-emitted acoustic cues to glean from vegetation or open ground; and bats that use echolocation to hunt prey near, but not on, vegetation (for a detailed review see [17]). For our study, we chose one representative from each group, each from the genus *Myotis*. We experimentally tested flexibility and learning performance in these species of closely-related, congeneric European bat species to test the hypothesis that foraging ecology predicts cognitive ability.

Water foraging bats use echolocation to hunt insects over water surfaces (open water foragers). Their foraging task should be the least demanding within the three groups because the echoes reflected from insects are not masked by echoes reflected from background structures. A water surface acts as an acoustic mirror reflecting almost all the sound energy away from the bats and returning only the echoes of the bat's prey [24–27]. Bodies of water tend to be uniform and unstructured and do not undergo large changes in an observable time span; they usually do not change from night to night. Insect abundance over water surfaces also does not seem to alter the stereotyped behavior of water foraging bats: analysis of hunting behavior of the open water foraging bat, *Myotis daubentonii*, under natural conditions showed no effect of food abundance on flight activity. It is possible that the high degree of stereotypy in behavior in this group of bats is due to uniformly high insect abundance in water foraging habitats [28,29]. We predict that associative learning between abundance of insect prey and specific locations or shapes plays little role in the hunting behavior of water foraging bats. As a representative of water foraging bats, we chose *M. capaccinii*, a species found in the Mediterranean region that hunts over water surfaces, with a preference for slow running rivers [30,31].

Bats from the second group glean arthropods from open ground or vegetation. In this situation insect echoes are masked by strong background-generated clutter echoes. Gleaning bats are specialized to find their prey by listening for prey-generated sounds such as rustling noises or communication sounds [32] and are therefore termed "passive listeners". Their foraging habitats are highly structured and undergo large changes over the course of a year. Consequently, these bats must be adept at learning to recognize specific landscape features, such as a freshly cut meadow, as a good foraging ground (e.g. [33]). They could also learn to associate prey profitability with specific sensory cues, such as rustling noise amplitude [34]. We predict that associative learning should play a more important role in the foraging behavior of passive listeners than open water foragers. To represent the passive listening gleaners, we chose *M. myotis*, a species which hunts arthropods from open accessible ground in forests or fields [35–37].

The third group, termed clutter specialists, consists of bats using echolocation to forage for insects in close proximity to vegetation. This foraging task is extremely challenging as it requires the ability to distinguish clutter echoes from the echoes of insects (e.g. [38,39]). Like passive listening gleaners, the foraging environment of clutter specialists changes rapidly: small plants and flowers appear and disappear, while trees grow leaves, blossoms and fruits and rapidly loose them again. Most important for bats, insect abundance changes in space and time as a function of plant phenology (e.g. [40]). Because flowering plants attract insects, the bats likely have to constantly build and rebuild associations with certain plants and places that are linked to high prey abundance [22]. Associative learning of cues that indirectly indicate the presence of prey, and the flexibility to update these associations rapidly over time, should thus be more important in these bats than in the former two groups. As a representative of this group, we investigated *M. emarginatus*, a clutter specialist sympatric with the other two species in our study [41–44].

We specifically chose three closely related species (illustrated in Figure 1) to decrease the likelihood that species differences could be attributed to phylogenetic distance. Our species choice is conservative in that *M. myotis* and *M. emarginatus* are more similar in their foraging ecology, but more distantly related to each other than each is to the ecologically dissimilar *M. capaccinii* [45].

Based on the hypothesis that the complexity of the foraging habitat predicts cognitive ability, we expected open water foragers to display relatively stereotyped behavior, slow learning and low flexibility. In contrast, we predicted that passive listening gleaners and clutter specialists would be fast learners and highly flexible [22]. To compare learning performance and flexibility, we used a purposefully artificial and thus species-fair plastic maze in which the animals had to crawl and search for food. A similar crawling paradigm has been successfully used for bats in other simple learning tests [46,47]. We quantified the bats' behavior on four tasks: exploration, simple rule learning, a reversal learning task that tested for flexibility, and complex rule learning. We predicted that the two species hunting in or near vegetation would be faster learners and show greater flexibility than the water foraging species.

Materials and Methods

Ethics Statement

Capture and husbandry were conducted in accordance with the species-specific recommendations of the Canadian Council on Animal Care on bats [48] and were licensed by the responsible Bulgarian authorities (MOEWSofia and RIOSV-Ruse, permit numbers 193/01.04.2009 and 205/29.05.2009). Officials from the Bulgarian Ministry of Environment and Water (MOEW) inspected our work in accordance with Section 8, Article 23, Paragraph 3 and 4 of the Bulgarian Biodiversity Law. According to Bulgarian laws no further ethical approval by a committee is required for a non-invasive behavioral study. No bats were harmed. All bats were released in good health, at or above capture weight, at their respective capture sites after the experiments.

Animals

Bat capture and experiments were conducted in Bulgaria. We used experimentally naïve, wild-caught adult male bats of the species *M. myotis* (n = 7), *M. capaccinii* (n = 8) and *M. emarginatus* (n = 7). We captured *M. myotis* and *M. capaccinii* in or near the entrance of caves in northeastern Bulgaria. *M. emarginatus* were mist-netted in the central Balkan Mountains near Gabrovo. The animals were then transferred to the Tabachka Bat Research

Functional Group	Open water forager	Clutter Specialist	Passive Listener
Prey detection by...	Echolocation	Echolocation	Passive listenting
Species used	*Myotis capaccinii* (n=8)	*Myotis emarginatus* (n=7)	*Myotis myotis* (n=7)
Complexity of habitat	Less complex hunting habitat	More complex hunting habitat	
	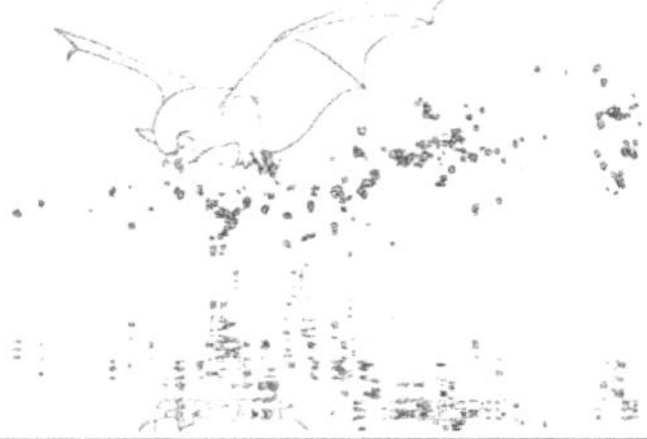	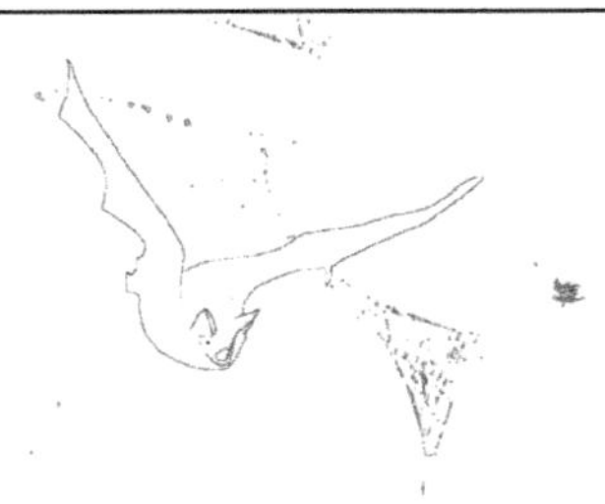	
Phylogeny according to [45]	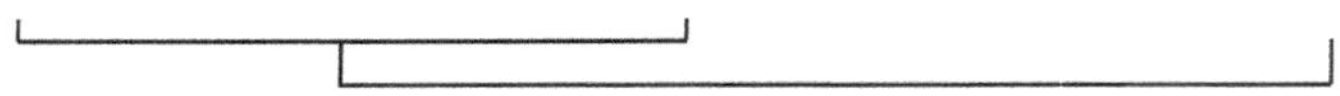		

Figure 1. The bat species used in the experiments were closely related European congeners representing three distinct foraging guilds.

Station (Bulgaria) of the Sensory Ecology Group (Max Planck Institute for Ornithology, Seewiesen, Germany), which is run in cooperation with the directorate of the Rusenski Lom Nature Park in the district of Ruse. *M. capaccinii* (7–10 g) and *M. emarginatus* (6–9 g) were housed together in a screen tent (2.2 m×0.9 m×1.1 m). *M. myotis* (20–27 g) were kept in a holding cage (50 cm×35 cm×40 cm). All animals had ad libitum access to water. For individual recognition, we gave all bats a within-species individual-specific haircut by cutting a small stripe of hair on one part of the back. After capture, the bats were hand-fed live mealworms (*Tenebrio molitor*) for two nights before starting training in the maze, to allow them time to adjust to the environment and the new food source. *M. myotis* received 4 g of mealworms per day, while *M. capaccinii* and *M. emarginatus* received 1.5 g and 1.7 g respectively. All bats readily accepted mealworms as food. When experiments began, bats were only fed in the experiment (with the exception of night 1, see below). The body mass of the bats was measured before and after each session to ensure that the animals maintained their weight. The dark-light-cycle and temperature in captivity mirrored ambient, natural conditions. The experiments with *M. capaccinii*, *M. myotis*, and four *M. emarginatus* were conducted in July and the first half of August. The experiments with the other three *M. emarginatus* took place in September of the same year. After the experiments all bats were released at their respective capture sites.

Experimental Setup

Experiments were conducted in a plastic maze. The simple form consisted of 4 plastic boxes (20 cm×13.5 cm×10 cm) connected to a large center box (24 cm×16.5 cm×12.5 cm) by plastic tubes (25 cm long, 7 cm internal diameter). All bats could easily crawl and turn in the tubes. One of the four boxes always served as the starting box while the other three contained mealworms (see Figure 2a). In all experiments, we placed the same amount of live mealworms in each feeding box to prevent the bats from receiving different acoustic, visual or olfactory cues from the three boxes. *M. myotis* was presented with normal-sized mealworms (approx. 0.1 g each) while the two smaller species received small mealworms (approx. 0.05 g each). The mealworms could crawl in the feeding boxes, but could not move back into the plastic tubes, as the entrance to the tubes was elevated by about 0.5 cm. The entrance to each box could be closed with a plastic slide. Additional arms could be added to the maze for the complex task (see Figure 2b). To remove possible olfactory cues, the mazes were cleaned after each experimental session with detergent and water and a clean set-up was used for each bat. Experiments were conducted in near darkness. The only light source was dim red light from the observer's headlamp (Tactikka plus, PETZL) and an infrared light (CONRAD, 1/3″ CMOS colour camera with IR) mounted on the ceiling to enable video recording. A camera (Watec, WAT-902H2 Ultimate) was placed above the maze to record all experiments. The videos were recorded on miniDV tapes with a camcorder (Sony DCR-TRV80E recorder).

Procedure

Experiment 1: Exploration. On night 1 we conducted an exploration experiment. The experiment began when a bat was placed into the maze and left there for 15 minutes. Prior to placing the bat in the maze, each box, including the center box, was baited with three mealworms. The bats were allowed to move freely and eat as many of the mealworms as they wanted to. Thus, the bats had the possibility to acclimate to the plastic maze and we were able to use video analysis of movement in the maze to assess the possible effect of species differences in body size and crawling performance on the results of the following experiments. After 15 minutes of exploration, the bat was removed from the maze and was additionally hand-fed up to the normal daily amount of food taking into account the number of mealworms eaten during exploration.

We analyzed the videos of the exploration experiment for several behavioral parameters to control for differences in crawling abilities. "Small scale exploration" was defined as the amount of time an animal spent exploring one of the boxes or exploring an

A

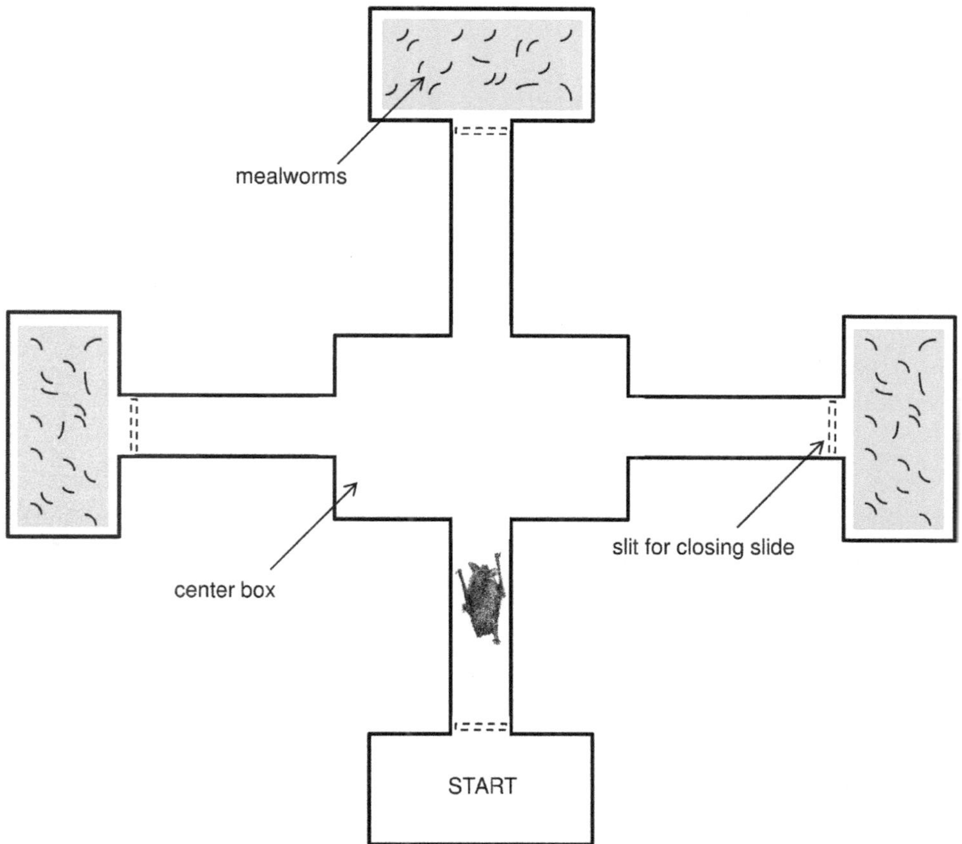
mealworms
center box
slit for closing slide
START

B

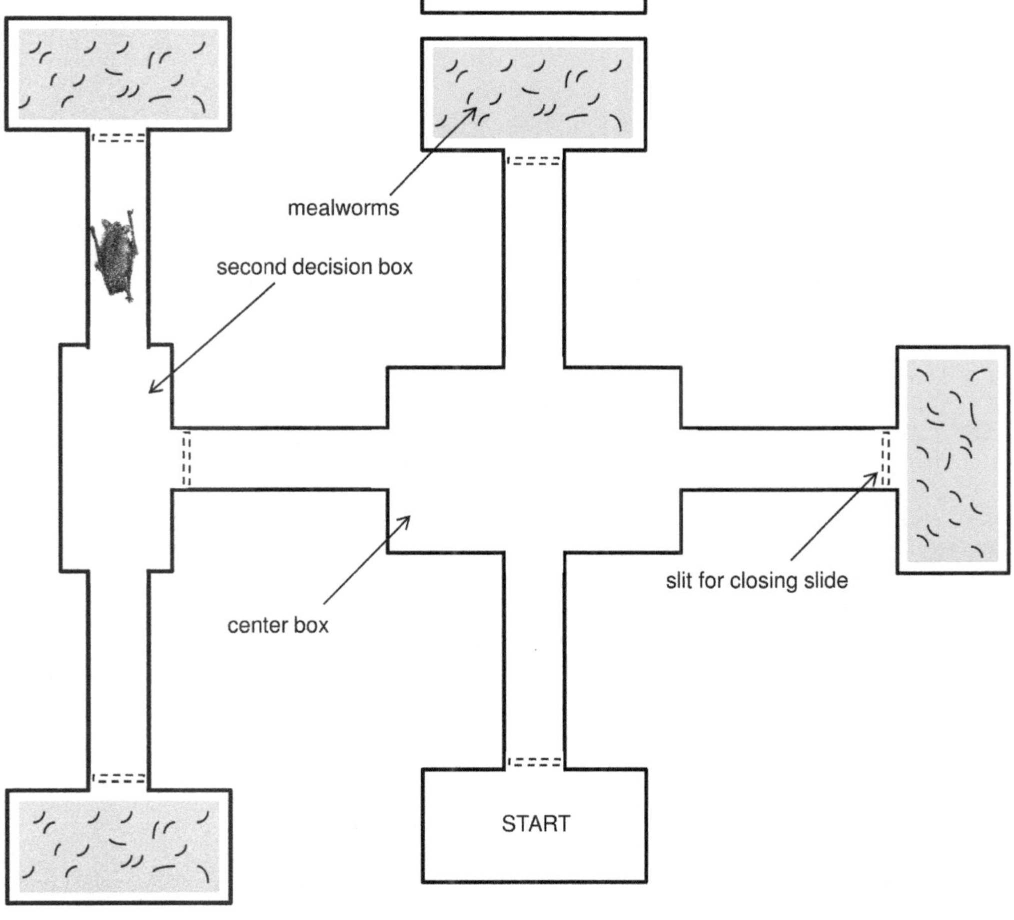
mealworms
second decision box
center box
slit for closing slide
START

Figure 2. Mazes used in experiments. a) Maze used in experiments 1 to 3; "Exploration", "Simple Rule Learning", "Flexibility and Re-learning" b) Extended Maze used in experiment 4; "Complex Rule Learning".

arm systematically by moving back and forth. In contrast "fast walking" was defined as the amount of time an animal was crawling straight through the maze; and "immobile" referred to times when the animal was not moving. We also quantified the "number of boxes visited" (excluding the start box), the "number of mealworms eaten", and the "latency to exit the start box". We additionally examined whether an animal explored the center box thoroughly or only crossed it when crawling from one arm to another. Two videos from *M. capaccinii* were lost due to technical failure. We analyzed the data using Anova, Kruskal-Wallis-test and Chi-square-test where appropriate.

Training. During the next two nights, the bats were trained to find rewards in one of the boxes (half of the bats were trained to the right box; half of the bats to the left box; balanced within species, fig. 2a). To train the bats to feed from the rewarded box, the other two arms were blocked by slides. In this training phase and all following experiments, each night, one session of 10 trials (one trial being one searching event) was conducted for each animal. At the beginning of each trial the bat was placed into the starting box. When it reached the target box it was allowed to eat two to four mealworms. Afterwards it was removed from the target box and the next trial started. If a bat did not move for more than five minutes (either did not leave the starting box or stopped in the center box), the trial was aborted. Then, after a break of two to five minutes, the next trial started.

Experiments 2–4. We conducted three additional experiments with all bats. All experiments were conducted in the same order. Each bat was tested in each experiment for at least three sessions (one session per night). Within each session after trial 4 and 7 the arms of the maze were interchanged to prevent the bat from following its own scent. After trial 7 the center box was rotated 180°.

If a bat made 8 out of 10 correct decisions, it was considered to have learned the task. To give the animals the opportunity to better consolidate the newly acquired information, individuals were required to make 8 out of 10 correct decisions on two consecutive nights before moving on to the next experiment. If a bat did not learn a task after 15 sessions, it was removed from the experiment. This rule only had to be applied for one *M. capaccinii*. The whole study required each bat to stay in captivity for a minimum of 14 nights (two nights of hand-feeding, one night in experiment 1, two nights training and three nights for each of the experiments 2–4). Theoretically, the study could take as long as 50 nights (2+1+2+15+15+15) for one individual, although no bat required the maximum amount of time in the experiments.

Experiment 2: Simple rule learning. On night 4 the second experiment started. For this and all subsequent experiments, all four arms of the maze were open. After reaching the center box, the bat was required to enter the arm in which it had formerly been trained to find mealworms (fig. 2a). In the target box it was again allowed to eat two to four mealworms before the next trial started. If a bat entered an unrewarded arm, the entrance to the box was blocked with a plastic slide. Subsequently, the bat was removed from the maze without receiving a food reward and the next trial started. We scored the number of days that an animal needed to meet the learning criterion of 8 out of 10 correct decisions.

Experiment 3: Flexibility and re-learning. To test flexibility and re-learning we conducted a reversal experiment in which the bats had to learn to visit the box opposite the formerly rewarded one (fig. 2a). If a bat entered the formerly rewarded arm it was denied access to the mealworms by closing the entrance to the box. We recorded how many trials were required for a bat to investigate a new arm; either straight ahead or the opposite box; we used this metric as a measure of flexibility. After visiting the newly rewarded arm for the first time, an animal could learn where to find the reward. We then analyzed the number of days it took an animal from the first correct visit until making 8 of 10 trials correct in one session, and used this as a measure of re-learning. One *M. capaccinii* did not learn the new position within 15 sessions and was therefore excluded from the second part of experiment 3 as well as from experiment 4.

Experiment 4: Complex rule learning. For experiment 4, two additional arms were added to the maze (Figure 2b). Upon choosing the same arm as was rewarded in experiment 3 the bat now entered a second decision box and had to turn in the opposite direction as before (either left-right or right-left) to obtain the mealworm reward. We used the same criterion and analysis as in experiment 3 to quantify how quickly the bats learned this more complex task. Due to the length of the study (see above) and constraints on the timing and use of the field station, logistical reasons made it necessary to release two *M. emarginatus* before they could participate in this experiment.

Statistical analysis was conducted in R [49]. For the post-hoc analysis following a Kruskal-Wallis-test, we followed Dunn's [50] suggestions for a test for multiple comparisons using rank sums.

Results

Experiment 1: Exploration

For most behavioral parameters we did not find any differences between species in the exploration experiment. We compared the latency to exit the start box ($F_{2;17} = 1.20$; $p = 0.325$), the time spent with small scale exploration ($F_{2;17} = 0.75$; $p = 0.489$) (see fig. 3), the time spent walking fast ($F_{2;17} = 0.06$; $p = 0.943$), and the time spent immobile ($F_{2;17} = 0.56$; $p = 0.581$). All box-and-whisker-plots show median, 25 percentile, 75 percentile, minimum, and maximum. Outliers have values at least 1.5 times the interquartile range (IQR) larger than the 75 percentile or 1.5 times the IQR smaller than the 25 percentile. On average it took the bats less than one minute to leave the start box and begin to explore the maze (mean = 0.58 min ± SD = 0.80). They spent ten minutes on small scale exploration (10.37 min ±2.71), two minutes on fast walking (1.8 min ±1.34), and three minutes immobile (2.75 min ±2.94). Three *M. capaccinii*, four *M. emarginatus*, and five *M. myotis* visited all three boxes. There was no difference in the number of boxes visited among the species (Kruskal-Wallis: df = 2; chi-squared = 0.51; p = 0.776). We only found differences in the absolute number of mealworms consumed (Kruskal-Wallis: df = 2; chi-squared = 8.714; p = 0.013) with *M. myotis* eating the most and *M. capaccinii* eating the least. To control for the effect of body mass we calculated the number of mealworms eaten per gram body mass of the bat and still found a difference in the relative number of mealworms consumed ($F_{2;17} = 7.73$; $p = 0.004$) with *M. emarginatus* eating the most and *M. capaccinii* eating the least.

Experiment 2: Simple Rule Learning

All species learned the task quickly (see fig. 4). After two days of training, there were no differences among any of the species in the number of days the bats needed to make 8 out of 10

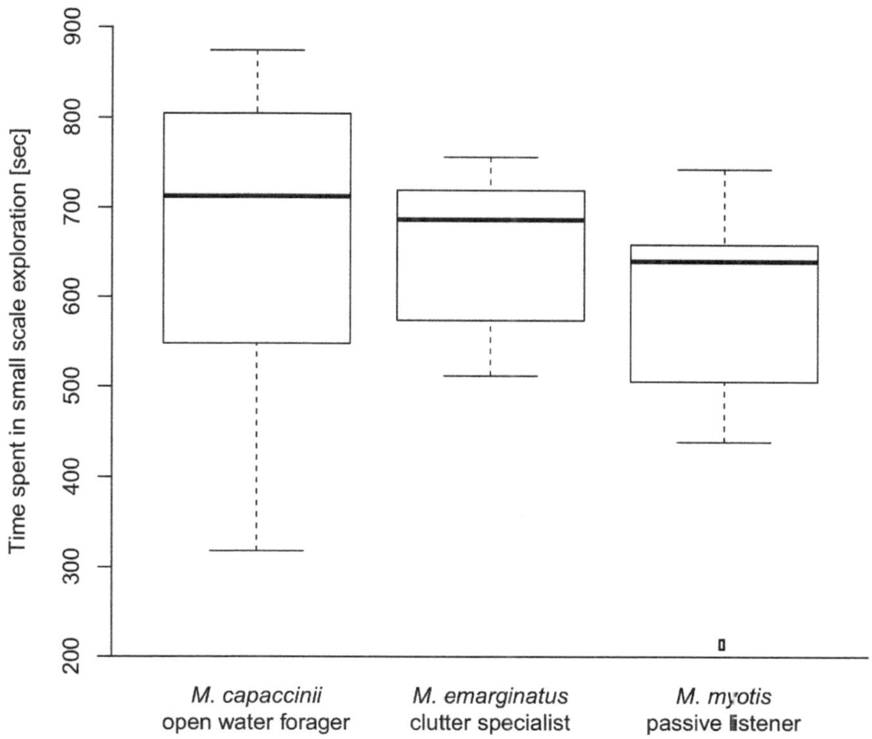

Figure 3. Experiment 1 ("Exploration"). Time spent with small scale exploration: There was no difference among the species in the time the bats spent with small scale exploration ($p = 0.49$). We found no difference in other parameters measured, except for the total number of mealworms eaten (not shown here).

correct decisions (Kruskal-Wallis: df = 2; chi-squared = 2.16; $p = 0.340$).

Experiment 3: Flexibility and Re-learning

Most *M. emarginatus* and *M. myotis* visited a new arm on the first day (trial 1 to 10), while most *M. capaccinii* required at least two days to visit a new arm for the first time (fig. 5 and supporting information Video S1). A Kruskal-Wallis-test showed significant differences between the species (df = 2; chi-squared = 7.68; $p = 0.022$). One *M. capaccinii* was an extreme outlier that required 78 trials to try a different arm. To ensure that the difference between the species was not due to this single individual, we tentatively excluded it from analysis and still found significant differences among the species (Kruskal-Wallis: df = 2; chi-squared = 6.30; $p = 0.04$). We did a post-hoc analysis following Dunn [50] to find out if there were differences between *M. capaccinii* and the other two species, as well as differences between *M. emarginatus* and *M. myotis*. *M. capaccinii* required significantly more trials to change its strategy and search in a new location than did the two other species. There was no difference in the number of trials required for *M. emarginatus* and *M. myotis* (fig. 5 and supporting information Video S1).

We found no difference among species in re-learning speed (Kruskal-Wallis: df = 2; chi-squared = 4.14; $p = 0.126$, fig. 6). However, when the *M. capaccinii* outlier (which needed six days to achieve 8 of 10 correct trials) was removed from the analysis, we found a significant difference among the species (Kruskal-Wallis: df = 2; chi-squared = 6.91; $p = 0.032$), with a faster re-learning speed in *M. capaccinii* than in the other species (Dunn: $p = 0.022$).

Experiment 4: Complex Learning

A Kruskal-Wallis-test showed a borderline significant difference among all species (df = 2; chi-squared = 5.81; $p = 0.055$) in the number of days needed to make 8 out of 10 correct choices. In the post-hoc analysis we conducted following Dunn [50] we found a clearly significant difference between *M. capaccinii* and the other two species ($p = 0.03$), implying that it took *M. capaccinii* longer to learn this more complex task of turning twice than it took either of the other species. The two *M. myotis* and *M. emarginatus* outliers (fig. 7) coupled with a low overall sample size (7 *M. capaccinii*, 5 *M. emarginatus*, 7 *M. myotis*) lead to only borderline significance in the Kruskal-Wallis-test (Kruskal-Wallis without outliers: df = 2; chi-squared = 8.65; $p = 0.013$). Again, there was no difference between *M. emarginatus* and *M. myotis* (fig. 7).

For the experiments 2–4 we investigated whether there were differences among the individuals that were first trained to turn right versus the ones that were trained to the left box. We found no differences for any parameter between these two groups (all $p > 0.1$).

Discussion

We find strong evidence that bat cognitive skills reflect their ecological niche. By comparing the ability to find food in a species-fair, artificial crawling maze we tested the learning performance of three closely related species that naturally forage in habitats differing in complexity, stability, and food predictability. We found that bat species hunting in more complex, less stable habitats with lower food predictability perform better in more complex learning tasks and are more flexible when the food source is relocated than those hunting in simple, more stable habitats with highly predictable food resources. Our results confirm the "Environ-

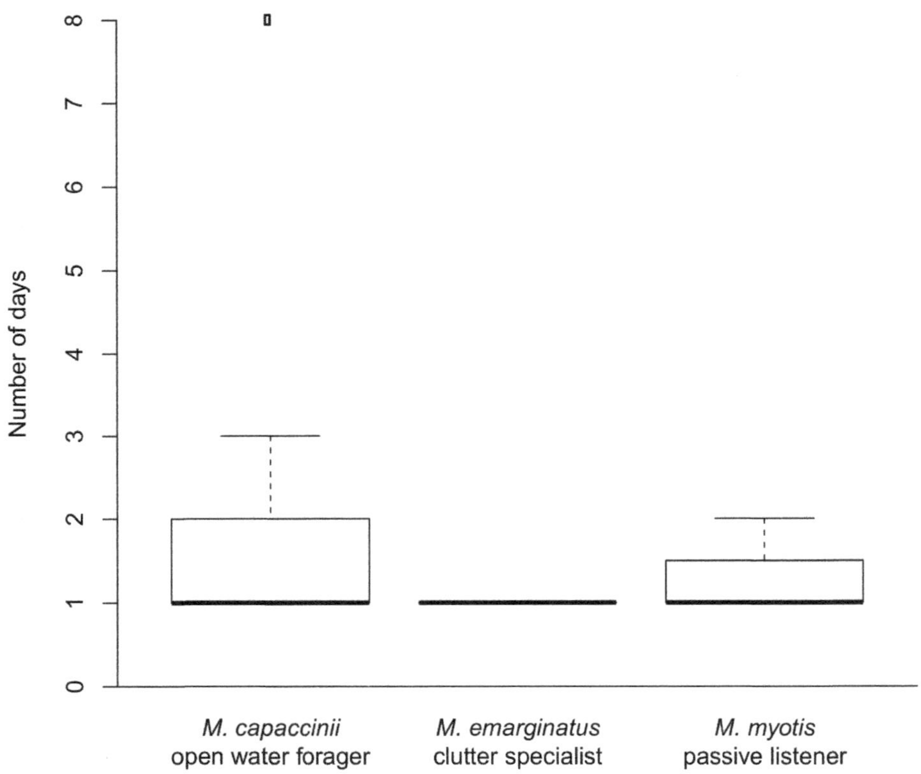

Figure 4. Experiment 2 ("Simple Rule Learning"). Number of days to reach criterion: There was no difference in learning performance among the three species (p = 0.34). Most animals reached the criterion on the first day after pre-training.

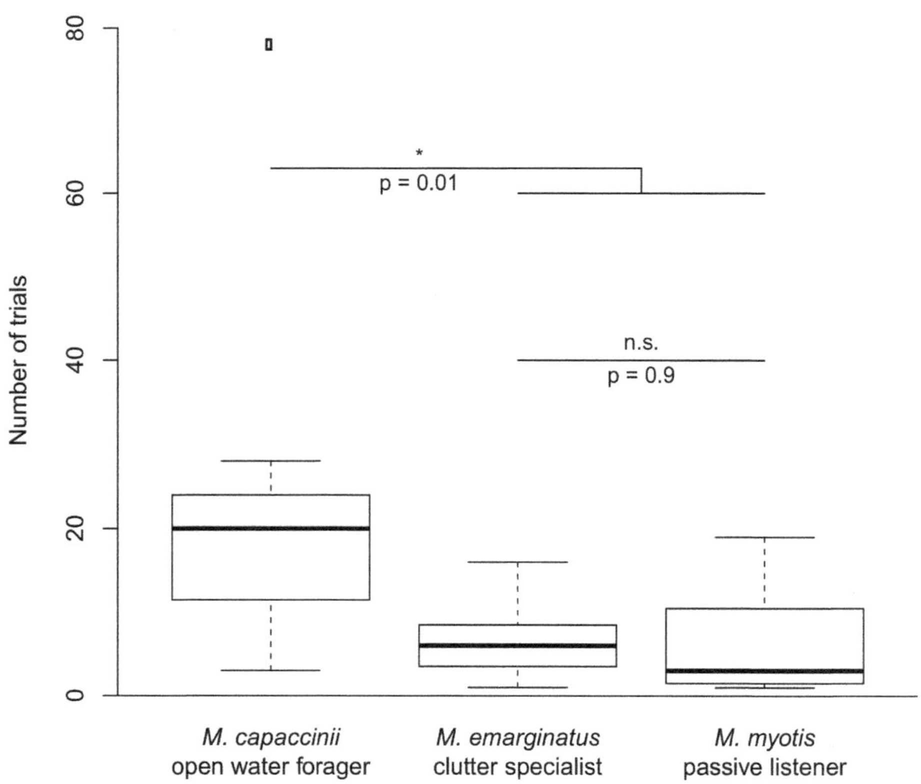

Figure 5. Experiment 3 ("Flexibility"). Number of trials until the animals visited a formerly unrewarded arm for the first time.

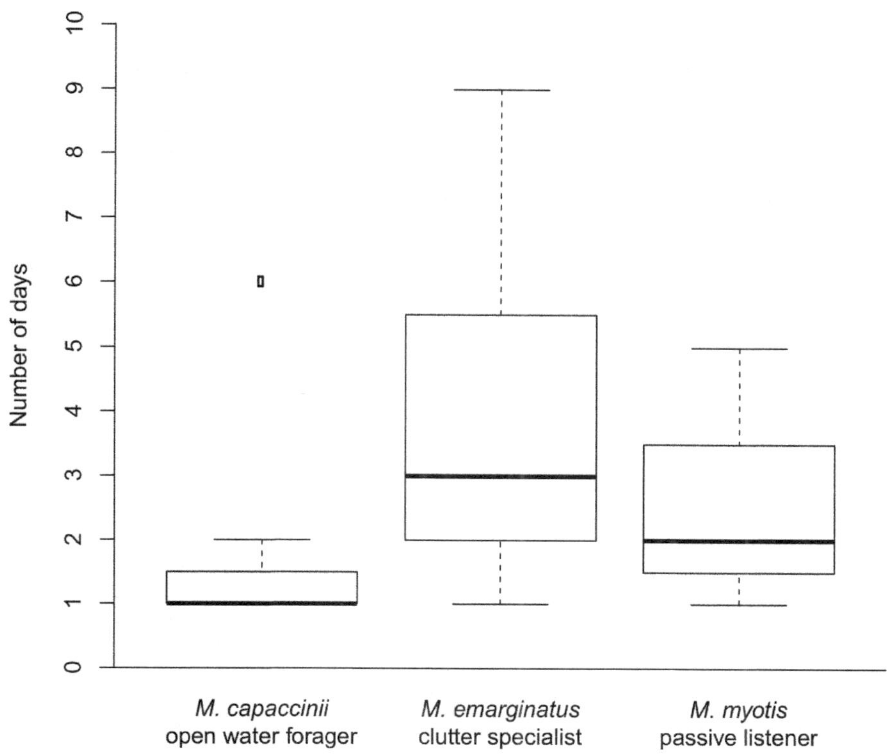

Figure 6. Experiment 3 ("Re-learning"). Number of days from visiting the newly rewarded box for the first time to reaching criterion (turning 8 out of 10 times in the opposite direction than before): There was no difference in learning performance among the species (p = 0.13).

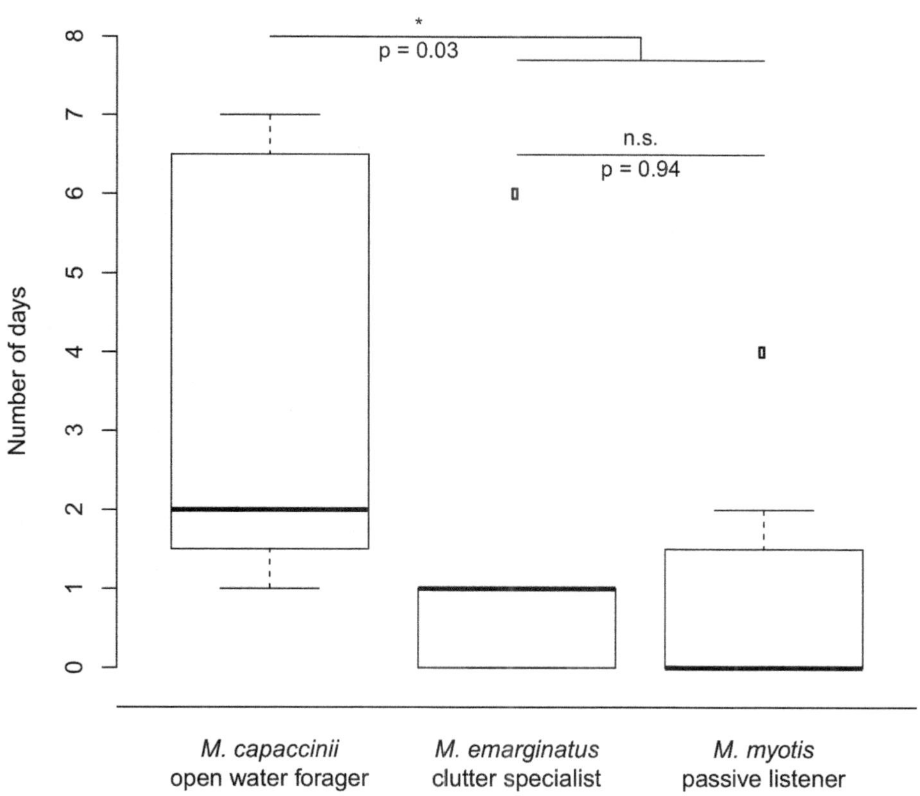

Figure 7. Experiment 4 ("Complex Rule Learning"). Number of days from visiting the newly rewarded box for the first time to reaching learning criterion.

mental Complexity Thesis" [10] and are the first to directly link bat learning and flexibility with ecological foraging niche.

Learning in an Artificial Environment

The comparison of learning performance of different animal species is extremely challenging [51]. To allow for comparison all species should be tested in the same setup, while each species unique set of sensorial and motor skills has to be accounted for to guarantee species-fairness. While a less artificial, more natural foraging task would have offered greater insight into the cognitive capabilities of these animals in nature, natural foraging tasks would have hindered cross-species comparisons. To compare learning and flexibility of taxa with widely disparate foraging ecologies (the goal of this study), we purposefully chose experimental tasks that were foreign to all. We used the exploration phase to assess possible differences in motor skills among the three species that might have hindered access and mobility within the maze. The lack of differences among the investigated species in the exploration experiment suggests that our results were not biased by any consistent species differences in the ability to cope with the maze. The difference in the number of mealworms eaten during exploration could be explained by body size differences or slight sensory inequalities. In the other experiments there were more mealworms in the boxes and the animals were given all the time they needed to consume two to four mealworms before the next trial started. This should have compensated for possible differences in detection and eating speed. We therefore conclude that the results of our experiments were not affected by differences in size or crawling performance among the species and that all animals had equal opportunities to learn about all the boxes and paths of the maze. The plastic maze was equally artificial for all bats and should not have posed a particular disadvantage to any one species. Because of its artificial novelty, it proved effective in elucidating consistent species differences in learning performance. Our results demonstrate that an artificial setup like ours can be useful to balance potentially biasing factors and to investigate cognitive abilities of different species in a species-fair manner.

Solving Simple and Complex Tasks

Two days of pre-training were sufficient to teach the bats to find food in one arm of the maze. These results show that the process of learning a simple rule (experiment 2) is quick, and suggests that simple rule learning is not sufficiently challenging to elicit differential responses in the species tested. It suggests that differences in cognitive abilities might only be detectable in more complex tasks.

Indeed, when the bats were required to exhibit more complex behavior to obtain a food reward – to turn twice to find the rewarded box in the complex learning task (experiment 4) – we found differences among the species. *M. myotis* easily completed this task, and most individuals met the learning criterion on the first day of testing. *M. emarginatus* also quickly learned to find and remember the new feeding box, most individuals learning within days 1–2. *M. capaccinii* showed the greatest inter-individual variation. Of the seven *M. capaccinii*, four learned the task rather quickly within days 1–2 and three learned the task very slowly, requiring six to seven days. Intermediate learning performance was not observed in this species. Interestingly, the one *M. capaccinii* that required 78 trials to investigate a new arm in experiment 3 (flexibility and re-learning) was the fastest learner in this task and succeeded in finding the food reward in 8 out of 10 trials during the first night. Thus, the least flexible individual was the fastest learner in the complex learning task. Even though the sample size in the complex rule learning experiment was smaller than in the other experiments and the results only show borderline significance ($p = 0.055$), this trend suggests that *M. capaccinii* needed the greatest number of days to remember the newly rewarded box. A larger dataset would be needed to demonstrate clearly significant results. However, while *M. capaccinii* were able to learn a simple rule as quickly as the other species, a more complicated path was more difficult for them to learn and remember. Thus, as predicted, the species foraging in the least complex environment, open water, performed more poorly than the passive listening gleaner and the clutter specialist.

In experiment 4, a few individuals of each of the three species went straight through the center box several times, even though the end box of this arm had never been rewarded. This behavior rarely occurred in the other experiments. It would be interesting to investigate if the bats had developed a cognitive map of the maze and would have turned in the direction of the rewarded box at the end of the arm or would have been able to perform shortcuts or use a more direct route, had this been available. Further investigations are necessary to determine whether insectivorous bats use cognitive maps in foraging as has been recently shown for fruit-eating bats on a larger scale [52], and if so, the conditions under which they do so.

Flexibility and Re-learning

As expected, we found a clear difference among the three species in the flexibility test (experiment 3). While most *M. emarginatus* and *M. myotis* tried new ways to find food on the first day when the familiar route was blocked, the open water forager, *M. capaccinii*, was persistent in visiting the formerly rewarded box. This supports the "Environmental Complexity Thesis" [10], which predicts that bats hunting in more complex and unpredictable habitats (for example, in or near vegetation) will show more flexible behavior than bats hunting in less complex and more predictable habitats (for example, over water surfaces). Game theoretical modeling suggests that individuals can enhance their probability of finding food by specializing on one food type when the location of their food sources is predictable in space and time [1]. The stereotypic behavior of *M. capaccinii* together with what we know from natural history and dietary studies [30,31] suggests that rather than specializing on a single food type, *M. capaccinii* has evolved to specialize on a specific and simple foraging habitat in which its particular foraging skills excel. The obstinacy of this species suggests that stereotypic behavior under natural conditions is beneficial and could be genetically determined. It is interesting that once the bats discovered the new food source in our artificial maze, there was no difference in the re-learning speed of a simple rule (e.g., always turn right or always turn left). There even was a trend that *M. capaccinii* learned the new position faster than the other species. This might be due to their greater persistence. This trend suggests that persistence can have evolutionary benefits. It is possible that in nature *M. emarginatus* and *M. myotis* continuously sample their surroundings for more and (potentially) better food sources, thereby making mistakes, while *M. capaccinii* persists in repeatedly visiting locations with a high probability of containing food. It is interesting to note that there is evidence that *M. capaccinii* in Israel have recently begun hunting fish adding this prey to their diet only within the last century. Especially during winter they forage to a great extent on *Gambusia affinis*, a species that was introduced in the area in the 1920s [53]. Occasional piscivory is known in this species from other areas and seems to be a common foraging strategy in other trawling bat species in times of high fish abundance [31,54,55]. In an experimental setup with *M. capaccinii* in the flight cage, dips into the water were not directionally targeting fish, but were carried out at random and seemed to follow stereotyped patterns [54], similar perhaps to the stereotyped foraging behavior we found in *M. capaccinii* in our maze experiment.

One *M. capaccinii* could not be included in the full set of experiments because it failed to complete the re-learning task within 15 sessions (experiment 3). This individual visited a new arm in the second night of experiment 3 (trial 20) and found the new food source for the first time in the fifth night of experiment 3 (trial 42), but even after ten additional nights of testing, never learned to consistently visit the newly rewarded box. This particular individual also required the most time of all bats to complete the simple rule learning task (see fig. 4; one *M. capaccinii* requiring eight days). This behavior differs from the other *M. capaccinii* which showed learning and re-learning within one or two nights once they found the food source, and highlights the potential for high inter-individual variation even within a species. Apart from this individual we did not find that any other bat consistently showed a different performance from its conspecifics (i.e. the outliers in different experiments were different individuals).

The relationship between flexibility and stereotypy and food predictability has been studied on an intraspecific level in the context of animal personality research. Studies of intraspecific variation in exploration speed, flexibility, and novelty response in birds [56] and small mammals [57] show that individuals that are more persistent and do not change their behavior when presented with changes in food positioning site have the advantage of finding more food under stable conditions. On the other hand, once they find a reliable food source they develop inflexible, routine-based behavior. Therefore, under more variable conditions, they have a disadvantage compared to less persistent individuals. Depending on how variable a specific environment of an individual or a population is, we predict a selective pressure to shape foraging behavior either towards more stereotypy or towards more flexibility. Because in our maze experiments we see predictable differences across species despite favorable conditions to the contrary, we further predict a genetic basis for stereotyped versus more flexible behavior, similar to the genetic basis for stress and novelty response [58], exploration propensity [59], and aggressive behavior [57]. It is possible that within-species differences on the flexibility-stereotypy gradient ultimately translate into species differences in these traits, as shown in the present study, through allopatric or even sympatric speciation with differential microhabitat selection.

Brain-size and Cognitive Abilities

Ratcliffe et al. [60] categorized predatory bat species into groups based on their natural foraging modes. They found that bat species they categorized as more flexible in their hunting strategies and ground gleaning species like *M. myotis* have larger relative brain sizes and a larger neocortex than less flexible species, such as open space aerial hawking bats. Eisenberg and Wilson [61] also found that aerial insectivores have smaller relative brain sizes than other insectivores, which in turn have smaller relative brain sizes than frugivorous bats. Several studies in birds and mammals show that the relative size of the neocortex as well as the relative size of the whole brain are strongly related to cognitive abilities and enhanced novelty response (e.g. [62,63] also reviewed in [64]). We thus predicted that *M. emarginatus* and *M. myotis* should have larger relative brains and neocortices than *M. capaccinii*. As data on brain sizes of *M. emarginatus* and *M. capaccinii* are not available yet, this prediction remains to be studied. However, available data on skull morphology indicate that *M. capaccinii* indeed has the smallest relative skull and hence potentially the smallest relative brain of our three species followed by *M. emarginatus* and then *M. myotis* (condylobasal length divided by forearm length and zygomatic width divided by forearm length; data from [65]).

Ecology and Sociality

Animal cognition researchers generally attribute the evolution of brains and intelligence to either ecology (e.g. [2]) or to social structure (e.g. [66]). Recently, however, there are models that combine the two (e.g. [1]). In our case, we have chosen bat species that differ widely in ecology, but are quite similar in their social systems. In all three species, females congregate in large maternity groups of up to several thousand individuals while males roost singly or in smaller bachelor colonies in summer; in winter, males and females from all three species hibernate singly or in small clusters ([67] *M. capaccinii* p. 213; *M. emarginatus* p. 244; *M. myotis* p. 254). In some cases pregnant or lactating females of all three species can be found in the same cave at the same time in Bulgaria (TC, personal observation). By choosing species with very similar social structures and other aspects of their biology, but disparate foraging ecologies, we can infer that the differences we find in learning and flexibility reflect ecology and not sociality.

Conclusion

Results from this study confirm our prediction that open water foragers are more stereotyped and hence less flexible than species hunting in more complex habitats. Contrary to our expectations, there was a trend that open water foragers learn simple rules more quickly than clutter specialists and passive listening gleaners. Given a more complex task involving two decisions, however, passive listening gleaners and clutter specialists showed a tendency to outperform bats hunting over water. Due to their unstructured foraging habitat, we would expect results from bats hunting in open space to be similar to those of open water hunters. More subtle differences between passive listening gleaners and clutter specialists might only be revealed in a yet more complex task.

Our data support the hypothesis that cognitive abilities of animals are shaped by the demands of their ecological background. Our results concur with those of recent studies on nectar-feeding bats that differ in their spatial working memory performance depending on their degree of dietary specialization, in which a specialized nectarivorous bat foraged more efficiently at artificial flower patches than a generalist that also includes fruits and insects in its diet [68]. They are also in line with recent findings for birds (e.g. [5,6]) and mammals [9].

Supporting Information

Video S1 Video clips demonstrate *M. capaccinii* navigating the maze in the flexibility and re-learning task and in the complex learning task. For comparisons of crawling performance, video clips demonstrate *M. myotis*, a much larger species, in the flexibility and re-learning task.

Acknowledgments

We acknowledge Angela Clarin and Katharina Clarin for help with the logistic flow of the experiment, Renate Heckel and Leonie Baier for drawing the bats in figure 1 and 2, the Tabachka Bat Research Station Team 2010 for support, and John Ratcliffe and Irmgard Teschke for helpful comments on the manuscript. Our work was supported by the International Max Planck Research School (IMPRS) for Organismal Biology. Sadly, PD Dr. Björn Siemers, head of the Sensory Ecology Group in Seewiesen, Germany, passed away in May 2012. We are deeply thankful for his support, friendship, and inspiration

Author Contributions

Conceived and designed the experiments: TMAC IR RAP BMS. Performed the experiments: TMAC. Analyzed the data: TMAC. Contributed reagents/materials/analysis tools: BMS. Wrote the paper: TMAC IR RAP BMS.

References

1. Overington SE, Dubois F, Lefebvre L (2008) Food unpredictability drives both generalism and social foraging: a game theoretical model. Behavioral Ecology 19: 836–841.
2. Parker S, Gibson K (1977) Object manipulation, tool use and sensorimotor intelligence as feeding adaptations in Cebus monkeys and great apes. Journal of Human Evolution 6: 623–641.
3. Rozin P, Kalat JW (1971) Specific hungers and poison avoidance as adaptive specializations of learning. Psychological Review 78: 459–486.
4. Dukas R, Ratcliffe JM (2009) Introduction. In: Dukas R, Ratcliffe JM, editors. Cognitive Ecology II. Chicago: University Of Chicago Press.
5. Mettke-Hofmann C, Gwinner E (2003) Long-term memory for a life on the move. Proceedings of the National Academy of Sciences of the United States of America 100: 5863–5866.
6. Gibson BM, Kamil AC (2005) The fine-grained spatial abilities of three seed-caching corvids. Learning & behavior 33: 59–66.
7. Micheli F (1997) Effects of experience on crab foraging in a mobile and a sedentary species. Animal behaviour 53: 1149–1159.
8. Day L, Crews D, Wilczynski W (1999) Spatial and reversal learning in congeneric lizards with different foraging strategies. Animal behaviour 57: 393–407.
9. Haupt M, Eccard J, Winter Y (2010) Does spatial learning ability of common voles (*Microtus arvalis*) and bank voles (*Myodes glareolus*) constrain foraging efficiency? Animal cognition 13: 783–791.
10. Godfrey-Smith P (2001) Environmental complexity and the evolution of cognition. In: Sternberg R, Kaufman J, editors. The evolution of intelligence. Mahwah, New Jersey: Lawrence Erlbaum Associates London. pp.223–250.
11. Tebbich S, Stankewitz S, Teschke I (2012) The relationship between foraging, learning abilities and neophobia in two species of Darwin's finches. Ethology 118: 135–146.
12. Simmons NB (2005) Order Chiroptera. In: Wilson DE, Reeder DM, editors. Mammal species of the world: a taxonomic and geographic reference, vol 1. Baltimore: John Hopkins University Press. pp.312–529.
13. Simmons NB (2005) An Eocene big bang for bats. Science 307: 527–528.
14. Norberg UM, Rayner JM (1987) Ecolocigal morphology and flight in bats (Mammalia; Chiroptera): wing adaptations, flight performance, foraging strategy and echolocation. Philosophical Transactions of the Royal Society B 316: 335–427.
15. Jones G, Holderied MW (2007) Bat echolocation calls: adaptation and convergent evolution. Proceedings of the Royal Society B 274: 905–912.
16. Jones G, Teeling EC (2006) The evolution of echolocation in bats. Trends in Ecology & Evolution 21: 149–156.
17. Schnitzler H-U, Moss C, Denzinger A (2003) From spatial orientation to food acquisition in echolocating bats. Trends in Ecology & Evolution 18: 386–394.
18. Weinbeer M, Kalko EKV (2007) Ecological niche and phylogeny: the highly complex echolocation behavior of the trawling long-legged bat, *Macrophyllum macrophyllum*. Behavioral Ecology and Sociobiology 61: 1337–1348.
19. Meyer CFJ, Weinbeer M, Kalko EKV (2005) Home-range size and spacing patterns of *Macrophyllum macrophyllum* (Phyllostomidae) foraging over water. Journal of Mammalogy 86: 587–598.
20. Kalko EKV, Schnitzler H-U (1989) The echolocation and hunting behavior of Daubenton's bat, *Myotis daubentoni*. Behavioral Ecology and Sociobiology 24: 225–238.
21. Schnitzler H-U, Kalko EKV, Kaipf I, Grinnell AD (1994) Fishing and echolocation behavior of the greater bulldog bat, *Noctilio leporinus*, in the field. Behavioral Ecology and Sociobiology 35: 327–345.
22. Siemers BM (2001) Finding prey by associative learning in gleaning bats: experiments with a Natterers bat *Myotis nattereri*. Acta Chiropterologica 3: 211–215.
23. Safi K, Dechmann DKN (2005) Adaptation of brain regions to habitat complexity: a comparative analysis in bats (Chiroptera). Proceedings of the Royal Society B 272: 179–186.
24. Boonman AM, Boonman M, Bretschneider F, Van de Grind WA (1998) Prey detection in trawling insectivorous bats: duckweed affects hunting behaviour in Daubenton's bat, *Myotis daubentonii*. Behavioral Ecology and Sociobiology 44: 99–107.
25. Siemers BM, Stilz P, Schnitzler H-U (2001) The acoustic advantage of hunting at low heights above water: behavioural experiments on the European "trawling" bats *Myotis capaccinii*, *M. dasycneme* and *M. daubentonii*. The Journal of experimental biology 204: 3843–3854.
26. Siemers BM, Baur E, Schnitzler H-U (2005) Acoustic mirror effect increases prey detection distance in trawling bats. Die Naturwissenschaften 92: 272–276.
27. Greif S, Siemers BM (2010) Innate recognition of water bodies in echolocating bats. Nature communications 1: DOI: 10.1038/ncomms1110. doi:10.1038/ncomms1110.
28. Ciechanowski M, Zając T, Zielińska A, Dunajski R (2010) Seasonal activity patterns of seven vespertilionid bat species in Polish lowlands. Acta Theriologica 55: 301–314.
29. Ciechanowski M, Zając T, Biłas A, Dunajski R (2007) Spatiotemporal variation in activity of bat species differing in hunting tactics: effects of weather, moonlight, food abundance, and structural clutter. Canadian Journal of Zoology 85: 1249–1263.
30. Almenar D, Aihartza J, Goiti U, Salsamendi E, Garin I (2006) Habitat selection and spatial use by the trawling bat *Myotis capaccinii* (Bonaparte, 1837). Acta Chiropterologica 8: 157–167.
31. Biscardi S, Russo D, Casciani V, Cesarini D, Mei M, et al. (2007) Foraging requirements of the endangered long-fingered bat: the influence of micro-habitat structure, water quality and prey type. Journal of Zoology 273: 372–381.
32. Jones PL, Page RA, Hartbauer M, Siemers BM (2010) Behavioral evidence for eavesdropping on prey song in two Palearctic sibling bat species. Behavioral Ecology and Sociobiology 65: 333–340.
33. Arlettaz R (1996) Feeding behaviour and foraging strategy of free-living mouse-eared bats, *Myotis myotis* and *Myotis blythii*. Animal Behaviour 51: 1–11.
34. Goerlitz HR, Siemers BM (2007) Sensory ecology of prey rustling sounds: acoustical features and their classification by wild grey mouse lemurs. Functional Ecology 21: 143–153.
35. Arlettaz R, Perrin N, Hausser J (1997) Trophic resource partitioning and competition between the two sibling bat species *Myotis myotis* and *Myotis blythii*. Journal of animal ecology 66: 897–911.
36. Russo D, Jones G, Arlettaz R (2007) Echolocation and passive listening by foraging mouse-eared bats *Myotis myotis* and *M. blythii*. The Journal of Experimental Biology 210: 166–176.
37. Siemers BM, Güttinger R (2006) Prey conspicuousness can explain apparent prey selectivity. Current biology 16: R157–R159.
38. Schnitzler H-U, Kalko EKV (2001) Echolocation by insect-eating bats. BioScience 51: 557–569.
39. Siemers BM, Schnitzler H-U (2004) Echolocation signals reflect niche differentiation in five sympatric congeneric bat species. Nature 429: 657–661.
40. Wang J, Kanwal J, Zhang C, Jiang T, Lu G, et al. (2010) Seasonal habitat use by greater horseshoe bat *Rhinolophus ferrumequinum* (Chiroptera: Rhinolophidae) in Changbai Mountain temperate forest, Northeast China. Mammalia 74: 257–266.
41. Flaquer C, Puig-Montserrat X, Burgas A, Russo D (2008) Habitat selection by Geoffroys bats (*Myotis emarginatus*) in a rural Mediterranean landscape: implications for conservation. Acta Chiropterologica 10: 61–67.
42. Krull D, Schumm A, Metzner W, Neuweiler G (1991) Foraging areas and foraging behavior in the notch-eared bat, *Myotis emarginatus* (Vespertilionidae). Behavioral Ecology and Sociobiology 28: 247–253.
43. Schumm A, Krull D, Neuweiler G (1991) Echolocation in the notch-eared bat, *Myotis emarginatus*. Behavioral Ecology and Sociobiology 28: 255–261.
44. Zahn A, Bauer S, Kriner E, Holzhaider J (2009) Foraging habitats of *Myotis emarginatus* in Central Europe. European Journal of Wildlife Research 56: 395–400.
45. Stadelmann B, Lin L-K, Kunz TH, Ruedi M (2007) Molecular phylogeny of New World Myotis (Chiroptera, Vespertilionidae) inferred from mitochondrial and nuclear DNA genes. Molecular phylogenetics and evolution 43: 32–48.
46. Page RA, Von Merten S, Siemers BM (2012) Associative memory or algorithmic search: A comparative study on learning strategies of bats and shrews. Animal cognition 15: 495–504.
47. Ruczynski I, Siemers BM (2011) Hibernation does not affect memory retention in bats. Biology letters 7: 153–155.
48. Canadian Council On Animal Care (2003) CCAC species-specific recommendations on: BATS: 1–9. Available: http://www.ccac.ca/Documents/Standards/Guidelines/Add_PDFs/Wildlife_Bats.pdf. Accessed 2013 April 24.
49. R Development Core Team (2012) R: A language and environment for statistical computing. Available: http://www.r-project.org/Accessed 2013 April 24.
50. Dunn OJ (1964) Multiple comparisons using rank sums. Technometrics 6: 241–252.
51. Bitterman ME (1975) The comparative analysis of learning - are the laws of learning the same in all animals? Science 188: 699–709.
52. Tsoar A, Nathan R, Bartan Y, Vyssotski A, Dell'Omo G, et al. (2011) Large-scale navigational map in a mammal. Proceedings of the National Academy of Sciences of the United States of America 108: 718–724.
53. Levin E, Barnea A, Yovel Y, Yom-Tov Y (2006) Have introduced fish initiated piscivory among the long-fingered bat? Mammalian Biology - Zeitschrift für Säugetierkunde 71: 139–143.
54. Aihartza J, Almenar D, Salsamendi E, Goiti U, Garin I (2008) Fishing behaviour in the long-fingered bat *Myotis capaccinii* (Bonaparte, 1837): an experimental approach. Acta Chiropterologica 10: 287–301.

55. Aihartza J, Goiti U, Alemar D, Garin I (2003) Evidences of piscivory by *Myotis capaccinii* (Bonaparte, 1837) in southern Iberian Peninsula. Acta chiropterologica 5: 193–198.
56. Verbeek M, Drent P, Wiepkema P (1994) Consistent individual differences in early exploratory behaviour of male great tits. Animal Behaviour 48: 1113–1121.
57. Benus R, Bohus B, Koolhaas J, Van Oortmerssen GA (1991) Heritable variation for aggression as a reflection of individual coping strategies. Experientia 47: 1008–1019.
58. Ruiz-Gomez MDL, Huntingford FA, Øverli Ø, Thörnqvist P-O, Höglund E (2011) Response to environmental change in rainbow trout selected for divergent stress coping styles. Physiology & behavior 102: 317–322.
59. Dingemanse NJ, De Goede P (2004) The relation between dominance and exploratory behavior is context-dependent in wild great tits. Behavioral Ecology 15: 1023–1030.
60. Ratcliffe JM, Fenton MB, Shettleworth SJ (2006) Behavioral flexibility positively correlated with relative brain volume in predatory bats. Brain, behavior and evolution 67: 165–176.
61. Eisenberg J, Wilson D (1978) Relative brain size and feeding strategies in the Chiroptera. Evolution 32: 740–751.
62. Reader SM, Laland KN (2002) Social intelligence, innovation, and enhanced brain size in primates. Proceedings of the National Academy of Sciences of the United States of America 99: 4436–4441.
63. Sol D, Duncan RP, Blackburn TM, Cassey P, Lefebvre L (2005) Big brains, enhanced cognition, and response of birds to novel environments. Proceedings of the National Academy of Sciences of the United States of America 102: 5460–5465.
64. Lefebvre L, Sol D (2008) Brains,lifestyles and cognition: are there general trends? Brain, behavior and evolution 72: 135–144.
65. Krapp F, Niethammer J, editors (2011) Die Fledermäuse Europas. 1st ed. Wiebelsheim: AULA-Verlag GmbH.
66. Byrne RW, Bates LA (2007) Sociality, evolution and cognition. Current biology 17: R714–723.
67. Dietz C, Von Helversen O, Nill D (2009) Bats of Britain, Europe and Northwest Africa. 1st ed. London: A & C Black Publishers Ltd.
68. Henry M, Stoner KE (2011) Relationship between spatial working memory performance and diet specialization in two sympatric nectar bats. PLoS ONE 6: e23773. doi:10.1371/journal.pone.0023773.

Investigating Behaviour and Population Dynamics of Striped Marlin (*Kajikia audax*) from the Southwest Pacific Ocean with Satellite Tags

Tim Sippel[1,2*¤], John Holdsworth[2], Todd Dennis[1], John Montgomery[1]

1 School of Biological Sciences, University of Auckland, Auckland, New Zealand, **2** Blue Water Marine Research, Whangarei, New Zealand

Abstract

Behaviour and distribution of striped marlin within the southwest Pacific Ocean were investigated using electronic tagging data collected from 2005–2008. A continuous-time correlated random-walk Kalman filter was used to integrate double-tagging data exhibiting variable error structures into movement trajectories composed of regular time-steps. This state-space trajectory integration approach improved longitude and latitude error distributions by 38.5 km and 22.2 km respectively. Using these trajectories as inputs, a behavioural classification model was developed to infer when, and where, 'transiting' and 'area-restricted' (ARB) pseudo-behavioural states occurred. ARB tended to occur at shallower depths (108±49 m) than did transiting behaviours (127±57 m). A 16 day post-release period of diminished ARB activity suggests that patterns of behaviour were affected by the capture and/or tagging events, implying that tagged animals may exhibit atypical behaviour upon release. The striped marlin in this study dove deeper and spent greater time at ≥200 m depth than those in the central and eastern Pacific Ocean. As marlin reached tropical latitudes (20–21°S) they consistently reversed directions, increased swimming speed and shifted to transiting behaviour. Reversals in the tropics also coincided with increases in swimming depth, including increased time ≥250 m. Our research provides enhanced understanding of the behavioural ecology of striped marlin. This has implications for the effectiveness of spatially explicit population models and we demonstrate the need to consider geographic variation when standardizing CPUE by depth, and provide data to inform natural and recreational fishing mortality parameters.

Editor: Daphne Soares, University of Maryland, United States of America

Funding: This research was funded primarily by the New Zealand Marine Research Foundation, with contributions from the University of Auckland, Tagging of Pacific Pelagics Program (a subprogram of the Census of Marine Life), Blue Water Marine Research and the National Geographic Society. TS was financially supported by a University of Auckland Doctoral Scholarship and Blue Water Marine Research. The funders had no role in study design, data collection and analysis, decision to publish, or preparation of the manuscript.

Competing Interests: The authors have declared that no competing interests exist.

* E-mail: tsippel@gmail.com

¤ Current address: Pelagic Fisheries Research Program, University of Hawaii, Honolulu, Hawaii, United States of America

Introduction

Striped marlin (*Kajikia audax*) [1] (Family: Istiophoridae) are important components of targeted (recreational and commercial) and non-targeted (bycatch) fishing activity throughout the Indo-Pacific. Their highly migratory nature means tagging is a critical component of understanding both their behavioural ecology and population biology. Two kinds of electronic tags, 1) archival, which collect high frequency environmental data (ambient temperature, depth, light-level); and 2) satellite-linked radio telemetry (SLRT), which provide high spatial resolution movement data, have become important tools for investigating the movement ecology of highly migratory pelagic species. Time-series models developed to approximate movements of tagged animals from archival tag data have provided important new biological insights, but both theoretical and practical limitations constrain the spatial resolution of this approach. Double-tagging enables the strengths of each technology to be utilized simultaneously [2,3]. Recently developed methods for electronic double-tagging of billfish [4] can provide very good inputs for behavioural classification models, and with the high cost and effort invested in collecting marine telemetry data, methods for extracting as much information as possible from the data are needed. Research into techniques for classifying behaviours from telemetry data is increasing rapidly [5–8], and analytical approaches are evolving, but standardized methods for classifying behaviour are yet to emerge [9]. These methods may contribute valuable insights into factors affecting foraging and migration ecology, and provide opportunities to examine important aspects including post-release behavior, depth distribution, and linkages between individual movements and geographic distribution.

Satellite tagging has been used to characterize the geographic distribution and general movement patterns of adult striped marlin from the Pacific Ocean over 1–9 month periods [10], and acoustic tracking has shown movement and behaviour trends at local scales over brief periods (up to 52 hours) [11,12]. However, research into relationships among the behaviour of individuals, spatial distributions of the population, and environmental conditions over longer periods can greatly improve biological and ecological understanding, as well as management of the species.

Electronic tags have been used widely to investigate the survivorship of billfish following release from recreational [13–

15] and commercial fishing gear [16–18]. However, very little information is available about potential capture and tagging effects on the behaviour of wild marine animals, despite the potentially important consequences for interpretation of telemetry studies, and management measures derived from them. The assumption implicit in most animal telemetry research is that the behaviours of animals that survive capture and tagging are representative of their broader untagged populations. The degree to which this assumption is valid may vary among species and tagging methods, but clearly has implications for interpretation of data derived from electronic tagging.

Information about regional variation of striped marlin occupancy of water column can significantly impact standardization of catch-per-unit-effort (CPUE) data and efforts to mitigate bycatch. Although known to spend the majority of their time near the surface, less is known about their utilization of other water column strata. Utilization of the water column by pelagic fishes tends to be shallower in the eastern tropical Pacific (ETP) than in most of the rest of the Pacific because of a sharp vertical oxycline and shallow thermocline [19]. As a result, maximum depths of striped marlin in the ETP are shallower [14] than the southwest Pacific [20]. Additional tagging can add to these datasets and help define water column use by striped marlin.

The geographic distribution of Pacific striped marlin has been inferred from commercial longline CPUE and conventional tagging data [21,22]. Characterized as 'horse-shoe shaped', it is continuous across the equator in ETP, but becomes meridionally discontinuous from the central Pacific and westward. However, insufficient spatial and/or temporal resolution in tagging data [11,12,23] have precluded the ability to link individual movements and behaviour with this distinct pattern.

Double-tagging data are used here to investigate the movement patterns of striped marlin within the southwest Pacific Ocean. A model is developed to classify pseudo-behavioural states and then to infer the frequency and duration of area-restricted and transiting behaviours. As a result, a basis for investigating their response to the capture/tagging process is also established. We show how different behavioural modes are related to swimming depths. Relationships among the behaviour of individuals and the distribution of the population are illustrated for the first time. This study provides the detailed analysis of long-term behaviour reported thus far, describing new insights into the biology and ecology of striped marlin. We discuss how this research can inform population dynamics models including spatio-temporal stratification, depth standardized CPUE, recreational fishing mortality, and natural mortality.

Materials and Methods

Animal ethics

This research was approved under permit AEC-R431, issued by the Animal Ethics Committee of University of Auckland School of Biological Sciences.

Capture and tagging

Striped marlin were caught between 2005–2008 from recreational fishing vessels as described in Holdsworth *et al* [4]. Tagging was completed while fish remained in the water alongside the boat during 2005, 2007, 2008; in 2006 fish were brought aboard and onto a padded deck mat via a stern ramp. A combination of pop-off satellite archival tags (PSAT)s and SLRT were used to investigate their movements and behaviour patterns. PSATs were either model PAT3 (2005) or PAT4 (2006–2008), SLRT tags were model SPOT5, and all were manufactured by Wildlife Computers (Redmond, Washington, U.S.A.). PSATs weighed 75 g in air and were 750 mm long (excluding antenna) with a maximum diameter around the float of 40 mm, while SPOT tags weighed 32 g in air with the dimensions 80×19.5×10.5 mm. PSAT tags were tethered by 300 lb monofilament fishing line to plastic (vinyl) intramuscular anchors which were implanted between dorsal pterygiophores of the fish, with a secondary anchoring point created by a looped conventional tag at the base of the float to hold the PSAT closer to the fish body. PSATs recorded water temperature, depth and sunlight intensity every 30–60 seconds and were programmed to summarize their data into 12 summary temperature and depth bins, divided into 3, 6, or 12 hour intervals at transmission. SPOT tags were mounted inside vinyl sleeves which were slipped over the upper tail lobe and stapled or bolted on [4].

Data analysis

All data analyses were conducted using the statistical environment **R** [24]. Use of functions from external packages are cited as 'function(**package**)'. Unless stated otherwise, averages are noted as mean ± standard deviation.

Location estimates

SLRT locations were computed directly by Argos satellites and properties of these positions (indices of quality and associated error structures) have been assessed extensively [6,25]. Light-based geopositions from PSATs were approximated using the tag manufacturer's proprietary software WC-GPE, which implements threshold light-level geolocation methods [26,27]. When estimating light-based geolocations from double-tagged animals, Argos quality mid-point positions from SLRT tags were used to calibrate these estimates in WC-GPE. The initial positions and sea-surface temperatures (SST) recorded simultaneously by the PSATs were used as inputs into an SST cross-referenced Kalman filter in the package **uKFSST** [28,29]. The magnitude and distribution of errors from Kalman filter location estimates for striped marlin has been assessed previously [4].

Temporal regularization of trajectories

Most telemetry studies record observations at irregular intervals, and it is common to use inferential models which require regular time-steps to analyze latent pseudo-behaviours [6,30,31]. We also took this approach. A regular time-series of locations was estimated from irregular observations for each animal using a continuous-time correlated random walk Kalman filter (CTCRW hereafter) from the package **crawl** [32,33]. From the continuous time-series, locations were extracted at 12-h intervals (00:00 and 12:00) over the course of the entire animal track, or similar to the average 2.3 locs/day of these data [4]. Appendix S1 includes further detail about regularization.

Classification of behaviour modes

A movement model was formulated to infer four discrete behavioural states; "slow-transiting", "fast-transiting", "slow-ARB", "fast-ARB" (ARB is short for area restricted behaviour). Trajectory segments with high turning angles have been referred to as 'area-restricted search' (ARS) previously [34,35]. The term 'search' is suggestive of a specific kind of behavior (searching), but the animal may or may not have found what it 'searches' for at any moment and classifying resting as a form of 'search' seems undesirable. As a result, the term 'area-restricted behavior' (ARB) is used as a more generic reference to area-restricted patterns. Appendix S1 details further the procedure of classifying behaviours.

Statistical analysis

Comparisons were made among the behaviours inferred by the model and distributions of maximum depths reported by PSAT tags. Profiles of depth and temperature (PDT)s, were transmitted by PSAT tags at 8 discrete depths (and mean temperature at depth) distributed between the minimum and maximum during pre-programmed summary periods. Maximum depths were chosen for analysis because they are the most commonly observed discrete measure of position in the water column across all animals. Maximum depths were non-normally distributed, so non-parametric tests for asymptotic distributions (analogous to t-tests for normally distributed data) were used to compare maximum depths with transiting and ARB behaviours modes (without respect to speed). Probability-density plots of time spent within discrete depth bins summarized over the same temporal periods described above were used to illustrate changes in depth distributions over time. Gaps in the depth data bins were interpolated using the function loess(**fields**) [36].

The probability of occurrence of ARB with respect to the number of days at liberty was estimated by fitting a proportional hazards model to produce a Kaplan-Meier survival curve using the function Surv(**survival**) [37]. Behaviour was coded as a binary variable (0 = not ARB, 1 = ARB) using days at liberty (or days since release, DAL hereafter) as the time variable and ordinal date as the predictor.

Results

Twenty-eight striped marlin were tagged, but three were excluded from analysis due to poor data quality or post-release mortality. Twenty-five datasets collected during the austral summers of 2005 ($n = 3$), 2006 ($n = 15$), 2007 ($n = 5$), and 2008 ($n = 2$) were included in the analysis (Table 1). Fifteen datasets were collected from individuals fitted with PSAT and SLRT tags simultaneously, and 10 SLRT only datasets were obtained. SLRT tracks averaged 32 days in length (range 1–114 days) and a mean of 25 PSAT geolocation days (range 0–133 days at liberty) per fish was also recorded. DAL ranged from 15 to 133 days, totaling 1432. A total of 1042 locations from SLRT and PSAT tags were observed on 398 different days.

Data quality

Temporal coverage of individual marlin was variable, with maximum duration for missing data from individual fish ranging from 1–41 days (7.6±8.8). The highest quality location data generally occurred within the first 4–8 weeks of the tracks when SLRT locations were available. Exceptions to this included STM06.8 and STM07.2, which had gaps of 14 and 20 days at the beginning of their tracks. STM07.1 and STM08.1 provided exceptionally good SLRT data for 103 and 115 days (max gaps 6.7 and 6.2 days) respectively (Table 1). Root mean squared (RMS) error for CTCRW smoothed uKFSST location estimates from transmitted PSAT data were 12.5 km and 87.8 km for longitude and latitude respectively. RMS errors for CTCRW smoothed uKFSST locations from archival PSAT data were 214.6 km and 30.0 km for longitude and latitude, respectively. The large overall RMS errors for archival longitudes was due to a bias over the initial 10 days at liberty for STM06.1 (longitude RMS = 264.6 km), but were much better and consistent with expectations for STM06.12 (longitude RMS = 31.8 km).

Behaviour classification

All four behavioural modes were predicted, but not all were observed in each individual (Figure 1). The number of behavioural modes inferred per individual ranged from 2–4. Mean travel speeds for slow-transit and fast-transit modes were 0.74±0.56 km/hr (range 0.0–3.5 km/hr) and 1.93±1.04 km/hr (range 0.0–8.5 km/hr), respectively. Continuous durations of segments of ARB lasted from 1 to 19.5 days (5.5±4.8 days). The number of days between ARB events ranged from 1.5 to 50 (14.3±13.3). Fast-ARB (mode 4) is probably heavily influenced by location error, as traveling rapidly in a highly tortuous manner for prolonged periods is not biologically realistic. We do not consider fast-ARB to be a distinct behaviour, but rather a mode which is useful for identifying trajectory segments more heavily influenced by high location error.

Representative tracks

Results for at least one marlin from each tagging season (2005–2008) are discussed. Two archival datasets were recovered from fish tagged in 2006 (Table 1), and analysis from one of these is included.

STM05.4 was tagged in the western Bay of Plenty on 18 March 2005 (Figure 2A). It departed the tagging location in fast-transiting mode with ARB first observed 5 days after release and a total of 4 times before ending on 17 May 2005 approximately 500 km southwest of Fiji. During the first two weeks at liberty depth oscillated between the surface and *c.* 75–150 m. During the ensuing two weeks STM05.4 spent the majority of its time at or near the surface, with only brief forays below 50 m (Figure 2B). From 24 April 2005, its depths frequently oscillated between the surface and *c.* 75–150 m. The highest latitude reached was 20.2°S on 10 May 2005 which coincided with dives ≥250 m on 10 and 12 May before turning south and shifting from ARB to fast-transiting mode.

STM06.10 was tagged on 31 March 2006 in the Tasman Sea at the Wanganella Banks (Figure 2C). It spent the first 14 days post-release in fast-transiting mode before changing to ARB for 5 days in mid-April. ARB was inferred 4 separate times during the trajectory. STM06.10 spent the following four months moving north/northwest, switching between transiting and ARB as it moved west of New Caledonia until the track terminated on 11 August 2006. The highest latitude reached was 20.5°S on 25 July 2006, with dives ≥300 m on 4 August 2006, before turning south and changing from slow-transit mode to fast-transit mode. Swimming depth varied between the surface and 150 m throughout the trajectory (Figure 2D).

STM06.1 (Figure 3A) was double-tagged on 10 January 2006 in the Tasman Sea, and 25 days of SLRT and 68 days of PSAT data were recorded before mortality occurred on 18 March 2006. After death the marlin sank to 596 m depth and drifted deeper to 644 m over the following two days. The descent rate over 300 m was 3.3 meters/minute. Its PSAT began transmitting 97 km away from its tagging location and its archival data were recovered in early 2008, when the tag was discovered on a beach in eastern Australia. STM06.1 departed the tagging location in fast-transiting mode for the initial three days, followed by switching between fast-ARB and fast-transiting for 17 days before spending about 2 weeks circling a small area, eventually returning to very near its capture location. During the first 10 days after release its diving activity was irregular, with crepuscular patterns not well defined. Night diving frequencies were higher during this period than the rest of the record, while spending almost all of its time in the upper 100 m. From day 11 onwards its behavioural patterns became more characteristic of day/night activity. In the early part of the record, ARB was commonly associated with high dive frequencies (within upper 100 m) at night, and during daylight very close to the surface. Transiting behaviour was associated with regular diving throughout both day and night.

Table 1. Striped marlin included in behavioural analysis.

Marlin ID	Weight (kg)	Start Date	Start Lat.	Start Lon.	End Lat.	End Lon.	Days At Liberty	SLRT Days	PSAT Days	Max Gap Days
STM05.2	85	25 Feb 05	−37.244	176.103	−31.245	173.185	21	9	21	11.0
STM05.3	100	26 Feb 05	−37.295	176.273	−29.506	180.460	21	21	21	2.5
STM05.4	70	18 Mar 05	−37.069	176.039	−21.703	176.134	60	25	60	2.0
STM06.1*	74	10 Jan 06	−32.635	167.563	−31.897	167.068	68	25	68	1.0
STM06.2	80	11 Jan 06	−31.683	167.833	−31.368	168.824	15	1	15	4.1
STM06.3	81	12 Jan 06	−31.707	167.835	−33.013	167.582	103	21	103	11.0
STM06.4**	66	13 Jan 06	−31.744	167.880	−33.443	180.484	28	28	0	4.1
STM06.5	110	4 Feb 06	−36.513	173.629	−24.105	167.376	80	21	80	41.0
STM06.6	66	20 Feb 06	−37.646	177.813	−24.191	177.850	65	23	65	2.3
STM06.7	66	1 Mar 06	−37.397	176.374	−24.552	182.792	85	43	85	19.0
STM06.8**	65	31 Mar 06	−31.706	167.846	−20.217	168.878	28	28	0	14.1
STM06.9**	81	13 Feb 06	−34.858	173.757	−32.638	173.053	17	17	0	6.4
STM06.10	75	31 Mar 06	−31.706	167.846	−21.717	161.812	133	36	133	7.0
STM06.11	95	31 Mar 06	−31.707	167.852	−23.757	175.176	68	25	68	12.0
STM06.12*	80	2 Apr 06	−31.709	167.833	−26.22	170.119	34	17	34	2.0
STM06.13	90	3 Apr 06	−31.667	167.817	−26.275	172.453	29	15	29	2.0
STM06.14	80	3 Apr 06	−31.683	167.850	−21.817	170.404	100	26	100	3.0
STM06.15**	76	4 Apr 06	−31.699	167.845	−27.179	166.796	18	18	0	2.3
STM07.1***	85	19 Feb 07	−37.485	177.982	−20.111	225.358	103	103	0	6.7
STM07.2***	60	19 Feb 07	−37.408	178.344	−30.682	182.985	59	59	0	19.8
STM07.3***	75	20 Feb 07	−37.509	177.847	−25.628	169.307	50	50	0	2.0
STM07.4	100	20 Feb 07	−37.560	177.660	−27.422	202.586	91	33	91	7.0
STM07.5***	80	21 Feb 07	−37.373	177.754	−34.448	180.691	18	18	0	1.6
STM08.1**	80	6 Mar 08	−37.449	177.964	−22.119	182.035	115	115	0	6.2
STM08.2**	75	8 Mar 08	−37.509	178.073	−27.701	179.716	22	22	0	1.5
Mean	79.8						57	32	39	8
Total							1431	799	973	

*Archival PSAT data were recovered.
**PSAT didn't transmit.
***Animal was single-tagged with SLRT only.

Later in the record transiting was associated with many deeper and colder dives (up to 350 m) during the day, with ARB again correlated with high dive frequencies to shallow depths (≤100 m) during the day (Figure 3B).

STM07.1 was tagged on 19 February 2007 at Waihau Bay in the eastern Bay of Plenty (Figure 3C). Upon departing the capture location it moved slowly to the northeast for 5 days before entering slow-ARB mode and spending the following 18 days within 100 km of the New Zealand coast in the eastern Bay of Plenty. On 13 March 2007 it made a sudden directed movement away from the New Zealand coast, spending the following 2.5 months in transiting mode as it moved into the central Pacific Ocean. From mid May (*c.* 145°W, *c.* 22°S) its northerly progress slowed, with the final SLRT location observed on 2 June 2007.

STM08.1 was tagged on 6 March 2008 at Waihau Bay (Figure 3D). It departed its tagging location in fast transiting mode; ARB was first detected 16 days after release, and 7 times during the trajectory. From the beginning of April this fish began moving east/northeast, traveling in a figure-eight loop before continuing north through May. The highest latitude reached was 20.2°S on 12 June 2006 before changing from slow-transit to fast-transit mode while turning south, with the track terminating on 29 June 2006.

Capture effects

On average, a lower proportion of ARB was predicted during the first 16 days after release. Survival analysis indicated the probability of observing transiting and ARB behaviours was not equal (50:50) until after 16 days post-release (Figure 4A). Proportions of fast-transiting behaviours were highest over a similar period, with proportions of slow-transiting and ARB behaviour increasing after approximately 16 days at liberty (Figure 4B). Travel speeds for tagged striped marlin during the initial 10 days post-release were significantly faster (1.97±1.20 km/h) than all subsequent periods (1.28±.93 km/h) ($p \leq 0.001$, $\theta = 11.8$). Relative turning angles (angles between consecutive locations) were significantly lower during the first 10 days at liberty than all subsequent periods ($p = 0.014$, $t = 2.45$). Trajectories were aligned significantly more northwards during the first 10 days at liberty than all subsequent days ($p \leq 0.001$, $t = 12.83$). In addition to the strong tendency to move away from tagging locations, a seasonal trend was observed in the likelihood

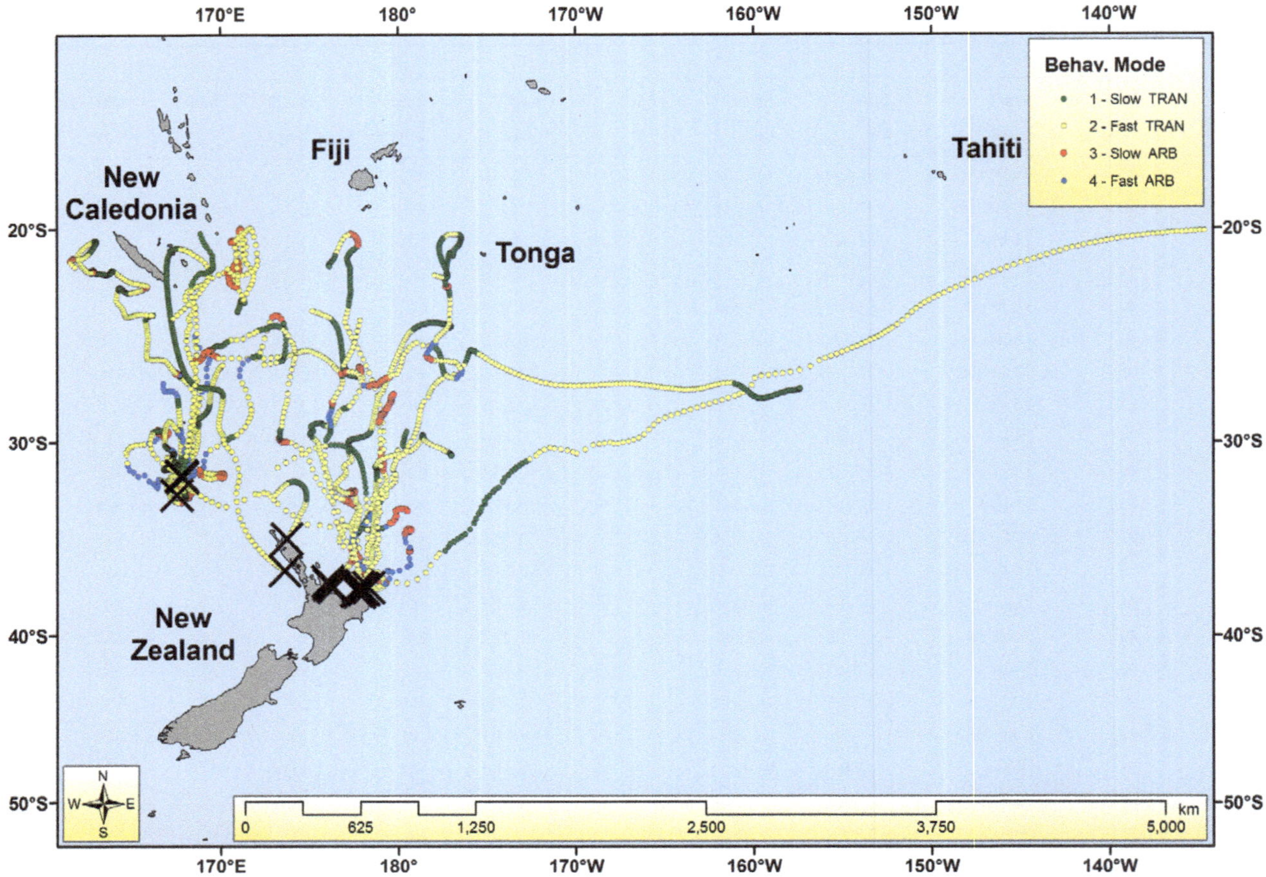

Figure 1. Twenty-five striped marlin trajectories colour coded by inferred modes of behaviour (X represents initial capture location).

of marlin returning to their capture vicinity within the same tagging year. Of the marlin tagged before 1 March, 5/14 circled back to within 500 km of their capture location (Figure 4C). None tagged after 1 March returned to within 500 km of their capture locations during the same season (Figure 4D).

Depth profiles

Maximum depths were significantly greater ($p \leq 0.001$) for transiting behaviours (127±57 m) than for ARB (108±49 m). Across all modes, they spent 51.4% of their time between within 1 m of the surface during the day, and 64.0% of their time in this range at night (Figure 5). However, they spent more time during night in the 50.5–200 m depth range (19.1%) than during the day (13.7%). The maximum depth observed overall was 460 m, where temperature at depth was 10.8°C, and surface temperature was 23.4°C.

Northerly extent

Five marlin (2005 = 1, 2006 = 3, 2008 = 1), initially tagged and captured in different months (February–April) and different locations (eastern NZ, Western NZ, Tasman Sea) moved towards the tropics and suddenly reversed directions at 20–21°S latitude upon arrival during the months April–August (Figure 6). These directional reversals coincided with a switch to fast-transiting mode in five fish. A sixth fish showed northerly progress leveling off at this latitude as the fish continued on a period 61 days of fast-transit (only trajectory not inset in Figure 6). A shift in depth distributions was apparent when PSAT data were available (n = 1, 2005; n = 3, 2006). At lower latitudes, depth plots show striped marlin spent most of their time at or near the surface (0–5 m). Approaching 20°S latitude movements tended toward subsurface depths of 5–10 m, or coincided with a deep spike dive below 250 m (Figures 2B and 2D).

Discussion

Data quality

Dealing with error-prone geolocations can be challenging, but Kalman filters enable the errors of marine telemetry techniques to be quantified and minimized [29,33,38–40] (see Appendix S1, Figures S1, S2 and S3, Tables S1 and S2). On average, CTCRW longitudes were 44 km closer to Argos longitudes than uKFSST longitudes were, with mean deviations from Argos of 12.5 km and 56.8 km respectively. Likewise, CTCRW latitudes were 24 km closer to Argos latitudes, than uKFSST latitudes were, on average (mean deviations were 87.8 km and 112.2 km respectively for uKFSST). The marked improvements in geolocation error distributions over those previously reported were due to geolocation processing at three levels. Firstly, calibrating approximations of light-level geolocations with Argos mid-points improved the initial estimates. With these initial estimates as inputs, uKFSST was then used to improve subsequent geolocation estimates,

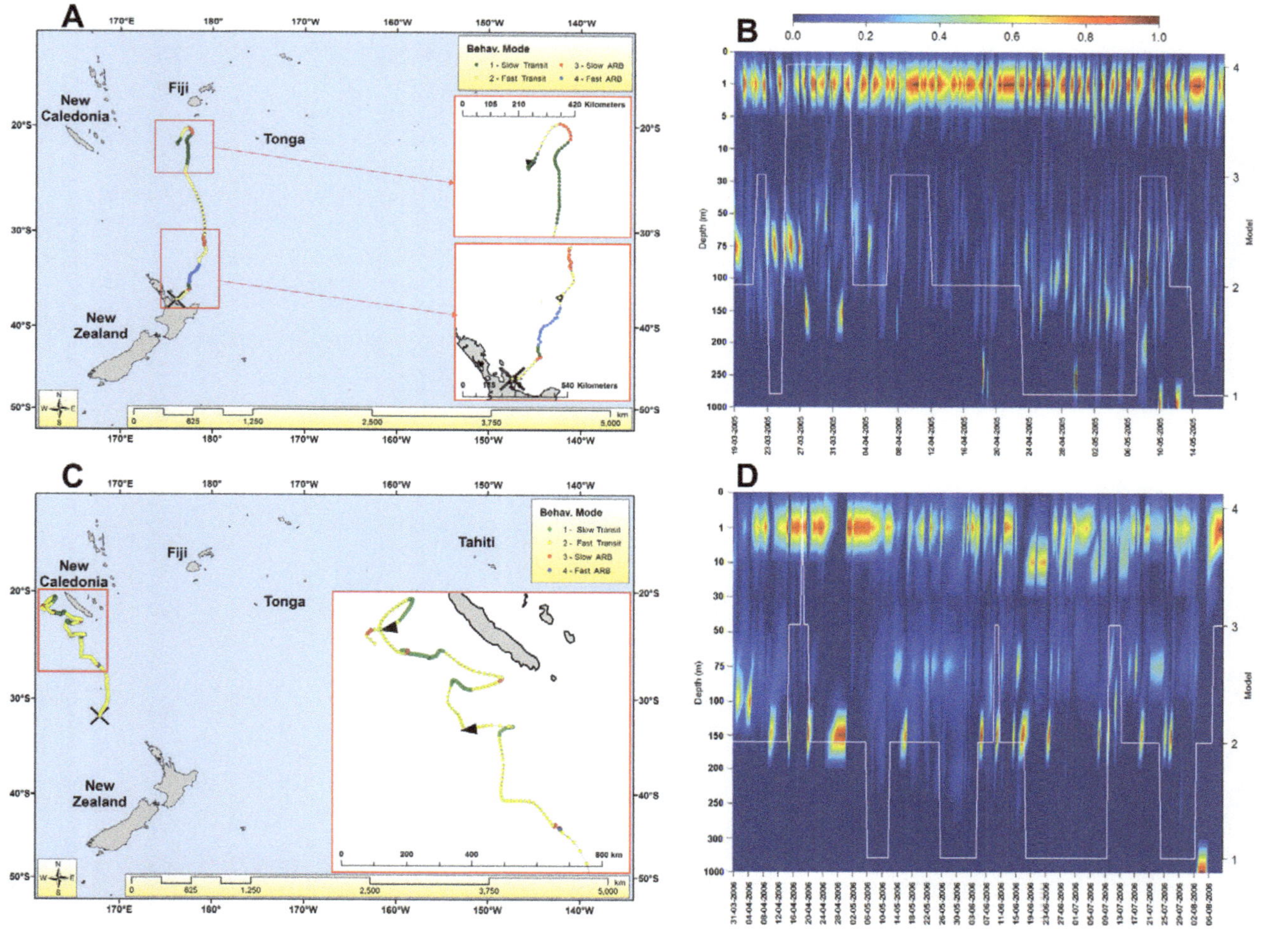

Figure 2. Individual striped marlin from 2005 and 2006 tagging seasons, representative of overall patterns. A. STM05.04 was tagged on 18 March 2005 and provided 60 days of data. Estimated behavioural modes are colour-coded and legend is at top-right. B. Time-at-depth profile for STM05.04. X-axis is time, primary Y-axis is depth (metres) of the water column, secondary Y-axis is estimated behavioural mode indicated with white trend-line, and colours are proportion of time (scaled 0–1, legend at top) spent at depth. C. STM06.10 was tagged on 31 March 2006 and provided 133 days of data. D. Time-at-depth profile for STM06.10.

particularly with respect to latitude. Finally, the ability to parameterize the CTCRW model with error terms for all location classes enabled further reduction of errors and estimation of trajectories on continuous time-scales. RMS longitude errors were reduced to near theoretical limits [26,27,41] and significantly below errors reported in other studies [2–4]. These studies did not calibrate light-level geolocations because their aim was to evaluate quality in the absence of calibration data. Furthermore, other studies have not used the third tier of location processing, state-space time-series regularization with algorithms such as the CTCRW [33]. The tendency of striped marlin to spend most of their time at the surface makes SLRT possible and maximizes the quality of light-level data for geolocation, however the effects of sea-state on the frequency and quality of SLRT are not well understood. However, the time-series of data available for behavioural modeling was significantly increased (doubled) by the inclusion of both Argos based SLRT and PSAT geolocation data.

STM06.1 made rapid and uncharacteristic west-to-east movements immediately upon release (Figure 3A). The magnitude of initial longitudinal movements (particularly in the initial 10 days) may have fallen outside the bounds normally expected by the light level geolocation algorithms used [26,27], inducing a longitude estimation bias which could not be resolved by further refinement or uKFSST. However, archival latitude geolocation RMS errors for STM06.12 were substantially smaller, again by nearly a half an order of magnitude from raw uKFSST estimates. Some datasets were particularly challenging to analyze, and occasionally it is not possible to resolve these issues. However, double-tagging with SLRT and PSAT tags shows that these problems are uncommon, but should be acknowledged.

Behavioural classification

In spite of the improved accuracy of our longitudes derived from light level geolocation, significant error remains for latitude estimates, which may lead to erroneous predictions of fast-ARB. Fast-ARB may not represent a real biological state, but rather a temporally autocorrelated series of location estimates of higher-than-average spatial error. Biologically relevant signals almost certainly lie within these segments, but the magnitude of measurement error precluded making biological inferences. Recently developed methods can reduce, or even eliminate bias in light level latitudes [39,40], which will improve the ability of behavioural models to represent real biological patterns. Data of

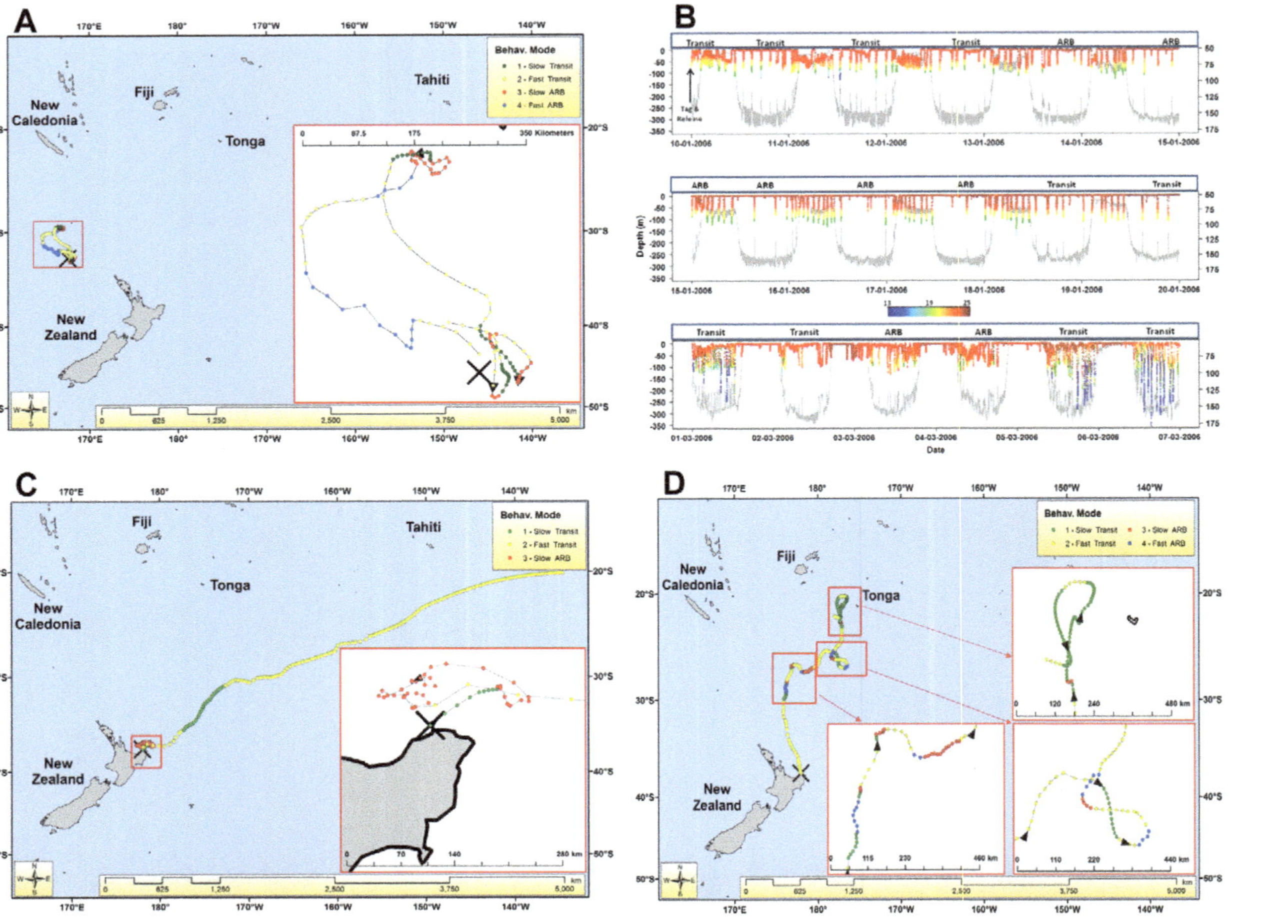

Figure 3. Representative individual striped marlin from 2006, 2007 and 2008 tagging seasons. A. STM06.01 was tagged on 10 January 2006 and provided 68 days of data. B. Archival time-series of temperature, depth and light-levels for STM06.01. Colored line represents water temperature with colors (legend in middle of plot) and depth (m) within the water column on the primary Y-axis; grey line shows light-level scaled in dimensionless units on the secondary Y-axis. C. STM07.01 was tagged on 19 February 2007 and provided 103 days of data. D. STM08.01 was tagged on 6 March 2008 and provided 115 days of data.

regular time-series are required by the segmentation algorithms used in this study, however interpolation of positions between observed locations introduces additional uncertainty and longer gaps between observations likely cause the model to over-estimate the proportion of transiting-mode behaviour.

ARB is representative of all behaviours such as foraging, resting and breeding that occur in spatially confined areas [42]. Data from the austral striped marlin breeding season [43,44] were not available in this study, which excludes ARB being interpreted as breeding. Striped marlin are obligate ram ventilators [45] with high metabolic rates [46], thus are unlikely to remain in localized resting states for extended periods (days). A substantial proportion of ARB is likely to be foraging behavior, but this mode also probably includes activities such as searching patterns, social activity, and periods of reduced activity. Transiting states can represent searching patterns, migration, and flight from predators among other possibilities. The frequent correspondence of changes in inferred behavioural states with variations in depth is encouraging (Figures 2B and 2D) because this provides a measure of confidence that meaningful patterns are discernible. Correspondence of behavioural states with depth distributions has been observed in Atlantic leatherback turtles as well (*Dermochelys coriacea*) using switching state-space models (SSM) [47]. The ability to corroborate behavioural inferences derived from 2-dimensional models with data from a third dimension provides a measure of confidence in detection of behavioural state changes. Furthermore, this confidence provides impetus to incorporate as much information as possible in future models to generate more realistic behavioural estimates.

The segmentation model we used calculates the likelihood of a trajectory segment belonging to a specific behavioural mode, but does not estimate a posterior probability as do other approaches to classification [6,47]. However, our model utilized more information (speed and turning angles, see Table S3 for movement parameter estimates), than indices such as first passage time [35] and straightness indices [48], and can be extended to incorporate other measures of behaviour including diving patterns. Schick *et al.* [49] considered the current status of modeling animal movement and where some hierarchical Bayesian modeling approaches might lead to further progress. Essentially, they envisioned a behavioural SSM with a resource selection function nested within the framework to govern switching probabilities. The approach used here, combined with resource selection analysis can provide the kind of a priori information to better inform an SSM as they envisaged.

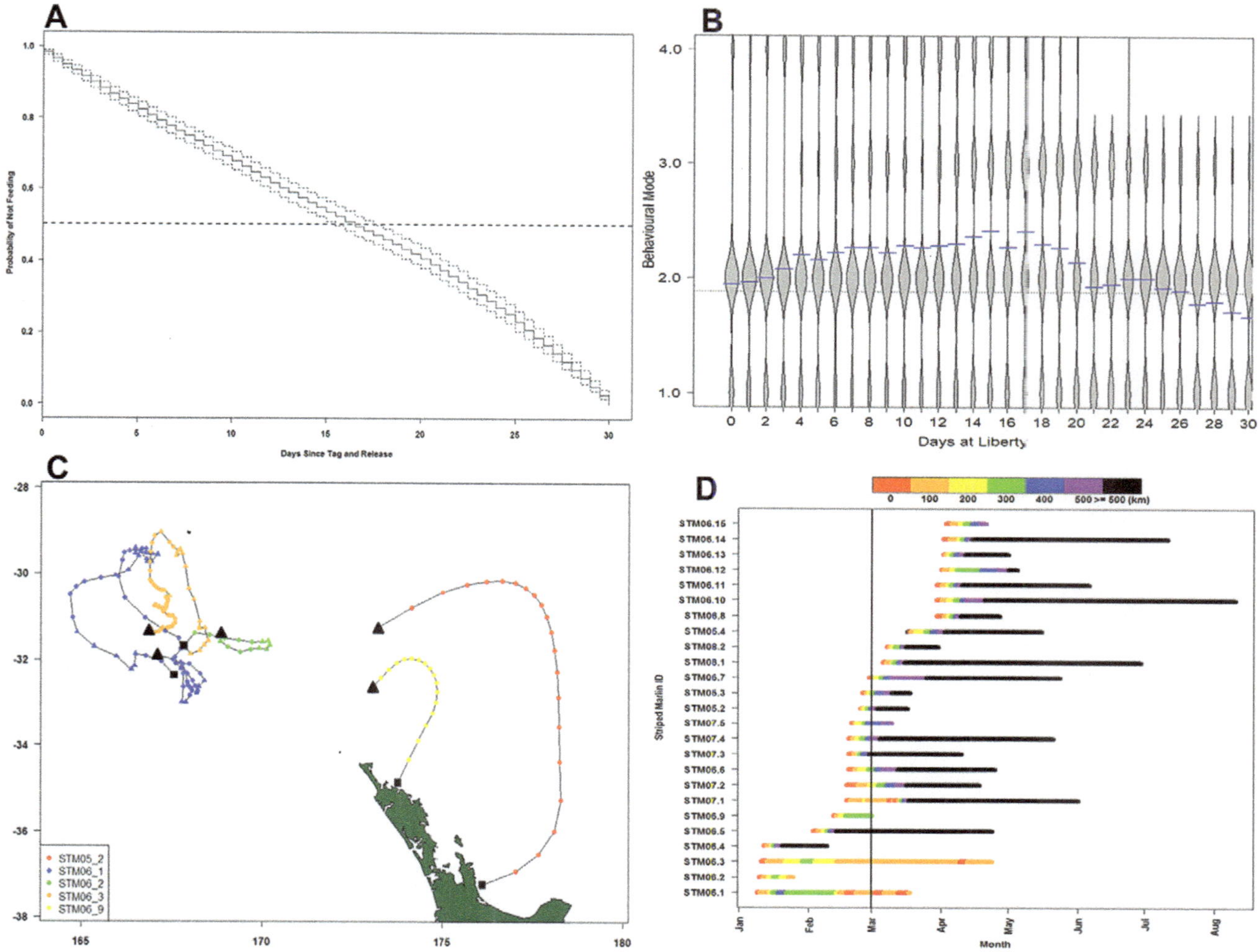

Figure 4. Capture effects on behaviour. A. Kaplan-Meier survival curve over first 30 days at liberty (solid stepped line) with 95% confidence intervals (dotted stepped lines). The probability of feeding through time is 1 - probability of not feeding (1 minus y-axis), with the odds of feeding or not estimated to be even (50:50) at day 16 since release (straight horizontal dashed line). B. Proportions of striped marlin in behavioural modes 1–4 by days at liberty. For each day, width of bubbles (called 'beans') represents proportion of all pooled striped marlin in each behavioural mode and blue dashes are mean behavioural mode for all pooled marlin. C. Five striped marlin tagged before 1 March (2005 and 2006) which circled back towards their tagging locations. Black squares represent tagging locations and black triangles are final transmitted locations. D. Distance away from initial capture location for each marlin, ordered by calendar tagging date. Color blocks (legend at top) denote straight-line distance (km) from initial release location, and the vertical black line marks 1 March, denoting 'early' and 'late' season.

Foraging patterns

Transiting behaviour tended to correspond with changes from surface to deeper waters, while ARB mode generally was associated with shallower depths. Opportunities to feed can be sparsely distributed through space and time for pelagic predators, so foraging frequencies observed here probably reflect these conditions. Shoals of prey (*ie.* sardines and other planktivorous fish) may sometimes persist for days [50] and even weeks, which would promote occurrence of ARB. However, diurnal vertical migrations of prey such as squid may not lead to discernible patterns of ARB. Transiting behaviour classified by the model likely includes foraging periods characterized by different movement patterns. Stomach contents of striped marlin from New Zealand waters have revealed prey common to surface-to-mid water depths, benthic teleosts and squid [51]. Stomach-content analysis of North Pacific striped marlin showed surface dwelling prey are dominant components of the diet, but occurrence of prey from deeper reaches of the water column indicate periodic changes in foraging patterns [52]. Squid, sardines and mackerel have previously been identified as primary components of their diet in North American waters [53–55]. Difference among these prey types complicate detection of all foraging habits using behavioural models comprised on only two-dimensional location data.

Jonsen *et al.* [47] found that leatherback sea-turtles, when foraging, tended to reverse directions frequently. We did not often observe such behaviour in this study. Disparity in patterns of foraging between these two studies may reflect different strategies in how prey is obtained at the different trophic levels. An effect of these different strategies could be different spatio-temporal scales at which foraging occurs. As patchy prey are encountered, prey-capture behaviour might occur over smaller spatio-temporal scales than grazing. Predatory foraging might be reflected by lower degrees of trajectory autocorrelation than grazing. The outcome of these strategic trade-offs are that turtles feed on more abundant and accessible prey (*ie.* jellyfish) which have lower nutritive value, while marlin feed on less abundant but higher nutritive value

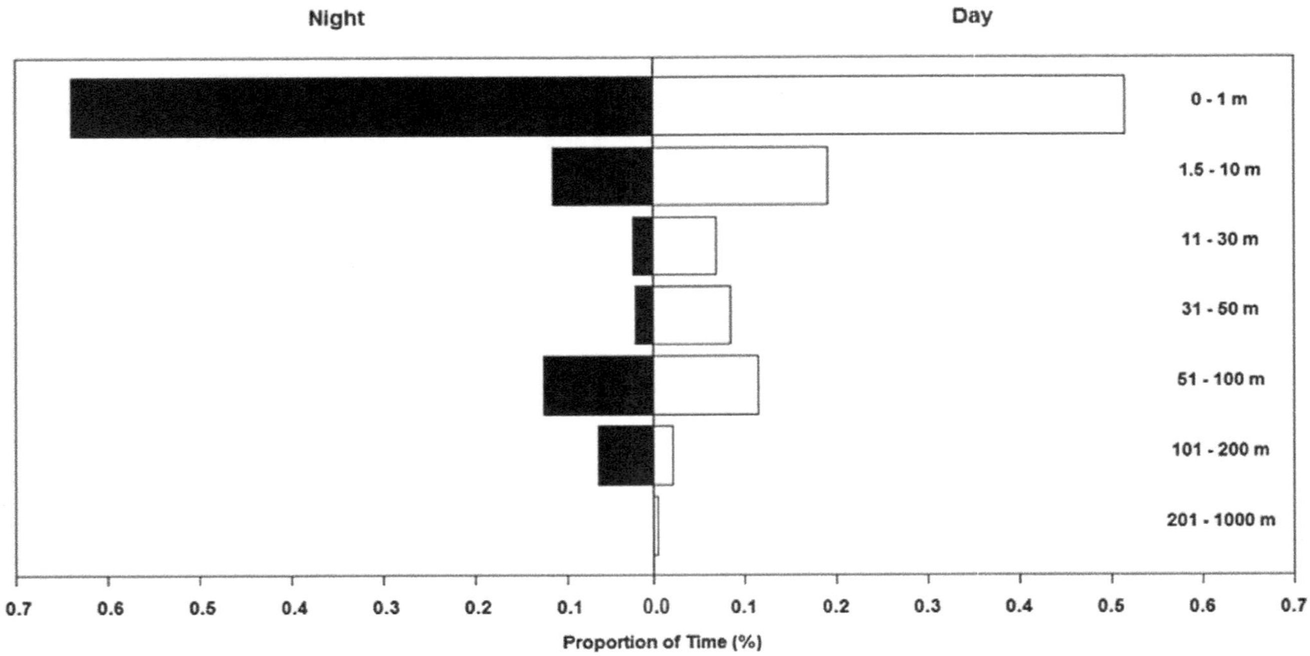

Figure 5. Proportion of time spent at depth (in meters) by day and night for all marlin tags reporting PSAT data between 2005 and 2008. Different tags were programmed to summarize data at 12 discrete depth intervals, and the seven bins shown here are aggregates of the bins that were common to all tags.

targets like saury (Family: Scomberesocidae) and mackerels (Family: Scombridae). Accordingly, models designed to identify analogous behaviours (ie. foraging) which driven by different strategies (*ie.* predatory vs. grazing) probably require unique configurations.

Across all striped marlin, a mean of nearly 3 weeks separated periods of ARB activity. Although up to 50 days separated these periods it should not be assumed that this is truly reflective of foraging frequency given the range of prey types commonly pursued. From visceral warming in archival tagging data, Bestley *et al.* [56] noted southern bluefin tuna (*Thunnus maccoyii*) fed on 76% of days, with periods between confirmed feeding events ranging from 3.9–24.4 days (mean 11 days). The ability to detect feeding through such physiological data improves confidence in estimates of foraging success. The intervals of feeding success observed from endothermic animals such as tunas are probably quite reliable, and provide some indication of reasonable feeding/fasting intervals in other highly migratory pelagic fishes.

Effects of capture and tagging

New Zealand's recreational catch rates of striped marlin peak around March and begin to diminish around April [57]. Japanese longline CPUE from the southwest Pacific Ocean shows relatively higher striped marlin abundance during summer/autumn and lower abundance during winter/spring around New Zealand [58]. Squire and Suzuki [21] proposed from these patterns that the population's movement trended towards southerly latitudes during summer/autumn. If these assemblages are motivated by foraging opportunities around New Zealand and the Tasman Sea, it could be expected that they would remain in these areas during these periods. A tendency of striped marlin to immediately depart capture locations in the southwest Pacific Ocean was initially detected by Sippel *et al.* [20] and multiple seasons of electronic tagging data show this to be a consistent pattern [4]. Tagged striped marlin tended to move north following release, while approximately 1/3 of those tagged during the summer circled back towards their capture locations, but none of those tagged in the autumn exhibited this behaviour. Similar patterns of departure in single tagged fish (PSAT only) where either documented or are apparent in [10,20], indicating that double-tagging did not have a greater effect on movement patterns.

Movement out of range of domestic fishing vessels may also contribute to the particularly low recapture rates (0.52%) of striped marlin in the New Zealand cooperative gamefish tagging program [23]. Of the 28 tagged during 2005–2008, two mortalities (7%) occurred within 48 hours of capture and release. This confirms high post-release survivorship, and other factors such as tag shedding probably contribute more to low conventional-tag return rates. Three mortalities of the 25 surviving capture and tagging (12%) were observed more than 30 days after release (33, 53, 67 days). Transmitted temperature at depth records from two marlin showed temperatures of greater than 25°C, which remained constant at depths ranging from the surface to 600 m. These temperature and depth patterns are similar to an apparent predation of a PSAT tagged Opah (*Lampris guttatus*) by a warm bodied shark (*ie.* a lamnid shark) [59]. Predation is probably an important reason for non-reporting of PSAT tags in billfish telemetry studies.

Time-series analysis indicates that the probability of occurrence of ARB and transiting behaviours was not equal (50:50) until 16 days after tagging. Bestley *et al.* [56] detected a 19±10 (mean ± SD) day period of post-release fasting (range 5–38 days). This is comparable to the 12.7 day period of diminished foraging probability detected from behavioural analysis of Atlantic leatherback sea turtle telemetry data [60], and are not unlike the 5–20 days of diminished body condition observed from short-term recaptures of conventionally tagged southern bluefin tuna [61]. Evidence for capture effects in other Istiophorid tagging data can

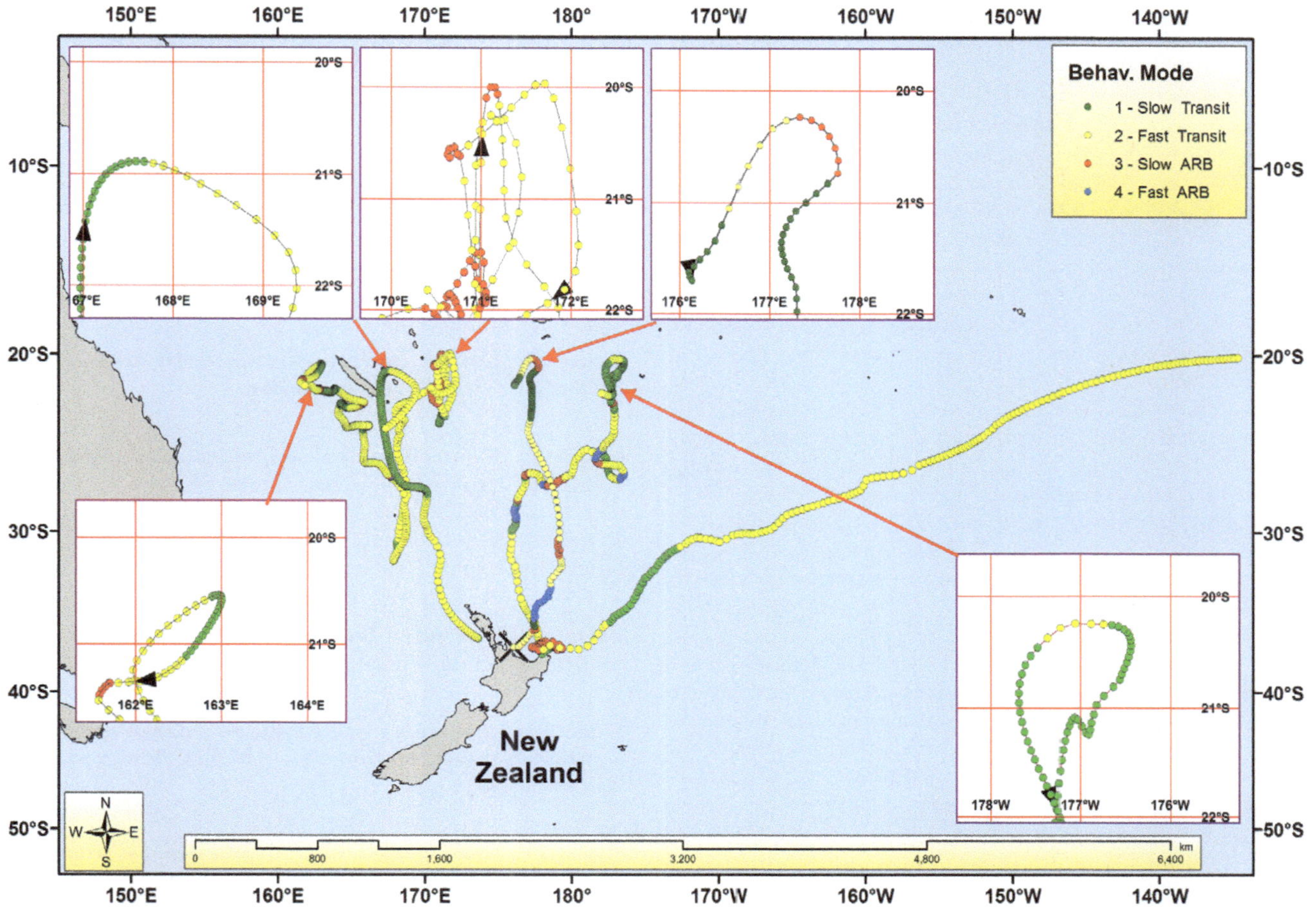

Figure 6. Six striped marlin trajectories (5 insets and one not inset) displaying either directional reversals or stopping northerly progress between 20–21°S latitude.

be deduced from departures of white marlin (*Kajikia albidus*) [15], black marlin (*Makaira indica*) [62], and Pacific blue marlin (*Makaira nigricans*) [17] from their capture/tagging locations.

We hypothesize that capture and/or tagging triggered migration in at least half of striped marlin sooner than would be expected if they were not captured and tagged. Since it is not possible to know the long-term movements of individual untagged fish, the primary difficulty in assessing the effects of capture and tagging is the challenge of gathering behavioural control data. An approach to obtaining such data is free-tagging uncaptured fish (wild, freely swimming fish). If substantial differences in behaviour are observed between captured and uncaptured striped marlin, this would suggest that the capture process contributes to departures. Two striped marlin have been free-tagged with a PSAT by blue-water spear fishers to investigate movements of uncaptured fish. One provided short-term data which are suggestive of different movement patterns (see Appendix S2, Figure S4), and data were not received from the other tag. There are numerous important factors to consider when testing this hypothesis which include regional, temporal, and ontogenetic effects.

A comprehensive investigation into the effects of exhaustive exercise on post-release survivorship and blood-borne stress indicators on sharks, tunas and marlin found compelling evidence that recreationally caught gamefish species are physiologically compromised after release [63]. Wells *et al.* [64] measured elevated plasma electrolytes, blood glucose, lactate and haematocrit levels in moribund striped marlin sampled dockside at recreational fishing competitions. They provided useful measures of peak of stress indicators, but it is very difficult to get baseline measurements of these indicators in unstressed striped marlin.

Because the effects of capture and tagging on animals behaviour and fitness are not well understood, further investigation is merited. Furthermore, capture effects that trigger early departure from the tagging region could bias spatio-temporal data used in management and stock assessment.

Depth and catchability

Utilization of the water column by striped marlin is mediated heavily by two primary oceanographic conditions, oxygen saturation and temperature relative to the surface. The maximum depth reported from PSAT tagging in the ETP was 192 m, with only 9% (4/45) of individuals reported to exceed 150 m depth [14], while no descents below 150 m were observed from acoustic tracking there and in the central Pacific [11,12]. Shallow oxyclines and thermoclines are uncommon in the southwest Pacific [65], and consequently our data show all striped marlin exceeded 150 m depth, descents to deeper than 200 m were common, and the maximum depth observed was 460 m. Like elsewhere, these fish spent more than half of their time within 10 m of the surface, but a second smaller peak (15%) was observed between 50–100 m (Figure 6) which has not been previously observed. Geographic variation of depth distribution has important consequences.

Among the uncertainties of the first southwest-Pacific stock assessment was vertical distribution which effects vulnerability to fishing activity [66]. Data cited about utilization of the water column was sourced from telemetry data in the ETP where striped marlin spend most of their time in the upper 40 m. We show occupancy of the water column differs substantially in the southwest Pacific, which should alter assumptions about susceptibility to fishing activity in stock-assessment. Regionally specific telemetry data can be useful for understanding gear selectivity and standardizing CPUE for billfish stock-assessments [67]. For example, relative CPUE for striped marlin from the Hawaii longline fishery was quite small at depths ≥120 m compared to shallower depths [68], but our data indicate these trends would probably be different in the southwest Pacific. Regionally standardized CPUE might be an improvement for future stock-assessments. Furthermore, it has been estimated that commercial bycatch accounts for around 90% of adult marlin mortality, and attempts to mitigate this will be affected by geographically variable vulnerability to gear at depth [69].

Gear depth is a consideration in recreational catch rates as well. Recreational billfishers usually troll for billfish at the surface, but we show how fishing activity targeted there would probably fail to attract the attention of marlin when they move deeper. This is a likely explanation for the marlin bite turning on and off, and innovations to recreational fishing techniques and gear might provide new opportunities to increase catches at depth.

Northerly extent

A 10–20° latitudinal band north and south of the equator is commonly recognized as a break in the geographic distribution of adult striped marlin in the central and western Pacific [21]. It is striking that a pattern of direction reversal or apparent termination of northerly progress between 20–21°S latitude was repeatedly observed in each season (2005–2008). Changes to fast-transiting mode coinciding with directional reversals suggest an individual response. Striped marlin moving further north than 20°S latitude have previously been documented [20], primarily during spring [10]. We did not acquire data during the spring, but oceanographic conditions probably help explain this pattern. Any number of combined oceanographic variables (*ie.* increased mixed layer depth and temperature, decline in oxygen saturation, changing upwelling/downwelling conditions, etc.) might cue these behaviour changes. The probable influence of oceanography on population structure and stratification demonstrates how environmental factors can be important inputs in population dynamics models [70]. The discontinuity in adult distribution observed in the southwest Pacific ocean is worth considering for spatial stratification of future stock-assessments.

Supporting Information

Figure S1 Joint probability densities of errors for CTCRW regularized uKFSST geolocations from transmitted PSAT data.

Figure S2 Effects of CTCRW regularization on double-tagged striped marlin STM06.14. The black line and points are raw location data, with triangles representing uKFSST location estimates and circles representing Argos locations from the SLRT tag. Red represents the smoothed CTCRW pathway.

Figure S3 Striped marlin trajectories by season 2005–2008 after CTCRW regularization.

Figure S4 Controlling for striped marlin capture effects. Red vector (red inset box): Travel distance and direction over 8 day PSAT deployment from speargun. Black vectors: Travel distance and direction over initial 8 days at liberty for six individuals captured with standard recreational fishing methods.

Table S1 Multiplier function K values used as error model parameters used in CTCRW regularization.

Table S2 Estimates of t-distribution parameters from non-linear minimization for uKFSST longitude and latitude errors.

Table S3 Density function parameters estimated by maximum likelihood for each individual.

Acknowledgments

We thank the many boats, captains, anglers and volunteers who helped over the years, particularly from the New Zealand Big Game Fishing Council. Special thanks to the Enterprise Motor Group and the Ultimate Lady for supporting the program. Pete Saul's wealth of fishing experience was a particular asset. Special thanks to Clement Calenge for helping with use of the adehabitat package, Toby Patterson for developing code to visualize depth data, Ian Jonsen for insightful discussions on modeling animal movement data, and Barbara Block for support.

Author Contributions

Conceived and designed the experiments: TS JH. Performed the experiments: TS JH. Analyzed the data: TS. Contributed reagents/materials/analysis tools: TS JH TD JM. Wrote the paper: TS. Edited the paper: JM TD JH.

References

1. Collette BB, McDowell JR, Graves JE (2006) Phylogeny of recent billfishes (Xiphioidei). Bulletin of Marine Science 79: 455–468.
2. Teo SLH, Boustany A, Blackwell S, Walli A, Weng KC, et al. (2004) Validation of geolocation estimates based on light level and sea surface temperature from electronic tags. Marine Ecology Progress Series 283: 81–98.
3. Wilson SG, Stewart BS, Polovina JJ, Meekan MG, Stevens JD, et al. (2007) Accuracy and precision of archival tag data: a multiple-tagging study conducted on a whale shark (Rhincodon typus) in the Indian Ocean. Fisheries Oceanography 16: 547–554.
4. Holdsworth J, Sippel T, Block B (2009) Near real time satellite tracking of striped marlin (Kajikia audax) movements in the Pacific Ocean. Marine Biology 156: 505–514.
5. Morales JM, Haydon DT, Frair J, Holsinger KE, Fryxell JM (2004) Extracting more out of relocation data: building movement models as mixtures of random walks. Ecology 85: 2436–2445.

6. Jonsen ID, Flemming JM, Myers RA (2005) Robust state-space modeling of animal movement data. Ecology 86: 2874–2880.
7. Tremblay Y, Roberts AJ, Costa DP (2007) Fractal landscape method: an alternative approach to measuring area-restricted searching behavior. Journal of Experimental Biology 210: 935–945.
8. Patterson T, Basson M, Bravington M, Gunn J (2009) Classifying movement behaviour in relation to environmental conditions using hidden Markov models. Journal of Animal Ecology 78: 1113–1123.
9. Nathan R, Getz WM, Revilla E, Holyoak M, Kadmon R, et al. (2008) A movement ecology paradigm for unifying organismal movement research. Proceedings of the National Academy of Sciences 105: 19052–19052.
10. Domeier ML (2006) An analysis of Pacific striped marlin (Tetrapturus audax) horizontal movement patterns using pop-up satellite archival tags. Bulletin of Marine Science 79: 811–825.
11. Holts D, Bedford D (1989) Activity patterns of striped marlin in the southern California bight. Mar Rec Fish 13: 81–93.
12. Brill RW, Holts DB, Chang RKC, Sullivan S, Dewar H, et al. (1993) Horizontal and vertical movements of striped marlin (Tetrapturus audax) near the Hawaiian Islands based on ultrasonic telemetry, with simultaneous measurements of ocean currents. Marine Biology 117: 567–574.
13. Graves JE, Luckhurst BE, Prince ED (2002) An evaluation of pop-up satellite tags for estimating postrelease survival of blue marlin (Makaira nigricans) from a recreational fishery. Fish Bull 100: 134–142.
14. Domeier ML, Dewar H, Nasby-Lucas N (2003) Mortality rate of striped marlin (Tetrapturus audax) caught with recreational tackle. Marine and Freshwater Research 54: 435–446.
15. Horodysky AZ, Graves JE (2005) Application of pop-up satellite archival tag technology to estimate postrelease survival of white marlin (Tetrapturus albidus) caught on circle and straight-shank (" J") hooks in the western North Atlantic recreational fishery. Fishery Bulletin 103: 84–96.
16. Kerstetter DW, Graves JE (2006) Post-release survival of white marlin released from commercial pelagic longline gear in the western North Atlantic. Fish Bull 104: 434–444.
17. Kerstetter DW, Luckhurst BE, Prince ED, Graves JE (2003) Use of pop-up satellite archival tags to demonstrate survival of blue marlin (Makaira nigricans) released from pelagic longline gear. Fishery Bulletin 101: 939–948.
18. Poisson F, Gaertner J, Taquet M, Durbec J, Bigelow K (2010) Effects of lunar cycle and fishing operations on lonline-caught pelagic fish: fishing performance, capture time and survival of fish. Fishery Bulletin 108: 268–281.
19. Prince E, Goodyear C (2006) Hypoxia-based habitat compression of tropical pelagic fishes. Fisheries Oceanography 15: 451–464.
20. Sippel T, Davie P, Holdsworth J, Block B (2007) Striped marlin (Tetrapturus audax) movements and habitat utilization during a summer and autumn in the Southwest Pacific Ocean. Fisheries Oceanography 16: 459–472.
21. Squire JL, Suzuki Z (1988) Migration trends of striped marlin (Tetrapturus audax) in the Pacific Ocean. In: Proceedings of the Second International Billfish Symposium, Kailua-Kona, Hawaii, August. pp 67–80.
22. Squire JL (1987) Striped marlin, Tetrapturus audax, migration patterns and rates in the Northeast Pacific Ocean as determined by a cooperative tagging program: Its relation to resource management. Marine Fisheries Review 49: 26–43.
23. Ortiz M, Prince ED, Serafy JE, Holts DB, Davy KB, et al. (2003) Global overview of the major constituent-based billfish tagging programs and their results since 1954. Marine and Freshwater Research 54: 489–507.
24. R (2008) R: A language and environment for statistical computing. Available: http://www.R-project.org.
25. Vincent C, Mcconnell BJ, Ridoux V, Fedak MA (2002) Assessment of Argos location accuracy from satellite tags deployed on captive gray seals. Marine Mammal Science 18: 156–166.
26. Hill RD, Braun MJ (2001) Geolocation by light level. In: Electronic tagging and tracking in marine fisheries: proceedings of the Symposium on Tagging and Tracking Marine Fish with Electronic Devices, February 7–11, 2000, East-West Center, University of Hawaii. pp 315–315.
27. Hill RD (1994) Theory of geolocation by light levels. Elephant Seals. Population Ecology, Behavior, and Physiology. pp 227–236.
28. Nielsen A, Sibert J (2009) ukfsst: Kalman Filter tracking including Sea Surface Temperature. www.nielsenweb.org. Accessed 1 June 2009.
29. Lam CH, Nielsen A, Sibert JR (2008) Improving light and temperature based geolocation by unscented Kalman filtering. Fisheries Research 91: 15–25.
30. Bailey H, Shillinger G, Palacios D, Bograd S, Spotila J, et al. (2008) Identifying and comparing phases of movement by leatherback turtles using state-space models. Journal of Experimental Marine Biology and Ecology 356: 128–135.
31. Brillinger DR, Stewart BS (1998) Elephant-seal movements: Modelling migration. Canadian Journal of Statistics 26: 431–443.
32. Johnson D (2008) crawl: Fit continuous-time correlated random walk models for animal movement data. Available: R package version 1.0-1.
33. Johnson DS, London JM, Lea MA, Durban JW (2008) Continuous-time correlated random walk model for animal telemetry data. Ecology 89: 1208–1215.
34. Kareiva P, Odell G (1987) Swarms of predators exhibit" preytaxis" if individual predators use area-restricted search. The American Naturalist 130: 233–270.
35. Fauchald P, Tveraa T (2003) Using first-passage time in the analysis of area-restricted search and habitat selection. Ecology 84: 282–288.
36. Furrer R, Nychka D, Sain S (2010) fields: Tools for spatial data. http://cran.r-project.org. Accessed 1 June 2009.
37. Therneau T, Lumley T (2009) survival: Survival analysis, including penalised likelihood. http://cran.r-project.org. Accessed 1 June 2009.
38. Sibert JR, Musyl MK, Brill RW (2003) Horizontal movements of bigeye tuna (Thunnus obesus) near Hawaii determined by Kalman filter analysis of archival tagging data. Fisheries Oceanography 12: 141–151.
39. Nielsen A, Sibert JR (2007) State-space model for light-based tracking of marine animals. Canadian Journal of Fisheries and Aquatic Sciences 64: 1055–1068.
40. Sibert JR, Nielsen A, Musyl MK, Leroy B, Evans K (2009) Removing Bias in Latitude Estimated from Solar Irradiance Time Series. Tagging and Tracking of Marine Animals with Electronic Devices. pp 311–322.
41. Musyl MK, Brill RW, Curran DS, Gunn JS, Boggs CH, et al. (2001) Ability of archival tags to provide estimates of geographical position based on light intensity. In: Electronic tagging and tracking in marine fisheries: proceedings of the Symposium on Tagging and Tracking Marine Fish with Electronic Devices, February 7–11, 2000, East-West Center, University of Hawaii. pp 343–343.
42. Turchin P (1991) Translating foraging movements in heterogeneous environments into the spatial distribution of foragers. Ecology 72: 1253–1266.
43. Hanamoto E (1977) Fishery oceanography of striped marlin I. Fishing season, fishing ground and movement pattern of the fish in the southern Coral Sea. Bulletin of the Japanese Society for Scientific Fisheries 43: 649–657.
44. Ueyanagi S (1959) Larvae of the striped marlin (Makaira misukurii). Report to the Nankai Regional Fisheries Research Laboratory 11: 130–146.
45. Wegner NC, Sepulveda CA, Graham JB (2006) Gill specializations in high-performance pelagic teleosts, with reference to striped marlin (Tetrapturus audax) and wahoo (Acanthocybium solandri). Bulletin of Marine Science 79: 747–759.
46. Brill RW (1996) Selective advantages conferred by the high performance physiology of tunas, billfishes, and dolphin fish. Comparative Biochemistry and Physiology Part A: Physiology 113: 3–15.
47. Jonsen ID, Myers RA, James MC (2007) Identifying leatherback turtle foraging behaviour from satellite telemetry using a switching state-space model. Marine Ecology Progress Series 337: 255–264.
48. Benhamou S (2004) How to reliably estimate the tortuosity of an animal's path: straightness, sinuosity, or fractal dimension? Journal of Theoretical Biology 229: 209–220.
49. Schick RS, Loarie SR, Colchero F, Best BD, Boustany A, et al. (2008) Understanding movement data and movement processes: current and emerging directions. Ecology Letters 11: 1338–1350.
50. Makris NC, Ratilal P, Symonds DT, Jagannathan S, Lee S, et al. (2006) Fish population and behavior revealed by instantaneous continental shelf-scale imaging. Science 311: 660–660.
51. Baker AN (1966) Food of marlins from New Zealand waters. Copeia 1966: 818–822.
52. Shimose T, Yokawa K, Saito H (2010) Habitat and food partitioning of billfishes (Xiphioidei). Journal of Fish Biology 76: 2418–2433.
53. Abitia LA, Galvan F, Muhlia A (1998) Trophic spectrum of striped marlin Tetrapturus audax (Philippi, 1887) off the coast of Cape San Lucas, Baja California Sur, Mexico. Revista de biologia marina y oceanografia 33: 277–290.
54. Abitia-Cardenas LA, Muhlia-Melo A, Cruz-Escalona V, Galvan-Magana F (2002) Trophic dynamics and seasonal energetics of striped marlin Tetrapturus audax in the southern Gulf of California, Mexico. Fisheries Research 57: 287–295.
55. Abitia-Cardenas LA, Galvan-Magana F, Rodriguez-Romero J (1997) Food habits and energy values of prey of striped marlin, Tetrapturus audax, off the coast of Mexico. Fishery Bulletin 95: 360–368.
56. Bestley S, Patterson TA, Hindell MA, Gunn JS (2008) Feeding ecology of wild migratory tunas revealed by archival tag records of visceral warming. Journal of Animal Ecology 77: 1223–1233.
57. Kopf RK, Davie PS, Holdsworth JC (2005) Size trends and population characteristics of striped marlin, Tetrapturus audax caught in the New Zealand recreational fishery. New Zealand Journal of Marine and Freshwater Research 39: 1145–1156.
58. Bromhead D, Pepperell J, Wise B, Findlay J (2004) Striped marlin: biology and fisheries. Bureau of Rural Sciences, Canberra.
59. Kerstetter DW, Polovina JJ, Graves JE (2004) Evidence of shark predation and scavenging on fishes equipped with pop-up satellite archival tags. Fish Bull 102: 750–756.
60. Sherrill-Mix SA, James MC (2008) Evaluating potential tagging effects on leatherback sea turtles. Endangered Species Research 4: 187–193.
61. Hampton J (1986) Effect of tagging on the condition of southern bluefin tuna, Thunnus maccoyii. Australian Journal of Marine and Freshwater Research 38: 699–705.
62. Gunn J, Patterson T, Pepperell J (2003) Short term movement and behaviour of black marlin Makaira indica in the Coral Sea as determined through a pop-up satellite archival tagging experiment. Marine and Freshwater Research 54: 515–525.
63. Skomal G (2007) Evaluating the physiological and physical consequences of capture on post-release survivorship in large pelagic fishes. Fisheries Management and Ecology 14: 81–89.
64. Wells R, McIntyre R, Morgan A, Davie P (1986) Physiological stress responses in big gamefish after capture: observations on plasma chemistry and blood factors. Comparative Biochemistry and Physiology Part A: Physiology 84: 565–571.

65. Longhurst A (2006) Ecological geography of the sea. Elsevier. 542 p.
66. Langley A, Molony B, Bromhead D, Yokawa K, Wise B (2006) Stock assessment of striped marlin (Tetrapturus audax) in the Southwest Pacific Ocean. WCPFC SC2 SA WP-6, Manila, Philippines. pp 7–18.
67. Uozumi Y (2003) Historical perspective of global billfish stock assessment. Marine and Freshwater Research 54: 555–566.
68. Boggs C (1992) Depth, capture time and hooked longevity of longline-caught pelagic fish: Timing bites of fish with chips. Fishery Bulletin 90: 642–658.
69. Kitchell J, Martell S, Walters C, Jensen O, Kaplan I, et al. (2006) Billfishes in an ecosystem context. Bulletin of Marine Science 79: 669–682.
70. Bertignac M, Lehodey P, Hampton J (1998) A spatial population dynamics simulation model of tropical tunas using a habitat index based on environmental parameters. Fisheries Oceanography 7: 326–334.

21

Movements of Blue Sharks (*Prionace glauca*) across Their Life History

Frederic Vandeperre[1,2]*, Alexandre Aires-da-Silva[3], Jorge Fontes[1,2], Marco Santos[1,2], Ricardo Serrão Santos[1,2], Pedro Afonso[1,2]

1 Centre of IMAR of the University of the Azores; Department of Oceanography and Fisheries, Horta, Portugal, **2** LARSyS Associated Laboratory, Lisboa, Portugal, **3** Inter-American Tropical Tuna Commission, La Jolla, California, United States of America

Abstract

Spatial structuring and segregation by sex and size is considered to be an intrinsic attribute of shark populations. These spatial patterns remain poorly understood, particularly for oceanic species such as blue shark (*Prionace glauca*), despite its importance for the management and conservation of this highly migratory species. This study presents the results of a long-term electronic tagging experiment to investigate the migratory patterns of blue shark, to elucidate how these patterns change across its life history and to assess the existence of a nursery area in the central North Atlantic. Blue sharks belonging to different life stages (n = 34) were tracked for periods up to 952 days during which they moved extensively (up to an estimated 28.139 km), occupying large parts of the oceanic basin. Notwithstanding a large individual variability, there were pronounced differences in movements and space use across the species' life history. The study provides strong evidence for the existence of a discrete central North Atlantic nursery, where juveniles can reside for up to at least 2 years. In contrast with previously described nurseries of coastal and semi-pelagic sharks, this oceanic nursery is comparatively vast and open suggesting that shelter from predators is not its main function. Subsequently, male and female blue sharks spatially segregate. Females engage in seasonal latitudinal migrations until approaching maturity, when they undergo an ontogenic habitat shift towards tropical latitudes. In contrast, juvenile males generally expanded their range southward and apparently displayed a higher degree of behavioural polymorphism. These results provide important insights into the spatial ecology of pelagic sharks, with implications for the sustainable management of this heavily exploited shark, especially in the central North Atlantic where the presence of a nursery and the seasonal overlap and alternation of different life stages coincides with a high fishing mortality.

Editor: Athanassios C. Tsikliras, Aristotle University of Thessaloniki, Greece

Funding: This work was conducted in the framework of the EU FP7 project MADE, Mitigating adverse ecological impacts of open ocean fisheries (FP7 KBBE/2007/1/210496). FCT/MCTES-MEC provided individual funding to FV, JF and PA (SFRH/BD/46891/2008, SFRH/BPD/66532/2009, and Ciência 2008/POPH/QREN). IMAR-DOP/UAz is Research and Development Unit no. 531 and LARSyS-Associated Laboratory no. 9 funded by the Portuguese Foundation for Science and Technology (FCT) through FCT - Pest/OE/EEI/LA0009/2011–2014 and by the Azores Fund for Science and Technology (FRCT), funded by OE, COMPETE, QREN and ProConvergencia. The Open Access of this paper is funded by FCT - Pest/OE/EEI/LA0009/2014 (OE, COMPETE and QREN). The funders had no role in study design, data collection and analysis, decision to publish, or preparation of the manuscript.

Competing Interests: The authors have declared that no competing interests exist.

* Email: vandeperre@uac.pt

Introduction

Sharks are generally characterised by a complex spatial organisation of their populations resulting from trade-offs between components of their life history, social and environmental interactions [1–5]. This complex organisation is reflected by their sexual segregation and the existence of discrete locations for key events during their life history, like pupping, nursing and mating. However, the existence and delineation of boundaries within shark populations and in particular the movements of individual sharks within and across these boundaries remain largely unknown. In the context of declining shark populations (e.g., [6–8]) and their deleterious ecological effects [9], unravelling such spatial organisation and, with it, accurately identifying Essential Fish Habitats, is key to develop appropriate management plans for the protection of the most vulnerable life stages [10]. This need is most compelling considering our current limitations in understanding the impacts of heterogeneously distributed fishing pressure on spatially structured shark populations [5,11].

Blue shark is an oceanic predator with a global distribution [12] and a major constituent of the by-catch in pelagic longline fisheries [13–17]. The present status of blue shark stocks is the subject of much debate [18–21]. Blue sharks can be very abundant locally and have a productive life history strategy with a faster growth and a higher number of smaller offspring when compared to other pelagic sharks [21–24]. They are placental viviparous and females can give birth to up to 82 pups ([25]; mean litter size of 37 pups: [26]; up to 70 pups: [27]). Males and females mature at around 183 cm FL and 180–185 cm FL respectively [25,26] at an age of 5–6 years [25]. However, sub-adult females (145 to185 cm FL) can engage in copulation and store spermatozoa in oviducal glands for later insemination, even if they still possess underdeveloped reproductive organs [25].

Results from conventional tagging studies have supported the conclusion, and concomitant management strategy, of a single blue shark stock in the North Atlantic (NA) [28,29]. Yet, blue shark populations are known to segregate by sex and life stages and to make use of ecologically important areas [12]. Pupping areas are generally thought to be located off the western coasts of Portugal and North Africa [30,31], although alternative locations near the Mid-Atlantic ridge in the central NA [17] and the Gulf of Guinea [26] have also been proposed. Although juveniles can be found in large parts of the NA, juvenile females appear to dominate in the eastern NA [32,33] whereas juvenile males dominate in the western NA [18,25,34]. Mating has been recorded in the western NA [25], while observations of fresh bite marks indicating recent mating events [17,25,35] and the presence of dense male aggregations [36] suggests that mating also occurs in other areas.

These areas have all been identified based on analyses of fisheries dependent data and conventional tagging data. While these studies are fundamental in understanding the distribution across life stages, they tell us little about the dynamics and connectivity of these areas and are necessarily limited to those regions and seasons where fishing occurs. Furthermore, the definition of EFHs requires an evaluation of their ecological functionality. For example, nursery areas are generally considered to provide juveniles with an advantage, e.g. shelter or abundant food, thus increasing their possibilities to recruit to the adult population [37]. In practice, this is often difficult to evaluate [3,37], particularly for pelagic sharks. Heupel et al. [3] therefore proposed a practical definition based on three criteria: (1) young-of-the-year sharks are more abundant than in other areas, (2) they have the tendency to remain or return for extended periods and (3) the area is repeatedly used across years.

Shark nurseries are typically closed bays or sheltered coastal areas that mainly provide protection from predators (e.g., [38–41]). In addition, nursery areas can be considered as one component of a shark's life history that, as a consequence of trade-offs between these components, would particularly benefit species with low breeding frequency, small size-at-birth, small litter sizes and/or low juvenile growth rates [3,40]. It is therefore pertinent to test the practical definition proposed by Heupel et al. [3] on highly migratory, productive, oceanic species like blue shark [22–24].

Previous telemetry studies conducted on blue shark focussed on relatively short term (up to 210 days) horizontal and vertical movements [42–46]. In the Northwest NA, acoustic telemetry of mainly adult males showed consistent offshore and southerly movements between August and March and diel vertical niche expansion, extending hundreds of meters during daytime [42]. Juvenile and adult blue sharks of both sexes tagged in late summer and autumn off the eastern coast of Canada with PSAT tags were observed to expand their vertical niche as they entered the gulfstream and moved south and east to offshore overwintering grounds [46]. Juveniles of both sexes tagged in the English Channel and off southern Portugal during summer and autumn, displayed a general southward movement, with increased residence in frontal areas and behavioural plasticity in relation to their diel depth preference [44,45]. Yet, because of the design and limited duration of the deployments, these studies provide us with little information about the seasonal movements and residency of different life stages and the complex spatial organisation of the NA blue shark population. In particular, longer-term data series are needed to validate the nursery assumptions.

Analyses of pelagic longline fishery data have indicated that the Azores region, in the central NA, is seasonally and alternately visited by immature, male and female adult blue shark, that it is important as a nursery ground and, potentially, also a pupping and mating ground for this species [17,47]. The seasonal presence of all these life stages in a region with a central location in the NA offers an exceptional opportunity to investigate their movements, gain insights into the complex structuring of the NA blue shark population, and test the assumptions of the nursery concept in an oceanic shark.

This study addresses two objectives. The first was to investigate long-term migratory patterns of different life stages of blue shark, to study how their movements change across their life history, and to assess the connectivity between different areas they occupy. The second goal was to verify the existence of a nursery area for NA blue shark in the central NA, and to determine its boundaries and stability over time. A tagging experiment was set up in the Azores archipelago in which 37 blue sharks belonging to different life stages were tagged with satellite transmitters set for long-term deployments.

Materials and Methods

Ethics Statement

This study was performed according to national Portuguese laws for the use of vertebrates in research, and the work and tagging protocol approved by the Azorean Directorate of Sea Affairs of the Azores Autonomous region (SRAM 20.23.02/Of. 5322/2009), which oversees and issues permits for scientific activities. All procedures followed the guidelines for the use of fishes in research of the American Fisheries Society. The field studies did not involve endangered or protected species, no animals were sacrificed, and procedures for reduction, replacement and refinement were thoroughly adopted.

1. Tagging experiment

Blue sharks were captured onboard the R/V 'Arquipelago' (25 m) using a commercial style American longline (monofilament mainline and wire leaders) geared with 200 to 400 hooks, light sticks and baited with mackerel and squid. Fish selected for telemetry were then tagged off an auxiliary 7 m, low gunnel, fibreglass boat. Sharks were tagged at the surface after immobilization and induction of tonic immobility [48]. A total of 37 animals were tagged with different models of satellite tags, i.e. SPOT (Wildlife Computer SPOT5) and PSAT tags (Wildlife Computer MK10-PATs and Mini-PATs). SPOT tags were attached to the dorsal fin of male and female sharks measuring 127 to 211 cm FL through four nylon threaded rods fixed through stainless steel nuts and programmed to emit continuously on alternating days (one day on, one day off). MK10-PAT and Mini-PAT tags were attached intramuscularly under the first dorsal fin using a stainless steel tether and one of three types of anchors (small and large Wilton anchors, conventional titanium darts), and equipped with a guillotine to prevent descents bellow ca. 1800 m. MK10-PAT tags were deployed on males and females measuring 142 to 202 cm FL and programmed to release after 180 days. Mini-PATs were deployed on smaller individuals of 112 to 127 cm FL of both sexes and programmed to release after 180 or 270 days. Three individuals were double tagged with a SPOT and MK10-PAT tag.

The experimental design focussed on immature individuals in order to test the nursery hypothesis. However, it was also designed to address the movement patterns of different sexes and life stages of blue shark seasonally occurring in the area [17]. For this study, size limits of population segments, based on available knowledge of the species' biology [25] and seasonal occurrence [17], were as follows: (i) Small juveniles (SJ) up to 120–130 cm FL; (ii) Large juveniles (LJ) from 120 to 183 and 185 cm FL for males and females (i.e. including sub-adults females) respectively; and (iii) Adults (AD) above this latter thresholds. Four dedicated tagging cruises were performed during the four different seasons. Each cruise specific population segments were targeted for tagging, reflecting their seasonal availability (Table 1; Fig. S1): (i) SJ males and females (112–127 cm FL, n = 8) during summer and autumn; (ii) LJ females (127–178 cm FL, n = 9) during autumn and winter; (iii) LJ males (133–183 cm FL, n = 14) during summer and autumn; (iv) AD females (201–202 cm FL, n = 2) during winter and spring; and (v) AD males (200–211 cm FL, n = 4) during summer and autumn.

2. Analytical methods

Geographical positions of the SPOT tag transmissions and the popup locations of the PAT tags were obtained through the Argos system. Geographical positions from the PAT tags were reconstructed from archived light intensity curves transmitted by the tags after popup or retrieved after physical recovery of the tag. Geolocation (GLS, Global Location Sensing) was performed with the WC-GPE2 software, a program provided by the tag manufacturer (Wildlife Computers).

The raw tracks were post-processed using the IKNOS-WALK model [49]. In this approach, tracks are corrected and interpolated at fixed intervals through a non-state based random walk using a forward particle filter. In contrast with state-space models, no inferences are made about the unknown state of an animal in order to calculate subsequent positions [49]. Instead, locations are estimated from a cloud of weighted particles. The weighting can be manipulated to apply corrections based on known constraints (e.g. maximum speed) or available data. This flexible and intuitive approach has the additional advantage of being able to deal with both Argos and GLS positions and doing so taking into account the specific error distributions of each data type. In particular, longitude information is still preserved for position estimation during equinox, when latitude estimation is problematic for light based geolocation [50]. Another advantage over some methodologies is that, in case of data gaps, positions are interpolated in a straight line without over-fitting artefacts.

The tracks were interpolated to obtain one daily position after imposing a maximum speed and impeding tracks from crossing land. The speed constraint is important for modelling the GLS positions, which have very large error distributions. It was set to 3 km/h based on speed estimates calculated from quality 2 and 3 Argos positions separated by time intervals of 1 to 4 days (Fig. S2) and movement information from the literature [42,46,51]. The error around estimated positions is represented by the dispersion of 50 time-matching alternative positions (represented as a cloud in Fig. 1). Fifty percent of the distances between modelled locations and 'real' GPS locations were reported to be less then 20 km and 104.8 km for Argos and GLS locations respectively [49]. The accuracy of the modelled tracks is further dependent on the quality of the track [49]. Consequently, an additional filter was applied on the modelled tracks in which interpolated points were considered to belong to a data gap when the dataset contained no raw location estimates within 72 hours from its time stamp (represented as yellow dots in Fig. 1).

In order to quantify and visualise the seasonal space use of the blue shark population segments, we calculated home range parameters for each of four quarters: from February to March, April to June, July to October and November to January. Fixed quarterly Kernel Utilisation Distributions (KUDs) were estimated for the different life stages based on aggregated locations of individuals within a given segment with deployments exceeding 90 days. KUDs were only calculated from estimated positions that did not belong to a data gap and for quarters for which more than 100 aggregated positions were available. All KUD calculations were performed using the adehabitatHR package [52] on the R statistical platform [53]. The remotely sensed SST data that we used in the analysis were obtained from NOAA's CoastWatch program and were provided as a monthly blended product from microwave and infrared sensors carried on multiple platforms (Japan's Advanced Microwave Scanning Radiometer AMSR-E, NOAA Advanced Very High Resolution Radiometer AVHRR, the Imager on NOAA's GOES spacecrafts, and the Moderate Resolution Imaging Spectrometer MODIS aqua) with complete coverage and a spatial resolution of 0.1 degrees. Subsequently, the SST data were averaged to a resolution of 1° using the arithmetic mean.

Results

1. Tag performance

In total we tagged 37 blue sharks with 40 tags, of which only three MK10-PAT tags deployed on LJ males failed to report (Table 1). SPOT tags (n = 21) reported for periods between 31 and 952 days (median = 220 days), eight of which for over one year (> 365 days), resulting in a cumulative transmission period of 6429 days. While these SPOT transmission periods are long, some tracks showed considerable data gaps. The reporting MK10-PAT tags (n = 7) released after 3 to 180 days (median = 91 days). The number of light intensity curves received through the Argos system and the actual number of GLS positions that WC-GPE2 was able to calculate varied greatly between tags (2–41 GLS locations). Access to the full archive of one recaptured tag (Shark 32) allowed for the calculation of 100 GLS positions. Mini-PATs generally performed better than the MK10-PAT tags, resulting in longer deployments of 102 to 273 days (mean = 201 days). Transmission and data quality were also higher than for the MK10-PAT tags resulting in 20 to 220 GLS locations per tag.

2. Recaptures

At least 8 sharks (21.6%) were recaptured by commercial fishermen anywhere from 3 to 952 days after tagging (Table 1). Recaptures were either reported by the fishermen or confirmed by the detection of Argos tracks typical of fishing operations by commercial longliners. An additional two sharks were possibly recaptured, but this could not be verified.

3. Movements

In general, blue sharks tagged in this study displayed wide ranging movements (Fig. 1). The longest deployment was registered for a LJ female (Shark 13, 145 cm FL) that travelled an estimated 28139 km over a period of 952 days. All sharks were oceanic, with individuals exploring shelf break areas, but none apparently moving onto continental shelves. While the tracks revealed important individual variability (Fig. 1), the distance travelled as a function of time at liberty was not significantly different between individuals or life stages (Fig. 2A). The only outlier was shark 9, of which we lost track during 389 days.

3.1 Small Juveniles (SJ). SJ sharks of both sexes were tagged with Mini-PAT tags providing low resolution location data (GLS). In general, no differences are apparent between the quarterly movements of SJ males and females (Fig. 3). Both sexes used a relatively limited area, comprised between 25° and 45°N and 22°

Table 1. Summary data for the 34 reporting blue sharks tagged off the Azores with satellite tags.

Shark	Deployment	Size (cm FL)	Sex	Life Stage	Class	Tag	Programmed Release (Days)	Recorded Release (Days)	Track Duration (Days)	Track Length (Km)	Argos Locations (n)	GLS Locations (n)	Data gaps (Days)	Recapture
1	17-10-2009	116	F	Juvenile	SJ	Mini-PAT	180	179	179	5316	0	54	74	
2	20-08-2010	122	F	Juvenile	SJ	Mini-PAT	270	102	102	3958	0	74	36	
3	20-08-2010	120	F	Juvenile	SJ	Mini-PAT	270	235	235	9346	0	208	17	
4	17-10-2009	116	M	Juvenile	SJ	Mini-PAT	180	180	180	5379	0	64	53	
5	20-08-2010	112	M	Juvenile	SJ	Mini-PAT	270	174	174	6295	0	124	36	
6	20-08-2010	127	M	Juvenile	SJ	Mini-PAT	270	193	193	6348	0	20	35	Y
7	20-08-2010	120	M	Juvenile	SJ	Mini-PAT	270	273	273	8761	0	193	37	
8	21-08-2010	112	M	Juvenile	SJ	Mini-PAT	270	270	270	11533	0	220	38	
9	19-02-2009	127	F	Juvenile	LJ	SPOT			877	14494	236	0	389	
10	19-02-2009	139	F	Juvenile	LJ	SPOT			616	16907	779	0	21	
11	26-02-2009	175	F	Sub-adult	LJ	Double	180	91	226	8275	293	10	20	Y
12	26-02-2009	142	F	Juvenile	LJ	Double	180	30	30	911	40	26	3	(Y)
13	26-02-2009	145	F	Juvenile	LJ	Double	180	89	952	28139	699	22	82	Y
14	26-02-2009	148	F	Sub-adult	LJ	SPOT			90	2637	100	0	8	(Y)
15	06-03-2009	165	F	Sub-adult	LJ	MK10 PAT	180	3	3	76	0	2	3	Y
16	06-03-2009	156	F	Sub-adult	LJ	MK10 PAT	180	179	179	5307	0	41	31	
17	02-12-2009	178	F	Sub-adult	LJ	SPOT			579	15498	314	0	137	
18	16-10-2009	168	M	Juvenile	LJ	SPOT			36	1213	56	0	4	
19	16-10-2009	133	M	Juvenile	LJ	SPOT			206	5041	152	0	24	
20	16-10-2009	156	M	Juvenile	LJ	SPOT			82	2245	87	0	8	
21	02-12-2009	130	M	Juvenile	LJ	SPOT			42	973	41	0	6	
22	20-08-2010	164	M	Juvenile	LJ	SPOT			244	6892	196	0	22	
23	20-08-2010	180	M	Juvenile	LJ	SPOT			381	13066	387	0	16	
24	20-08-2010	140	M	Juvenile	LJ	SPOT			126	2364	189	0	6	Y
25	20-08-2010	183	M	Juvenile	LJ	SPOT			116	3846	160	0	9	
26	20-08-2010	183	M	Juvenile	LJ	SPOT			524	15066	612	0	32	Y
27	20-08-2010	159	M	Juvenile	LJ	SPOT			212	5994	332	0	10	
28	20-08-2010	172	M	Juvenile	LJ	SPOT			369	12451	610	0	12	Y
29	06-03-2009	202	F	Adult	AD	MK10 PAT	180	143	143	3577	0	2	89	
30	18-05-2012	201	F	Adult	AD	SPOT			33	1297	33	0	7	
31	16-10-2009	201	M	Adult	AD	SPOT			161	5108	133	0	22	
32	17-10-2009	200	M	Adult	AD	MK10 PAT	180	103	103	4338	0	100	13	Y
33	20-08-2010	207	M	Adult	AD	SPOT			228	6710	341	0	29	
34	20-08-2010	211	M	Adult	AD	SPOT			526	14500	929	0	24	

F – female; M – male; SJ – Small Juveniles; LJ – Large Juvenile, including sub-adults for females; AD – Adult; Double – Double tagged with SPOT and MK10 PAT tags; (Y) not confirmed recapture.

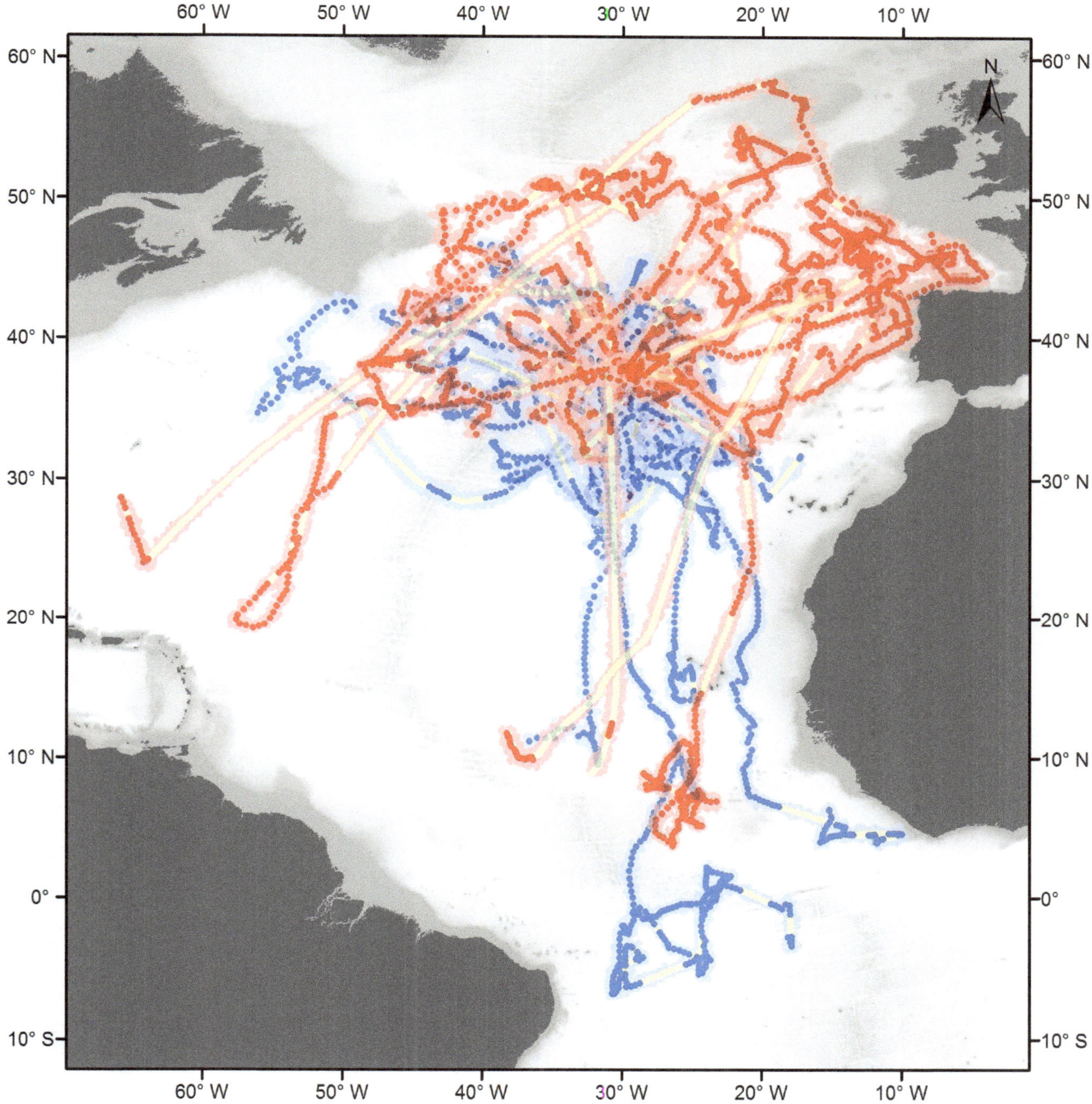

Figure 1. Reconstructed migratory pathways of blue sharks tagged in the Azores. Red and blue dots represent the most probable daily position estimates of respectively female and male sharks, with pink and light blue clouds representing the respective errors around the position estimates. Yellow dots indicate data gaps (i.e. days without position information within 72 h intervals).

and 56°W for up to 235 days. Two sharks (sharks 1 and 4), a male and a female tracked for a period of 6 months between October 2009 and May 2010 remained mainly in an area south-southwest of the Azores, near the MIDAR and the Atlantis – Great Meteor seamount complex. The six sharks, four males and two females tagged in August 2010 and tracked for periods between 101 and 270 days, spent most time in this same general area. Two males (sharks 7 and 8) initiated a westward movement during January. Shark 8 moved as far west as the eastern edges of the New England seamounts, before we lost its track near the southern slopes of the Grand Banks at the end of May. Shark 7 moved to an area close to 36.5°N 43°W, before returning to the same general area south-southeast of the Azores in May. Shark 5 moved to the Newfoundland Basin during autumn and returned by January.

3.2 Large Juveniles (LJ). Large juvenile females were tagged in autumn (2009, n = 1) and winter (2009, n = 8) with SPOT and/or MK10-PAT tags, providing high resolution location data (Argos data) for most individuals (sharks 9–15 and 17). In contrast with the SJ of both sexes, LJ females performed extensive latitudinal movements, utilising an area comprised between 31°N and 59°N and 3°W and 50°W (Fig. 4). LJ females displayed marked seasonal migrations, residing in the southern part of their range during

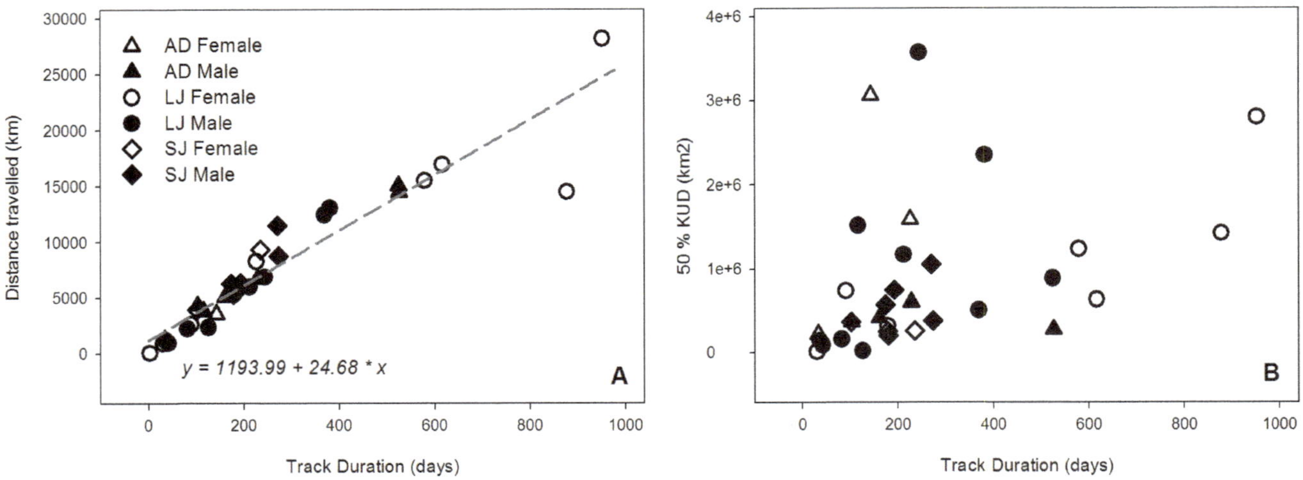

Figure 2. Distance travelled (A) and 50% fixed Kernel Utilisation Distributions (KUD) (B) in function of the track duration for the blue sharks belonging to different life stages. AD – Adults; LJ – Large juvenile; SJ – Small juvenile.

winter (i.e. between 31°N and 45°N) and moving northward during summer (i.e. between 45°N and 59°N). During the latter season, LJ females explored mainly seamount and slope areas in the north-eastern NA, from the Bay of Biscay to the Hatton bank (sharks 9, 10, 13 and 16), near the Charlie-Gibbs Fracture Zone (sharks 13 and 17) and east and south of the Flemish cap (sharks 13 and 17). Four individuals (three juveniles and one sub-adult, sharks 9, 10, 13, 17), were tracked for over a year allowing the observation of a full seasonal cycle (Fig. 5). All four sharks returned to the Azores approximately one year after release, despite having moved to different areas during summer. Whereas shark 10 moved to the eastern NA during two consecutive summers, in the same period shark 13 explored almost the entire longitudinal range between the Flemish cap and the Bay of Biscay. Sharks 12 and 15 were probably both recaptured close to the tagging site 30 and 3 days after release, respectively. Shark 14 moved to the Iberian coast.

LJ males were tagged with SPOT tags during summer (2010, n = 7) and autumn (2009, n = 4), providing high resolution location data for all sharks. This population segment also displayed extensive movements, but the seasonal nature of these movements appeared less pronounced as for LJ females (Fig. 6). Nonetheless, by December, 10 of 11 LJ males started moving south after spending up to 105 days (median = 37 days) in an area close to the tagging location. The core of this area, located south-southwest of the islands of Faial and Pico, is constituted by a complex of seamounts and offshore banks (e.g. Condor Seamount, Açores Bank and Princess Alice Bank) that joins the MIDAR further to the southwest. We will refer to this core area as the Princess Alice complex. Four sharks (sharks 21, 18, 20 and 24) were lost between 21 and 46 days after they initiated their southward movement. At that time these fish were located approximately at the latitude of Great Meteor Seamount, 8 degrees south of the tagging location. After the initial southward movement, four of the remaining individuals (sharks 22, 23, 25 and 27), dispersed further south to tropical latitudes with one shark eventually moving to the South Atlantic, where it remained 8 months. In contrast, two sharks that were tracked for 524 and 369 days (sharks 26 and 28, respectively), remained in the area south-southwest of the Azores after their initial southward movement in autumn. At the end of January, both sharks initiated a westward movement to areas in the north-western NA, where they remained for approximately 1.5 months before returning towards the Azores during May. The following autumn, shark 26 moved again to the area south-southwest of the Azores and initiated a new westward movement in December towards the Corner Rise Seamounts. In contrast with the other LJ

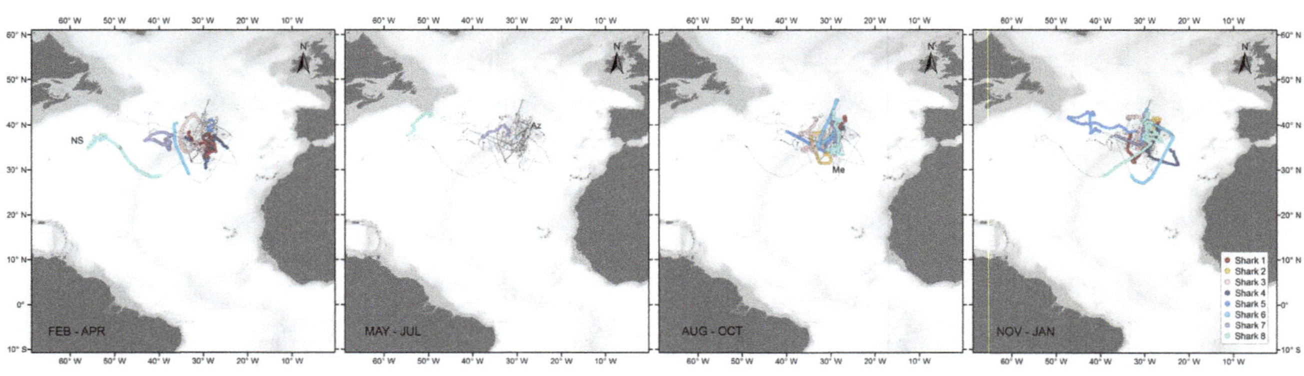

Figure 3. Reconstructed movements of small juvenile male (green – blue) and female (yellow – red) blue shark tagged in the Azores by quarter. Grey lines represent track segments of the individuals during other quarters. Az – Azores, Me – Great Meteor Seamount; NS – New England Seamounts.

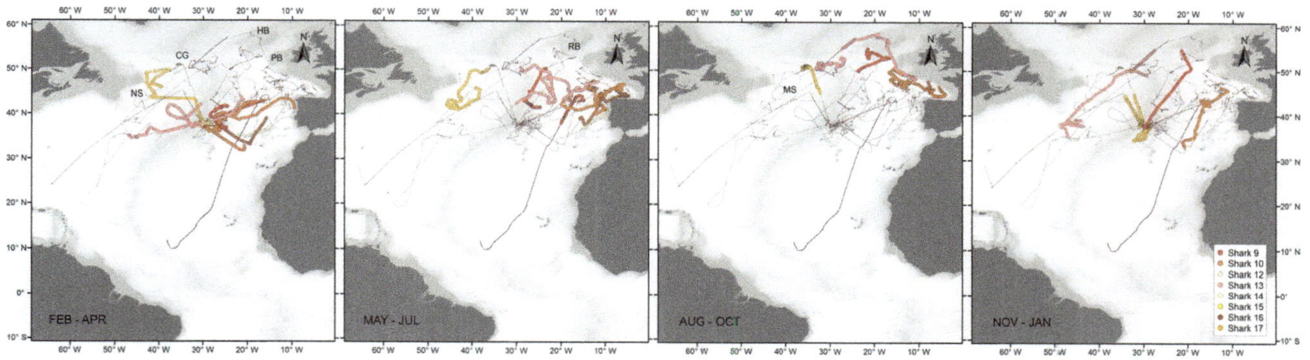

Figure 4. Reconstructed movements of large juvenile and sub-adult (LJ) female blue shark tagged in the Azores by quarter. Grey lines represent track segments of the individuals during other quarters. NS – Newfoundland Seamount; CG – Charlie-Gibbs Fracture Zone; HB – Hatton Bank; PB – Porcupine Bank; RB – Rockall Bank; MS – Milne Seamount.

males, shark 19 (140 cm FL) exhibited movements similar to SJ sharks, remaining in a smaller area close to the Archipelago during the entire duration of the deployment (206 days).

3.3 Adults. Two AD females (shark 29, 202 cm FL; shark 30, 201 cm FL) and one large LJ female (shark 11, 175 cm FL) were tagged in winter (2009, n = 2) and spring (2012, n = 1) with SPOT and/or MK10-PAT tags, providing high resolution location data for two individuals (sharks 11 and 30). Shark 30 released fully developed pups during capture, indicating late stage pregnancy at the time of tagging. All three individuals migrated south within 33 days after tagging (Fig. 7). After remaining 135 days (June to October) in tropical waters, in an area comprised between 3.5°N–11°N and 23°W–28.5°W, shark 11 was recaptured by a fisherman who reported that the shark was carrying embryos at the time. Similar southward migrations towards tropical and sub-tropical waters were also observed for 3 juvenile and sub-adult females that were tracked for over a year. After 830 days of tracking, shark 9 reappeared for a period of 29 days in tropical waters near 10°N 36°W in June (2011) after its signal had been lost for 389 days. Sharks 13 and 17 both migrated to the southern part of the Sargasso Sea by June 2011 after approximately 826 and 526 days, respectively. Shark 13 moved back north towards the Corner Rise Seamounts and then east towards the MIDAR near 33°N and 40.5°W, where we lost its track in October 2011. The southward migrations of sharks 9, 13, 17 and 29 were accompanied by large periods (137, 82, 389 and 89 days respectively) when no Argos locations or good GLS locations were obtained and reconstructed migratory pathways were difficult to produce.

The movements of the AD males were similar to those observed for LJ males (Fig. 7). Four adult males were tagged during summer (2010, n = 2) and autumn (2011, n = 2) with SPOT or MK10-PAT tags, providing high resolution location data for three sharks (sharks 31, 33 and 34). By December all four individuals started moving south. Before that time, three sharks (sharks 31, 33, 34) spent between 14 and 117 days within the Princes Alice complex. Shark 32 also remained in the vicinity of the archipelago. Thereafter, all four sharks stayed in the area south-southwest of the Azores, mainly near the MIDAR and the Atlantis – Great Meteor seamount complex, but moving as far south as 24°N, at least until the end of January. Shark 31 returned to the Archipelago by the middle of March, when we lost its signal. Shark 33 moved east until we lost its track between Madeira and the Canary Islands at the beginning of April. Shark 34 moved west towards the north-western NA during the month of February, reaching an area south of the Newfoundland Seamount in March. The second summer the shark moved again to the Azores, spending most time within the Princes Alice complex, initiating a new southward movement during November.

4. Space use

Distance travelled as a function of time at liberty was similar across individuals (Fig. 2A) but the concomitant variation on home range size changed dramatically (Fig. 2B). The 25% and 50% KUD areas were highly variable both between sexes and life stages (Fig. 8). The latitudinal and longitudinal displacements over time (Fig. 9) further illustrate the contrast in spatio-temporal dynamics across the species' life history. SJ of both sexes were contained within a narrow latitudinal range and mainly utilised an area delimited by the Azores archipelago to the North, the Atlantis – Great Meteor seamount complex to the South, and the MIDAR to the South-West. There was little difference between sexes, except that males extended their range more towards the West. In contrast, LJ of both sexes showed a wider range in both latitude and longitude. The segregation between both sexes becomes apparent during summer and autumn, when females extended their range to the North and East and males to the South. Conversely, AD females moved South and South-West while AD males used a more limited range, mainly extending their distribution South and West of the tagging location. The figures also clearly illustrate the intermediate geographical position of the tagging location as the distributions of all life stages spatially overlap in this area, albeit in different quarters.

5. Time at SST

The spatio-temporal segregation patterns are also reflected in the pooled time-at-SST profiles for the different population components (Fig. 10). No differences were apparent between SJ males and females, which show two preferred maxima, i.e. approximately 17°C and 25°C, corresponding to the most common temperatures within the narrow latitudinal range they occupied. The figure also shows a discrepancy between the SST recorded by the Mini-PATs and that obtained from remote sensing, which is probably due to heating of the surface layer as a consequence of increased sun exposure and low wind speeds during summer. In contrast, the temperature niche segregation between male and female LJ was evident, with males seemingly preferring warmer waters between 16°C and 30°C. This SST range was roughly identical as for AD males, whereas AD females used a wider temperature range (15°C to 30°C) and, apparently,

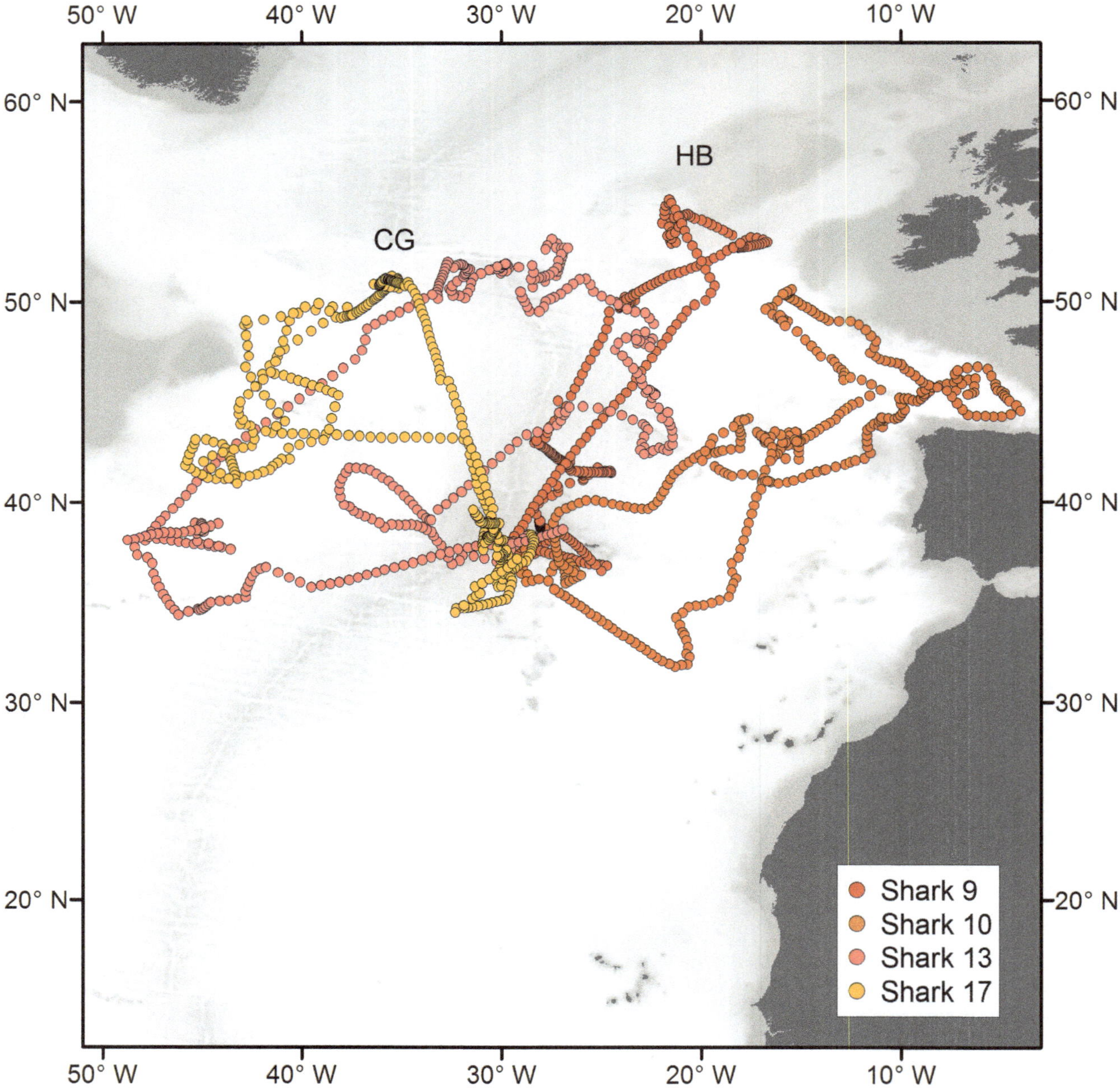

Figure 5. One year reconstructed tracks of four large juvenile and sub-adult (LJ) female blue shark tagged in the Azores. CG – Charlie-Gibbs Fracture Zone; HB – Hatton Bank.

preferred warmers waters within it (25°C to 30°C). However, this pattern is based on a smaller dataset and seasonal coverage.

Discussion

The blue sharks tagged in the Azores displayed extensive movements over vast areas of the North Atlantic, with one individual eventually moving into the southern hemisphere. The sharks appeared to be mainly oceanic and to explore shelf break areas, but did not venture onto the continental shelves, although many studies about blue shark catch data and conventional tagging are from shelf areas (e.g., [18,28,30,34,54]). This finding supports the conclusions of previous studies that pointed out the potential bias that release sites from conventional tagging studies can introduce in the interpretation of movements [12]. Furthermore, the sharks in this study exhibited considerable individual variability in their movements, both among individuals and life stages as well as individually across time.

1. Testing the nursery concept

Our data showed that both male and female SJ blue shark tend to remain for extended periods of time (up to 235 days) in a general area delimited by the Azores, the Atlantis – Great Meteor seamount complex and the Mid-Atlantic Ridge (MIDAR). Additionally, the movement patterns and SST preference did not show any indications of segregation between the two sexes at the spatial scale considered in this study. This pattern is in agreement with available data from fisheries-dependent conventional tagging programs, which suggest that juveniles of 100–

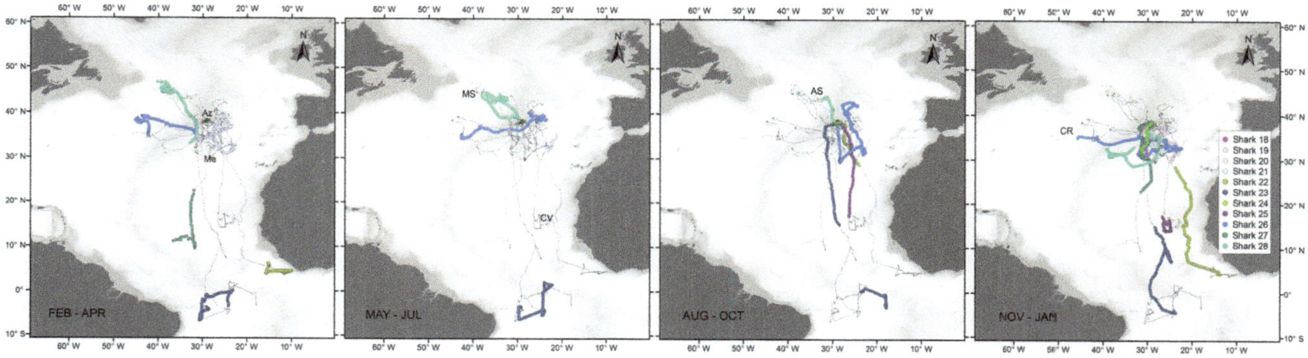

Figure 6. Reconstructed movements of large juvenile (LJ) male blue shark tagged in the Azores by quarter. Grey lines represent track segments of the individuals during other quarters. Az – Azores; Me – Great Meteor Seamount; MS – Milne Seamount; AS – Altair Seamount; CR – Corner Rise Seamounts.

130 cm FL do not participate in extensive latitudinal migrations [12,28,30,32,55]. Nonetheless, Queiroz et al. [44] found movements of juvenile blue shark closer to continental shelves to be more extensive, and Litvinov [36] concluded that male and female blue shark segregate during their first year, at lengths smaller than 70 cm FL. This is in disagreement with the results from the present study and a recent demographic analysis of fisheries data from the Azores region [17]. Both studies indicate that segregation does not occur before, at least, the second year, when blue sharks start to take part in larger scale migrations.

Our study does not present indications of connectivity between the central NA and other juvenile grounds, such as the continental shelves of the Iberian Peninsula and Northern Africa [12,32], reinforcing the hypothesis that parturition may also take place in this area of the NA [17]. Moreover, our study provides additional evidence of the existence of a nursery area for blue shark in the central NA. Not only are juvenile blue sharks locally abundant across years [17,47], but individual blue shark have the tendency to remain and return to the area for extended periods, meeting the three criteria proposed by Heupel et al. [3]. Still, in light of the classical notion of shark nursery areas [40], the classification of such a vast offshore area as a nursery could be disputed. Even so, oceanic nurseries have previously been suggested for some oceanic species, but solely based on observations of small individuals and/or pregnant females in offshore areas, e.g. salmon shark, *Lamna ditropis* [56,57], bigeye thresher, *Alopias superciliosus* [58], and oceanic whitetip, *Carcharhinus longimanus* [59]. Despite the consistent spatial patterns observed, our study was not designed to clarify whether the area provides juveniles with increased survival probability by providing them shelter from predators or some

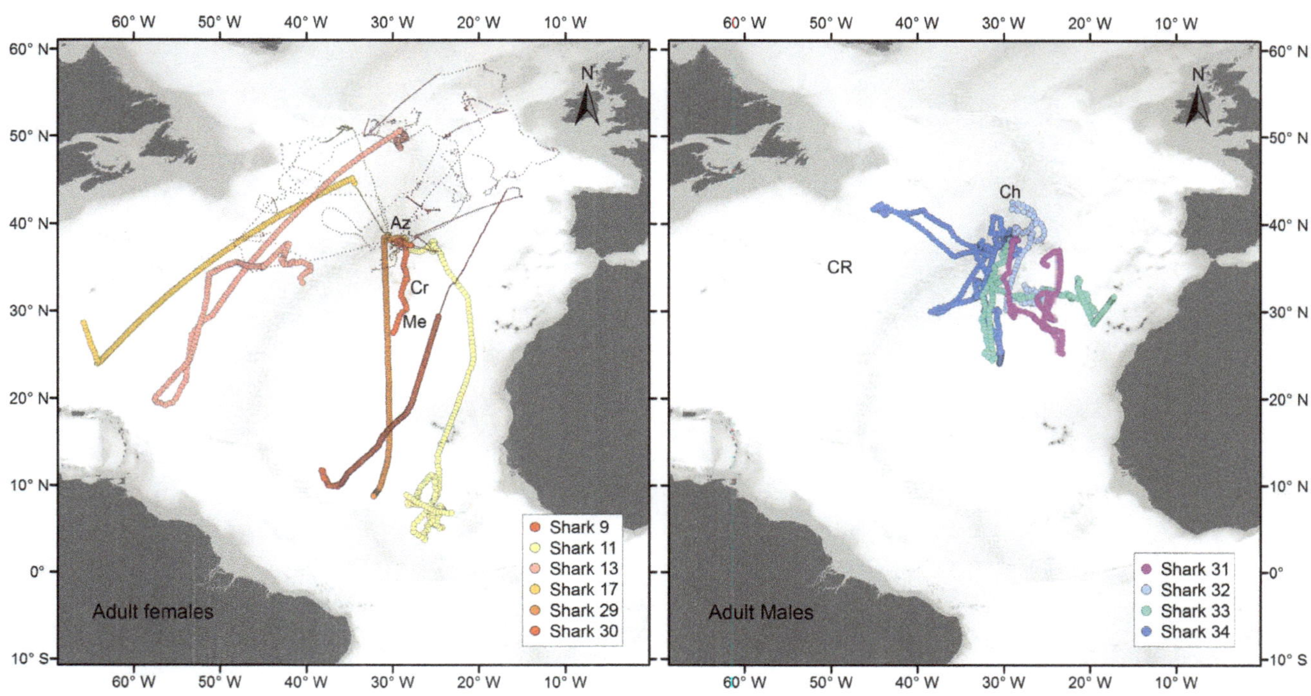

Figure 7. Reconstructed movements of adult (AD) male (right) and female (left) blue shark tagged in the Azores by quarter. Grey lines represent track segments of the individuals during other quarters. Az – Azores; Me – Great Meteor Seamount; Ch – Chaucer Seamount; Cr – Cruiser Seamount; CR – Corner Rise Seamounts.

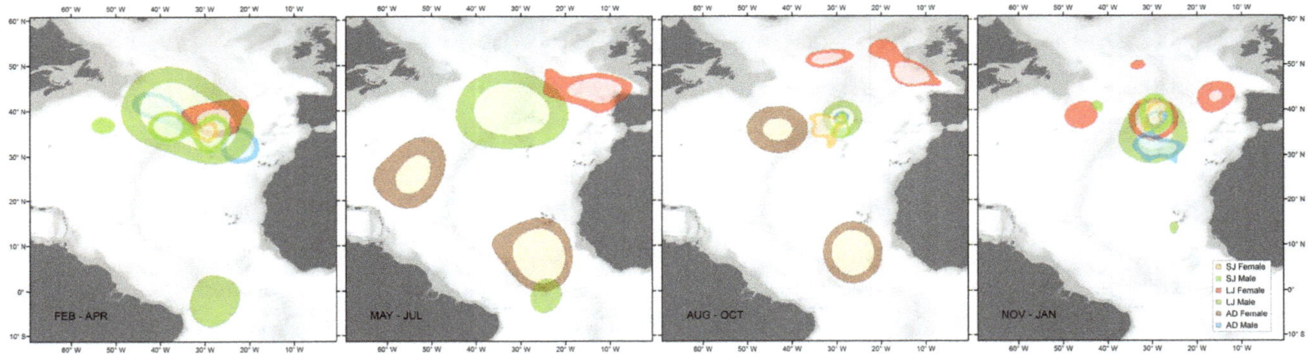

Figure 8. Quarterly 25% and 50% Kernel Utilisation Distributions (KUD) for the different life stages of blue sharks tagged in the Azores. Orange – Small Juvenile (SJ) females; Green – SJ males; Red – Large Juvenile and Sub-adult females; Dark green - Large Juvenile (LJ) males; Brown – Adult (AD) females; Blue – AD males.

other advantages [3,37]. Oceanic sharks have typically larger litters of smaller young than their coastal counterparts, which has been interpreted as an adaptive strategy of the offspring as a whole to thrive on patchy oceanic resources, benefiting from the supposedly lower predation pressure in the open ocean [60]. Lower predation risk in oceanic waters is also seen as an

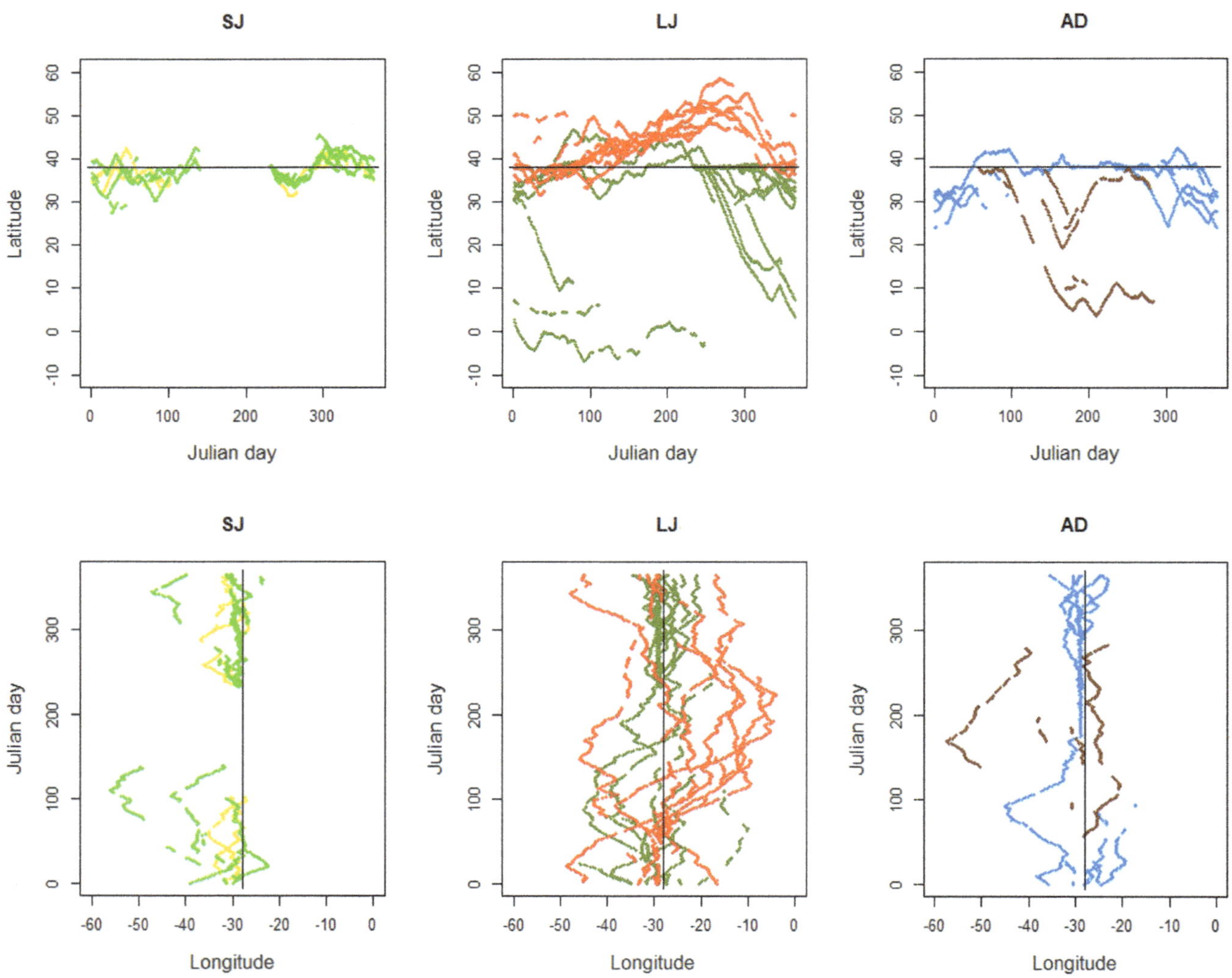

Figure 9. Latitudinal and longitudinal displacements by day of the year of different life stages of blue sharks tagged in the Azores. Orange – Small Juvenile (SJ) females; Green – SJ males; Red – Large Juvenile and Sub-adult females; Dark green - Large Juvenile (LJ) males; Brown – Adult (AD) females; Blue – AD males.

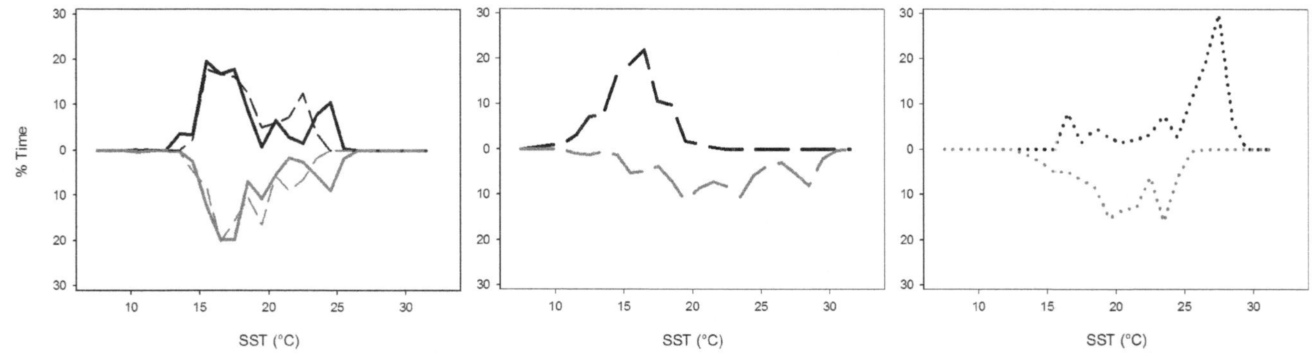

Figure 10. Aggregated long-term time at Sea Surface Temperature (SST, °C) profiles for male (grey) and female (black) blue sharks belonging to different life stages. Top panel: Small Juveniles (SJ); Centre panel: Large Juveniles and Sub-adults (LJ); Bottom panel: Adults (AD). SST values were obtained from averaged (1°) monthly SST images for NOAA's CoastWatch program, except for the dotted curves in the top panel that were obtained directly from the Mini-PAT tags.

evolutionary driver for the oceanic juvenile stage in some sea turtles [61,62]. A favourable habitat providing increased growth rates and thus a reduced time at vulnerable sizes [3] therefore appears to be a plausible advantage for an oceanic nursery habitat of blue shark.

Juvenile survival has been shown to have great impact on blue shark population growth [24]. Due to their small size-at-birth, juvenile blue sharks are probably highly vulnerable during their first year of life [22], which could be balanced through rapid juvenile growth [24,40]. While food is generally not considered a limiting factor for juvenile sharks in typically prey rich estuarine and coastal nursery areas [40], it has been shown to limit the survival of juvenile scalloped hammerheads in a Hawaiian nursery [41] and may play an important role in oceanic and oligotrophic environments. It can therefore be argued that the area provides juvenile blue shark with optimal growth conditions supported by ample food resources associated with the diversity of topographic features (seamounts and islands) [63] and localised oceanographic processes [64]. In addition, the intermediate geographical location of the wider Azores ensures a favourable temperature niche throughout the year. This combination of factors is probably what confers the region an important role for other highly migratory species, e.g. blue and fin whales [65] and juvenile loggerhead sea turtles [66,67]. Remarkably, the distance travelled by these SJ sharks as a function of time is similar to that of other life stages. This finding strongly suggests that their limited distribution when compared to that of larger life stages is unlikely to be merely a consequence of limited swimming capacities. In conclusion, our study provides strong support for the existence of an oceanic shark nursery, although further research is needed to clarify its primary drivers. The designation of the wider Azores as a nursery area also seems meaningful from a management perspective, considering the spatial scale at which the pelagic longline fishery operates.

2. Movements into adulthood

LJ females are known to undertake large scale latitudinal migrations, moving to northern latitudes during summer [12,30,32,33]. In the NA, they dominate the summer catches off south-west England and are present off the US and Canada [18,25,34] whereas in winter high abundances are mainly recorded off Portugal [55] and around the Azores [17,47]. This putative migratory pattern was directly observed in all of our tagged LJ females with periods at liberty exceeding 90 days (n = 6). Although the summer patterns revealed considerable individual variability as individuals explored almost the entire longitudinal range across the NA from the Flemish Cap to the Gulf of Biscay, LJ females displayed a strong connectivity with the eastern NA as indicated by conventional tagging [54]. Regardless of this variability, the four sharks that retained the transmitters for at least one full year were all tracked back to the Azores region during winter, indicating a high degree of site fidelity to the region. LJ females stayed within colder temperatures than younger or older sharks (10°–20°C) and showed a well marked preference for surface waters with an SST between 15° and 16°C throughout the year. The affinity of blue shark for certain water temperatures has been extensively described in the literature although these studies rarely made the same distinction between life stages as we do, making it difficult to compare findings. Nonetheless, highest abundances of juvenile and sub-adult females in the Northwest and central NA were recorded within the same range [17,18].

Tagging AD female blue shark was not the prime objective of the present study. Nonetheless, the few large sub-adult and adult female that we tagged all displayed directed southern movements to tropical latitudes in sharp contrast with LJ females. One AD female was later recaptured in tropical waters carrying embryos, indicating that these movements are related to the reproductive cycle. Additionally, we were able to observe a behavioural switch in three females up to 2.5 years after release, when they reached the theoretical adult size between 170 and 200 cm FL [68]. Up to that point, these sharks had remained in cold waters migrating between sub-tropical and temperate regions, after which all of them moved to southern tropical grounds in June and July utilising pathways on both sides of the ocean. Female blue sharks are thought to move offshore from coastal areas in the north-western NA after mating and to migrate towards the eastern NA to mature and deliver pups [18,25]. However, our results also show that a proportion of mature females move South towards the south-western NA, where their arrival in early summer coincides with the increasing seasonal catch rates of mature and pregnant females [27]. This finding supports the hypothesis of a southbound movement of mature blue sharks into the south-western NA as inferred from conventional tagging data [12,29]. In the eastern NA, adult and pregnant females are seasonally observed in different stages of the reproductive cycle. Mature and pregnant females are found during winter near the Canary Islands and African coast, with mating and pupping believed to occur off Portugal [12] and near the Azores during spring [17]. Castro and

Mejuto [26] found a large proportion of adult females in the Gulf of Guinea, many of which were pregnant. The present study indicates that small adult females also move to offshore areas further south (3°–12°N) into the south-eastern NA. This is in agreement with their presence in tropical latitudes [26,27], a behaviour that was hypothesised to facilitate fertilization and embryonic development [69].

LJ and AD males displayed similar movements and do not appear to segregate spatially. After spending up to 117 days within the Princes Alice seamount complex, both groups started moving south to an area south-southwest of the Azores during autumn. This autumn migration coincides with a decrease in abundance of both life stages in local catches [17]. Accordingly, Litvinov [70] observed high abundance of adult males above the seamounts south of the Azores during autumn and labelled these concentrations as "male clubs". In fact, three sharks (2 LJ and 1 AD) remained in that area until January before initiating a westward movement to offshore areas in the western Atlantic. By the end of spring these fish returned to the Azores, eventually resuming a new southward movement at the end of summer. Such a clear cyclic pattern of migration appears to be common throughout the life history of male blue sharks, as it was also observed in small juveniles. It demonstrates the strong connectivity between the north-western and central NA for males, as previously hypothesised [28,71].

In the northwest Atlantic, adult and sub-adult blue shark enter coastal waters during late spring and summer and engage in mating [25]. Thus, the subsequent migration of males from the central NA to waters off the north-western NA in late spring could well be related to such mating. The behavioural shift in spring of two tagged females in these off-shore areas appears to support this hypothesis, but their capacity to store sperm would not require such a temporal matching of events. In addition to the western migration cycle, LJ males also exhibited an important southward dispersal, which is in accordance with their dominance in tropical waters [26,27].

3. Underlying mechanisms of blue shark spatial ecology

The long-term tracking of blue shark in this study revealed the existence of a high degree of site fidelity with the central NA across almost all life stages, except for adult females for which no long-term tracks were available. Such site fidelity appears to be common in migratory shark species (e.g. great white shark, *Carcharodon carcharias* [72], tiger shark, *Galeocerdo cuvier* [48], broadnose sevengill shark, *Notorynchus cepedianus* [73]) and other pelagic predators like tuna [74] and cetaceans [75] and is generally associated with feeding, parturition or mating. The identification of these specific areas is considered an important step for the management of highly mobile species [76] and the fact that site fidelity towards the central NA is present throughout its life history emphasises the importance of the area for blue shark.

The cyclic migrations and dispersal to tropical latitudes of LJ and AD males suggest some degree of behavioural polymorphism within the male population supported by what seems to be a high tolerance and adaptation to habitat conditions. In fact, the wide geographical range occupied by these males was also reflected in their broad SST niche (12–30°C and 15–30°C, respectively), in agreement with fisheries catch analysis [17]. Such behavioural polymorphism, also observed in many fish (e.g., [77]) and sharks [48], may have ecological reasons, such as the existence of different feeding or reproductive strategies [48]. Given the segregation between AD and LJ females, one hypothesis to explain the behavioural polymorphism in male blue shark could therefore be the co-existence of different reproductive strategies aimed at mating with either of the two female life stages. Indeed, AD females are seldom observed at the mating areas in the north-western NA, where LJ females occur (see above) which are visited by males exhibiting the western migration cycle. Conversely, LJ females are virtually absent in tropical latitudes to where both AD females and (some) males appear to converge. This possibility casts a previously unknown complexity on the behavioural ecology of pelagic sharks and blue shark in particular.

The patterns of spatial segregation between sexes, which are considered a common trait in sharks [1,2,38,39], change over the course of the blue shark's life cycle indicating that the relative importance of the drivers behind the segregation changes along with it. This is particularly so for females, which undergo pronounced ontogenetic shifts in distribution and movement patterns during their juvenile phase as well as upon reaching maturity. Contrastingly, male blue sharks expand their range, mainly to warmer waters during their juvenile phase, but typical movement patterns are observed throughout their life history. As a result, LJ females segregate more clearly from other life stages, occupying colder waters and mainly occurring together with smaller sized individuals of both sexes [17,55]. Because this is a juvenile life stage and growth is similar for males and females, this pattern is likely a means for LJ females to avoid aggressive courtship behaviour until approaching sexual maturity [2,5], which is also why females develop an epidermis twice as thick as that of males [25]. The behavioural shift observed in females upon reaching maturity could then be related to the search of warmer thermal habitat to aid fertilisation and embryonic developments as proposed by Hazin et al. [69]. Other drivers in shaping the segregation patterns in blue shark can not be excluded (see [2,5] for reviews), and it is possible that certain patterns were not observed because they act at different spatial scales than could be detected in the present study.

Conclusion

The long term tracking of different life stages performed in this study demonstrates the complex structuring of the NA blue shark population. The study builds upon previous studies [17,47], providing strong evidence for the existence of a discrete central NA nursery roughly delimited by the Azores archipelago in the North, the Atlantis–Great Meteor seamount complex in the South and the MIDAR in the South-West. This oceanic nursery is used by juveniles of both sexes, at least during their first year, after which they start to show the typical segregation pattern that can be seen in the older juvenile component. Upon reaching maturity, females move to southern latitudes utilising migration pathways on both sides of the NA. Males extended their range, displaying a high connectivity with the western NA and dispersal to tropical latitudes. Moreover, there is a high degree of site fidelity with individuals of nearly all life stages returning to the central NA, emphasizing the importance of this region and their possible role as pupping grounds. Such complex spatial structuring and philopatric behaviour will have important consequences for the management and conservation of blue shark [11,78] in the Atlantic, especially considering that the seasonal overlap and alternation of different life stages in the central NA nursery coincides with high levels of fishing mortality [79].

Supporting Information

Figure S1 Overview of the tagging experiment.

Figure S2 Frequency distribution of blue shark speeds (km/h). Speed (km/h) was calculated as the displacement between quality 2 and 3 Argos locations separated by time intervals of 1–4 days.

Acknowledgments

We would like to thank the crew of the RV Arquipelago and the following individuals who were instrumental in helping with the fieldwork to tag the blue sharks: N. Serpa, M. Schmiing, C. Pham, G. Graça, R. Guedes and A. Morais. JA Musick and one anonymous reviewer provided comments on the manuscript.

Author Contributions

Conceived and designed the experiments: FV PA JF MS. Performed the experiments: FV PA JF MS. Analyzed the data: FV. Contributed reagents/materials/analysis tools: FV PA RSS. Contributed to the writing of the manuscript: FV PA JF AA.

References

1. Klimley AP (1987) The determinants of sexual segregation in the scalloped hammerhead shark, Sphyrna lewini. Env Biol Fish 18: 27–40.
2. Sims DW (2005) Differences in habitat selection and reproductive strategies of male and female sharks. In Ruckstuhl KE and Neuhaus P (Eds), Sexual Segregation in Vertebrates: Ecology of the Two Sexes. Cambridge University Press, Cambridge, UK, pp. 127–147.
3. Heupel MR, Carlson JK, Simpfendorfer CA (2007) Shark nursery areas: concepts, definition, characterization and assumptions. Mar Ecol Prog Ser 337: 287–297.
4. Robbins RL (2007) Environmental variables affecting the sexual segregation of great white sharks *Carcharodon carcharias* at the Neptune Islands South Australia. J Fish Biol 70: 1350–1364.
5. Wearmouth VJ, Sims DW (2008) Sexual segregation in marine fish, reptiles, birds and mammals: Behaviour patterns, mechanisms and conservation implications. Adv Mar Biol 54: 107–170.
6. Ward P, Myers RA (2005) Shifts in Open-Ocean Fish Communities Coinciding With the Commencement of Commercial Fishing. Ecology 86: 835–847.
7. Dulvy NK, Baum JK, Clarke S, Compagno LJV, Cortes E, et al. (2008). You can swim but you can't hide: the global status and conservation of oceanic pelagic sharks and rays. Aquatic Conserv: Mar Freshw Ecosyst 18: 459–482.
8. Ferretti F, Myers RA, Serena F, Lotze HK (2008) Loss of large predatory sharks from the Mediterranean Sea. Conserv Biol 22: 952–964.
9. Ferretti F, Worm B, Britten G, Heithaus MR, Lotze HK (2010) Patterns and ecosystem consequences of shark declines in the ocean. Ecol Lett 13: 1055–1071.
10. Grüss A, Kaplan DM, Guenette S, Roberts CM, Botsford LW (2011) Consequences of adult and juvenile movements for marine protected areas. Biol Conserv 144: 692–702.
11. Mucientes GR, Queiroz N, Sousa LL, Tarroso P, Sims DW (2009) Sexual segregation of pelagic sharks and the potential threat from fisheries. Biol Lett 5: 156–159.
12. Nakano H, Stevens J (2008) The biology and ecology of the blue shark *Prionace glauca*. In: In: Camhi M, Pikitch E, Babcock E (Eds), Sharks of the Open Ocean: Biology, Fisheries and Conservation. Blackwell Publishing, Oxford, UK, pp. 140–148.
13. Mejuto J (1985) Associated catches of sharks, *Prionace glauca*, *Isurus oxyrinchus*, and *Lamna nasus*, with NW and N Spanish swordfish fishery, in 1984. International Council for the Exploration of the Sea, Council Meeting 1985/H 42:16 pp.
14. Hurley PCF (1998) A review of the fishery for pelagic sharks in Atlantic Canada. Fish Res 39: 107–113.
15. Castro J, de la Serna JM, Macias D, Mejuto J (2000) Estimaciones cientificas preliminares de los desembarcos de especies asociados realizados por la flota Espanola de palangre de superficie en 1997 y 1998. ICCAT Coll Vol Sci Pap 51: 1882–1893.
16. Mejuto J, García-Cortés B, Ramos-Cartelle A, de la Serna JM (2009) Scientific estimations of by-catch landed by the Spanish surface longline fleet targeting swordfish (Xiphias gladius) in the Atlantic ocean with special reference to the years 2005 and 2006. ICCAT Coll Vol Sci Pap 64: 2455–2468.
17. Vandeperre F, Aires-da-Silva A, Santos M, Ferreira R, Bolten AB, et al. (2014) Demography and ecology of blue shark (*Prionace glauca*) in the central North Atlantic. Fish Res 153: 89–102.
18. Simpfendorfer CA, Hueter RE, Bergman U, Connett SMH (2002) Results of a Fishery-Independent Survey for Pelagic Sharks in the Western North Atlantic, 1977–1994. Fish Res 55: 175–192.
19. Burgess GH, Beerkircher LR, Cailliet GM, Carlson JK, Cortes E, et al. (2005). Is the Collapse of Shark Populations in the Northwest Atlantic Ocean and Gulf of Mexico Real? Fisheries 30: 19–26.
20. Aires-da-Silva A, Hoey JJ, Gallucci VF (2008b) A Historical Index of Abundance for the Blue Shark (*Prionace glauca*) in the Western North Atlantic. Fish Res 92: 41–52.
21. Anonymous (2009) Report of the 2008 shark stock assessment meeting. ICCAT Coll Vol Sci Pap 64: 1343–1491.
22. Cortés E (2000). Life history patterns and correlations in sharks. Reviews in Fisheries Science 8: 299–344.
23. Frisk MG, Miller TJ, Fogarty MJ (2001) Estimation and Analysis of Biological Parameters in Elasmobranch Fishes: a Comparative Life History Study. Can J Fish Aquat Sci 58: 969–981.
24. Aires-da-Silva A, Gallucci VF (2007) Demographic and Risk Analyses Applied to Management and Conservation of the Blue Shark (*Prionace glauca*) in the North Atlantic Ocean. Mar Freshwater Res 58: 570–580.
25. Pratt HL (1979) Reproduction in the Blue Shark, *Prionace glauca*. Fish Bull 77: 445–470.
26. Castro J, Mejuto J (1995) Reproductive parameters of blue shark *Prionace glauca*, and other sharks in the Gulf of Guinea. Mar Fresh Res 46: 967–973.
27. Tavares R, Ortiz M, Arocha F (2012) Population Structure, Distribution and Relative Abundance of the Blue Shark (*Prionace glauca*) in the Caribbean Sea and Adjacent Waters of the North Atlantic. Fish Res 129: 137–152.
28. Kohler N, Turner P, Hoey J, Natanson L, Briggs R (2002) Tag and recapture data for three pelagic shark species, blue shark (*Prionace glauca*), shortfin mako (Isurus oxyrinchus), and porbeagle (Lamna nasus) in the North Atlantic Ocean. ICCAT Coll Vol Sci Pap 54: 1231–1260.
29. Kohler N, Turner P (2008) Stock structure of the blue shark (*Prionace glauca*) in the North Atlantic Ocean based on tagging data. In: Camhi M, Pikitch E, Babcock E (Eds), Sharks of the Open Ocean: Biology, Fisheries and Conservation. Blackwell Publishing, Oxford, UK, pp. 339–348.
30. Stevens JD (1990) Further Results From a Tagging Study of Pelagic Sharks in the North-East Atlantic. J Mar Biol Assoc UK 70: 707–720.
31. Buencuerpo V, Rios S, Moron J (1998). Pelagic Sharks Associated With the Swordfish, Xiphias gladius, Fishery in the Eastern North Atlantic Ocean and the Strait of Gibraltar. Fish Bull 96: 667–685.
32. Stevens JD (1976) 1st Results of Shark Tagging in Northeast Atlantic, 1972–1975. J Mar Biol Assoc UK 56: 929–937.
33. Henderson AC, Flannery K, Dunne J (2001) Observations on the Biology and Ecology of the Blue Shark in the North-East Atlantic. J Fish Biol 58: 1347–1358.
34. Campana SE, Marks L, Joyce W, Kohler NE (2006) Effects of Recreational and Commercial Fishing on Blue Sharks (*Prionace glauca*) in Atlantic Canada, With Inferences on the North Atlantic Population. Can J Fish Aquat Sci 63: 670–682.
35. Stevens JD (1974) Occurrence and Significance of Tooth Cuts on Blue Shark (*Prionace glauca* L) From British Waters. J Mar Biol Assoc UK 54: 373–378.
36. Litvinov FF (2006) On the Role of Dense Aggregations of Males and Juveniles in the Functional tructure of the Range of the Blue Shark *Prionace glauca*. J Ichthyol 46: 613–624.
37. Beck MW, Heck KL, Able KW, Childers DL, Eggleston DB, et al. (2001) The identification, conservation, and management of estuarine and marine nurseries for fish and invertebrates. BioScience 51: 633–641.
38. Springer S (1967) Social organisation of shark populations. In Gilbert PW, Mathewson RF, Rall DP (Eds), Sharks, Skates and Rays. The Johns Hopkins Press, Baltimore, pp. 149–174.
39. Bass AJ (1978) Problems in studies of sharks in the Southwest Indian Ocean. In: Hodgson ES, Mathewson RF (Eds) Sensory biology of sharks, skates and rays. Office of Naval Research, Department of the Navy, Arlington, VA, pp. 545–594.
40. Branstetter S (1990) Early life-history implications of selected carcharhinoid and lamnoid sharks of the northwest Atlantic. In Pratt HL, Gruber SH and Taniuchi T (Eds), Elasmobranchs as Living Resources: Advances in the Biology, Ecology, Systematics, and the Status of the Fisheries. NOAA Technical Report NMFS 90. pp. 17–28. (US Department of Commerce: Washington, DC.)
41. Duncan KM, Holland KN (2006) Habitat use, growth rates and dispersal patterns of juvenile scalloped hammerhead sharks *Sphyrna lewini* in a nursery habitat. Mar Ecol Prog Ser 312: 211–221.
42. Carey FG, Scharold JV (1990) Movements of blue sharks (*Prionace glauca*) in depth and course. Mar Biol 106: 329–342.
43. Klimley AP, Beavers SC, Curtis TH, Jorgensen SJ (2002) Movements and swimming behaviour of three species of sharks in La Jolla. Env Biol Fish 63: 117–135.
44. Queiroz N, Humphries NE, Noble LR, Santos AM, Sims D (2010) Short-term movements and diving behaviour of satellite-tracked blue sharks *Prionace glauca* in the northeastern Atlantic Ocean. Mar Ecol Prog Ser 406: 265–279.
45. Queiroz N, Humphries NE, Noble LR, Santos AM, Sims DW (2012) Spatial Dynamics and Expanded Vertical Niche of Blue Sharks in Oceanographic Fronts Reveal Habitat Targets for Conservation. PloS ONE 7: e32374.
46. Campana SE, Dorey A, Fowler M, Joyce W Wang Z, et al. (2011) Migration Pathways, Behavioural Thermoregulation and Overwintering Grounds of Blue Sharks in the Northwest Atlantic. PloS ONE 6: e16854.

47. Aires-da-Silva A, Ferreira R, Pereira J (2008a) Case study: blue shark catch patterns from Portuguese swordfish longline fishery in the Azores. In: Camhi M, Pikitch E, Babcock E (Eds), Sharks of the Open Ocean: Biology, Fisheries and Conservation. Blackwell Publishing, Oxford, UK, pp. 230–234.
48. Meyer CG, Papastamatiou YP, Holland KN (2010) A multiple instrument approach to quantifying the movement patterns and habitat use of tiger (*Galeocerdo cuvier*) and Galapagos sharks (*Carcharhinus galapagensis*) at French Frigate Shoals, Hawaii. Mar Biol 157: 1857–1868.
49. Tremblay Y, Robinson PW, Costa DP (2009) A Parsimonious Approach to Modeling Animal Movement Data. PLoS ONE 4, e4711.
50. Nielsen A, Bigelow KA, Musyl MK, Sibert JR (2006) Improving light-based geolocation by including sea surface temperature. Fish Oceanogr 15(4): 314–32.
51. Stevens JD, Bradford RW, West GJ (2010) Satellite tagging of blue sharks (*Prionace glauca*) and other pelagic sharks off eastern Australia: depth behaviour, temperature experience and movements. Mar Biol 157: 575–591.
52. Calenge C (2006). The package adehabitat for the R software: a tool for the analysis of space and habitat use by animals. Ecol Model 197: 516–519.
53. R Core Team (2013) R: A language and environment for statistical computing. R Foundation for Statistical Computing, Vienna, Austria. (http://www.R-project.org/).
54. Fitzmaurice P, Green P, Keirse G, Kenny M, Clarke M (2005) Stock discrimination of the blue shark, based on Irish tagging data. ICCAT Coll Vol Sci Pap 58: 1171–1178.
55. Queiroz N, Lima FP, Maia A, Ribeiro PA, Correia JP, et al. (2005) Movement of Blue Shark, *Prionace glauca*, in the North-East Atlantic Based on Mark - Recapture Data. J Mar Biol Assoc UK 85: 1107–1112.
56. Nakano H, Nagasawa K (1996) Distribution of Pelagic Elasmobranchs Caught by Salmon Research Gillnets in the North Pacific. Fish Sci 62(5): 860–865.
57. Goldman KJ, Musick JA (2008) The biology and ecology of the salmon shark, Lamna ditropis. In: Camhi M, Pikitch E, Babcock E (Eds), Sharks of the Open Ocean: Biology, Fisheries and Conservation. Blackwell Publishing, Oxford, UK, pp. 95–104.
58. Matsunaga H, Yokawa K (2013) Distribution and ecology of bigeye thresher Alopias superciliosus in the Pacific Ocean. Fish Sci 79:737–748.
59. Bonfil R, Clarke S, Nakano H (2008) The biology and ecology of the Oceanic whitetip shark, Carcharhhinus longimanus. In: Camhi M, Pikitch E, Babcock E (Eds), Sharks of the Open Ocean: Biology, Fisheries and Conservation. Blackwell Publishing, Oxford, UK, pp.128–139.
60. Snelson FF, Roman BL, Burgess GH (2008) The reproductive biology of pelagic elasmobranchs. In: Camhi M, Pikitch E, Babcock E (Eds), Sharks of the Open Ocean: Biology, Fisheries and Conservation. Blackwell Publishing, Oxford, UK, pp. 24–45.
61. Musick JA, Limpus CJ (1997) Habitat utilization and migration in juvenile sea turtles. In: Lutz PL, Musick JA (Eds) The biology of sea turtles. CRC Press, Boca Raton, Florida, USA, pp. 137–164.
62. Bolten AB (2003) Variation in sea turtle life history patterns: neritic vs. oceanic developmental stages. In: Lutz PL, Musick JA, Wyneken J (Eds) The biology of sea turtles, Vol II. CRC Press, Boca Raton, Florida, USA, pp. 243–257.
63. Morato T, Varkey DA, Damaso C, Machete M, Santos M, et al. (2008) Evidence of a seamount effect on aggregating visitors. Mar Ecol Prog Ser 357: 23–32.
64. Mailly T, Blayo E, Verron J (1997) Assessment of the ocean circulation in the Azores region as predicted by a numerical model assimilating altimeter data from Topex/Poseidon and ERS-1 satellites. Ann. Geophysicae 15: 1354–1368.
65. Silva MA, Prieto R, Jonsen I, Baumgartner MF, Santos RS (2013) North Atlantic Blue and Fin Whales Suspend Their Spring Migration to Forage in Middle Latitudes: Building up Energy Reserves for the Journey? PLoS ONE 8: e76507.
66. Bolten AB (2003) Active swimmers passive drifters: the oceanic juvenile stage of loggerheads in the Atlantic system. In: Bolten AB and Witherington BE (Eds). Loggerhead Sea Turtles. Smithsonian Institution Press, Washington, DC. p.63–78.
67. Santos MA, Bolton AB, Martins HR, Riewald B, Bjorndal KA (2007) Air-breathing visitors to seamounts: sea turtles. In: Pitcher TJ, Morato T, Hart PJB, Clarke MR, Haggan N, Santos RS (Eds), Seamounts: Ecology, Fisheries and Conservation. Blackwell Publishing, Oxford, UK, pp. 239–244.
68. Skomal GB, Natanson LJ (2003) Age and Growth of the Blue Shark (*Prionace glauca*) in the North Atlantic Ocean. Fish Bull 101: 627–639.
69. Hazin FHV, Couto AA, Kihara K, Otsuka K, Ishino M, et al. (1994) Reproduction of the blue shark *Prionace glauca* in the south-western equatorial Atlantic Ocean. Fisheries Sci 60: 487–491.
70. Litvinov FF (2007) Aggregations of large pelagic sharks above seamounts. In: Pitcher TJ, Morato T, Hart PJB, Clarke MR, Haggan N, Santos RS (Eds), Seamounts: Ecology, Fisheries and Conservation. Blackwell Publishing, Oxford, UK, pp. 202–206.
71. Mejuto J, García-Cortés B, Ramos-Cartelle A (2005) Tagging-recapture activities of large pelagic sharks carried out by Spain or in collaboration with the tagging programs of other countries. ICCAT Coll Vol Sci Pap 58: 974–1000.
72. Jorgensen SJ, Reeb CA, Chapple TK, Anderson S, Perle C, et al. (2009) Philopatry and migration of Pacific white sharks. Proc R Soc B 277: 679–688
73. Barnett A, Abrantes KG, Stevens JD, Semmens JM (2011) Site fidelity and sex-specific migration in a mobile apex predator: implications for conservation and ecosystem dynamics. Anim Behav 81: 1039–1048.
74. Walli A, Teo SLH, Boustany A, Farwell CJ, Williams T, et al. (2009) Seasonal movements, aggregations and diving behavior of Atlantic bluefin tuna (Thunnus thynnus) revealed with archival tags. PloS One 4: e6151.
75. Foote AD, Simila T, Vikingsson GA, Stevick PT (2010) Movement, site fidelity and connectivity in a top marine predator, the killer whale. Evol Ecol 24: 803–814.
76. Martin TG, Chades I, Arcese P, Marra PP, Possingham HP, et al. (2007) Optimal conservation of migratory species. PloS One 2: e751.
77. Afonso P, Fontes J, Holland KN, Santos RS (2009) Multi-scale patterns of habitat use in a highly mobile reef fish, the white trevally *Pseudocaranx dentex*, and their implications for marine reserve design. Mar Ecol Prog Ser 381: 273–286.
78. Hueter RE, Heupel MR, Heist EJ, Keeney DB (2005) Evidence of Philopatry in Sharks and Implications for the Management of Shark Fisheries. J Northw Atl Fish Sci 35: 239–247.
79. Aires-da-Silva A, Maunder NM, Gallucci VF, Kohler NE, Hoey JJ (2009) A spatially structured tagging model to estimate movement and fishing mortality rates for the blue shark (*Prionace glauca*) in the North Atlantic Ocean. Mar Fresh Res 60: 1029–1043.

22

White-Nose Syndrome Fungus: A Generalist Pathogen of Hibernating Bats

Jan Zukal[1,2], Hana Bandouchova[3], Tomas Bartonicka[2], Hana Berkova[1], Virgil Brack[4], Jiri Brichta[3], Matej Dolinay[2], Kamil S. Jaron[5], Veronika Kovacova[3], Miroslav Kovarik[6], Natália Martínková[1,5], Karel Ondracek[3], Zdenek Rehak[2], Gregory G. Turner[7], Jiri Pikula[3]*

1 Institute of Vertebrate Biology, Academy of Sciences of the Czech Republic, Brno, Czech Republic, **2** Department of Botany and Zoology, Masaryk University, Brno, Czech Republic, **3** Department of Ecology and Diseases of Game, Fish and Bees, University of Veterinary and Pharmaceutical Sciences, Brno, Czech Republic, **4** Environmental Solutions & Innovations Inc., Cincinnati, Ohio, United States of America, **5** Institute of Biostatistics and Analysis, Masaryk University, Brno, Czech Republic, **6** Administration of the Moravian Karst Protected Landscape Area, Nature Conservation Agency of the Czech Republic, Blansko, Czech Republic, **7** Pennsylvania Game Commission, Harrisburg, Pennsylvania, United States of America

Abstract

Host traits and phylogeny can determine infection risk by driving pathogen transmission and its ability to infect new hosts. Predicting such risks is critical when designing disease mitigation strategies, and especially as regards wildlife, where intensive management is often advocated or prevented by economic and/or practical reasons. We investigated *Pseudogymnoascus [Geomyces] destructans* infection, the cause of white-nose syndrome (WNS), in relation to chiropteran ecology, behaviour and phylogenetics. While this fungus has caused devastating declines in North American bat populations, there have been no apparent population changes attributable to the disease in Europe. We screened 276 bats of 15 species from hibernacula in the Czech Republic over 2012 and 2013, and provided histopathological evidence for 11 European species positive for WNS. With the exception of *Myotis myotis*, the other ten species are all new reports for WNS in Europe. Of these, *M. emarginatus, Eptesicus nilssonii, Rhinolophus hipposideros, Barbastella barbastellus* and *Plecotus auritus* are new to the list of *P. destructans*-infected bat species. While the infected species are all statistically phylogenetically related, WNS affects bats from two suborders. These are ecologically diverse and adopt a wide range of hibernating strategies. Occurrence of WNS in distantly related bat species with diverse ecology suggests that the pathogen may be a generalist and that all bats hibernating within the distribution range of *P. destructans* may be at risk of infection.

Editor: Justin G. Boyles, Southern Illinois University, United States of America

Funding: This study was supported by the Grant Agency of the Czech Republic (Project No. P506/12/1064) and by the National Speleological Society of the USA. The funders had no role in study design, data collection and analysis, decision to publish, or preparation of the manuscript.

Competing Interests: V.B. is currently affiliated with a commercial company Environmental Solutions & Innovations Inc. He provided expertise in bat ecology in this paper. This does not alter the authors' adherence to PLOS ONE policies on sharing data and materials.

* E-mail: pikulaj@vfu.cz

Introduction

Host-pathogen dynamics represent a balance between the pathogen's ability to infect and the host's ability to resist, with an intensive arms race between the two reflected in co-evolutionary adaptations. With host switching, pathogens temporarily escape the arms race. New, naive hosts may show lower resistance and other characteristics favourable to the pathogen. Overlapping distribution of a pathogen and its potential host(s) is key to host switching driven by opportunity [1]. The spread of emerging wildlife pathogens may have economic consequences, even in species indirectly linked to humans [2]. Fungal infections in amphibians and bats that result in population declines [3], for example, can lead to increased agricultural costs where humans chemically compensate for ecosystem services provided by these organisms in terms of insect control.

White-nose syndrome (WNS) is an emerging disease of hibernating bats associated with skin infection by *Pseudogymnoascus [Geomyces] destructans*, a recently recognised fungal pathogen [4–7]. Severe skin damage results in disruption of torpor pattern, premature depletion of fat reserves and mortality in affected bats in North America [8]. High mortality rates at affected localities and the rapid spread of the infection since 2006 continues to threaten bat diversity [9–11].

While WNS has characteristics of an epizootic, gradually expanding through North American hibernacula from its original detection site [12,13], *P. destructans* is pan-European in distribution [14,15]. Aside from seasonality in the appearance of white fungal growth [15], detailed spatio-temporal data for *P. destructans* infection in Europe are lacking. Fortunately, mass mortality has not been observed in European bats to date [14].

WNS can be transmitted either directly through bat-to-bat contact or indirectly through contact with pathogen propagules in the environment [6,16] and the infection's spread is assumed to be both density- and frequency-dependent [10]. Multiple factors, such as hibernation in large assemblages or length of hibernation season, play a role in the epidemiology of this fungal disease and ecological and behavioural characteristics of bat species may affect the risk of infection [3]. Traits such as selection of hibernaculum roost sites with differing microclimatic conditions and solitary versus gregarious hibernation behaviour may also influence the impact of WNS [10]. Other risk factors associated with hibernating bat mortality in North America include distance from the first WNS-affected site, cluster size, species diversity and composition and type of hibernaculum [13].

Prior to 2012, bat species positive for *P. destructans* in North America included *Myotis austroriparius, M. grisescens, M. leibii, M. lucifugus, M. septentrionalis, M. sodalis, M. velifer, Perimyotis subflavus* and *Eptesicus fuscus* [11,16,17]. Dermatohistopathology has revealed fungal infection with cupping erosions and skin invasion diagnostic for WNS in a *M. myotis* specimen hibernating in the Moravian Karst, Czech Republic [18] and eight species (*M. myotis, M. oxygnathus/M. blythii, M. brandtii, M. daubentonii, M. dasycneme, M. mystacinus, M. nattereri* and *M. bechsteinii*) have been reported positive for the WNS fungus in Europe based on direct microscopy of characteristic *P. destructans* conidia, fungal culture and genetic analysis [14,15,19–22]. Photographic evidence of fungal growth suggests that *M. emarginatus, E. nilssonii* and *Rhinolophus hipposideros* may also prove positive for *P. destructans* [14].

In summary, a total of 17 vespertilionid bat species had been reported positive for the WNS fungus in North America and Europe prior to 2012 and, as the epizootic spreads through North America and surveillance continues in Europe, it is expected that the number of infected species will increase. Hereinafter, the term *P. destructans*-infected or -positive relates to those species for which the fungal pathogen has been confirmed by laboratory methods such as fungal culture and genetic analysis. WNS-positive represents those species where the infection has been diagnosed through characteristic histopathological lesions, such as fungal hyphae densely packed in so-called cupping erosions and/or invasion of the dermis [23].

Little is known about *P. destructans* infection in European bat species less abundant or less commonly observed in hibernacula. Knowledge of pathological effects associated with the WNS fungus in European bat species is even poorer. While it is a commonly held view that European bats are more resistant or resilient than those in North America [17], our monitoring revealed three further species positive for *P. destructans* infection in 2012. Differences in their hibernation behaviour and taxonomy inspired us to examine the ecological and behavioural traits and phylogeny of European and North American species reported positive for *P. destructans* in order to identify any similarities in behaviour and habitat use and to identify any other species that may be at risk.

First, we examined the hypothesis that more bat species are positive for WNS in Europe than currently reported via histopathology, considered as the 'gold standard' for diagnosing WNS [23,24]. Second, we hypothesised that ecology, behaviour and phylogenetic relationships of hibernating bat species influence risk of infection by *P. destructans*. Aside from ecological similarities, those species most often found positive for *P. destructans* and WNS belong to the genus *Myotis*, indicating that phylogenetic relatedness of hosts may facilitate invasion by the fungus. To test this, we compared the ecological and behavioural traits of hibernating bats from Europe and North America. We grouped species with similar behaviour and habitat use and used confirmed positive species to propose possible additional species susceptible to infection. We constructed a phylogeny of vespertilionids and rhinolophids from Europe and North America, to test the hypothesis that infected bats are phylogenetically closely related. Finally, we screened species of unknown infection status in Czech hibernacula to test the validity of our models and predictions.

Here, we provide histopathological evidence of multiple European species positive for WNS. We found that infected bat species are ecologically diverse, utilising a range of hibernating and feeding strategies. Although bat species previously described as being *P. destructans*-positive have been phylogenetically related, the pattern begins to break down with the newly diagnosed taxa; the data presented herein demonstrating that the host range for this fungal pathogen is more diverse than previously realized.

Materials and Methods

Ethics statement

The Czech Academy of Sciences' Ethics Committee has reviewed and approved Animal Use Protocol No. 169/2011 in compliance with Law No. 312/2008 on Protection of Animals against Cruelty, as adopted by the Parliament of the Czech Republic. Bats were monitored for WNS and presence of the causative agent *P. destructans* in the spring of 2012 and 2013 in caves of the Moravian Karst, mines near Mala Moravka in the Jeseniky Mountains, and in the Podyji National Park, all in the Czech Republic. Non-lethal sampling was in compliance with Law No. 114/1992 on Nature and Landscape Protection and was based on permits 01662/MK/2012S/00775/MK/2012, 866/JS/2012 and 00356/KK/2008/AOPK issued by the Nature Conservation Agency of the Czech Republic. Bats were handled so as to minimise stress and duration of sampling procedures between capture and release. Numbered aluminium rings were attached around the forearm for long-term identification prior to release at the site.

Screening bat species in Czech hibernacula for *P. destructans* infection

When screening bats for WNS and *P. destructans* we 1. captured bats emerging from hibernacula at the end of the hibernation season using mist nets, 2. swabbed the wing membrane for fungal culture using the Fungi-Quick transport system (Copan Innovation, Italy), 3. briefly illuminated the bats with a flashlight to detect any visible fungal growth, 4. trans-illuminated the wing membrane using ultraviolet light (UV; wavelength 368 nm) to detect any WNS lesions, 5. photographed wing membranes of each bat under both visible and UV light, 6. took a wing punch biopsy from all WNS-suspected skin lesions (i.e. areas of orange-yellow fluorescence) using a sterile and disposable 4 mm skin biopsy punch (Kruuse, Denmark), 7. used polymerase chain reaction (PCR) to confirm *P. destructans* from fungal cultures or skin swabs using the FLOQSwabs system (CopanFlock Technologies, Italy), and 8. undertook complete histopathological examinations of skin samples.

A total of 276 bats were screened for WNS and *P. destructans* and 123 skin biopsies were taken for histological examination from 15 bat species (Table 1).

Formalin-fixed punch biopsy samples were embedded in paraffin and serial 5 μm tissue sections were prepared and stained for fungi with periodic acid–Schiff stain. Histopathological findings were classified as WNS based on previously described diagnostic criteria [23]. Samples collected to cultivate fungi were transferred onto Petri dishes containing Sabouraud agar, sealed with parafilm, inverted and incubated in the dark at 10 °C. Pure fungal cultures were established from fungal growth developing at 14 days or later. *Pseudogymnoascus destructans* was confirmed through characteristic asymmetrically curved conidia via direct microscopy [5]. Fungal isolates or skin swabs were further identified using PCR and follow-up sequencing of amplicons [25] and real-time PCR [24]. A pure culture of *P. destructans* isolate grown at 10 °C on Sabouraud agar and genetically confirmed (EMBL-Bank accession number: HE588133; [18]) served as a PCR control.

Table 1. Bats examined for white-nose syndrome and *Pseudogymnoascus destructans* infection in Czech hibernacula (Europe).

Species	Screened	Biopsied	Histo+	WNS prevalence (%)	St. error
Myotis myotis[a]	67	56	37	55.22	6.08
Myotis daubentonii[b]	25	13	4	16.00	5.76
Myotis bechsteinii[b]	21	7	2	9.52	6.78
Myotis nattereri[b]	20	8	3	15.00	7.10
Myotis brandtii[b]	17	1	1	5.88	8.23
Myotis alcathoe[b]	8	7	0	0	15.94
Myotis emarginatus[b]	39	7	5	12.82	5.35
Rhinolophus hipposideros[b]	28	5	1	3.57	5.18
Eptesicus nilssonii[b]	4	1	1	25.00	26.89
Plecotus auritus[c]	23	11	5	21.73	8.60
Barbastella barbastellus[c]	17	3	3	17.64	8.24
Plecotus austriacus	3	1	0	0	32.22
Eptesicus serotinus	2	1	0	0	39.61
Pipistrellus pipistrellus	1	1	0	0	48.47
Myotis dasycneme[b]	1	1	1	100	48.47
Total	**276**	**123**	**63**	**22.82**	**2.53**

[a]= species reported positive for WNS fungus prior to 2012, [b] = species recognised as positive in 2012, [c] = bat species recognised as positive in 2013. Screened = numbers of bats captured and examined using UV light trans-illumination to detect WNS lesions, biopsied = numbers of bats biopsied due to WNS-suspected skin lesions viewed under UV light, histo+ = specimens positive for WNS diagnostic features under histopathological examination (i.e. cupping erosions and fungal invasion of dermis), WNS prevalence = percentage of bats positive on histopathology out of the total number screened.

Analysis of European and North American bat ecological traits

The list of European and North American bat species was prepared according to Simmons [26]; those species included in the study being those with complete or partial distribution in continental Europe or North America for which data are available. In total, we reviewed ecological and behavioural variables for 87 species. Of these, 47 were assessed for all variables and were subjected to ecological modelling analysis.

Eleven traits were chosen to describe bat species: 1. Infection status (*P. destructans*-positive/*P. destructans*-negative), 2. Cave or non-cave hibernation, 3. Region (Palearctic/Nearctic distribution), 4. Clustering during hibernation (clustering/non-clustering; i.e. hibernating in groups where multiple individuals touch), 5. Temperature preference (thermophilic/cryophilic; mainly according to Webb *et al.* [27]), 6. Preferred roost type during hibernation (exposed/hidden/both), 7. Size of clusters during hibernation (no clustering/small clusters < 50 bats/medium clusters of 51 to 500 bats/large clusters > 500 bats), 8. Distribution range (very large area/large area of approximately half a continent/moderate size/small area/very small area; according to Horáček *et al.* [28]), 9. Food (dominant insect food group represented by Diptera/Lepidoptera/Coleoptera/generalists), 10. Foraging habitat (open/edge/closed), 11. Body size (small - up to 5 g/medium - 5 to10 g/large - over 10 g).

These traits, which were assessed based on a primary literature review and expert evaluation, were chosen as those most likely to influence susceptibility of bats to WNS, the spread of *P. destructans* infection or survival rate during hibernation [10]. As the disease can inflict long-term damage to affected bats surviving the hibernation season, other medically relevant factors may also influence survival and reproduction in the active season [29]. Categorical variables were coded as $n - 1$ binary, dummy variables, where n is the number of categories. Data for the traits of each bat species are provided in Table S1. Grouping of ecologically similar species was performed via neighbour-joining clustering of squared Euclidean distances using ape in R language [30,31].

Phylogenetic reconstruction of European and North American bats

The phylogeny of bats from Europe and North America was extracted from a maximum likelihood phylogenetic tree using Phylocom version 4.2 [32]. The complete phylogeny of the Vespertilionidae, Miniopteridae and Cistugidae families (from which the tree used here was pruned) was reconstructed from a concatenated DNA sequence matrix of 13 mitochondrial and nuclear genes with 64% of missing data (Table S2). Three *Rhinolophus* species were used as an outgroup. Phylogeny was reconstructed using the partitioning scheme suggested by the greedy algorithm, utilising the Bayesian Information Criterion assessment in PartitionFinder [33]. The tree space was searched using maximum likelihood analysis with automatic majority-rule bootstopping option [34]. By extracting the target species' phylogeny from a comprehensive tree, we were able to obtain a phylogeny that exploits currently available diversity to optimise relationships and branch lengths, and thus mitigate possible analysis artefacts.

Statistical analysis and hypothesis testing of *P. destructans* infection occurrence

We explored the distribution pattern of *P. destructans* infection on a tree based on bat trait variation and molecular phylogeny. Occurrence of *P. destructans* infection represents a presence/absence variable, rather than a continuous trait, meaning that it is suitable for community structure analysis. The phylogenetic signal for explanatory variables and for *P. destructans* infection was calculated in Phylocom using the comstruct function. In order to assess relatedness of species that share a specific characteristic, mean phylogenetic distance (MPD) and mean nearest phylogenetic taxon distance (MNTD) were compared to the null model, which assumes random dispersal of the trait on the tree. MPD measures the mean branch length between two randomly selected taxa from a sample, and is calculated as the sum of branch lengths to the node representing their most recent common ancestor. MNTD is the mean branch length between a taxon within the sample and its nearest relative. The null model randomised samples across phylogeny in 9,999 replicates. The distribution of the trait on a tree is clustered if values of MPD and MNTD obtained are higher than 95% of values obtained from the null samples standardised by the standard deviation of the null samples. The comparison is expressed as net relatedness index (NRI) and nearest taxon index (NTI) greater than zero [32]. Clustered distribution of a trait or phylogenetic signal means that species that share the trait are more closely related to one another than to a random taxon sampled from the tree.

Species' ecological and behavioural characteristics often show a heritable component such that close relatives have similar traits [35]. Such characteristics might then be adaptive and their evolution further decoupled from the assumption of sample independence needed for general statistical approaches. The evolutionary relationships of traits in our dataset were removed from comparisons by using phylogenetic generalised least squares (PGLS) in the Caper package of R [36]. We used a variance-covariance matrix calculated from the phylogeny with branch lengths transformed according to the Ornstein-Uhlenbeck model in geiger [37]. The PGLS model was developed via a step-down procedure, using the Akaike Information Criterion (AIC) to compare alternative models.

Results

Screening species with unknown infection status in Czech hibernacula

We tested a broad diversity of European hibernating bats for *P. destructans* infection and skin lesions pathognomonic for WNS. Analysis of 123 skin biopsy samples collected in 2012 and 2013 revealed histopathological findings matching criteria used for diagnosis of WNS in 63 bats (22.82% prevalence; Table 1) of 11 species, i.e. *M. myotis*, *M. daubentonii*, *M. bechsteinii*, *M. nattereri*, *M. brandtii*, *M. emarginatus*, *M. dasycneme*, *E. nilssonii*, *R. hipposideros*, *B. barbastellus* and *P. auritus* (Figure 1). With the exception of *M. myotis*, the other ten species are all new reports of WNS in Europe. Of these, *M. emarginatus*, *E. nilssonii*, *R. hipposideros*, *B. barbastellus* and *P. auritus* are new to the list of *P. destructans*-infected bat species. Fungus isolates or skin swabs from histopathologically positive bats were identified as *P. destructans* using PCR.

Risk of *P. destructans* infection in bat species of unknown infection status

Testing the hypothesis of phylogenetic relatedness. *P. destructans*-infected species were clustered together by molecular phylogeny (MPD = 0.212, NRI = 2.913, $p < 0.001$), meaning that pairs of infected species were, on average, more closely related than random species pairs from Europe and North America. When sister species or nearest relatives were considered, however, our results indicated that infection of both, either or neither was random (MNTD = 0.111, NTI = 1.556, $p = 0.06$; Table 2, Figure 2). Nine explanatory variables showed NRI and/or NTI not equal to zero, indicating that phylogenetic comparative methods were needed due to relatedness of taxa with shared traits (Table 2).

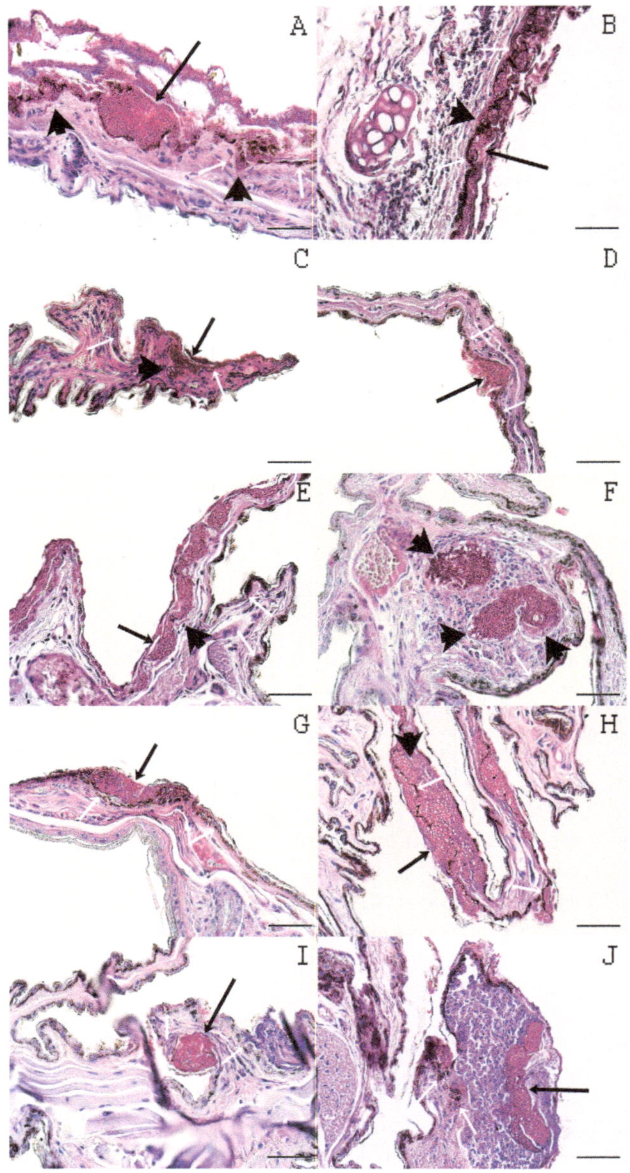

Figure 1. Histopathological skin lesions consistent with white-nose syndrome in ten European bat species. (A) *Myotis emarginatus*, (B) *Eptesicus nilssonii*, (C) *Rhinolophus hipposideros*, (D) *Plecotus auritus*, (E) *Barbastella barbastellus*, (F) *M. dasycneme*, (G) *M. nattereri*, (H) *M. daubentonii*, (I) *M. bechsteinii*, (J) *M. brandtii*. The photographs illustrate i) extensive infection of the wing membrane and cup-shaped epidermal erosions (A, E, H, J; long black arrow); ii) cup-like epidermal erosions in the pinna (B; long black arrow), iii) *Pseudogymnoascus destructans* hyphae obscuring the basement membrane and invading the dermis (A, B, C, E, H; black arrow); iv) a single cupping erosion packed with fungal hyphae in the wing membrane (C, D, G, I; long black arrow); v) colonisation of a hair follicle by *P. destructans*, fungal hyphae present in the associated sebaceous gland and regional connective tissue (F; black arrow); vi) marked signs of inflammation (B, F, J); and vii) a cellular inflammatory crust that sequesters fungal hyphae (A, J). White arrows within each photograph indicate the interface between epidermis and dermis. Periodic acid-Schiff stain; scale bar = 50 μm. *M. myotis* not shown because WNS lesions in this species have already been documented elsewhere [18].

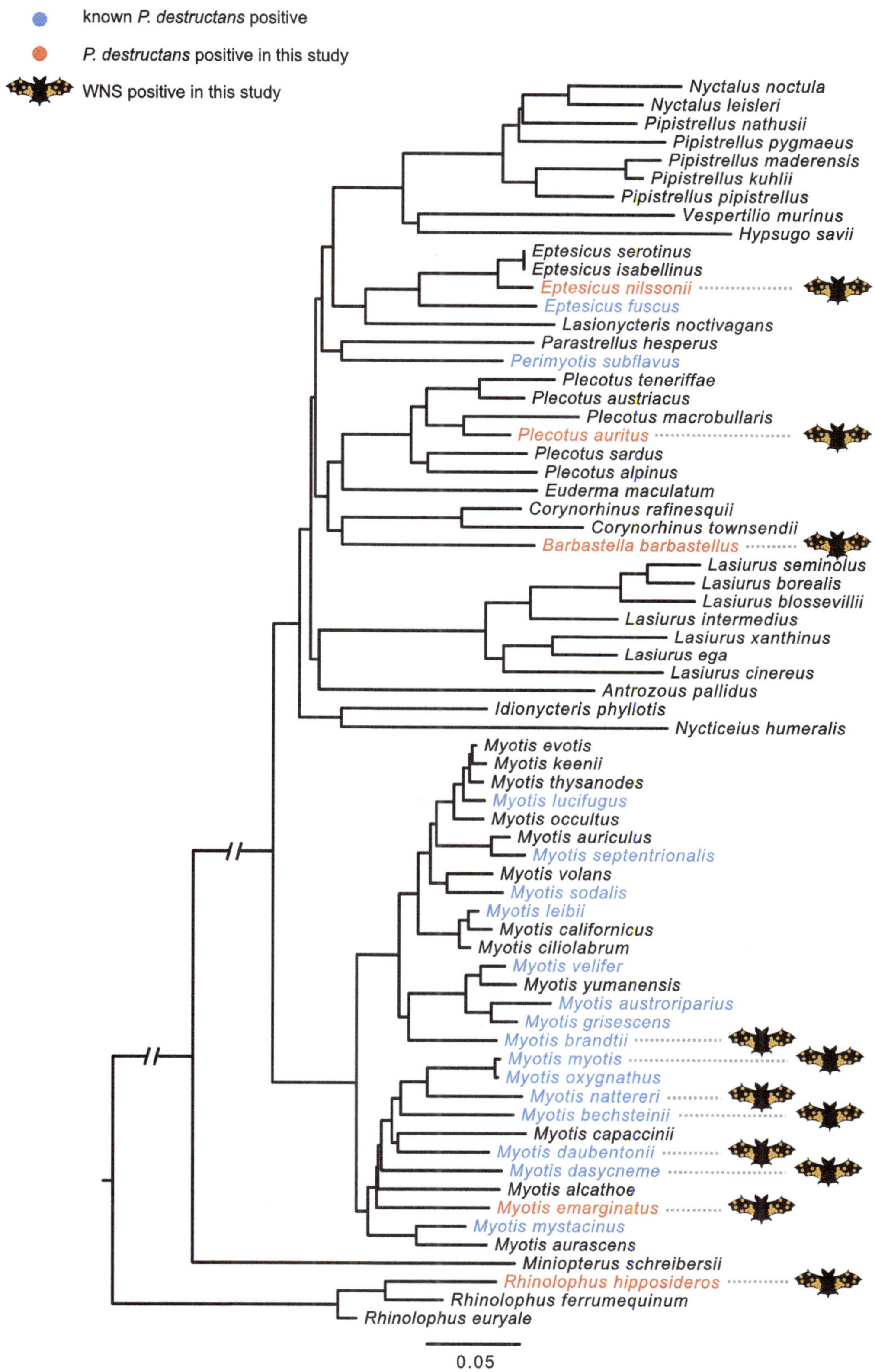

Figure 2. Phylogenetic reconstruction of bats from Europe and North America. The reconstruction was based on a concatenated DNA sequence matrix from 13 loci, purged from a maximum likelihood vespertilionid phylogeny rooted on *Rhinolophus*. Blue = species reported positive for WNS fungus prior to 2012, red = species recognised positive in this study, bat image = bat species diagnosed as WNS positive in this study.

Table 2. *Pseudogymnoascus destructans* infection in relation to chiropteran phylogeny and ecological similarity.

Variable	*N*	MPD	NRI	p_{NRI}	MNTD	NTI	p_{NTI}
Explanatory							
CAVE	42	0.270	1.300	0.105	**0.091**	**1.984**	**0.025**
REGION	23	0.332	−1.606	0.944	0.114	1.125	0.135
CLUSTER	27	**0.241**	**2.151**	**0.012**	0.109	1.180	0.124
TEMPERATURE	21	0.318	−1.020	0.841	0.128	0.569	0.292
SHELTERhidden	31	**0.246**	**2.147**	**0.013**	0.121	0.115	0.450
SHELTERexposed	21	0.316	−0.929	0.814	0.119	1.004	0.165
SHELTERboth[a]	5	0.224	1.051	0.106	0.170	0.754	0.229
CSIZEno	20	0.337	−1.639	0.947	0.125	0.736	0.240
CSIZEsmall	7	**0.126**	**3.025**	**0.001**	**0.082**	**2.474**	**0.004**
CSIZEmedium	9	0.280	0.201	0.474	0.162	0.404	0.358
CSIZElarge	11	0.265	0.592	0.306	0.166	0.064	0.488
RANGEverylarge	15	0.333	−1.217	0.875	0.188	−1.328	0.902
RANGElarge	13	0.311	−0.565	0.708	**0.213**	**−1.872**	**0.970**
RANGEmoderate	14	0.246	1.189	0.112	0.133	0.859	0.203
RANGEsmall	5	**0.098**	**2.865**	**0.001**	**0.059**	**2.675**	**0.001**
FOODcolleoptera	5	0.258	0.482	0.330	0.169	0.706	0.249
FOODdiptera	9	0.237	1.093	0.117	0.144	0.902	0.188
FOODgeneralist	18	0.313	−0.763	0.770	0.128	0.772	0.232
FOODlepidoptera	14	0.286	0.117	0.475	0.147	0.355	0.368
FOODother	1	n/a					
HABITATclosed	11	0.280	0.228	0.458	0.142	0.802	0.220
HABITATopen	13	0.314	−0.623	0.724	0.178	−0.652	0.739
HABITATedge	23	0.267	0.855	0.211	**0.093**	**2.321**	**0.007**
BODYsmall	10	0.292	−0.097	0.576	0.142	0.855	0.204
BODYmedium	21	**0.211**	**2.829**	**0.001**	**0.105**	**1.698**	**0.048**
BODYlarge	16	**0.368**	**−2.259**	**0.989**	0.149	0.070	0.475
Response							
Pd+ on phylogeny (Figure 2)	20	**0.212**	**2.914**	**0.001**	0.111	1.555	0.058
Pd+ on 'eco' tree (Figure 3)	20	7.839	0.862	0.201	4.356	0.277	0.384

[a] = species using both types of shelters are also included in the previous categories.

Phylogenetic signal of explanatory variables on a phylogeny of bats from Europe and North America and of *P. destructans* infection on both phylogeny and a neighbour-joining tree based on squared Euclidean distances of ecological and behavioural traits of hibernating bat species. Values in bold indicate significant clustering or over-dispersion of *P. destructans* infection on the tree. Note that all categories of explanatory variables were tested here, but n - 1 dummy variables were included in the PGLS model. *N* = number of species scored positive for the given variable, MPD = mean phylogenetic distance, NRI = net relatedness index, MNTD = mean nearest taxon phylogenetic distance, NTI = nearest taxon index.

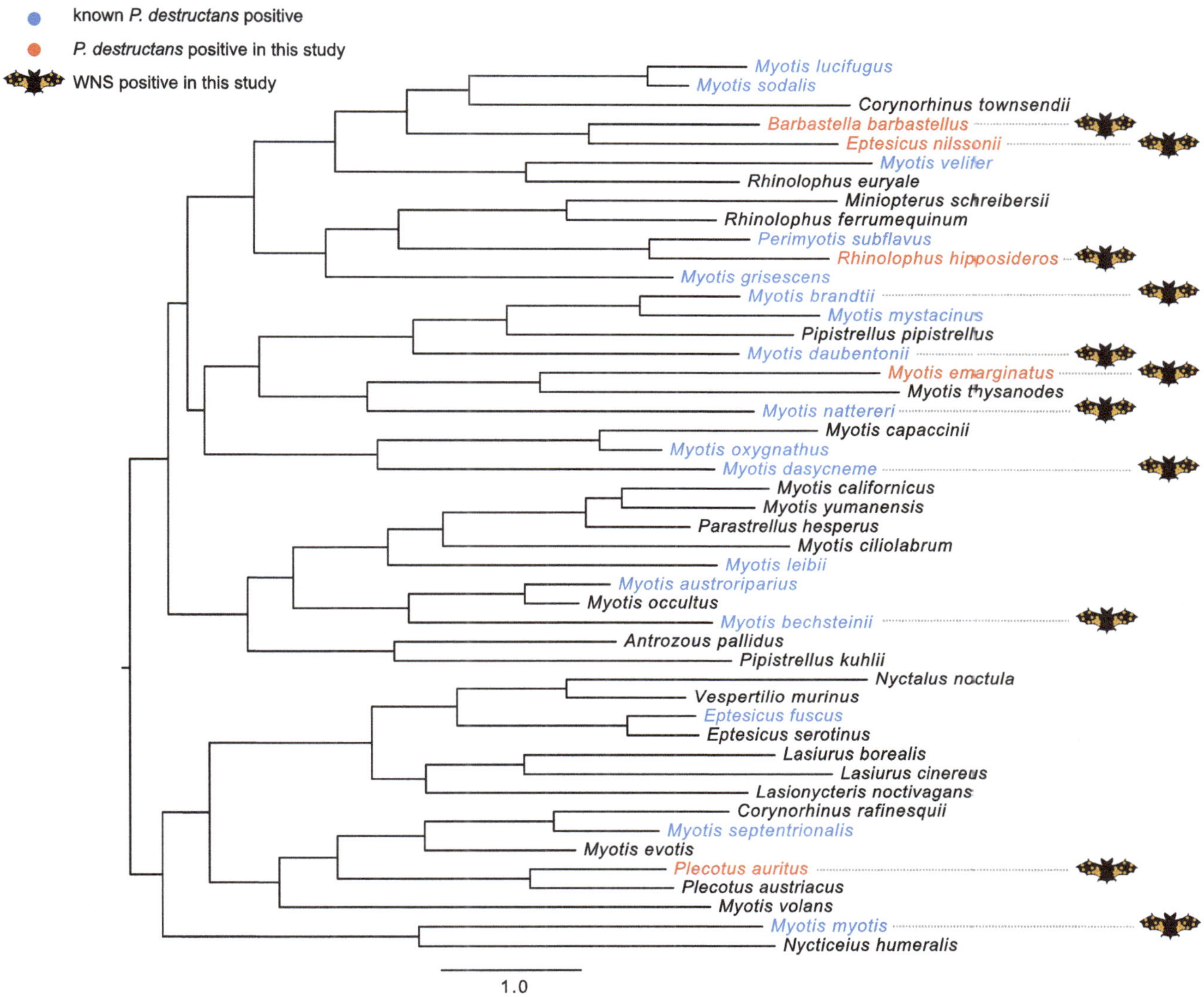

Figure 3. Neighbour-joining tree based on ecological and behavioural traits of bats from Europe and North America (rooted at midpoint). See Figure 2 for a description of the colour scheme.

Testing the hypothesis of ecological similarity. The ecological similarity tree for European and North America bats was constructed using the squared Euclidean distances of the traits dataset (Figure 3). Analysis of *P. destructans*-infected species distribution indicated that infected species were randomly distributed (MPD = 7.839, NRI = 0.862, p = 0.201) across the ecological diversity of bats from these two continents. The most ecologically similar species were also infected randomly (MNTD = 4.356, NTI = 0.277, p = 0.384; Table 2, Figure 3).

Predicting species at risk from *P. destructans* infection. We explored relationships between ecological traits after removing the effects of bat species relatedness using PGLS. The final model, displaying lowest AIC (AIC = 69.08, F-statistic = 8.98, df = 7 and 39, $p < 0.001$, adjusted R^2 = 0.55), differed from two more complex models by ΔAIC < 3 (Table S3). The addition of the variables did not markedly alter results of the analysis as reported below, and therefore we used the model with the lowest AIC. It describes the relationship between *P. destructans* infection in bat species and Temperature preference during hibernation (β = −0.207, SE = 0.085); Roost Shelter during hibernation: Hidden (β = −0.109, SE = 0.211), Exposed (β = 0.560, SE = 0.192); Cluster Size during hibernation: Small (β = −0.386, SE = 0.130), Medium (β = −0.309, SE = 0.194); Distribution range size: Moderate (β = −0.201, SE = 0.089); and Feeding habitat: Closed (β = 0.319, SE = 0.113). Shapiro-Wilks' normality test indicated that model residuals were normally distributed (W = 0.981, p = 0.63).

The fitted values of *P. destructans* infection in bats based on the PGLS model showed overlap between the *P. destructans*-positive and -negative bats (Figure 4). Bat species currently recognised as *P. destructans*-negative with highest fitted PGLS values were (in descending order): *Corynorhinus townsendii*, *Lasiurus cinereus*, *Plecotus austriacus*, *Rhinolophus ferrumequinum*, and *Miniopterus schreibersii*.

Discussion

At the end of the hibernation seasons of 2012 and 2013, we screened 276 bats of unknown infection status and biopsied all bats with WNS-suspected skin lesions. Both the number of bats and the number of taxa (n = 15) examined make this the most extensive

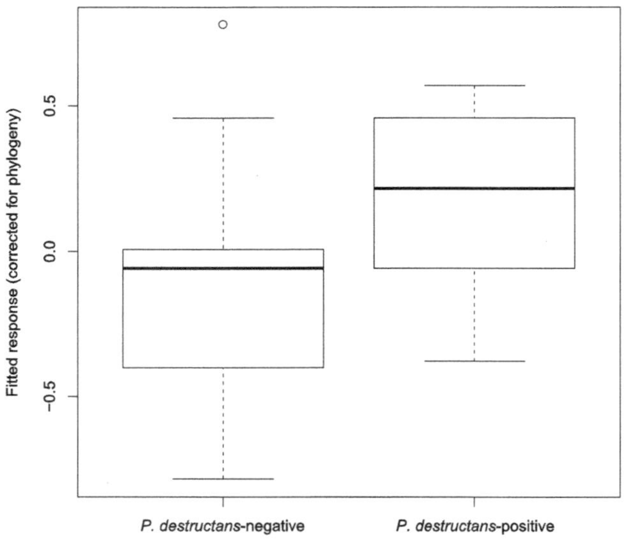

Figure 4. Boxplot of fitted values from the phylogenetic generalised least squares model of *P. destructans* infection. Predictions for infected and non-infected species overlap.

and species-rich study of *P. destructans* infection in Europe to date. Earlier European studies have provided data from bats originally sampled for fungal microscopy, culture and genetic analysis when they exhibited obvious fungal growth during hibernation, numbers of species examined ranging from 1 to 12 and numbers of specimens from 1 to 107 [14,15,19–22].

This study documented five additional bat species as positive for *P. destructans* infection and added ten species from the genera *Myotis*, *Eptesicus*, *Plecotus*, *Barbastella* and *Rhinolophus* to the list of European bat species showing histopathological findings consistent with WNS [23]. Species-specific prevalence of WNS-diagnostic skin lesions ranged from 0 to 100% (4–55% for species with $n > 20$; Table 1). Highest prevalence of WNS lesions was observed in *M. myotis* (after excluding species for which just one specimen was caught, i.e. *M. dasycneme*). Note, however, that with prevalence differing by an order of magnitude, detection of positive specimens may have been biased by small sample size in rarer species and species less frequently visible in hibernacula. With our experimental design (i.e. trapping of winter survivors leaving hibernacula and non-destructive UV fluorescence screening), we were able to confirm that *M. myotis* had highest prevalence of WNS lesions (based on histopathology), with the possible exception of *M. dasycneme*. Our results confirmed WNS skin lesions in 11 bat species, which is in contrast to a previous study that found no invasive growth of *P. destructans* in European bats [38]. The latter study, however, examined a low number of bats and used skin biopsy methods lacking in sensitivity and this may explain the failure to detect lesions diagnostic of WNS.

Bat species with *P. destructans* infection exhibited diverse hibernation behaviours. Among these, *R. hipposideros* was special as it is an exclusively solitary hibernator and hangs free in exposed places with the wing membranes covering the body. It is capable of hibernation at higher temperatures but requires the microclimate stability ensured by using the inner parts of caves [39]. *Rhinolophus hipposideros* is also the most abundant species in winter-monitoring counts, amounting to more than 50% of all bats registered in Czech hibernacula (unpublished data, Czech Bat Conservation Trust). Moreover, as a member of the suborder Pteropodiformes, it is phylogenetically distantly related to all other species infected by *P. destructans* (Figure 2). As documented by the low infection prevalence (3.57%), environmentally mediated indirect and density-dependent transmission probably does not result in higher risk for this rhinolophid species [10]. Similarly, *E. fuscus* and *M. leibii*, both solitary hibernators from North America, were the least impacted species. In comparison, higher declines were observed in large winter colonies of two species that roost solitarily or in small groups, i.e. *P. subflavus* and *M. septentrionalis* [10]. Disease risk, however, was not related to conspecific transmission only. When multiple co-occurring species can host the pathogen, density-dependent transmission can be amplified [40,41].

We confirm here that bat species previously known to be positive for *P. destructans* may later show as WNS-positive based on histopathology. As the lists of bats positive for *P. destructans* or WNS lesions in North American and European species are nearly equal, conclusions drawn from analysis of infection or disease risk should be similar. Traits describing ecological and behavioural characteristics of bats occurring within the known distribution of *P. destructans* indicate that species belonging to additional genera may also be found positive for the infection in the future. Our screening of Czech underground hibernacula, however, demonstrates that the initial, relatively low, number of bat species positive for *P. destructans* infection is more likely the result of sampling bias than a biological phenomenon. Currently, affected species are ecologically diverse, to the point where predictions for infected and non-infected species overlap. We therefore assume that more species will be revealed as WNS-positive with increased sensitivity of detection methods [42].

Phylogenetic representation of *P. destructans* infection indicates that closely related species are most likely to be infected. In terms of field surveys, therefore, some *Myotis* bats are universally likely to be WNS-positive and most effort should be devoted to these species. Interestingly, *M. alcathoe* was free of WNS-positive skin lesions in the present study (but note the low sample size). *Myotis* species typically form clusters during hibernation and such behaviour promotes frequency-dependent transmission of the infection, independent of population size, and may yet drive the species to extinction [9] unless they change their social behaviour, as documented in *M. lucifugus* and *M. sodalis* [10]. In light of our new data, clustering behaviour is not a descriptor common to infected species. Rather, a suite of characteristics, including cluster size, type of shelter during hibernation, temperature at hibernation, as well as size of the distribution range and feeding in closed habitats, play a role in characterising bats with *P. destructans* infection.

Based on the list of species currently known to be affected by WNS or *P. destructans*, it is clear that the fungus is neither species-, genus- nor family-specific. The multi-host occurrence of the pathogen might make the disease less predictable using ecologically- and phylogenetically-based analysis [43]; however, this is likely to change in the future as additional species are revealed as susceptible. Hibernation in contaminated caves and mines under conditions favourable for fungal growth [5,44] appears to be the main risk factor [4,12]. Distribution of *P. destructans* is also correlated with disease in hibernating bats [45]. Importantly, 25 species of insectivorous bats presently hibernate in the United States and Canada, all of which represent possible hosts of the fungal pathogen should the disease spread to their geographic range [46]. This scenario is predicted to happen in most counties with caves in the contiguous United States by the winter of 2105-2106 [47].

The reason bats in North America have been so hard-hit, with millions dying, while bats in Europe apparently cope better with the infection, has not yet been explained. Likewise, the pathogenesis of WNS still remains unclear [12]. Behavioural aberrations, physiological disruption and immunosuppression during hibernation are, however, considered key pathomechanisms [4,48,49]. On the other hand, restoration of immune responses in WNS-positive bats early in post-hibernation may result in immune-mediated destruction of infected tissues and death [50].

The fact that our samples were mostly collected from bats emerging from hibernacula at the end of the hibernation season indicates that European bat species can survive *P. destructans* infection and highlights the need for a comparison of European and North American bat population responses to this fungal disease. As all European bat species are strictly protected and any thorough pathological study of *P. destructans* infection would be controversial [18,22], implementation of non-lethal sampling methods is necessary, such as the wing membrane biopsy used in this study [42]. While a detailed comparison of histopathological findings in European and North America bats represents a valid approach to the better understanding of WNS mortality [17,48], it was outside the objectives of this ecological study. We are however, planning a comprehensive study to investigate the extent of WNS wing lesions in hibernating bats from the two continents.

Conclusions

This hypothesis-driven study explored clustering of *P. destructans* infection in relation to chiropteran ecological and behavioural trait variation and phylogeny and supported this with field data. Extension of the surveillance to a broader number of species to test the study's hypotheses identified multiple European species positive for WNS. The increased number of positive bat species resulted in random dispersion of *P. destructans* infection across trait trees and weakened the pattern of phylogenetic clustering of *P. destructans* infection. Distantly related bat species, characterised by diverse life histories, were infected and all hibernating bats may, therefore, be at risk from *P. destructans* infection. Ecological and evolutionary constraints on hibernating bats do not pose a barrier to this generalist fungal pathogen, with WNS occurring in both suborders of bats. Our findings indicate that a wider focus is needed in studying the ecology and epidemiology of this fungal disease of major conservation concern

Supporting Information

Table S1 Ecological and behavioural traits of European and North American bat species. The dataset includes the most common behavioural or trait value for each bat species.

Table S2 Accession numbers for phylogeny reconstruction. A total of 13 available mitochondrial and nuclear genes were used for phylogeny reconstruction of bat species from Europe and North America.

Table S3 Phylogenetic generalized least squares model selection. The model predicting *P. destructans* infection based on ecological and behavioural characteristics of bats was selected with the step-down procedure, where the full model is given on the first line and removed variables are listed subsequently.

Acknowledgments

The authors would like to thank Kevin Roche for his correction of the English text. Comments from Justin Boyles and two anonymous reviewers helped to improve the manuscript.

Author Contributions

Conceived and designed the experiments: JZ H. Bandouchova NM JP. Performed the experiments: H. Bandouchova JP JZ TB H. Berkova JB MD KSJ VK MK KO ZR. Analyzed the data: JZ MD NM KSJ VB GGT. Contributed reagents/materials/analysis tools: JZ H. Bandouchova NM JP GGT. Wrote the paper: JZ H. Bandouchova NM JP. Compiled the ecological and behavioural traits of bats: JZ. Reconstructed the phylogeny of European and North American bats: MD NM. Performed the statistical analysis: JZ KSJ NM. Examined the histopathological findings: H. Bandouchova JP. Collected field data: JZ H. Bandouchova TB H. Berkova JB MD KSJ VK MK KO ZR. Cultured and diagnosed the fungus: JB VK KO. Reviewed compiled data on North American bats: VB GGT. Provided a new method to detect WNS lesions: GGT.

References

1. Irwin NR, Bayerlová M, Missa O, Martínková N (2012) Complex patterns of host switching in New World arenaviruses. Mol Ecol 21: 4137–4150.
2. Boyles JG, Cryan PM, McCracken GF, Kunz TH (2011) Economic importance of bats in agriculture. Science 332: 41–42.
3. Eskew EA, Todd BD (2013) Parallels in amphibian and bat declines from pathogenic fungi. Emerg Infect Dis doi: http://dx.doi.org/10.3201/eid1903.120707.
4. Blehert DS, Hicks AC, Behr M, Meteyer CU, Berlowski-Zier BM, et al. (2009) Bat white-nose syndrome: An emerging fungal pathogen? Science 323: 227–227.
5. Gargas A, Trest MT, Christensen M, Volk TJ, Blehert DS (2009) *Geomyces destructans* sp nov associated with bat white-nose syndrome. Mycotaxon 108: 147–154.
6. Lorch JM, Meteyer CU, Behr MJ, Boyles JG, Cryan PM, et al. (2011) Experimental infection of bats with *Geomyces destructans* causes white-nose syndrome. Nature 480: 376–378.
7. Minnis AM, Lindner DL (2013) Phylogenetic evaluation of *Geomyces* and allies reveals no close relatives of *Pseudogymnoascus destructans*, comb. nov., in bat hibernacula of eastern North America. Fungal Biol 117: 638–649.
8. Warnecke L, Turner JM, Bollinger TK, Lorch JM, Misra V, et al. (2012) Inoculation of bats with European *Geomyces destructans* supports the novel pathogen hypothesis for the origin of white-nose syndrome. Proc Natl Acad Sci USA 109: 6999–7003.
9. Frick WF, Pollock JF, Hicks AC, Langwig KE, Reynolds DS, et al. (2010) An emerging disease causes regional population collapse of a common North American bat species. Science 329: 679–682.
10. Langwig KE, Frick WF, Bried JT, Hicks AC, Kunz TH (2012) Sociality, density-dependence and microclimates determine the persistence of populations suffering from a novel fungal disease, white-nose syndrome. Ecol Lett 15: 1050–1057.
11. Turner GG, Reeder DM (2009) Update of white-nose syndrome in bats, September 2009. Bat Research News 50(3): 47–53.
12. Blehert DS (2012) Fungal disease and the developing story of bat white-nose syndrome. PLoS Pathog 8: e1002779.
13. Wilder AP, Frick WF, Langwig KE, Kunz TH (2011) Risk factors associated with mortality from white-nose syndrome among hibernating bat colonies. Biol Lett 7: 950–953.
14. Martínková N, Bačkor P, Bartonička T, Blažková P, Červený J, et al. (2010) Increasing incidence of *Geomyces destructans* fungus in bats from the Czech Republic and Slovakia. PLoS One 5: e13853.
15. Puechmaille SJ, Wibbelt G, Korn V, Fuller H, Forget F, et al. (2011) Pan-European distribution of white-nose syndrome fungus (*Geomyces destructans*) not associated with mass mortality. PLoS One 6: e19167.
16. Foley J, Clifford D, Castle K, Cryan P, Ostfeld RS (2010) Investigating and managing the rapid emergence of white-nose syndrome, a novel, fatal, infectious disease of hibernating bats. Conserv Biol 25: 223–231.
17. Puechmaille SJ, Frick WF, Kunz TH, Racey PA, Voigt CC, et al. (2011) White-nose syndrome: is this emerging disease a threat to European bats? Trends Ecol Evol 26: 570–576.
18. Pikula J, Bandouchova H, Novotny L, Meteyer CU, Zukal J, et al. (2012) Histopathology confirms white-nose syndrome in bats in Europe. J Wildl Dis 48: 207–211.
19. Kubátová A, Koukol O, Nováková A (2011) *Geomyces destructans*, phenotypic features of some Czech isolates. Czech Mycol 63: 65–75.
20. Puechmaille SJ, Verdeyroux P, Fuller H, Gouilh MA, Bekaert M, et al. (2010) White-nose syndrome fungus (*Geomyces destructans*) in bat, France. Emerg Infect Dis 16: 290–293.

21. Šimonovičová A, Pangallo D, Chovanová K, Lehotská B (2011) *Geomyces destructans* associated with bat disease WNS detected in Slovakia. *Biologia* 66: 562–564.
22. Wibbelt G, Kurth A, Hellmann D, Weishaar M, Barlow A, et al. (2010) White-nose syndrome fungus (*Geomyces destructans*) in bats, Europe. Emerg Infect Dis 16: 1237–1243.
23. Meteyer CU, Buckles EL, Blehert DS, Hicks AC, Green DE, et al. (2009) Histopathologic criteria to confirm white-nose syndrome in bats. J Vet Diagn Invest 21: 411–414.
24. Muller LK, Lorch JM, Lindner DL, O'Connor M, Gargas A, et al. (2013) Bat white-nose syndrome: a real-time TaqMan polymerase chain reaction test targeting the intergenic spacer region of *Geomyces destructans*. Mycologia 105: 253–259.
25. Lorch JM, Gargas A, Meteyer CU, Berlowski-Zier BM, Green DE, et al. (2010) Rapid polymerase chain reaction diagnosis of white-nose syndrome in bats. J Vet Diagn Invest 22: 224–230.
26. Simmons NB (2005) Chiroptera. In: Wilson DE, Reeder DM, editors. Mammal species of the world: A taxonomic and geographic reference. Smithsonian series in comparative evolutionary biology, 3rd edition. Washington DC: Smithsonian Inst. Press. pp. 312–529.
27. Webb PI, Speakman JR, Racey PA (1996) How hot is a hibernaculum? A review of the temperatures at which bats hibernate. Can J Zool 74: 761–765.
28. Horáček I, Hanák V, Gaisler J (2000) Bats of Palearctic region. In: Woloszyn BW, editor. Proceedings of the VIIIth European bat research symposium: 23–27 August 1999. Krakow, Poland: Institute of Systematics and Evolution of Animals PAS. pp. 11–157.
29. Reichard JD, Kunz TH (2009) White-nose syndrome inflicts lasting injuries to the wings of little brown myotis (*Myotis lucifugus*). Acta Chiropt 11: 457–464.
30. Paradis E, Claude J, Strimmer K (2004) APE: Analyses of phylogenetics and evolution in R language. Bioinformatics 20: 289–290.
31. R Core Team: R: a language and environment for statistical computing. Vienna, Austria: R Foundation for Statistical Computing. Available: http://www.R-project.org. Accessed 2013 Oct 10.
32. Webb CO, Ackerly DD, Kembel SW (2008) Phylocom: software for the analysis of phylogenetic community structure and character evolution. Bioinformatics 24: 2098–2100.
33. Lanfear R, Calcott B, Ho SYW, Guindon S (2012) PartitionFinder: Combined selection of partitioning schemes and substitution models for phylogenetic analyses. Mol Biol Evol 29: 1695–1701.
34. Pattengale ND, Alipour M, Bininda-Emonds ORP, Moret BME, Stamatakis A (2010) How many bootstrap replicates are necessary? J Comput Biol 17: 337–354.
35. Felsenstein J (1985) Phylogenies and the comparative method. Am Nat 125: 1–15.
36. Orme D, Freckleton R, Thomas G, Petzoldt T, Fritz S, et al. (2013) Caper: Comparative analyses of phylogenetics and evolution in R. R package version 0.5.2. Available: http://CRAN.R-project.org/package=caper. Accessed 2014 Feb 24.
37. Harmon LJ, Weir JT, Brock CD, Glor RE, Challenger W (2008) GEIGER: investigating evolutionary radiations. Bioinformatics 24: 129–131.
38. Wibbelt G, Puechmaille SJ, Ohlendorf B, Mühldorfer K, Bosch T, et al. (2013) Skin lesions in European hibernating bats associated with *Geomyces destructans*, the etiologic agent of white-nose syndrome. PLoS ONE 8: e74105.
39. Zukal J, Berková H, Řehák Z (2005) Activity and shelter selection by *Myotis myotis* and *Rhinolophus hipposideros* hibernating in the Kateřinská cave (Czech Republic). Mamm Biol 70: 271–281.
40. Keesing F, Holt RD, Ostfeld RS (2006) Effects of species diversity on disease risk. Ecol Lett 9: 485–498.
41. Ostfeld RS, Keesing F (2012) Effects of host diversity on infectious disease. Annu Rev Ecol Evol Syst 43: 157–182.
42. Turner GG, Meteyer CU, Barton H, Gumbs JF, Reeder DM, et al. (2014) Nonlethal screening of bat-wing skin with the use of ultraviolet fluorescence to detect lesions indicative of white-nose syndrome. J Wildl Dis, doi: 10.7589/2014-03-058.
43. Woolhouse MEJ, Taylor LH, Haydon DT (2001) Population biology of multihost pathogens. Science 292: 1109–1112.
44. Verant ML, Boyles JG, Waldrep W, Wibbelt G, Blehert DS (2012) Temperature-dependent growth of *Geomyces destructans*, the fungus that causes bat white-nose syndrome. PLoS ONE 7: e46280.
45. Lorch JM, Muller LK, Russell RE, O'Connor M, Lindner DL, et al. (2013) Distribution and environmental persistence of the causative agent of white-nose syndrome, *Geomyces destructans*, in bat hibernacula of the Eastern United States. Appl Environ Microbiol 79: 1293–1301.
46. U.S. Geological Survey (2012) White-Nose Syndrome Threatens the Survival of Hibernating Bats in North America. Available: http://www.fort.usgs.gov/WNS/Default.asp. Accessed 2014 Jan 05.
47. Maher SP, Kramer AM, Pulliam JT, Zokan MA, Bowden SE, et al. (2012) Spread of white-nose syndrome on a network regulated by geography and climate. Nat Commun 3: 1306.
48. Cryan PM, Meteyer CU, Boyles JG, Blehert DS (2010) Wing pathology of white-nose syndrome in bats suggests life-threatening disruption of physiology. BMC Biology 8: 135.
49. Cryan PM, Meteyer CU, Blehert DS, Lorch JM, Reeder DM, et al. (2013) Electrolyte depletion in white-nose syndrome bats. J Wildl Dis 49: 398–402.
50. Meteyer CU, Barber D, Mandl JN (2012) Pathology in euthermic bats with white nose syndrome suggests a natural manifestation of immune reconstitution inflammatory syndrome. Virulence 3: 583–588.

A Comparison of Measures of Boldness and Their Relationships to Survival in Young Fish

James R. White[1]*, Mark G. Meekan[2], Mark I. McCormick[1], Maud C. O. Ferrari[3]

1 School of Marine and Tropical Biology, James Cook University, Townsville, Queensland, Australia, **2** Australian Institute of Marine Science, Botany Building, The University of Western Australia, Crawley, Western Australia, Australia, **3** Department of Biomedical Sciences, Western College of Veterinary Medicine, University of Saskatchewan, Saskatoon, Canada

Abstract

Boldness is the propensity of an animal to engage in risky behavior. Many variations of novel-object or novel-environment tests have been used to quantify the boldness of animals, although the relationship between test outcomes has rarely been investigated. Furthermore, the relationship of outcomes to any ecological aspect of fitness is generally assumed, rather than measured directly. Our study is the first to compare how the outcomes of the same test of boldness differ among observers and how different tests of boldness relate to the survival of individuals in the field. Newly-metamorphosed lemon damselfish, *Pomacentrus moluccensis*, were placed onto replicate patches of natural habitat. Individual behavior was quantified using four tests (composed of a total of 12 different measures of behavior): latency to enter a novel environment, activity in a novel environment, and reactions to threatening and benign novel objects. After behavior was quantified, survival was monitored for two days during which time fish were exposed to natural predators. Variation among observers was low for most of the 12 measures, except distance moved and the threat test (reaction to probe thrust), which displayed unacceptable amounts of inter-observer variation. Overall, the results of the behavioral tests suggested that novel environment and novel object tests quantified similar behaviors, yet these behavioral measures were not interchangeable. Multiple measures of behavior within the context of novel environment or object tests were the most robust way to assess boldness and these measures have a complex relationship with survivorship of young fish in the field. Body size and distance ventured from shelter were the only variables that had a direct and positive relationship with survival.

Editor: Howard I. Browman, Institute of Marine Research, Norway

Funding: Funding was provided by an ARC discovery grant, ARC Centre of Excellence for Coral Reef Studies (MIM & MGM), Australian Institute of Marine Science (MGM), the Yulgilbar Foundation and Natural Sciences and Engineering Research Council of Canada (MCOF). The funders had no role in study design, data collection and analysis, decision to publish, or preparation of the manuscript.

Competing Interests: The authors have declared that no competing interests exist.

* E-mail: james.ryan.white@gmail.com

Introduction

The propensity of an animal to take a risk is often described along an axis of boldness and shyness, where high likelihood of risk-taking is defined as boldness and low likelihood is defined as shyness. This behavior is important on both ecological and evolutionary time scales. Individuals can display various levels of boldness or shyness that can influence the outcome of everyday ecological challenges, such as competition for females [1] or food [2], foraging under predation pressure [3–5] and habitat selection [6,7]. Consequently, boldness and shyness can influence reproduction, survival and thus ultimately affect fitness. Boldness may have underlying physiological components and may be heritable [8–10], so can be subject to evolution following natural selection in subsequent generations [11].

Measurements of some aspect of behavior on the boldness-shyness axis dominate research on animal personality, termed 'behavioral syndromes' [12,13]. These syndromes refer to behavioral differences among individuals or species that are consistent over time or across situations [14,15]. Unfortunately, attempts to generalize the results of this work are hampered by a lack of common language and methodology [16,17]. For instance, some studies have defined boldness as the tendency of an individual to move through or explore an unfamiliar space (i.e. a novel environment) [18,19,7], while others consider it the propensity to forage under predation risk [20] or alternatively, reaction to a novel object [21]. Additionally, researchers have used a variety of behavioral attributes to measure boldness, such as latency to emerge into a novel environment, frequency of predator inspection [3,22], propensity to enter traps [19], or flight response to a novel object [19,23]. These measures may have some relation to one another (i.e. correlated behavioral measures within or across certain contexts), but do not necessarily quantify the same behavioral trait [24]. Recent attempts have been made to address this issue with proposed standardized terminology [25,26], however this has yet to be adopted universally.

The techniques used to measure boldness are almost as numerous as the studies that have assessed this trait in different taxa. Some researchers have argued that boldness should be tested in familiar, rather than novel environments [26] and to date, only a few studies have attempted to quantify behavior using multiple tests of boldness among individuals. For example, Wilson and Goden (2009) assessed individual differences in exploratory behavior, activity, and anti-predator behavior of juvenile sunfish using novel object and environment tests in the laboratory [27], while an earlier aquaria study by Brown et al. (2007) found a strong correlation between two independent assays of boldness (time to emerge into a novel environment and propensity to inspect a novel object) in a peociliid fish [28].

Due to the great variety of techniques used to quantify boldness, it remains unclear how studies compare in terms of the trait that they actually measure. Additionally, given that few assessments of behavioral syndromes have been conducted within an organism's natural environment, it is also difficult to determine how the results of these tests predict the likelihood of real ecological consequences for the subject animals.

Clearly, there is a need to clarify the relationships among the various measures of and tests for behavior on the boldness-shyness axis on subject animals in the field. Here, we focus on this task using a tropical reef fish model. Young reef fish can be collected at the end of their larval phase immediately prior to settlement on the reef, when they are naïve to reef-based predators and behaviors learned after settlement [29]. Also, by collecting fish from a single recruitment pulse, we control for gross variations in size and age [30]. In this phase of their life cycle, reef fishes typically experience high mortality [31], with rates within the first 48 hours of benthic life averaging 57% [32,31] but sometimes >90% [33]. The distributions that are established through differential mortality often set the pattern for abundances of juveniles and later life stages. Because experience can influence behavioral phenotypes [7,34,35], the use of naïve study organisms allows us to control for variation and consistency in behavior and to examine ecologically important behavioral traits at a critical ontogenetic boundary [36]. Here, we use short-term (48 hours) survival as a measure of the ecological consequences of differences in boldness, assayed using a variety of techniques. For juvenile coral reef fish, short-term survival immediately following settlement is a critical selective bottleneck for populations and is relatively straightforward to measure, making it ideal for use in our study. While our survival estimate is just one of a number of possible estimates of fitness that are ecologically relevant, because of the magnitude of mortality at this stage, the trait of survivorship is likely to be very important. For these young reef fish, we aimed to determine: 1) if different types of boldness measurements quantified a similar behavioral trait, 2) which of the commonly-used methods of assessing boldness (variants of novel object and novel environment tests) was the most closely correlated with an ecological outcome (survival), and 3) which behavioral measures were easiest to conduct *in situ* with low variability among multiple observers. Based on our previous experience with this system and study species, we predicted that novel object and environment tests would not covary in how they quantified boldness, with novel environment activity measures more likely to predict survivorship. We expected that correlations among behaviors would show that bold fish tended to be larger overall, spend more time actively foraging in ways that left them more exposed to predators, while being less reactive to any sort of novel object test than shy fish.

Methods

Ethics Statement

This study was carried out in strict accordance with the recommendations under James Cook University (JCU) ethics protocols and approved by the JCU Animal Ethics Committee (Permit Number: A1067). All efforts were made to minimize animal handling and stress.

Study Site and Species

This study was conducted on the shallow reef (2–4 m depth) offshore from the Lizard Island Research Station (14°40′S, 145°28′E) on the northern Great Barrier Reef, Australia. Our study species, the lemon damsel, *P. moluccensis*, is common on Indo-Pacific coral reefs [37]. Juveniles settle from the plankton at night [38], between October and January around the time of the new moon [39], preferentially settling on live coral [40]. Larvae recruit onto the reef after approximately 20 days in the plankton, at about 11 mm standard length [41]. *P. moluccensis* has a relatively small home range [42], moving only small distances (<1 m) during the first few months after settlement [40]. Due to its high abundance, small size, rapid development, and sedentary nature, *P. moluccensis* is an ideal model organism for field and laboratory based behavioral studies [43].

Experimental Design

Collection. We collected newly-metamorphosed juveniles of *P. moluccensis* using moored light traps (see small light trap of Figure 1 in Meekan et al. 2001 for design, [44]) during November 2010. Traps were anchored approximately 100 m from the nearest reef in ~10 m of water at dusk and left overnight. Catches were emptied from the traps the next morning between 05:30–07:00 h. Fish collected from the traps were transported to the laboratory where *P. moluccensis* was separated from all other species and maintained in a 25 L aquarium of aerated seawater for at least 24 h to acclimatize to local conditions and reduce handling stress before experiments began. Fish were fed *Artemia* nauplii twice daily while in captivity. After acclimation, each *P. moluccensis* was placed into a clip-seal polyethylene bag containing aerated seawater and were measured for total length (to the nearest mm) with calipers, photographed, and then transported to the field in individually-labeled plastic bags. After final observations, study organisms were released unharmed on nearby natural habitat. Fish collection locations/activities and handling protocols were approved by the Great Barrier Reef Marine Park Authority (Permit Number: G10/33784.1) and JCU Animal Ethics Committee (Permit Number: A1067).

Observational protocol. All behavioral observations were made on individual fish in the field. Divers released a single fish onto a small patch reef (30×30×30 cm) haphazardly chosen from 35 that were constructed from live and dead pieces of the bushy hard coral *Pocillopora damicornis* on the shallow (3–4 m water depth) sand flat. *P. moluccensis* recruits occur naturally in this habitat. Reefs were deployed in rows, 5 m apart and approximately 10 m from the nearest area of natural reef. Means and ranges of temperatures did not vary among reefs (M. McCormick unpubl. data) and care was taken in reef construction to ensure that patch reefs had only very minor differences in habitat structure. Previous studies have shown that such minor variation in topographic complexity of patch reefs has no effect on behavior of young fish [36,43]. Before introduction of the study fish, patch reefs were cleared of any resident fishes using hand nets. These were released on nearby natural reef far enough away to prevent their return (approx. 10 m). Individual study fish were then released onto their respective patch reefs and the first behavioral variable (latency to enter a novel environment; see description below) was recorded. Immediately afterwards, small wire cages (about 40×40×40 cm, 12 mm mesh size) were placed over the patch to allow the fish to acclimate to the new surroundings while being protected from predation. Cages were left a minimum of 20 min and carefully removed immediately before observations. Following established protocols, divers conducted observations from at least 1.5 m away (with the aid of a 2×magnifying glass) to avoid any effects that may have been caused by the proximity of the observer to the target fish [36,43]. A pilot study where estimates of distance were

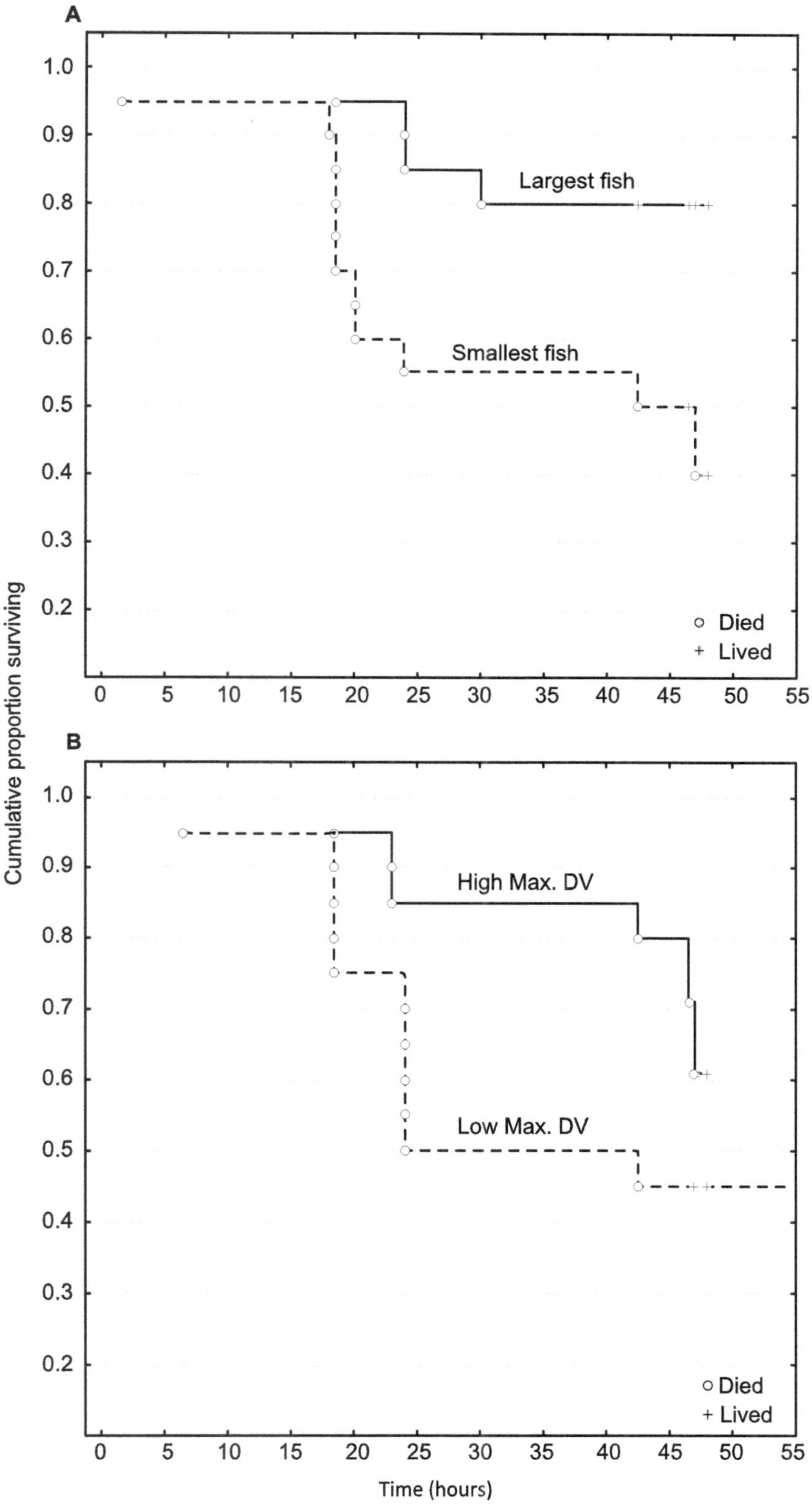

Figure 1. Survival over two nights in the field. Kaplan-Meier survival analysis with respect to: a) maximum distance moved and b) size (TL) of juvenile *Pomacentrus moluccensis* on patch reefs in the field. Fish were sequentially ranked for their scores on each trait and two groups (high and low ranked) of twenty fish (21.7% of total) were compared. Solid lines and dashed lines represent the two groups of highest and lowest ranked fish, respectively. Symbols represent presence or absence of individual fish during subsequent mortality surveys.

checked against a ruler found these estimates to be within 10% of the true value.

Behavioral traits were measured for a total of 92 fish during eight periods of observation spread over 5 days. The first six of these periods (n = 59 fish) were conducted by three experienced observers, each assessing the same fish simultaneously to quantify variance in measures among observers. All subsequent observations were conducted by JRW and MGM. Data from all observation periods were used for comparisons of behavioral traits among fish and data collected by three observers was used for a comparison of variability in estimates of behavior among observers. Each behavioral test was only trialed once with individual fish because *P. moluccensis* has been shown to recognize threats after a single exposure [45], which could have altered the outcomes of some boldness measures. In general, the behavioral responses of individuals have been shown to be very stable (e.g. a coefficient of variation for 3 consecutive observations ranging from 0–0.15; repeated measures ANOVA over 15 observations across 5 days, mean intraclass correlation coefficient = 0.69) at least over the time of our relatively short experiments (Mero 2009 unpublished thesis, [36]). In both a pilot study and this experiment, we found no relationship between observed behaviors of individual fish and specific patch reefs or time of day. This suggests differences in local environmental conditions such as minor variations in habitat, light conditions and food abundance across patch reefs did not noticeably influence behaviors.

The behavior of each fish was assessed using variations of two novel-object and two novel-environment tests that were composed of 12 behavioral measures:

1) Novel environment: release. After resident fish were cleared from the patch reefs, each damselfish was carefully released from the plastic bag onto the sand 10 cm from the patch reef. The amount of time it took for the fish to move onto refuge of the patch reef was termed 'latency at release'. This was timed from the moment the fish exited the bag, to the instant it reached the edge of the reef shelter. If the individual took more than 60 seconds to move to the reef, observations were discontinued and individuals were assigned a top value (~10% of fish).

2) Novel environment: overall activity. Six behavioral measures were recorded simultaneously over a 3 min observation interval for each fish: bite rate (number of feeding strikes towards objects floating in the water column); distance moved (total distance covered (cm) during 3 min); distance ventured (the maximum distance (cm) fish moved away from their patch reef; the distance ventured from the patch (categorized as % of time spent within 0, 2, 5, or 10 cm away from the patch); and position on the reef (categorized as a cumulative proportion of the time spent at varying heights over the 3 min observation period, with the top of the patch taken as height of 1, middle of the patch a height of 0.5, and bottom a height of 0). Mean distance ventured was calculated from the sum of the proportions of time spent in each of the distance categories multiplied by the distance that each category represented. Relative height on the patch was summarized as a cumulative proportion of the time spent at varying heights over the 3 min observation period, calculated from the sum of the proportions multiplied by the height categories (0, 0.5, or 1). Estimated distances were verified with a ruler after the 3 min observation period was completed.

3) Novel object: benign. Each fish was presented with a novel object (2.4×2.1×1.6 cm consistent assortment of blue and yellow Lego^{TM} blocks, with the same blocks used for each fish) that was gently placed 10 cm away from its location. Fish were not obviously disturbed by this action. Over a 60 s observational period, minimum approach distance (cm) and a visual estimate of mean approach distance (cm) were recorded.

4) Novel object: threat. The reaction of each damselfish to the thrust (~120 cm/s over 20 cm) of an observer's probe (pencil 13 cm long) towards them was recorded as the minimum distance from the tip of the probe (cm) before fleeing, the maximum distance traveled (cm) by the fish after the presentation of the threat, and the latency (seconds) of the fish to leave shelter of a particular part of the coral patch and return to its original location. Latency was limited to a 60 s observation time. A reaction score was quantified as a continuous variable on a 0–3 scale with 0.1 unit increments, where: **0**- hiding in refuge before or immediately after thrust and seldom emerging afterwards; **1**- retreating to refuge when scared and taking more than 5 s to re-emerge, then tentatively striking at food; **2**- retreating to refuge when scared but emerging quickly and striking at food; **3**- not hiding but continuing to explore or strike at food aggressively. The reaction score summarized the combination of overall individual behavior during the 3 min observation and reaction to the probe thrust.

Survival

The presence of fish on reefs was monitored twice daily (between 10:00–11:00 and 15:00–16:00 h) over two days (mean 44.9 h). Previous studies have shown that any migration of newly-settled fish from patch reefs in this location is negligible (<0.007% of 300 tagged fish in 3 days) so that the absence of fish from a reefs can most likely be attributed to predation [46].

Data Analysis

The overall variability of each behavioral measure was quantified using a coefficient of variation. The coefficient of variation and comparison of behavioral traits with survival were calculated using one score (from the most experienced observer, MGM) per fish. Behavioral responses were z-transformed to standardize differences in mean and variance while maintaining patterns of covariance.

In order to compare observers, the range of values (maximum-minimum scores) for each trait recorded by the three observers was compared across six observation periods (n = 59). Because the range values did not meet assumptions of normality, a Friedman test was used as a nonparametric alternative to one-way repeated measures ANOVA.

The influence of a single behavioral trait on survival was determined with Kaplan-Meier survival analysis and its significance with Cox's F-Test using multiple single-predictor models. In order to highlight the influence of behaviors at either high or low extremes, the twenty highest and twenty lowest scoring fish of each trait were compared. Traits identified as significant by the Kaplan-Meier test were further compared using phenotypic selection gradient analysis [47] as a more explicit test of the relationships between single and combinations of traits on fitness. This test was used to identify behavioral traits that best predicted survivorship, while accounting for direct and indirect selection. First, behavioral variables were z-transformed (standardized). Then, logistic regression was used to regress the standardized values, their squared terms, and the cross-products of the pairwise combinations on relative fitness (whether an individual lived or died, divided by average fitness of the population) to estimate directional, stabilizing, and correlation selection gradients, respectively [47,34].

Relationships between behavioral traits were analyzed using Pearson's product moment correlation. The statistical effect value (r) associated with these correlations are simply used as potential indicators of the strength of relationships rather than indicators of

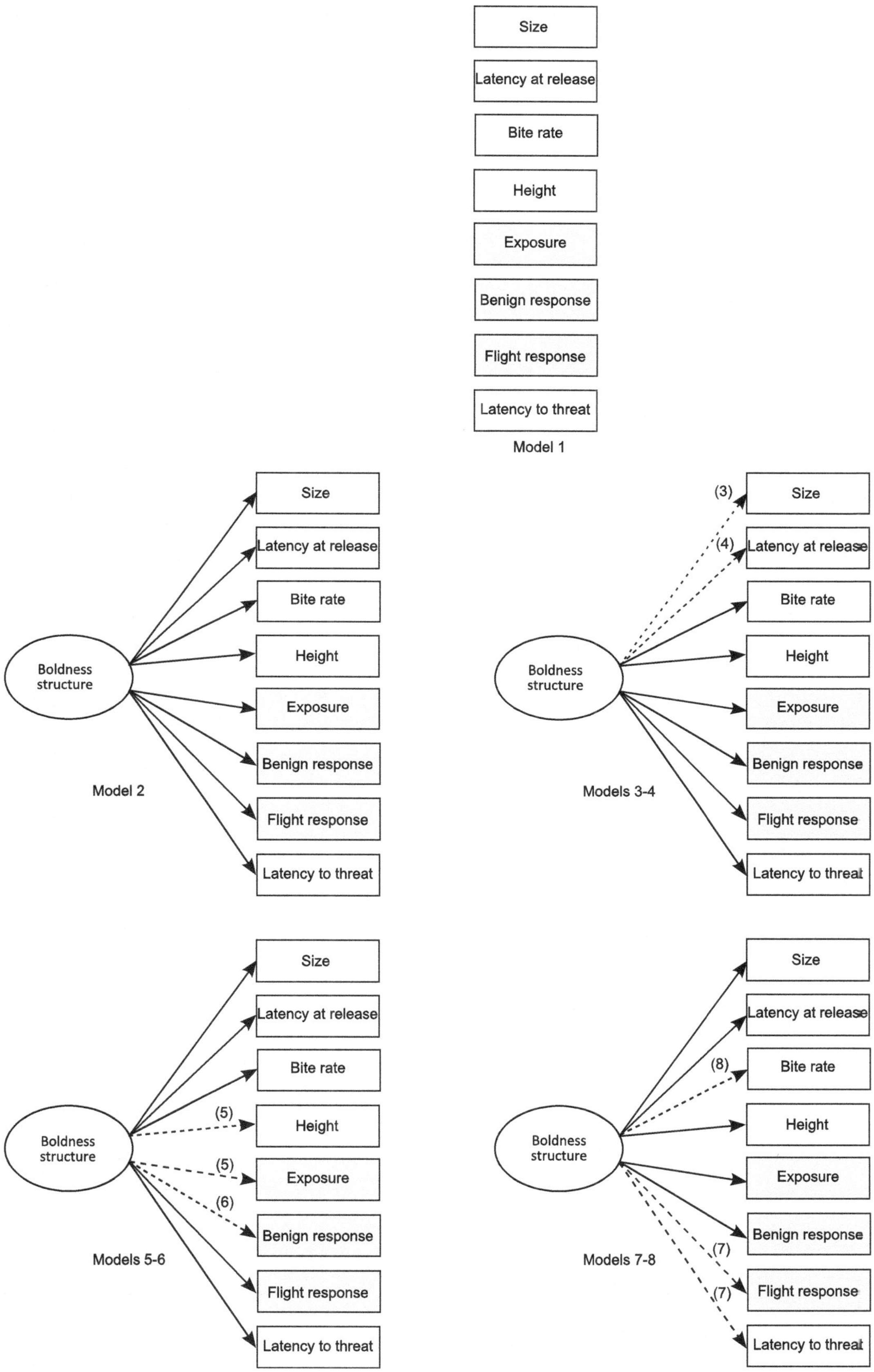

Figure 2. Eight models of boldness syndrome structure developed based on *a priori* hypotheses of boldness structure. Model 1 represents behavioral independence. Model 2 represents a domain-general model of syndrome structure while models 3–8 are more constrained, representing different types of boldness tests. The measured behaviors are represented in rectangular boxes, with shaded boxes representing

composite variables. Underlying causal connections (latent variables) resulting in boldness structure are represented in ovals [48]. In order to save space, multiple models are presented with alternative structures denoted by dashed lines labeled with model number (e.g. model 3 excluded size, as denoted with a dashed line labeled 3).

biological significance. However, sequential Bonferroni adjustments are included to account for multiple testing (Type I) errors.

Confirmatory factor analysis, a form of structural equation modeling [48], was used to determine the structure of a combination of behavioral measures used to assess boldness for the population during a 48 hr post-settlement period. We followed the proposed framework established by Dingemanse et al. (2010) for using structural equation modeling (SEM) to compare hypothesized patterns of behavioral covariance. Eight alternative models formulated *a priori* (as described below) for boldness syndrome structure were separately assessed and the relative fit of each model was compared. Models were compared using Akaike's Information Criterion (AIC), which was calculated from model discrepancies ($\hat{C}$) estimated by maximum likelihood using Bollen-Stine bootstrapping (2000 bootstraps). AIC values compare the fit of a model to data while rewarding parsimony, with lower values indicating greater model support [49,50]. Models were compared by AIC differences (ΔAIC) relative to the model with the lowest AIC value, with ΔAIC values greater than two suggesting less support [51]. The maximum convergence limit for data to fit to models was set at 50 iterations.

In order to increase parsimony of the structural equation models, the most similar behavior responses were combined into composite variables by extracting their factor scores using factor analysis. Distance ventured and maximum distance ventured were combined into a new variable termed 'Exposure'. Minimum and average distances to Lego blocks were combined to form the new variable 'Benign response', while minimum and maximum distances to the threatening object (probe) formed the new variable 'Flight response'. Rather than the traditional method of using factors with eigenvalues greater than one, parallel analysis was used to determine the number of factors to be extracted (using permutations of 1000 parallel generated datasets) as outlined in Budaev (2010). With the correct number of factors determined by the parallel analysis, factor scores were calculated using principle axis factoring with Varimax rotation and the regression method [52].

Prior to SEM analysis, Bartlett's test of sphericity and the Kaiser-Meyer-Olkin (KMO) index were calculated for the dataset. The sphericity test determined if the behavioral variance-covariance matrix differed from random [50], while the KMO index compared observed correlations and partial correlations among original variables [52]. In our data, the matrices differed from random ($\chi^2_{28} = 118.40$, $P<0.001$). The KMO values were above the 0.5 acceptable threshold [52] with KMO = 0.52. However, the efficacy of the KMO test for a confirmatory factor analysis with a single latent factor (as used in this study) is unknown [50] and both tests are unlikely to be necessary for simple models with few observed variables [53].

Eight *a priori* hypotheses of boldness structure were considered based on the different types of boldness tests in behavioral syndrome literature (models 1–8, Fig. 2). Model 1 was the null model, where there was an absence of covariance and behavioral responses varied independently [54]. Model 2 represented a domain-general model of boldness structure, where all types of novel environment and novel object tests were linked via an underlying factor. Models 3 and 4 represented a domain-general model where size and latency at release, respectively, were considered contextually different from the rest of the behavioral responses. Model 5 considered foraging and height contextually different. Model 6 removed the benign response from the other boldness measures. Model 7 removed the threatening novel object measures: flight response and latency to threat. Model 8 considered bite rate contextually different from other activity, novel object, and novel environment tests.

Because models were built on *a priori* hypotheses, models 2–8 were compared against the model of no boldness syndrome structure (model 1) to quantify the amount of variation explained by the different models. This was done by calculating D_x, which represented the proportion of variation in the behavioral variance-covariance matrix explained by each model, relative to the null model [55,50]. D_x was calculated as: $D_x = 1 - \hat{C}_x/\hat{C}_{null}$ where $\hat{C}_{null}$ was the discrepancy for the null model (i.e. model 1, Fig. 2) and $\hat{C}_x$ was the discrepancy for other hypothesized models (i.e. models 2–8, Fig. 2). D_x is interpreted similarly to an R^2 value [50].

Statistical analyses used SPSS version 20.0 (SPSS Inc., Chicago, IL, U.S.A.). Structural equation models were constructed using AMOS version 20.0 (SPSS, Inc.).

Results

Variability of Behaviors

Most traits showed high variability among individuals (Table 1), which allowed one or more traits to affect post-settlement mortality. Coefficients of variation ranged between 8–82% for most measures, with the exceptions of latency at release, time budget and escape latency to a probe thrust, which all had CVs over 100% of mean values. Latency at release had the highest CV (167%), but this was skewed due to a small number of fish (9 of 92 fish) that did not move to patch reefs within the 60 s observation period. The CV reduced to 102% when these slow-to-respond fish were excluded from the data set. The time budget had high CVs since few fish remained motionless or did activities other than feeding. Some fish (7 of 92) remained hidden within the refuge of the patch reefs after the probe thrust, skewing the CV for this measure.

Mortality was monitored for at least two nights in the field (mean 44.9 h). A total of 41.8% of all fish disappeared from reefs and were assumed to have died (Table 2). Of these, 84% died within the first 24 h, typically at sometime between the last observation in the afternoon and the next observation the following morning.

Variability among Observers

The threat test was the only measure that showed significant variability among observers across trials ($\chi^2(5) = 12.72$, $p = 0.026$). There was no pattern of improvement in observer consistency over time, with the variability in trials 1, 4, and 6 lower on average than trials 2, 3, and 5. The variability in threat test scores among observers ranged from a 0.5 to 1.0 difference (on 0–3 scale).

Other behavioral measures did not differ significantly among trials and observers, suggesting no major improvement or decline in observer consistency. Most variables had a low level of observer variance (Table 1), with a difference of only 1–3 cm or 1–3 seconds (~3% of maximum observation time). However, distance moved had a relatively large variance, with observers disagreeing by an average of 12 cm in most trials, although this improved to 5 cm by the end of the study. Estimates of bite rate were

Table 1. Summary statistics for various measures of novel object or novel environment tests of the Lemon damsel (*Pomacentrus moluccensis*).

Variable	N	Mean	SD	CV (%)	Mean Inter-Observer SD/Equivalent in units
Physical character					
Size (cm)	92	1.3	0.1	8	N/A
Novel environment: release					
Latency at release (s)	92	17.1	28.5	167	1.0/2
Novel environment: activity					
Bite rate	92	26.7	15.1	56	8.1/16
Distance moved (cm)	92	17.5	14.4	82	8.4/15
Distance ventured (% time index)	92	1.8	0.9	52	0.6/1
Max. distance ventured (cm)	92	3.3	1.8	55	1.8/3
Position on reef (height index)	92	2.5	0.7	26	0.1/0.2
Novel object: benign					
Minimum distance to Legos (cm)	92	4.2	1.8	44	1.0/2
Mean distance to Legos (cm)	92	7.2	2.7	37	1.2/2
Novel object: threat					
Minimum distance to threat (cm)	92	3.2	2.2	69	0.7/2
Max. distance travelled from threat (cm)	92	5.3	1.8	34	1.6/3
Latency to threat (s)	92	14.2	18.3	129	2.4/4
Threat test (0–3 score)	92	1.7	0.6	37	0.4/0.8

moderately variable but improved with time, with the average difference ranging between 6–22 strikes. Because of the high and inconsistent inter-observer variability in measures of the threat test and distance moved, these measures were omitted from subsequent analysis.

Individual Behavioral Traits and Survival

The ability to discriminate survivors from non-survivors on the basis of a single behavioral trait was poor. However, Kaplan-Meier survival analysis showed maximum distance ventured ($F_{12,22} = 2.42$, $p = 0.035$) and initial size ($F_{8,22} = 3.72$, $p = 0.007$) were good predictors of survival (Figures 1 a & b). Larger fish and those willing to venture further from the reef had better survival rates. Bag latency at release was suggestive of a trend ($F_{16,28} = 1.97$, $p = 0.056$), with fish that quickly moved to the patch reef having lower average mortality.

The phenotypic selection analysis showed a significant relationship between the behavioral traits and relative fitness (), however this model accounted for a relatively low amount of the variation with Cox & Snell $R^2 = 0.096$ (Table 3). Overall, larger fish survived better, with size as the only variable identified as significant directional ($\beta = 0.469$, $p<0.05$) and stabilizing ($\beta = 0.234$, $p<0.05$) selection gradients, even though the size range was only 1.1–1.6 mm total length. No other directional, stabilizing, or correlational selection gradients were found to be significant. The model was adequate and predicted 63% of the responses correctly.

Table 2. Survival (%) of newly settled Lemon damsel (*Pomacentrus moluccensis*) on patch reefs.

Field trial	Trial duration (h)	N	Survival (%)
1	47	9	60
2	47	7	57
3	42.5	9	67
4	42.5	12	67
5	42.5	7	43
6	42.5	14	43
7	48	14	64
8	47	20	65
Mean or total	44.9	92	58

Table 3. Directional, stabilizing and correlational standardized selection gradients (β) from logistic regression.

	β	SE	*P*- value	β *avggrad*
Size	0.469	0.240	0.050	0.170
Latency at release	−0.418	0.234	0.074	−0.152
Max. DV	0.072	0.229	0.754	0.026
Size^2	0.234	0.120	0.050	0.085
Latency at release^2	−0.209	0.117	0.074	−0.076
Max. DV^2	0.036	0.115	0.754	0.013
Size * Latency at release	0.256	0.280	0.360	0.093
Size * Max. DV	0.100	0.230	0.665	0.036
Latency at release * Max. DV	0.232	0.206	0.261	0.084

Model , *P*= 0.026, Cox & Snell $R^2 = 0.096$.

Correlations among Behavioral Traits

There were two significantly correlated relationships between behavioral traits (Table 4). Bite rate had a high positive correlation with exposure. Bite rate was also moderately negatively correlated with latency to a threat. This general lack of correlation suggests that each variable is quantifying a different aspect of behavior or space use.

Structure of Multiple Behavioral Traits

There was equal support for models in which response to the benign novel object (model 6, ΔAIC = 0; Table 5), size (model 3, ΔAIC = 0.10; Table 5), latency at release (model 4, ΔAIC = 0.29; Table 5) varied independently of other behavioral measures and also for the model in which all measures were included (model 2, ΔAIC = 0.85; Table 5). These models explained approximately 51% of the variance-covariance matrix variation in behavior (Table 5). In summary, four models fit the data equally well and accounted for about half the total variation.

The behavioral patterns were best explained by models that showed a similar pattern in variable loadings. Path coefficients for the best fit models (models 2–4, 6) all had negative loadings for bite rate, exposure, size and height and positive loadings for latency at release, latency to threat, benign novel object and flight responses (Fig. 3). Loadings with the same sign imply an unknown proximate factor or factors that affect the expression of behaviors in the same manner [50]. The SEM structure explained a high amount of variance in data sets for bite rate and exposure behaviors, suggesting these measures were better suited to assess boldness of juvenile fish in the field.

Table 5. Model comparison results for confirmatory factor analysis.

Model (x)	Ĉ (discrepancy)	k	AIC	ΔAIC	D_x
6	58.76	15	88.76	0	0.51
3	58.87	15	88.87	0.10	0.50
4	59.05	15	89.05	0.29	0.50
2	57.61	16	89.61	0.85	0.52
5	92.00	14	119.99	31.23	0.23
8	91.66	15	121.66	32.89	0.23
1	118.8	8	134.8	46.04	0

Structural equation models (SEMs) were evaluated based on difference in Akaike's information criterion (AIC) values. Small values represent an increased parsimony-informed fit to the data. AIC values were calculated based on the discrepancy between the statistical model for a hypothesis (Ĉ) and the number of parameters (k). D_x values represent the proportion of the variance explained by the focal model relative to null expectations of no boldness structure. D_x can be interpreted as analogous to R^2. Unlisted models were those where the data did not converge within 50 iterations.

Discussion

Individual Behavioral Traits and Survival

Single behavioral traits had limited ability to predict survivorship for our model species. Those fish that were larger or were willing to venture further from the edge of patch reefs had greater survivorship during this critical phase of the life cycle; a conclusion supported by studies of intra- and inter-specific behavioral interactions at this life stage [56,57]. There was a strong (though non-significant at p = 0.056) trend for fish that moved quickly to patch reefs when released to survive better than those that were slow to travel to the reef. Phenotypic selection analysis suggested only size had a significant effect on survivorship and that combinations of behavioral measures did not influence survival. Size and condition at settlement has previously been shown to be important for survival [58,59], with larger fish often having greater survivorship [60]. However, this pattern is not consistent at all times and places, with some studies showing that newly-settled individuals that were larger suffered higher mortality than smaller fish in some cases [43,61]. Additionally, earlier work has found no links between foraging behaviors and selective mortality at settlement [43], or a positive correlation between distance ventured from reefs and mortality [36]. Such differences in outcomes of studies may simply be a reflection of the temporal or spatial variability in predator/prey abundance [62,63] or a predator's individual preference of prey species [64]. These complex relationships between predator/prey abundance and predator behaviors could be a major driving force in shaping individual variation in the prey's behavior and ultimately, survival in the population. For example, Holmes & McCormick (2009) have shown that one of the major predators on newly-settled damselfish, *Pseudochromis fuscus*, which is common in shallow reefs adjacent to our patch reefs [61,65], preferentially targets larger recruiting fishes. If *P. fuscus* was more abundant in previous years, or selectively targets certain species [66], then spatial and temporal differences in the relationship of size or behavioral traits with mortality would be expected.

Table 4. Phenotypic correlations between seven behavioral traits for Lemon damselfish.

†Behavior	Relative fitness	Size	Latency at release	Bite rate	Height	Exposure	Benign response	Flight response	Latency to threat
Relative fitness	–	0.25*	−0.24*	0.00	0.04	0.18	0.09	−0.05	−0.12
Size		–	−0.22*	0.11	0.21*	−0.26	−0.05	0.10	0.08
Latency at release			–	−0.13	−0.15	0.01	0.14	−0.09	0.20
Bite rate				–	0.16	**0.61***	−0.12	−0.01	**−0.35***
Height					–	−0.05	−0.31**	0.19	−0.16
Exposure						–	0.11	0.13	−0.24*
Benign response							–	0.25	0.30**
Flight response								–	0.15
Latency to threat									–

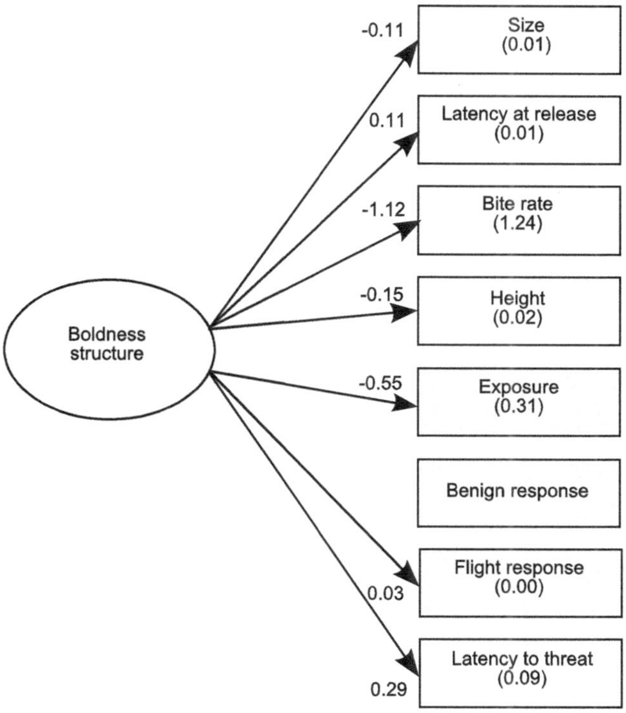

Figure 3. Best fitting structural equation model (SEM). This SEM shows how behaviors were related within the best fitting model for damselfish. Numbers in parentheses are variances of the different behaviors explained by the SEM structure (R^2) for 'model 6' (see Fig. 2). Numbers associated with arrows are standardized factor loadings for the effects of the underlying boldness structure on a particular behavior. These represent how behavioral responses are predicted to change based on changes to the underlying boldness structure (e.g. a shift of 1 SD along the distribution of boldness structure for the population would result in a 0.15 SD decrease in height).

We used short-term (over 2 nights) survival as an ecologically relevant measure of the consequences of behavioral decisions although other measures of fitness (e.g. long term survival, reproductive output, offspring quality, etc.) or some other aspect of an animal's ecology could be used as an equally valid trait against which behaviors could be compared. Indeed, the different measures of boldness might vary in relevance depending on the trait against which they are measured and ontogenetic stage [2]. The high and selective mortality that normally occurs during the settlement transition for organisms with complex life cycles such as fishes makes the short term mortality measured in the present study, and the behavioral correlations explored, ecologically relevant.

Correlations among Behavioral Traits

The limited number of correlations among behaviors found in our study suggests that the behavioral variables we assessed measured slightly different aspects of boldness and were not interchangeable. The positive relationship between the composite variable 'Exposure' and bite rate was expected because juvenile fish tend to actively swim and explore the vicinity of their habitat while foraging. Fish that had higher bite rates also tended to quickly resume feeding after being threatened with a probe. With size being the principal predictor of short-term survival, one viable strategy would be for these fish to prioritize behaviors that maximized growth rates. By growing quickly, juveniles would escape gape-limited predators and better compete for space and resources. In this case, it would be advantageous for juvenile pomacentrids to quickly learn to recognize and ignore false threats, a trait that is a feature of these fishes [45].

Structure of Multiple Behavioral Traits

Multiple SEM models could be fitted to the data for juvenile lemon damselfish. This suggests that there was considerable variability in the expression of boldness among individuals at the same life stage, in this case within the first few days of settling to the coral reef environment. Having a relatively adaptable expression of boldness at this time may allow individuals to properly assess and deal with the risks associated with the large assortment of predators that preferentially target fish recruits.

The use of a wild-caught population of juvenile fish rather than laboratory-bred individuals may account for a lower value for overall model fit ($D_x = 0.51$) compared to similar studies [50]. Previous work has shown similar species of juvenile damselfish are highly flexible in their behavioral responses across different situations (White et al. in review). Relatively large individuals also had relatively high bite rates and spent more time near the top of the reef (greater height) while being relatively quick to exit the bag at release, were more exposed, and less reactive to novel objects. This was in agreement with our predictions on how boldness would be structured. However, contrary to our predictions, novel object and novel environment tests did not vary independently, with the fit of the data lending equal support to the unrestricted domain general model (model 2). All measures were considered to be behavioral responses that were contextually similar in regards to boldness structure. In other words, all measures accounted for the structure of boldness.

Variability among Observers

Variability among observers measuring the same trait did not decline or increase over time for most behaviors, with the exception of the threat test. Variation in this measure increased during the study, probably reflecting the subjective nature of the measure, at least when multiple observers were involved in the work. Measures of bite rate, escape distance from a probe thrust and minimum distance from a probe thrust all showed some signs of reduced variation among observers over time. Observer variation in observed bite rate was initially high, but was reduced to acceptable levels after limited training. Overall, generation of consistent and accurate measures of distance moved and reaction to the threat test proved difficult when multiple observers were involved, however the recording of behavior using high resolution cameras may offer a means to further reduce this source of variation in these measurements.

Conclusion

Although we measured 12 behavioral variables, only one (distance from shelter) predicted short-term survival. Fish size (a physical character) was the most influential in determining survival. In the past, most studies have considered boldness as a binary trait that was that could be quantified with a single variable. However, our study suggested that multiple measures of behavior and habitat use were necessary to adequately quantify boldness in our study species, because all quantified slightly different and largely uncorrelated aspects of behavior. Additionally, our multivariate analysis suggested that both novel object and environment tests were related via some underlying causal factor to boldness structure, but the lack of correlations suggested that these behavioral measures were not interchangeable. For our study animal, a tropical reef fish, we argue that most of the behavioral variables measured that required little to no interaction with the study subject gave a good overall insight into boldness structure. Boldness measures that involve interaction (e.g. presentation of novel objects), while correlated with another measure (bite rate), provided only a small amount of additional predictive value with regards to boldness structure of the fish. Also, due to the ability of *P. moluccensis* [45] and other juvenile fishes [67,68] to learn rapidly, novel object tests may be less repeatable once fish have acclimated toward the stimuli [69]. We suggest that novel object tests may engender responses that have little relevance to the environments in which naïve young fish find themselves after settlement, so that the results may have no bearing on the likely behavior of individuals in response to natural predators, at least in the first few days after settlement. While our results show novel environment and object tests both give insight into boldness structure, the repeatability and ecological relevance should be considered when selecting the most appropriate boldness measure for a study organism.

Acknowledgments

We would like to thank M. Takahashi for assistance in the field as well as the staff of Lizard Island Research Station (Australian Museum) for logistical support. R. Beg, R. Jones, N. Dingemanse, and N. Dochtermann all provided advice on statistical analysis. C. Norris provided inspiration for this study.

Author Contributions

Conceived and designed the experiments: JRW MGM MIM MCOF. Performed the experiments: JRW MGM. Analyzed the data: JRW MIM MCOF. Contributed reagents/materials/analysis tools: JRW MGM MIM MCOF. Wrote the paper: JRW MGM MIM.

References

1. Dugatkin LA, Alfieri MS (2003) Boldness, behavioral inhibition and learning. Ethol Ecol Evol 15: 43–49. doi:10.1080/08927014.2003.9522689.
2. Dingemanse NJ, Both C, Drent PJ, Tinbergen JM (2004) Fitness consequences of avian personalities in a fluctuating environment. P R Soc B 271: 847–852.
3. Dugatkin LA (1992) Tendency to inspect predators predicts mortality risk in the guppy (*Poecilia reticulata*). Behav Ecol 3: 124–127. doi:10.1093/beheco/3.2.124.
4. Biro PA, Abrahams MV, Post JR, Parkinson EA (2006) Behavioural trade-offs between growth and mortality explain evolution of submaximal growth rates. J Anim Ecol 75: 1165–1171. doi:10.1111/j.1365-2656.2006.01137.x.
5. Stamps JA (2007) Growth-mortality tradeoffs and "personality traits" in animals. Ecol Lett 10: 355–363. doi:10.1111/j.1461-0248.2007.01034.x.
6. Wilson DS, Coleman K, Clark AB, Biederman L (1993) Shy-bold continuum in pumpkinseed sunfish (*Lepomis gibbosus*): An ecological study of a psychological trait. J Comp Psychol 107: 250–260. doi:10.1037/0735-7036.107.3.250.
7. Budaev SV (1997) " Personality" in the guppy (*Poecilia reticulata*): A correlational study of exploratory behavior and social tendency. J Comp Psychol 111: 399–411. doi:10.1037/0735-7036.111.4.399.
8. Boissy A (1995) Fear and fearfulness in animals. Q Rev Biol 70: 165–191.
9. Koolhaas JM, Korte SM, De Boer SF, Van Der Vegt BJ, Van Reenen CG, et al. (1999) Coping styles in animals: current status in behavior and stress-physiology. Neurosci Biobehav R 23: 925–935. doi:10.1016/S0149-7634(99)00026-3.
10. Brown C, Burgess F, Braithwaite VA (2007) Heritable and experiential effects on boldness in a tropical poeciliid. Behavioral Ecology and Sociobiology 62: 237–243. doi:10.1007/s00265-007-0458-3.
11. Réale D, Festa-Bianchet M (2003) Predator-induced natural selection on temperament in bighorn ewes. Animal Behaviour 65: 463–470. doi:10.1006/anbe.2003.2100.
12. Clark AB, Ehlinger TJ (1987) Pattern and adaptation in individual behavioral differences. In: Bateson PPG, Klopfer PH, editors. Perspectives in Ethology. New York: Plenum. 1–47.
13. Sih A (2004) Behavioral syndromes: an ecological and evolutionary overview. Trends in Ecology & Evolution 19: 372–378. doi:10.1016/j.tree.2004.04.009.
14. Dall SRX, Houston AI, McNamara JM (2004) The behavioural ecology of personality: consistent individual differences from an adaptive perspective. Ecol Lett 7: 734–739. doi:10.1111/j.1461-0248.2004.00618.x.
15. Sih A, Bell AM, Chadwick Johnson J (2004) Behavioral syndromes: an ecological and evolutionary overview. Trends Ecol Evol 19: 372–378. doi:10.1016/j.tree.2004.04.009.
16. Gosling SD (2001) From Mice to Men: What Can We Learn About Personality From Animal Research? Psychol Bull 127: 45–86. doi:10.1037/0033-2909.127.1.45.
17. Toms CN, Echevarria DJ, Jouandot DJ (2010) A Methodological Review of Personality-Related Studies in Fish: Focus on the Shy-Bold Axis of Behavior. Int J Comp Psychol 23: 1–25.
18. Fraser DF, Gilliam JF, Daley MJ, Le AN, Skalski GT (2001) Explaining leptokurtic movement distributions: intrapopulation variation in boldness and exploration. Am Nat 158: 124–135.
19. Wilson DS, Clark AB, Coleman K, Dearstyne T (1994) Shyness and boldness in humans and other animals. Trends Ecol Evol 9: 442–446. doi:10.1016/0169-5347(94)90134-1.
20. Brown C, Jones F, Braithwaite V (2005) In situ examination of boldness-shyness traits in the tropical poeciliid, *Brachyraphis episcopi*. Anim Behav 70: 1003–1009. doi:10.1016/j.anbehav.2004.12.022.
21. Wright D, Nakamichi R, Krause J, Butlin RK (2006) QTL Analysis of Behavioral and Morphological Differentiation Between Wild and Laboratory Zebrafish (*Danio rerio*). Behav Genet 36: 271–284. doi:10.1007/s10519-005-9029-4.
22. Budaev SV, Zworykin DD, Mochek AD (1999) Consistency of individual differences in behaviour of the lion-headed cichlid, *Steatocranus casuarius*. Behav Process 48: 49–56. doi:10.1016/S0376-6357(99)00068-6.
23. Frost AJ, Winrow-Giffen A, Ashley PJ, Sneddon LU (2007) Plasticity in animal personality traits: does prior experience alter the degree of boldness? P R Soc B 274: 333–339. doi:10.1098/rspb.2006.3751.
24. Réale D, Reader SM, Sol D, McDougall PT, Dingemanse NJ (2007) Integrating animal temperament within ecology and evolution. Biol Rev 82: 291–318. doi:10.1111/j.1469-185X.2007.00010.x.
25. Budaev SV, Brown C (2011) Personality Traits and Behaviour. In: Brown C, Laland K, Krause J, editors. Fish Cognition and Behavior. Fish and Aquatic Resources Series. Oxford, UK: Wiley-Blackwell.
26. Réale D, Reader SM, Sol D, McDougall PT, Dingemanse NJ (2007) Integrating animal temperament within ecology and evolution. Biological Reviews 82: 291–318. doi:10.1111/j.1469-185X.2007.00010.x.
27. Wilson ADM, Godin JGJ (2009) Boldness and behavioral syndromes in the bluegill sunfish, *Lepomis macrochirus*. Behav Ecol 20: 231. doi:10.1093/beheco/arp018.
28. Brown C, Jones F, Braithwaite VA (2007) Correlation between boldness and body mass in natural populations of the poeciliid Brachyrhaphis episcopi. Journal of Fish Biology 71: 1590–1601. doi:10.1111/j.1095-8649.2007.01627.x.
29. Lonnstedt OM, McCormick MI, Meekan MG, Ferrari MCO, Chivers DP (2012) Learn and live: predator experience and feeding history determines prey behaviour and survival. P R Soc B: 1–8. doi:10.1098/rspb.2011.2516.
30. Kerrigan BA (1996) Temporal patterns in size and condition at settlement in two tropical reef fishes (Pomacentridae: *Pomacentrus amboinensis* and *P. nagasakiensis*). Mar Ecol Prog Ser 135: 24–41. doi:10.3354/meps135027.
31. Almany GR, Webster MS (2006) The predation gauntlet: early post-settlement mortality in reef fishes. Coral Reefs 25: 19–22. doi:10.1007/s00338-005-0044-y.
32. Doherty PJ, Dufour V, Galzin R, Hixon MA, Meekan MG, et al. (2004) High mortality during settlement is a population bottleneck for a tropical surgeonfish. Ecol 85: 2422–2428. doi:10.1890/04-0366.
33. Gosselin LA, Qian P (1997) Juvenile mortality in benthic marine invertebrates. Mar Ecol Prog Ser 146: 265–282. doi:10.3354/meps146265.
34. Bell AM, Sih A (2007) Exposure to predation generates personality in threespined sticklebacks (*Gasterosteus aculeatus*). Ecol Lett 10: 828–834. doi:10.1111/j.1461-0248.2007.01081.x.
35. Dingemanse NJ, Van der Plas F, Wright J, Réale D, Schrama M, et al. (2009) Individual experience and evolutionary history of predation affect expression of heritable variation in fish personality and morphology. P R Soc B 276: 1285.
36. McCormick MI, Meekan MG (2010) The importance of attitude: the influence of behaviour on survival at an ontogenetic boundary. Mar Ecol Prog Ser 407: 173–185. doi:10.3354/meps08583.
37. Beukers J, Jones G (1998) Habitat complexity modifies the impact of piscivores on a coral reef fish population. Oecologia 114: 50–59. doi:10.1007/s004420050419.
38. Pitcher CR (1988) Validation of a technique for reconstructing daily patterns in the recruitment of coral reef damselfish. Coral Reefs 7: 105–111. doi:10.1007/BF00300969.
39. Meekan MG, Milicich MJ, Doherty PJ (1993) Larval production drives temporal patterns of larval supply and recruitment of a coral reef damselfish. Mar Ecol Prog Ser 93: 217–225.
40. McCormick MI, Weaver CJ (2012) It Pays to Be Pushy: Intracohort Interference Competition between Two Reef Fishes. PLoS ONE 7. Available: http://dx.plos.org/10.1371/journal.pone.0042590. Accessed 11 August 2012.
41. Wellington G, Victor B (1989) Planktonic larval duration of one hundred species of Pacific and Atlantic damselfishes (Pomacentridae). Mar Biol 101: 557–567. doi:10.1007/BF00541659.
42. Brunton B, Booth D (2003) Density- and size-dependent mortality of a settling coral-reef damselfish (*Pomacentrus moluccensis* Bleeker). Oecologia 137: 377–384. doi:10.1007/s00442-003-1377-2.
43. Meekan MG, von Kuerthy C, McCormick MI, Radford B (2010) Behavioural mediation of the costs and benefits of fast growth in a marine fish. Anim Behav 79: 803–809. doi:10.1016/j.anbehav.2009.12.002.

44. Meekan MG, Wilson SG, Halford A, Retzel A (2001) A comparison of catches of fishes and invertebrates by two light trap designs, in tropical NW Australia. Mar Biol 139: 373–381. doi:10.1007/s002270100577.
45. Mitchell MD, McCormick MI, Ferrari MCO, Chivers DP (2011) Friend or foe? The role of latent inhibition in predator and non-predator labelling by coral reef fishes. Anim Cogn 14: 707–714. doi:10.1007/s10071-011-0405-6.
46. Hoey AS, McCormick MI (2004) Selective predation for low body condition at the larval-juvenile transition of a coral reef fish. Oecologia 139: 23–29. doi:10.1007/s00442-004-1489-3.
47. Lande R, Arnold SJ (1983) The measurement of selection on correlated characters. Evolution: 1210–1226.
48. Grace JB (2006) Structural Equation Modeling and Natural Systems. Cambridge: Cambridge University Press.
49. Akaike H (1973) Information theory and an extension of the maximum likelihood principle. In: Petran BN, Csaki F, editors. Budapest: Akademaiai Kiadi. 267–281.
50. Dingemanse NJ, Dochtermann N, Wright J (2010) A method for exploring the structure of behavioural syndromes to allow formal comparison within and between data sets. Animal Behaviour 79: 439–450. doi:10.1016/j.anbehav.2009.11.024.
51. Burnham KP, Anderson DR (2002) Model Selection and Multimodel Inferences: a Practical Information-Theoretic Approach. New York: Springer-Verlag.
52. Budaev SV (2010) Using Principal Components and Factor Analysis in Animal Behaviour Research: Caveats and Guidelines. Ethol 116: 472–480. doi:10.1111/j.1439-0310.2010.01758.x.
53. Dochtermann NA, Jenkins SH (2011) Multivariate Methods and Small Sample Sizes. Ethol 117: 95–101. doi:10.1111/j.1439-0310.2010.01846.x.
54. Coleman K, Wilson DS (1998) Shyness and boldness in pumpkinseed sunfish: individual differences are context-specific. Anim Behav 56: 927–936. doi:10.1006/anbe.1998.0852.
55. Stamps J, McElreath R, Eason P (2005) Alternative models of conspecific attraction in flies and crabs. Behav Ecol 16: 974–980. doi:10.1093/beheco/ari083.
56. McCormick M (2009) Behaviourally mediated phenotypic selection in a disturbed coral reef environment. PLoS ONE 4: e7096. doi:10.1371/journal.pone.0007096.
57. McCormick MI (2012) Lethal effects of habitat degradation on fishes through changing competitive advantage. P R Soc B. Available: http://rspb.royalsocietypublishing.org/cgi/doi/10.1098/rspb.2012.0854. Accessed 13 August 2012.
58. Vigliola L, Meekan MG (2002) Size at hatching and planktonic growth determine post-settlement survivorship of a coral reef fish. Oecologia 131: 89–93. doi:10.1007/s00442-001-0866-4.
59. Gagliano M, McCormick MI, Meekan MG (2007) Survival against the odds: ontogenetic changes in selective pressure mediate growth-mortality trade-offs in a marine fish. P R Soc B 274: 1575. doi:10.1098/rspb.2007.0242.
60. McCormick MI, Hoey AS (2004) Larval growth history determines juvenile growth and survival in a tropical marine fish. Oikos 106: 225–242. doi:10.1111/j.0030-1299.2004.13131.x.
61. McCormick MI, Meekan MG (2007) Social facilitation of selective mortality. Ecology 88: 1562–1570. doi:10.1890/06-0830.
62. Holmes TH, McCormick MI (2006) Location influences size selective predation on newly settled reef fish. Marine Ecology Progress Series 317: 203–209.
63. Fuiman LA, Meekan MG, McCormick MI (2010) Maladaptive behavior reinforces a recruitment bottleneck in newly settled fishes. Oecologia 164: 99–108. doi:10.1007/s00442-010-1712-3.
64. Smith BR, Blumstein DT (2010) Behavioral types as predictors of survival in Trinidadian guppies (Poecilia reticulata). Behavioral Ecology 21: 919.
65. Holmes TH, McCormick MI (2009) Influence of prey body characteristics and performance on predator selection. Oecologia 159: 401–413. doi:10.1007/s00442-008-1220-x.
66. Almany GR, Peacock LF, Syms C, McCormick MI, Jones GP (2007) Predators target rare prey in coral reef fish assemblages. Oecologia 152: 751–761.
67. Colgan P, Gotceitas V, Frame J (1991) Individual variation, acquisition, and reacquisition in a foraging task by juvenile bluegill sunfish (*Lepomis macrochirus*). J Comp Psych 105: 78–84. doi:10.1037/0735-7036.105.1.78.
68. Kieffer J, Colgan P (1992) Differences in learning by foraging juvenile pumpkinseed and bluegill sunfish in a structured habitat. Environ Biol Fish 33: 359–366. doi:10.1007/BF00010948.
69. Wilson AD, Godin JG (2009) Boldness and behavioral syndromes in the bluegill sunfish, *Lepomis macrochirus*. Behav Ecol 20: 231.
70. Quinn GP, Keough MJ (2002) Experimental Design and Data Analysis for Biologists. Cambridge: Cambridge University Press. 537 p.

24

Context-Dependent Preferences in Starlings: Linking Ecology, Foraging and Choice

Marco Vasconcelos[1]*, Tiago Monteiro[2¤], Alex Kacelnik[2]*

1 School of Psychology, University of Minho, Braga, Portugal, 2 Department of Zoology, University of Oxford, Oxford, United Kingdom

Abstract

Foraging animals typically encounter opportunities that they either pursue or skip, but occasionally meet several alternatives simultaneously. Behavioural ecologists predict preferences using absolute properties of each option, while decision theorists focus on relative evaluations at the time of choice. We use European starlings (*Sturnus vulgaris*) to integrate ecological reasoning with decision models, linking and testing hypotheses for value acquisition and choice mechanism. We hypothesise that options' values depend jointly on absolute attributes, learning context, and subject's state. In simultaneous choices, preference could result either from comparing subjective values using deliberation time, or from processing each alternative independently, without relative comparisons. The combination of the value acquisition hypothesis and independent processing at choice time has been called the Sequential Choice Model. We test this model with options equated in absolute properties to exclude the possibility of preference being built at the time of choice. Starlings learned to obtain food by responding to four stimuli in two contexts. In context [AB], they encountered options A_5 or B_{10} in random alternation; in context [CD], they met C_{10} or D_{20}. Delay to food is denoted, in seconds, by the suffixes. Observed latency to respond (L_i) to each option alone (our measure of value) ranked thus: $L_A \approx L_C < L_B << L_D$, consistently with value being sensitive to both delay and learning context. We then introduced simultaneous presentations of A_5 vs. C_{10} and B_{10} vs. C_{10}, using latencies in no-choice tests to predict sign and strength of preference in pairings. Starlings preferred A_5 over C_{10} and C_{10} over B_{10}. There was no detectable evaluation time, and preference magnitude was predictable from latency differentials. This implies that value reflects learning rather than choice context, that preferences are not constructed by relative judgements at the time of choice, and that mechanisms adapted for sequential decisions are effective to predict choice behaviour.

Editor: Daniel Osorio, University of Sussex, United Kingdom

Funding: This work was supported by Biotechnology and Biological Sciences Research Council (BBSRC) Grant BB/G007144/1 to AK www.bbsrc.ac.uk; TM was supported by a Doctoral Grant from the Portuguese Foundation for Science and Technology (FCT) www.fct.pt/index.phtml.en. The funders had no role in study design, data collection and analysis, decision to publish, or preparation of the manuscript.

Competing Interests: The authors have declared that no competing interests exist.

* E-mail: mvasconcelos@psi.uminho.pt (MV); alex.kacelnik@zoo.ox.ac.uk (AK)

¤ Current address: Champalimaud Neuroscience Programme, Lisbon, Portugal

Introduction

We present a theoretical and experimental investigation of animal choice that links foraging ecology, state- and context-dependent learning and hypothetical cognitive mechanisms for choice. Foraging animals search, detect, pursue, catch, and consume prey of diverse kinds, each time experiencing the consequences of the whole cycle of events. Throughout these sequences they store information about prey physical attributes (e.g., search time, size, taste, handling time) and about experienced consequences of their decision to pursue such prey (e.g., hedonic impact, improvements in energetic state or other psychological correlates of survival and reproduction).

From a learning point of view, experienced consequences constitute biologically programmed signals that retrospectively give value to (reinforce) earlier links of each typical sequence (e.g., [1,2]) and hence affect future foraging decisions. In terms of learning theory, stimuli identifying prey types are conditional stimuli (CS) and the consequences of pursuing each prey type are unconditional stimuli (US). If, in infrequent occasions, the CSs corresponding to different outcomes occur simultaneously, for instance if two prey types are simultaneously in the field of view, the subject faces a choice. In such cases, preference between the stimuli that identify prey types will depend on the history of reinforcement, namely the returns with which each set of stimuli has been correlated in the past (e.g., [3–5]). This can produce apparently irrational preferences, because absolute properties of prey types (i.e. context and state independent properties) are not the only determinants of their subjective value. The point we are making for learning can underlie observed irrationalities such as the Concorde (in evolutionary biology) or Sunk Cost (in economics) fallacies [6–8] and 'within-trial contrast' in experimental psychology [9,10]. To elaborate this point, consider how an unwarranted preference can be caused by learning history: consuming two prey of equal body mass causes the same energy gain and should be treated indifferently, but if typically one of them has higher searching costs, the costly one would have been experienced under greater need and would elicit a memory for higher hedonic impact [11–18], leading to it being preferred. The rationale can be extended to the effect of context. If two prey of equal body mass are typically met in periods where they rank differently respect to the background of alternative prey types, then they can be mnemonically associated with positive (elation) or negative (frustration) contrasts, independently of energetic state. Such contrast effects are well known behaviourally and neurobiologically (e.g., [19–21]).

Optimal choice in simultaneous encounters should be free from such effects, but simultaneous encounters may be too infrequent to select for learning processes immune to these fallacious effects. The structure of most species' ecology means that state- and context-dependent reinforcement, evolved as adaptations to serial cycles, are likely to dominate learning-dependent valuation (e.g., [22–24]). Thus, we conclude that the subjective valuation of each option may be affected by factors other than their intrinsic objective value because of learning processes but these valuations must interact to control choices [25–29], and this we discuss next.

Learning, choice and decision making attract multidisciplinary interest, and it is not surprising that theories follow differing rationales and make different predictions. For instance, a comprehensive recent review ([30]; see also [31,32]) includes process-specific predictions in the definition of what a decision is, starting thus: 'A decision is a deliberative process that results in the commitment to a categorical proposition. An apt analogy is a judge or jury that *must take time to weigh evidence* for alternative interpretations and/or possible ramifications before settling on a verdict.' (p. 536, our italics).

The view that decisions take time and deliberation time reflects a tradeoff between speed of action and accuracy of decision is a logical expectation, but it is still subject to empirical research. We have argued that because sequential encounters are frequent and simultaneous choices rare, the latter are handled by mechanisms selected for their performance in the former, and this may mean an absence of deliberation time. The mechanism we propose is that in the (rare) simultaneous encounters, each option elicits its own typical latency to act reflecting its encoded subjective value, and a "choice" results from responding towards the option eliciting the shorter latency [28].

Our overall view is that subjective valuation reflects state- and context-dependent changes in wellbeing caused by each option in no-choice encounters, and that when options are met simultaneously preferences reflect a horse-racing competition for expression of each option treated independently. To emulate a stylized version of natural foraging, in our experiments the subjects learn by facing different alternatives sequentially, and in infrequent occasions face simultaneous choices between types. In particular, we trained European starlings (*Sturnus vulgaris*) in two contexts, [AB] and [CD], thus generating context-dependent differential utilities across foraging options [33,34]. In context [AB], they sequentially encountered options A_5 or B_{10} in random alternation; in context [CD], they met C_{10} or D_{20} in a similar fashion. Suffixes indicate delays to food, in seconds. The animals then faced infrequent simultaneous choice trials between options that belonged to different contexts, namely between A_5 and C_{10} and between B_{10} and C_{10} (Figure 1).

We make the following specific predictions: (i) in sequential encounters stimuli whose outcome causes more favourable state changes relative to their context will cause shorter latencies to respond. Notice that options equal in absolute properties but experienced in different contexts can then have different value; (ii) preferences in simultaneous encounters should be predictable by assuming cross-censorship between the latencies observed in sequential encounters, and (iii) decisions should not take extra time; responding to an option in a simultaneous choice should not take longer than when the same option is encountered alone.

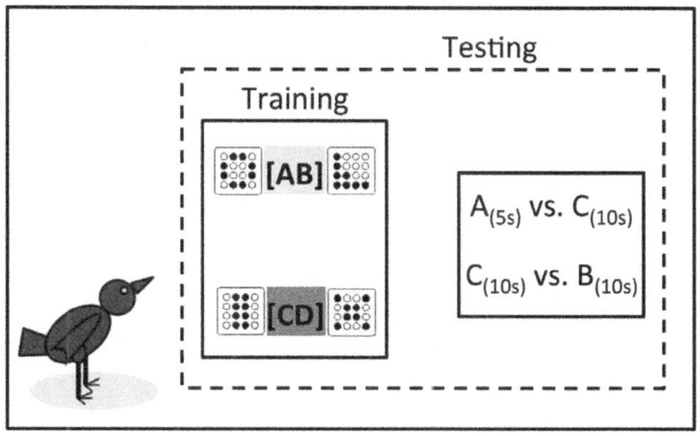

Figure 1. Experimental design schematics. Animals were trained in two alternating contexts. In context [AB], they encountered options A_{5s} or B_{10s} in random alternation; in context [CD], they met C_{10s} or D_{20s} (*Training*). The animals were then tested by introducing, infrequent simultaneous choice trials between options that belonged to different contexts, namely between A_{5s} and C_{10s} and between B_{10s} and C_{10s} (*Testing*). Delay to food in each option is denoted, in seconds, by the suffixes; symbols exemplify the type of stimuli presented.

Materials and Methods

Ethics statement

The study was approved and carried out in strict accordance with the recommendations of the Ethical Review Committee of the Department of Zoology of the University of Oxford. The starlings that participated in this experiment were obtained and kept under Natural England license No. 20083718 and were reintroduced into a communal aviary after the experiment, before being released back into the wild. All efforts were made to minimize suffering.

Subjects

Eight wild-caught adult European starlings (*Sturnus vulgaris*) participated in the experiment. All had laboratory experience unrelated to the contingencies they encountered here. The birds were kept outdoors before starting the experiment, when they were transferred to individual indoor cages (160 cm×45 cm×40 cm) that served as housing and testing chambers. Indoor temperatures ranged from 15 to 18°C and lights followed a 12:12 dark schedule with light from 0700 to 1900. Starlings were rewarded with 20 mg BioServ® precision pellets throughout the experimental sessions and provided with three hours of free access to turkey crumbs (14:30 to 17:30) and 10 mealworms daily. This regime allows the starlings' body weights to remain stable at approximately 90% of their free-feeding weight [35].

Apparatus

Each cage had an operant panel attached to the centre of the back wall. The panel had three sections: a middle subpanel, facing the cage (10 cm wide), and two side subpanels (each 10 cm wide) attached to the cage at a 110 degree angle from the centre subpanel. Each panel had a response key in the centre (11 cm from the bottom) that could be illuminated with a variety of hues and black and white symbols. The middle subpanel also had a food hopper (2 cm from the bottom) that was connected to a pellet dispenser (Campden Instruments®). Trials were controlled by custom software running on the Microsoft® Windows® operating system and attached to an Animal Behavior Environment Test System (Campden Instruments®) via WhiskerServer [36].

Procedure

Training. After brief preliminary training (detailed in Text S1) the training phase began. All subjects were exposed to four stimuli (options), implemented as coloured keys that, once pecked, led to a food reward being delivered after a pre-established waiting time, or delay. Two of the options were encountered in one context and the other two in a different context. Contexts are defined exclusively by the pair of options that could potentially be encountered. In context [AB] there were random sequential encounters with options A_5 or B_{10} and in context [CD] with options C_{10} or D_{20}.

Starlings received one training session per day for 12 days. Each session was divided into four 44-trial sub-sessions, each sub-session involving one of the two contexts, yielding up to 176 trials per day/bird (thus a maximum of 7.04 g of food could be received during testing by each subject). The order of presentation of contexts alternated across subjects and days. Half of the subjects started the first session with context [AB] and the other half with context [CD]. On each particular session, the subjects that began with context [AB] ended with context [CD], and on the next day, they began with context [CD] and ended with context [AB]. Contexts were separated by an interval of either 45 (first and third transition) or 60 minutes (second transition). Sessions ended after 176 trials or 6.5 hours from the session start, whichever came first.

Within each context, birds experienced 40 single-option trials (20 per option) and 4 peak trials (2 per option), all separated by a 30-s inter-trial-interval (ITI). Single-option trials emulated sequential encounters in natural settings and provided the birds information about the properties of each option. These trials began with the green centre key flashing (700 ms ON, 300 ms OFF). A single peck to this key switched its light off and caused one of the side keys to start flashing one of four possible symbols at the same rate used for the centre key. A peck to this key turned it steadily on and initiated the delay associated with the displayed symbol (5, 10 or 20 s). The time between pecking the green centre key and pecking the flashing side key is hereafter referred to as the latency to respond. The first peck after the delay elapsed turned the side key symbol off and produced two food pellets followed by the ITI. Peak trials were identical to single-option trials except that once the flashing side key was pecked, the symbol remained continuously on for 60 s, and then extinguished without reinforcement. These trials were used to determine the animals' knowledge of the delay to reward associated with each particular option [37,38]. The order and sides in which the options were presented were pseudo-randomised with the constraint that no two-peak trials could occur consecutively. Symbol/delay parings were partially counterbalanced across birds. To equate the rate of reward across contexts and thus minimize possible differences in satiation states, we adjusted the usual 30 s ITI every fifth single-option trial. We proceeded as follows: we summed the delays to food from the preceding five single-option trials, subtracted this value from its theoretical maximum of 100 s (5 delays of 20 s) and then added this value to the 30 s ITI.

Testing. This phase was structurally similar to the training phase except that choice trials were added. All such trials were cross-context, involving choices either between A_5 and C_{10} or between B_{10} and C_{10}. Half of the subjects were first tested with choices between A_5 and C_{10} and the remaining half with choices between B_{10} and C_{10}. When preference reached stability, choice types were swapped: subjects that previously chose between A_5 and C_{10} faced choices between B_{10} and C_{10}, and vice-versa. Both tests continued until the stability criteria were met: a standard deviation less than 10 for choice percentages and no visible trend in choice percentages, both during the last three sessions.

Sixteen simultaneous choice trials were presented daily. These were presented in pairs and separated by the ITI. Each pair of choice trials was presented at the beginning and the end of each context substituting the first and last two single-option trials that would normally occur during training. Choice trials were similar to single-option trials with the exception that pecking the flashing centre key caused two different symbols, one on each side key, to begin flashing simultaneously. A single peck to one of these keys turned the selected symbol continuously on, triggered the appropriate delay, and extinguished the unselected option. Thus, a bird committed itself to one of the options with its first peck to a flashing symbol. Again, the time between pecking the centre key and pecking one of the flashing side keys will be considered the as latency to choose. Once an option was selected, the trial proceeded as a regular single-option trial.

SCM simulations. Only latencies from single-option trials collected at stability were used to run the simulations. To compute predictions for preferences of A_5 vs. C_{10} and of B_{10} vs. C_{10}, we used latencies from single-option trials in the same test sessions in which choice proportions were measured. A separate Monte Carlo simulation using a nonparametric estimate of the cumulative distribution functions of the data yielded extremely similar predictions and thus is not reported.

Each simulation involved 10,000 experiments of 96 trials each [48 A_5 vs. C_{10} and 48 B_{10} vs. C_{10} trials]. For any given trial, the simulation drew one latency for each option [A_5 and C_{10} or B_{10} and C_{10}] from the population of observed latencies during single-option trials and predicted the option to be chosen in the simultaneous presentation as the one yielding the shortest latency.

Data analysis. Prior to analysis, all proportion and latency data were successfully normalized using an arcsine square root and a natural log transformation, respectively [39,40]. A Type-1 error rate of 0.05 was adopted for all reported statistical comparisons. Preliminary analyses (see Text S2 for details) revealed that the context in which test trials occurred had no significant effect on preference, thus we pooled data across testing contexts for all further analyses. One bird was dropped early on from the experiment due to showing a strong side preference.

Results

Preferences

When given a choice between two options with the same within-context ranking but signalling different delays to reward [A_5 vs. C_{10}], starlings chose C_{10}, the option with lower immediacy, 19.8% of the time (range: 2.1–47.9%). Conversely, when immediacy was the same but within-context ranking differed [B_{10} vs. C_{10}], starlings preferred the option with higher ranking, C_{10}, 69.7% of the time (range: 54.2–85.1%). One-sample t tests of magnitude of preference against indifference revealed that both these findings deviated significantly from chance [$t(6) = -3.828$, $p = .009$ and $t(6) = 5.086$, $p = .002$ for the A_5 vs. C_{10} and B_{10} vs. C_{10} tests, respectively]. Figure 2 shows both average and individual choice proportions.

Perception and encoding of delay to reward

A candidate explanation for the preferences observed in the B_{10} vs. C_{10} tests lies in possible distortions of memory for waiting times. If the representation of the temporal properties of each option (i.e., their immediacy) is affected by its context, then the starlings may in fact be choosing between options that, in spite of being objectively equivalent, are remembered as signalling subjectively different delays to reward. We used two techniques to test this hypothesis: one based on pecking patterns and the other

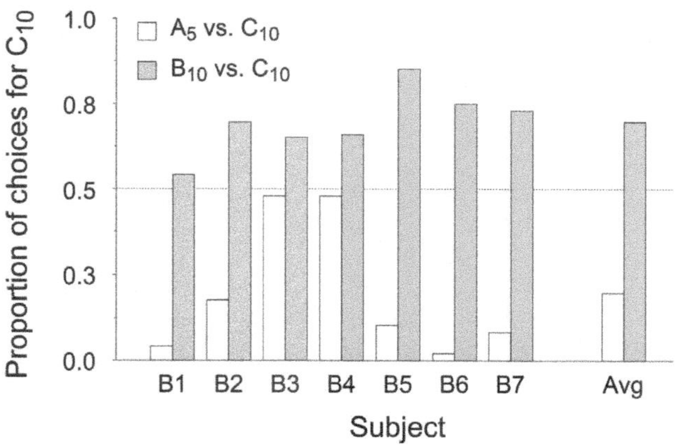

Figure 2. Proportion of choices for option C_{10} in both preference tests. White and grey bars show preference for C_{10} in the A_5 vs. C_{10} and B_{10} vs. C_{10} tests, respectively. The rightmost columns show average data. The horizontal dashed line indicates chance level.

based on peak time (i.e., the time with maximum response rate). Figure 3 shows the pecking pattern in peak trials, averaged across subjects. For all options, the average maximum response rates were near the programmed reward time while peak rates decreased and spreads increased with longer delays. A repeated-measures analysis-of-variance (ANOVA) with option and time as fixed factors, subjects as random factor and pecking rate as dependent variable confirmed a significant effect of option [$F(3,18) = 42.629$, $p<.001$], time [$F(59,354) = 121.546$, $p<.001$] and their interaction [$F(177,1062) = 18.919$, $p<.001$]. Of particular interest are the pecking patterns for options B_{10} and C_{10} as they provide a test for the hypothesis that contextual effects on preference were mediated by distortions of subjective temporal estimates. Post-hoc pairwise comparisons revealed that pecking rates did not differ systematically between options B_{10} and C_{10} ($p = .289$), but did differ for all other comparisons (largest $p = .018$), thus suggesting that the effect of within-context ranking was not mediated by distortions in interval timing. The analyses based on the location of peaks returned similar findings (see Text S3 for details).

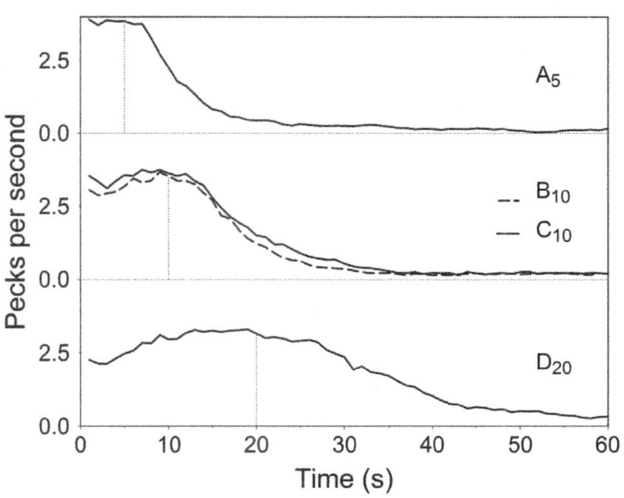

Figure 3. Mean pecks per second in peak trials for options A_5, B_{10}, C_{10} and D_{20}. The vertical dashed lines indicate the time at which reward was due in rewarded trials.

Sequential-Choice Model

Simulations. According to the SCM, preferences in simultaneous choice trials should be predictable from the latencies observed in sequential encounters. Predicting choice preferences from performance in single-option (no-choice) trials is an exclusive feature of the SCM, and for this reason it is of paramount importance for this study. The simulations predicted average preferences for C_{10} of 37.9% and 70.3% in A_5 vs. C_{10} and B_{10} vs. C_{10} tests, respectively. Figure 4 shows the observed preferences (black bars) and the predictions of the simulations (white bars). Qualitatively, the simulation predicted both kinds of choices accurately. However, while the SCM predicted the results of the B_{10} vs. C_{10} tests with high accuracy, it did not do so for the A_5 vs. C_{10} tests, where observed preferences were more extreme than those predicted. Figure 5 shows the observed versus predicted preferences for C_{10} for all subjects in both tests as well as an unconstrained regression line for each test. The best case model performance occurs when every data point falls in the main diagonal (where predictions of the model and empirical observations match). For the B_{10} vs. C_{10} tests, the regression line falls close to the theoretical optimum whereas for the A_5 vs. C_{10} tests the regression deviates considerably. One way to read the results of the A_5 vs. C_{10} tests in Figure 5 is that, based on the latencies of sequential trials, the simulations never predict preferences for A_5 stronger than 80%, but the data do show such cases. We will refer to a potential explanation for this dissociation in the ability of the SCM to quantitatively predict choice across tests in the discussion.

Latencies in sequential encounters. The SCM argues that latencies to respond to opportunities encountered sequentially reflect both the attributes of each option and those of the background opportunities, and that such latencies should be good predictors of the relative preferences of options when encountered simultaneously. Here we test the first part of these arguments. Figure 6 shows the average median latencies for each option during single-option trials. A repeated-measures ANOVA con-

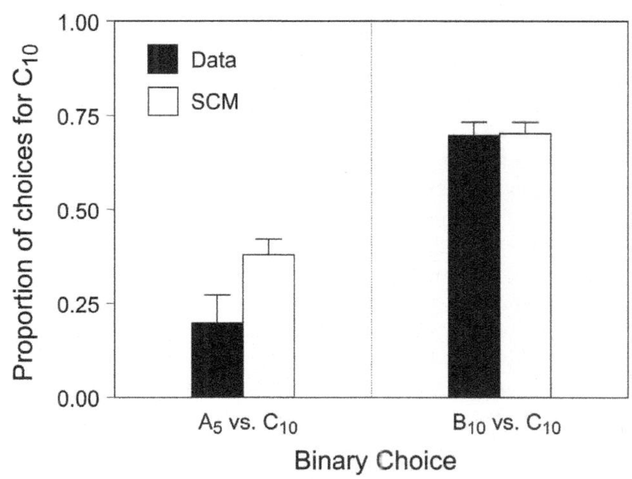

Figure 4. Observed and predicted proportion of choices for option C_{10} in both preference tests. Black bars represent the observed average preference for C_{10} in the A_5 vs. C_{10} (left) and B_{10} vs. C_{10} (right) preference tests. White bars represent the preference for C_{10} predicted by the Monte Carlo simulations in the same tests. Error bars represent one SE.

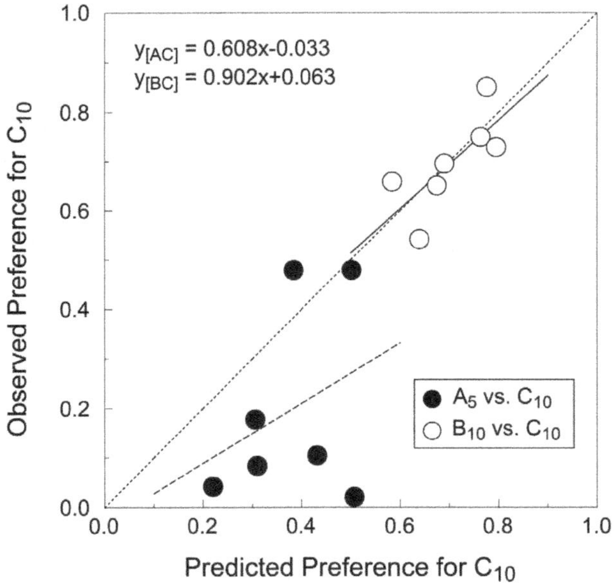

Figure 5. Overall fit of the SCM. Scatter plot of the preference for C_{10} predicted by the SCM versus the observed preference for C_{10}. Each dot represents one subject, with black and white dots referring to the A_5 vs. C_{10} and B_{10} vs. C_{10} tests. Both lines are linear regressions.

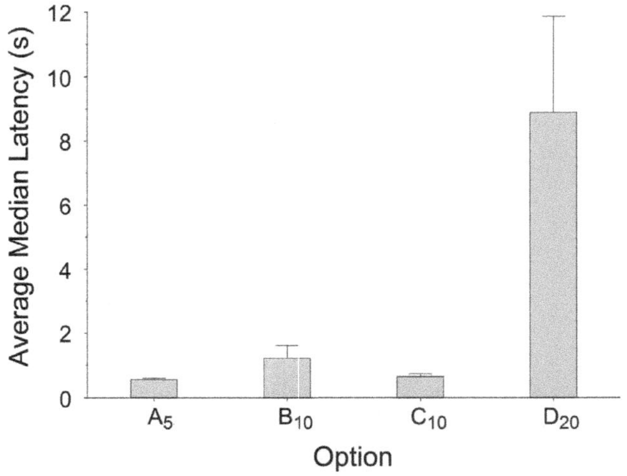

Figure 6. Latencies to accept each option in single-option trials. Median latencies to accept options A_5, B_{10}, C_{10} and D_{20} in single-option trials, averaged across subjects (+SE).

ducted on these data confirmed a significant effect of immediacy [Greenhouse-Geisser adjusted $F(1.236, 7.417) = 20.042$, $p = .002$]. Post-hoc pairwise comparisons showed that latencies differed significantly between options (largest $p = .019$) except for the comparison between options A_5 and C_{10} (p = .081). Further, these latencies reflected both the immediacy and ranking of each option in its own context: greater immediacy and better ranking both shortened latency to accept an option. In particular, the within-context ranking effect is demonstrated because latencies to accept option C_{10} were significantly shorter than latencies to accept option B_{10} despite both options signalling the same delay to reward. In contrast, despite A_5 and C_{10} signalling different delays in this case the difference between median latencies was short of conventional significance. These findings show that ranking did participate in the valuation of the options. The absence of a conventionally significant difference between A_5 and C_{10}, does not imply that absolute attributes do not have any effect, because these latencies were sufficiently short to approach psychophysical reaction times (cf. [41]) rather than strategically variable delays to accept individual options. If a subject's response is close to its reaction time, "floor effects" are likely to mask any possible difference in subjective value.

Since there is a danger that an analysis based on medians may miss a potential effect of differences in the tails of the distributions, we also performed an analysis based on the whole frequency distributions of latencies (see Text S4 and Figure S1). This analysis showed the same trends. In sum, the latencies collected during single-option trials confirm the SCM's proposal that such latencies should be sensitive both to the option's objective properties (i.e. its immediacy) and to the properties of the background alternatives available in the environment (i.e. its ranking).

Latencies in simultaneous choices. The comparison between latencies in single-option and choice trials pitches models based on deliberative comparison at the time of choice and on independent processing of options against each other: executing a cognitive comparison should make simultaneous-choice latencies longer than single-option latencies, while cross-censorship in the horse-racing SCM model should lead to choices taking less time than single option responses. The magnitude of the expected shortening depends on the similarity between the distributions of latencies collected during single-option trials, and shortening should be greater for the option that is less preferred (i.e., more censored) in choice trials.

Figure 7 shows the average median latencies for each option when chosen in simultaneous choice trials and when encountered alone in single-option trials. Paired-samples *t*-tests failed to reveal significant difference in latencies depending on whether an option was encountered by itself or chosen in a simultaneous choice situation [all $ts(6) \geq -1.577$ and ≤ 1.326, smallest $p = .166$].

This central tendency analysis was complemented by analyses of the complete frequency distributions of latencies (see Figure S2 for details). We summarized each individual latency distribution by

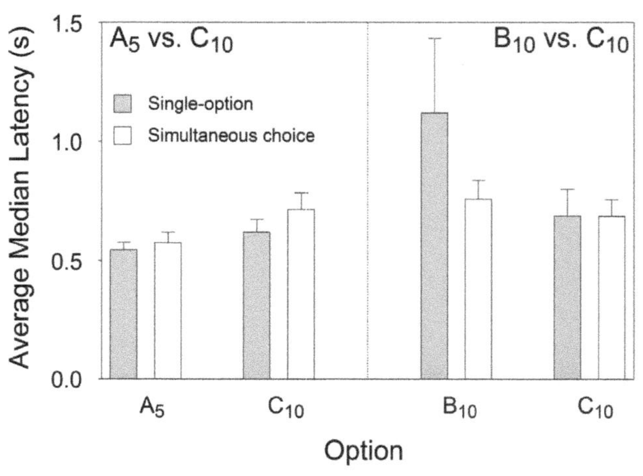

Figure 7. Latencies to accept each option in single-option and simultaneous choice trials. Latencies are separated according to collection time: either at stability in the A_5 vs. C_{10} preference tests or at stability in the B_{10} vs. C_{10} preference tests. Grey and white bars present median latencies averaged across subjects (+SE) from single-option and simultaneous choice trials, respectively.

means of a cumulative frequency score of latencies both in forced and simultaneous choice trials using the approach described in Text S4. We also calculated such scores for choice latencies in the SCM Monte Carlo simulations. The latter predicted a significant shortening only for option B_{10} [$t(6) = 2.571$, $p = .042$], but the observed shortening for this option, while having the expected trend, did not reach conventional statistical significance in a two-tail test [$t(6) = 2.155$, $p = .075$], even though it was the strongest shortening we observed (see panel B, Figure S2). We are aware that given the strength of the *a priori* directional prediction towards a shortening a one-tail test might have been considered more appropriate, but we chose to take the same conservative approach that we used throughout and treat this result as suggestive but not robustly demonstrated. All other predicted and observed shortenings were non-significant. No lengthening of latencies (which would support a deliberative comparison model) was either predicted by the Monte Carlo simulations or observed for any option.

Discussion

Our main objectives were to integrate ecological reasoning, knowledge about learning, and concepts of decision processes to advance towards a comprehensive model of foraging choices. We tested predictions about learning effects (basically that subjective value as acquired in no-choice encounters depends on state and context-dependent effects of each option's outcome) and whether the translation of these subjective values into preferences in simultaneous encounters involved a relative comparison at the time of choice or resulted from the expression of the same mechanisms used in no-choice encounters. We found support for our learning hypotheses because when facing simultaneous choices between options differing in either delay to reward or in within-context ranking, starlings' choices were better explained by remembered context-dependent utility than by remembered objective properties. Analysis of temporal responding showed that preferences were not driven by distorted memory for temporal properties of the options.

Consistently with the assumptions of the SCM, latencies to accept options presented singly were affected both by the options' objective properties and their within-context ranking: the higher the immediacy and the better the ranking the shorter the corresponding latencies. This supports the view that latencies expressed in sequential encounters reflect the operation of a valuation process and justifies their use in predicting simultaneous choice [25–29].

One of the unique predictions of the SCM is that we ought to be able to predict preferences in simultaneous choices from latencies observed during sequential encounters. The present experiment tested this prediction in a situation where valuation was context-dependent and thus significantly influenced by background alternatives. The model was qualitatively accurate in both choice tests and remarkably accurate quantitatively in one of them. Specifically, the model predicted preferences in simultaneous choices were almost identical to those observed when the options have the same immediacy and different rankings [B_{10} vs. C_{10}] but were directionally correct but less extreme than observed when options had the same ranking and different immediacies [A_5 vs. C_{10}]. This pattern of findings relates to our latency findings: both the analysis based on median latencies and that based on cumulative frequency distributions indicated that the latencies recorded for A_5 in single-option trials were somewhat shorter (but not reliably so) than those found for C_{10}. Thus, the results from the Monte Carlo simulations (that used latencies from the same distributions) did not reach the strength of preference for A_5 that was observed, despite predicting preference in the correct direction. Conversely, we did find significant differences between the distribution of latencies for B_{10} and C_{10}, which in turn yielded predictions that virtually matched the observed preferences.

What could have caused this differential success? The SCM's assumption is that differences in subjective value of an option relative to its background alternatives are detectable in the readiness to respond to it, expressed as a distribution of latencies to accept single options. The model does not claim that latencies cause choices but instead that latencies and choices are expressions of a common process acting both in sequential and simultaneous choice [28]. Tests of the SCM use latencies as 'windows' to the processes involved in the subjective valuation of each option. When there are clear differences in latency distributions such as observed between options B_{10} and C_{10} (cf. Figure S1) the model predicts preference very accurately. When latency distributions mostly overlap, such as for options A_5 and C_{10} (cf. Figure S1), the model is unable to generate extreme preferences such as those obtained in the A_5 vs. C_{10} preference tests.

Clearly, given that in the case of A_5 vs. C_{10} preferences were very strong but latencies did not differ much, comparable latency distributions are not diagnostic of equivalent subjective value. We believe that this may have been an expression of a floor effect. The average median latencies for options A_5 and C_{10} during sequential encounters were .565 and .654 sec, respectively, even though they had a 2:1 ratio in terms of immediacy. These extremely short response latencies together with the starlings' sensitivity to their immediacy [viz., they strongly preferred A_5 in the A_5 vs. C_{10} preference tests] suggest that latencies to accept these options were levelled by reaction times. The high subjective value of both options led to latencies so short as to reach the limits of reaction time in the present apparatus. The fact that the birds can generate strong choice preferences in the absence of differences in latencies argues for the existence of routes to drive choice other than those embodied in the SCM, but it would appear that these additional routes only kick in when latencies are very short and their distributions superimpose.

We also pointed out that while the conventional deliberative models of choice predict a lengthening of latencies in choices versus no-choice encounters, the SCM predicts the opposite [25,28]. Our simulations of the SCM choice mechanism indicated that a shortening should be observed when comparing latency to B_{10} in single-option trials and when in the B_{10} vs. C_{10} simultaneous-choice trials. As previously stated, the shortening should be particularly noticeable for the poorer options, as indeed it was. Conversely, options A_5 and C_{10} elicited short, virtually undistinguishable latencies during single-option trials, making any change in latency between sequential and choice trials very difficult to detect.

The experimental results yielded a shortening, albeit not conventionally significant in a two-tailed test ($p = 0.075$). In no case was there any indication for an increase in latencies consistent with a deliberation process at the time of choice. Taken together these results favour the SCM cross-censorship mechanism rather than of a deliberative process at the time of choice, particularly because the task both families of models face is asymmetric: latencies have floor but not ceiling and thus lengthening should be easier to observe than shortening.

In short, then, the confluence of ecological and mechanistic hypotheses that we endorse argues that because in natural circumstances animals typically meet feeding opportunities sequentially and seldom simultaneously, their learning and decision processes do not include cognitive mechanisms evolved due to their performance in simultaneous choice, and our observations endorse this view. In sequential encounters animals assign subjective value to prey types according to their state and the contrast with the background, and in simultaneous encounters, when by necessity state and background are unique, the learning past still controls choice. Preference thus is not built at the time of choice, by a cognitive deliberation that evaluates options' differences, but by the shadow of past state-dependent valuation learning.

Supporting Information

Figure S1 Average cumulative frequency distribution of latencies for each option during sequential encounters. The greater the area below each function the shorter the overall latencies to accept that particular option.

Figure S2 Average cumulative frequency distribution of latencies for each option in single-option and simultaneous choice trials. Triangles and dots represent data from single-option and simultaneous choice trials. Latencies are separated according to collection time: either at stability in the A_5 vs. C_{10} preference tests or at stability in the B_{10} vs. C_{10} preference tests. (A) Distributions for A_5 in the A_5 vs. C_{10} preference tests. (B) Distributions for B_{10} in the B_{10} vs. C_{10} preference tests. (C) Distributions for C_{10} in the A_5 vs. C_{10} preference tests. (D) Distributions for C_{10} in the B_{10} vs. C_{10} preference tests.

Acknowledgments

We are grateful to Simon and Jan Parry for their assistance in trapping our participants, Justine Aw for her help conducting the experiment and Esteban Freidin for fruitful discussions.

Author Contributions

Conceived and designed the experiments: MV TM AK. Performed the experiments: MV TM. Analyzed the data: MV TM AK. Contributed reagents/materials/analysis tools: MV TM AK. Wrote the paper: MV TM AK.

References

1. Bouton ME (2007) Learning and behavior: A contemporary synthesis. Sunderland, MA: Sinauer Associates.
2. Rescorla RA (1988) Behavioral Studies of Pavlovian Conditioning. Annual Review of Neuroscience 11: 329–352.
3. Gallistel CR, Gibbon J (2000) Time, rate, and conditioning. Psychological Review 107: 289–344.
4. Staddon JER, Cerutti DT (2003) Operant conditioning. Annual Review of Psychology 54: 115–144.
5. Sutton RS, Barto AG (1998) Reinforcement Learning: An Introduction. Cambridge, MA: MIT Press.
6. Arkes HR, Ayton P (1999) The sunk cost and Concorde effects: Are humans less rational than lower animals? Psychological Bulletin 125: 591–600.
7. Dawkins R, Carlisle TR (1976) Parental investment, mate desertion and a fallacy. Nature 262: 131–133.
8. Navarro AD, Fantino E (2005) The sunk cost effect in pigeons and humans. Journal of the Experimental Analysis of Behavior 83: 1–13.
9. Klein E, Bhatt R, Zentall T (2005) Contrast and the justification of effort. Psychonomic Bulletin & Review 12: 335–339.
10. Zentall TR (2010) Justification of Effort by Humans and Pigeons: Cognitive Dissonance or Contrast? Current Directions in Psychological Science 19: 296–300.
11. Aw J, Holbrook RI, Burt de Perera T, Kacelnik A (2009) State-dependent valuation learning in fish: Banded tetras prefer stimuli associated with greater past deprivation. Behavioural Processes 81: 333–336.
12. Aw J, Vasconcelos, Kacelnik A (2011) How costs affect preferences: Experiments on state-dependence, hedonic state and within-trial contrast in starlings. Animal Behaviour 81: 1117–1128.
13. Marsh B, Schuck-Paim C, Kacelnik A (2004) Energetic state during learning affects foraging choices in starlings. Behavioral Ecology 15: 396–399.
14. Pompilio L, Kacelnik A (2005) State-dependent learning and suboptimal choice: when starlings prefer long over short delays to food. Animal Behaviour 70: 571–578.
15. Pompilio L, Kacelnik A, Behmer ST (2006) State-dependent learned valuation drives choice in an invertebrate. Science 311: 1613–1615.
16. Schuck-Paim C, Pompilio L, Kacelnik A (2004) State-dependent decisions cause apparent violations of rationality in animal choice. Plos Biology 2: 2305–2315.
17. Vasconcelos M, Urcuioli PJ (2008) Deprivation level and choice in pigeons: A test of within-trial contrast. Learning & Behavior 36: 12–18.
18. Johnson AW, Gallagher M (2011) Greater effort boosts the affective taste properties of food. Proceedings of the Royal Society B: Biological Sciences 278: 1450–1456.
19. Schultz W, Dayan P, Montague PR (1997) A neural substrate of prediction and reward. Science 275: 1593–1599.
20. Reynolds GS (1961) Behavioral contrast. Journal of the Experimental Analysis of Behavior 4: 57–71.
21. Gutman A (1977) Positive contrast, negative induction and inhibitory stimulus control in the rat. Journal of the Experimental Analysis of Behavior 27: 219–233.
22. Stephens DW, Krebs JR (1986) Foraging theory. Princeton, NJ: Princeton University Press.
23. Stephens DW, Kerr B, Fernández-Juricic E (2004) Impulsiveness without discounting: the ecological rationality hypothesis. Proceedings of the Royal Society of London Series B: Biological Sciences 271: 2459–2465.
24. Stephens DW, Dunlap AS (2009) Why do animals make better choices in patch-leaving problems? Behavioural Processes 80: 252–260.
25. Shapiro MS, Siller S, Kacelnik A (2008) Simultaneous and sequential choice as a function of reward delay and magnitude: Normative, descriptive And process-based models tested in the European Starling (Sturnus vulgaris). Journal of Experimental Psychology-Animal Behavior Processes 34: 75–93.
26. Freidin E, Aw J, Kacelnik A (2009) Sequential and simultaneous choices: Testing the diet selection and sequential choice models. Behavioural Processes 80: 218–223.
27. Vasconcelos M, Monteiro T, Aw J, Kacelnik A (2010) Choice in multi-alternative environments: a trial-by-trial implementation of the sequential choice model. Behav Processes 84: 435–439.
28. Kacelnik A, Vasconcelos M, Monteiro T, Aw J (2011) Darwin's 'Tug-of-War' vs. Starlings' 'Horse-Racing': Choice without Comparisons. Behavioral Ecology and Sociobiology 65: 547–558.
29. Aw J, Monteiro T, Vasconcelos M, Kacelnik A (2012) Cognitive mechanisms of risky choice: Is there an evaluation cost? Behavioural Processes 89: 95–103.
30. Gold J, Shadlen M (2007) The neural basis of decision making. Annual Review of Neuroscience 30: 535–574.
31. Gold JI, Shadlen MN (2001) Neural computations that underlie decisions about sensory stimuli. Trends in Cognitive Sciences 5: 10–16.
32. Vlaev I, Chater N, Stewart N, Brown GDA (2011) Does the brain calculate value? Trends in Cognitive Sciences 15: 546–554.
33. Belke TW (1992) Stimulus preference and the transitivity of preference. Animal Learning and Behavior 20: 401–406.
34. Pompilio L, Kacelnik A (2010) Context-dependent utility overrides absolute memory as a determinant of choice. Proceedings of the National Academy of Sciences of the United States of America 107: 508–512.
35. Bateson M (1993) Currencies for decision making: the foraging starling as a model animal. Oxford: University of Oxford.
36. Cardinal RN, Aitken MRF (2010) Whisker: A client-server high-performance multimedia research control system. Behavior Research Methods 42: 1059–1071.
37. Catania AC (1970) Reinforcement schedules and psychophysical judgment: A study of some temporal properties of behavior. In: Schoenfeld WN, editor. The theory of reinforcement schedules New York: Appleton-Century-Crofts. pp. 1–42.
38. Roberts S (1981) Isolation of an internal clock. Journal of Experimental Psychology: Animal Behavior Processes 7: 242–268.

39. Zar JM (1999) Biostatistical analysis. Upper Saddle River, NJ: Prentice Hall.
40. Grafen A, Hails R (2002) Modern Statistics for the Life Sciences. New York: Oxford University Press.
41. Blough DS (2006) Reaction-time explorations of visual perception, attention, and decision in pigeons. In: Wasserman EA, Zentall TR, editors. Comparative Cognition: Experimental Explorations of Animal Intelligence. New York: Oxford University Press. pp. 89–105.

25

Stochastic Optimal Foraging: Tuning Intensive and Extensive Dynamics in Random Searches

Frederic Bartumeus[1,2]*, Ernesto P. Raposo[3], Gandhimohan M. Viswanathan[4], Marcos G. E. da Luz[5]

1 ICREA Movement Ecology Laboratory, CEAB-CSIC, Blanes, Spain, **2** CREAF, Cerdanyola del Vallès, Barcelona, Spain, **3** Laboratório de Física Teórica e Computacional, Departamento de Física, Universidade Federal de Pernambuco, Recife-PE, Brazil, **4** Departamento de Física Teórica e Experimental, Universidade Federal do Rio Grande do Norte, Natal-RN, Brazil, **5** Departamento de Física, Universidade Federal do Paraná, Curitiba-PR, Brazil

Abstract

Recent theoretical developments had laid down the proper mathematical means to understand how the structural complexity of search patterns may improve foraging efficiency. Under information-deprived scenarios and specific landscape configurations, Lévy walks and flights are known to lead to high search efficiencies. Based on a one-dimensional comparative analysis we show a mechanism by which, at random, a searcher can optimize the encounter with close and distant targets. The mechanism consists of combining an optimal diffusivity (optimally enhanced diffusion) with a minimal diffusion constant. In such a way the search dynamics adequately balances the tension between finding close and distant targets, while, at the same time, shifts the optimal balance towards relatively larger close-to-distant target encounter ratios. We find that introducing a multiscale set of reorientations ensures both a thorough local space exploration without oversampling and a fast spreading dynamics at the large scale. Lévy reorientation patterns account for these properties but other reorientation strategies providing similar statistical signatures can mimic or achieve comparable efficiencies. Hence, the present work unveils general mechanisms underlying efficient random search, beyond the Lévy model. Our results suggest that animals could tune key statistical movement properties (e.g. enhanced diffusivity, minimal diffusion constant) to cope with the very general problem of balancing out intensive and extensive random searching. We believe that theoretical developments to mechanistically understand stochastic search strategies, such as the one here proposed, are crucial to develop an empirically verifiable and comprehensive animal foraging theory.

Editor: Mark D. McDonnell, University of South Australia, Australia

Funding: The authors thank for financial support the Spanish Government and Ministry of Science (BFU-2010-22337), the Human Frontier Science Program (RGY0084-2011), and CNPq, CAPES, and FACEPE (Brazilian agencies). The funders had no role in study design, data collection and analysis, decision to publish, or preparation of the manuscript.

Competing Interests: The authors have declared that no competing interests exist.

* Email: fbartu@ceab.csic.es

Introduction

Optimal foraging is one of the most extensively studied optimization process in ecology and evolutionary biology [1–5]. To fully develop a comprehensive theory of animal foraging, one must understand separately the evolutionary trade-offs and the contribution of the different elements involved in the foraging dynamics [3,6], including pre-detection components such as search and taxis, and post-detection events such as pursuing and handling a prey or exploiting a patch [3,6]. Landscape behavioral ecology [7–9], for example, highlights the relevance of the search component by observing that animal perceptual scales are often much smaller than the exploration scales. It also shows that the degradation of sensory or memory-based information leads to stochastic motor outputs and decision making [8,10].

Despite the vast amount of literature quantifying movement in the context of foraging, only recently the focus has shifted to the *spatial search optimization problem*, aimed at explaining movement patterns under low information load. [11–13]. Here we develop in great depth a spatially explicit Stochastic Optimal Foraging Theory (SOFT, see [14,15]) to explore the underlying mechanisms responsible for random encounter success. Importantly, the assumption of a foraging animal as a random wanderer passes over a more comprehensive picture of animal search ecology, where distinct quantity and quality of information (present or past) is processed [16–18]. Simple stochastic models, however, are useful to tackle complex spatial search problems [19]. In this sense, random walk methods in general and SOFT [15] in particular should play a complementary role to biologically-detailed modeling and shed light on universal features of search strategies [13].

In its core formulation SOFT quantifies the distance traveled (or time spent) by a random walker that starts moving from a given initial position within a spatial region delimited by absorbing boundaries, which represent the targets to be found. Each time the walker reaches the boundaries, an encounter is computed and the search process starts all over again. Averages on the properties of many walk realizations are aimed to reproduce the dynamics of a forager looking for successive targets under specific environmental conditions [14].

SOFT differs from classical optimal foraging theory in two fundamental aspects. First, it is a spatially explicit theory focused on the exploration component of the classical patch exploration/exploitation trade-off [1,2,4]. Second, it considers inter-patch travel as a *random search* [11–13,20] rather than a *traveling salesman* [21] optimization process. In particular, doubt is cast on whether inter-patch motion should always be ballistic, or else, complex search patterns occur between patches. In consonance with previous works, e.g. [22–25], current search theory has identified reorientations (turns) as key behavioral events for random search optimization [11–13].

Using the SOFT framework, we quantify the impact of animals' reorientation strategies and spreading capacity on the search efficiency. Specifically, we unveil the main aspects associated with the tension between optimizing the encounters for nearby targets while balancing the ability to search for further areas. Assuming distinct movement strategies (represented by probability density functions, i.e. pdfs, for the move lengths) we study which common statistical characteristics of such pdfs can lead to higher encounter rates. Those pdfs yielding reorientation patterns with a proper compromise between local and far away searching (like Lévy walks) are the ones optimizing (or at least improving) search efficiency. In practice, in Lévy walk models the turning angles are drawn from a uniform probability distribution and the lengths of the moves are chosen from a power-law probability distribution [13,26].

Finally, one technical point deserves some considerations. The one-dimensional (1D) context approached in this work may not seem the most general case and certain particular characteristics of random walks might depend on the spatial dimension. Nevertheless, it is mathematically convenient and already contains the essentials of any random search in n D [20,27] (as for the optimization condition and behavior at low density of targets). In the simple but very general dynamics assumed here, the searcher chooses a random direction and draws a step length ℓ from a given pdf $p(\ell)$. It then moves in a straight line, looking for a target along the way. If the forager does not find any target after traveling ℓ, it changes direction and pick a new distance to go. This process is repeated many times during a full search. Hence, the statistics of the ℓ's, which depict rectilinear locomotion, represents one of the main aspects of the problem (e.g., related to the search efficiency). This ubiquitous 1D process embedded into an arbitrary n D search makes the simpler 1D search to be qualitatively representative of the more complex n D situation [14]. Therefore, the 1D formulation has the twofold advantage of providing very general results (the underlying mechanisms and trade-offs of efficient searches remain valid in more complex environmental scenarios) and at the same time being simple enough for a full analytical description.

Analysis

We focus here on a point-like searcher, with a radius of vision (or perceptual range) r_v, that looks for equally spaced point-like targets separated by a distance λ in a 1D search space (see Fig. 1a). The searcher starts at an arbitrary point x_0. It then scans the space between the two nearest (boundary) targets by choosing a direction (left or right) with equal probability and taking move lengths ℓ from a pdf $p(\ell)$ (Fig. 1b). Such normalized pdf, namely,

$$\int_{-\infty}^{+\infty} p(\ell)d\ell = 1, \qquad (1)$$

is taken to be symmetric $[p(-\ell)=p(\ell)]$ and kept unaltered during the whole search process.

An encounter takes place each time the searcher is at a detection distance r_v from a target. The search then restarts with the walker now placed at distance x_0 either to the left or to the right (with 50%–50% probability) of the last target found (Fig. 1c). This dynamics is repeated a large number of times, N_{found}, and averages are performed so as to lead to a proper statistical characterization of the full movement pattern. Note that the total traveled distance (L_{tot}) is composed by the concatenation of all the partial search trajectories performed by the searcher along consecutive encounters. The above mentioned averages will describe a typical random search process in an environment with global density of targets $\rho \propto 1/(\text{mean distance between targets}) = 1/\lambda$.

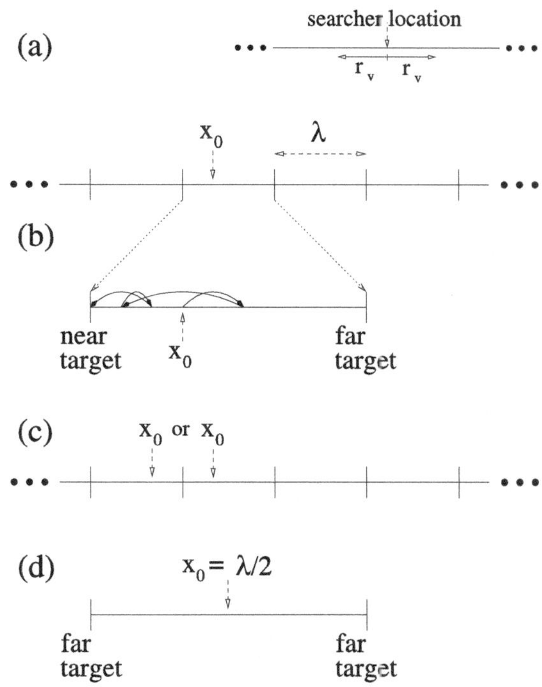

Figure 1. Diagrams showing the symmetric and asymmetric initial search conditions for the 1D stochastic search model. (a) The 1D searching environment with equally spaced point-like targets. The initial searcher location is x_0. A detail of the searcher perceptual range (or radius of vision r_v) is shown. (b) An example of the walk movement dynamics where a target is found after four steps. (c) In the asymmetric condition, each time the searcher finds a target, the search is re-initialized by placing it a distance x_0 to the right or to the left of the target found. (d) In the symmetric starting condition, $x_0 = \lambda/2$.

The present framework allows an important technical simplification. Due to the above features the search for the first target is statistically equivalent to the search for any subsequent target. In this way, we obtain exactly the same results by restricting our study to the situation where the searcher is always within the region $0 \leq x \leq \lambda$, with the targets (acting as absorbing boundaries, meaning that once a target is found that particular search dynamics ends and the process of looking for a target starts all over again) at $x=0$ and $x=\lambda$. Thus, we consider the walker movement until reaching a boundary (just as above) and repeat the procedure a number N_{found} of times. We should also observe that in terms of the general random walk theory, one event, namely, the finding of a single target either at $x=0$ or $x=\lambda$, corresponds to the so called first-passage-time problem [28].

In this work we focus mainly on two limiting initial conditions (see, e.g., Refs. [15,17,20]), even though the analysis for initial conditions intermediate between these two cases is also provided (see also [14]). Suppose first that the starting point is at the middle of the interval, $x_0 = a = \lambda/2$ (Fig. 1d). In the regime of low density of targets (large λ), this means that once a target is found, the searcher restarts its search faraway from the subsequent targets to be found. This symmetric condition corresponds to the foraging case where, once a food item is located and consumed, there are no other resources nearby, and thus the forager needs to travel relatively away from its present location in order to find targets. In contrast, the asymmetric condition $x_0 = a$, where a is either close to 0 or λ, represents the situation where, once a target is located, another one is always available nearby, so that the next search starts with both a close and a distant target.

Symmetric conditions are well represented by destructive (e.g. hunting) search processes in homogeneous resource landscapes [20], whereas asymmetric conditions can represent both non-destructive (e.g. pollinating) searches, regardless the landscape properties, and destructive (e.g. hunting) searches when targets are patchily distributed [20,27]). These two initial limiting conditions illustrate the fundamental competition between looking efficiently for nearby targets and exploring new regions to find distant targets [14,15,27]. It is also possible to take x_0 as a random variable distributed according to some given pdf and use the present framework to explore search processes in heterogeneous landscapes characterized by a distribution of initial conditions (see [14,15,17] for an explicit analysis of the role of x_0 to the search efficiency).

Some other features should also be considered. First, we set $p(\ell) = 0$ for $|\ell| < \ell_0$, so that there is a minimum move length ℓ_0. As an example, ℓ_0 could be taken as the maximum resolution of an empirical data set. Actually, here ℓ_0 can assume any arbitrarily small positive value. Moreover, although in principle the parameter ℓ_0 does not need to be related to r_v, we set for simplicity $r_v = \ell_0$ in our calculations. Since our interest is in the low density limit, where movement patterns may be an essential factor for survival, we also take $\lambda \gg r_v$ and $\lambda \gg \ell_0$. The stochastic character of the search is favored in this low-density regime.

We define a statistical search efficiency η as

$$\eta = \frac{N_{\text{found}}}{L_{\text{tot}}}. \tag{2}$$

By denoting $\langle L \rangle$ as the average distance traveled between two successive targets found, we write

$$L_{\text{tot}} = N_{\text{found}} \langle L \rangle, \tag{3}$$

which leads to

$$\eta = \frac{1}{\langle L \rangle}. \tag{4}$$

Likewise, the total number of moves in the search walk can be written as

$$N = N_{\text{found}} \langle n \rangle, \tag{5}$$

with $\langle n \rangle$ being the average number of moves between two successive targets.

Of great importance in our analysis is the quantification of the relative contributions to the search efficiency of the encounters of nearby and distant targets. We thus provide a factorization of the average distance between two successive targets in the form

$$\langle L \rangle = p_0 \langle L_0 \rangle + p_\lambda \langle L_\lambda \rangle, \tag{6}$$

where we assign p_0 (p_λ) to the probability of the searcher to find the nearest (farthest) target, with $p_0 + p_\lambda = 1$, and $\langle L_0 \rangle$ ($\langle L_\lambda \rangle$) denotes the corresponding average traveled distance. By further defining $\langle |\ell| \rangle(x_0)$ as the average length of a single step starting at position x_0, we have that $\langle |\ell| \rangle$, $\langle L \rangle$, p_0, p_λ, $\langle L_0 \rangle$ and $\langle L_\lambda \rangle$ are key functions to fully describe the random search process. Note that all of them depend on the walker initial position x_0. For instance, for the symmetric case ($x_0 = \lambda/2$) it follows that $\langle L_0 \rangle = \langle L_\lambda \rangle$ and $p_0 = p_\lambda = 1/2$. In contrast, in the asymmetric case $\langle L_0 \rangle < \langle L_\lambda \rangle$ and $p_0 > p_\lambda$. The necessary mathematical machinery to calculate the above quantities has been developed in [29,30] for the specific case of Lévy walks. To make the present contribution self-contained, we review in a very comprehensive manner all the steps necessary to obtain such quantities for any $p(\ell)$ (Appendix S1 in File S1), and particularize our general results for the biologically relevant cases of stretched exponential, log-normal, and gamma pdfs (Appendix S2 in File S1).

The above construction describes the essential dynamics of a non-informed stochastic foraging process, where search efficiency results only from proper choices for $p(\ell)$. In the following we review previously considered quantities [15] and propose new ones, all of them $p(\ell)$-dependent and aimed to characterize the full random search dynamics. These quantities constitute good figures of merit for an optimal theory.

The root-mean-square displacement

The searcher's root-mean-square (r.m.s.) displacement is defined in terms of averages over the position x of the searcher according to

$$R \equiv [\langle (\Delta x)^2 \rangle]^{1/2} = [\langle x^2 \rangle - \langle x \rangle^2]^{1/2}. \tag{7}$$

The quantity R is the square root of the second moment of the probability distribution of the position, and is one of the most important quantities related to the motion in general. Since x is a function of time t, so it is R. Hence, how R scales with t or with an operational time equivalent (like N_{found} or the number of moves N) tells us how "fast" diffusion is taking place. Indeed, for the scaling relations

$$R \sim t^\nu \quad \text{or} \quad R \sim N^\nu \quad \text{or} \quad R \sim N_{\text{found}}^\nu, \tag{8}$$

we identify that for normal (Brownian) diffusion $\nu = 1/2$, while for superdiffusion $\nu > 1/2$, and in the ballistic limit $\nu \to 1$.

Many reaction-diffusion processes have built-in multiple time scales, for example, distinct "slow" and "fast" dynamics with different relaxation time scales. Consider the case of random searches. We expect a fast regime at scales up to the encounter of the first targets, i.e., $N_{\text{found}} < N_{\text{crossover}}$ or $t < t_{\text{crossover}}$, where the diffusive properties strictly depend on the shape of the move length pdf, and therefore can sensibly differ from Brownian normal diffusion. Actually, in this regime if the pdf $p(\ell)$ is heavy-tailed, then the foraging process may appear as effectively superdiffusive [31]. On the other hand, there is also a long-term subsequent diffusive regime for the overall search trajectory where Brownian (normal) dynamical behavior pervades. At these large time scales, the correlations are lost and the process is essentially random and memory-free. These features might be shared among different move length pdfs, for some specific choices of their parameters. In between these two regimes, we expect a crossover to take place from one to the other.

Finding a target either at $x=0$ or $x=\lambda$ is essentially a mean first-passage-time problem [28]. Let $\langle x\rangle_{\rm fpt}$ and $\langle x^2\rangle_{\rm fpt}$ represent the average of the linear and quadratic searcher's position, x and x^2, over all walks departing from $x_0=a$ and ending either at $x=0$ or $x=\lambda$ by an encounter (first passage) event. By taking into account the radius of vision r_v, the detection of targets occurs at $x=r_v$ and $x=\lambda-r_v$. Recalling from Eq. (6) that $p_0(a)$ and $p_\lambda(a)$ are, respectively, the probabilities for a walker starting at $x_0=a$ to find the target at $x=0$ and $x=\lambda$ (with $p_0+p_\lambda=1$), then $\langle x\rangle_{\rm fpt}=r_v p_0+(\lambda-r_v)p_\lambda$ and $\langle x^2\rangle_{\rm fpt}=r_v^2 p_0+(\lambda-r_v)^2 p_\lambda$. Thus in analogy to Eq. (7) we find the corresponding r.m.s. of the first time passage at positions $x=0$ or $x=\lambda$,

$$R_{\rm fpt}=[\langle x^2\rangle_{\rm fpt}-\langle x\rangle_{\rm fpt}^2]^{1/2}=(\lambda-2r_v)(p_0p_\lambda)^{1/2}. \qquad (9)$$

It should be clear that the quantities R, Eq. (7), and $R_{\rm fpt}$, Eq. (9), are not the same because there is no first-passage restriction in the calculation of the standard r.m.s. distance R. Nevertheless, the dynamics of these two quantities are closely related, as we argue below. The minimum time required for the searcher with constant v to find (ballistically) the farthest target, say at $x=\lambda$, is $t_{\rm ballistic}=(\lambda-r_v-a)/v$ (note the target is detected when the searcher reaches the position $x=\lambda-r_v$). In this sense, by denoting $t_0=\langle L_0\rangle/v$ as the average time to find the closest target at $x=0$, we conclude that a searcher can go back and forth to the closest site typically a number $t_{\rm ballistic}/t_0$ of times during the time interval needed to reach the target at $x=\lambda$. At very short times, the searcher has not traveled far. So, for short times, $t<t_{\rm crossover}$, the closest target will influence much more the behavior of R, since the probability to find the nearby target is much higher. As time passes, the searcher increases its probability to move away from the closest target. Then, it is the characteristic time of finding the distant target which contributes the most to the behavior of R. In such time scales, $t_{\rm crossover}$ should have both lower and upper limits (respectively, the minimum and the average time to find the farthest target). Consequently,

$$\frac{(\lambda-r_v-a)}{v}<t_{\rm crossover}<\frac{\langle L_\lambda\rangle}{v}, \qquad (10)$$

once we can set the minimum time to find the farthest target as $t_{\rm ballistic}=(\lambda-r_v-a)/v$ and the average time to find the farthest target as $t_\lambda=\langle L_\lambda\rangle/v$ (t_λ marks the onset of the Brownian regime when revisits to both close and faraway targets becomes frequent). For an asymmetric initial condition ($a\approx r_v$) and low targets density ($\lambda\gg r_v$), we find that $\lambda/v<t_{\rm crossover}<\langle L_\lambda\rangle/v$. Specifically, if the pdf allows superdiffusivity we expect the crossover to approach the lower boundary value, also characterized by the density scale of the system, i.e., $t_{\rm crossover}\approx\lambda/v$. By dividing this expression by the average time $\langle L\rangle/v$ to reach either targets, we can also express the onset in terms of the number of targets found:

$$\frac{(\lambda-r_v-a)}{\langle L\rangle}<N_{\rm crossover}<\frac{\langle L_\lambda\rangle}{\langle L\rangle}. \qquad (11)$$

On the other hand, at long times the central limit theorem guarantees that if the move lengths have finite variance and if the correlations are of short-range (e.g., exponentially decaying), then the diffusion must converge to Brownian normal motion (upon coarse graining or renormalization). So, for $t>t_{\rm crossover}$ or $N_{\rm found}>N_{\rm crossover}$, R converges to a Brownian-type expression ($R_{\rm Brownian}$). For the asymmetric (non-destructive) case, in which the walker starts from a fixed distance $x_0=a=r_v+\Delta x$ to the closest target, with $\Delta x\ll r_v\ll\lambda$, the *actual* r.m.s. distance after $N_{\rm found}\gg 1$ "moves" (i.e., $N_{\rm found}\gg 1$ targets found) has been derived in [15], i.e. $R_{\rm Brownian}=\lambda p_\lambda^{1/2}N_{\rm found}^{1/2}$. Notice the presence of the Brownian dynamics (diffusion exponent $\nu=1/2$) in the long-term regime, in agreement with the central limit theorem. Further, by writing the total distance traversed until the encounter of $N_{\rm found}$ targets as $L_{\rm tot}=vt=N_{\rm found}\langle L\rangle$, Eq. (3), we obtain

$$R_{\rm Brownian}=\left(\frac{\lambda^2 v p_\lambda}{\langle L\rangle}\right)^{1/2} t^{1/2}. \qquad (12)$$

The close-to-distant ratio of target encounters

A fundamental indicator of the mechanisms of efficient searching (and surprisingly not discussed before) is the ratio Q between the average numbers of encounters of the closest and the farthest targets. Indeed, as we shall see from the numerical results below, it is the specific trade-off between visiting nearby and distant regions of targets that ultimately sets the general features (including the necessary degree of superdiffusivity) for optimal search strategies.

By definition we can write $Q=p_0/p_\lambda=p_0/(1-p_0)$. Nevertheless, a more interesting relation for Q can be derived as follows in terms of the degree of target revisitability. Consider first $\langle N_N\rangle$ to be the average number of distinct targets visited in a search walk with a total of $N_{\rm found}$ targets found. We also denote by $\langle N_0\rangle$ and $\langle N_H\rangle$ the average numbers of revisits of first (an event where a target just visited is immediately found in the next encounter) and higher (an event where a target previously visited is found again, but not in the sequence) orders, respectively. From these definitions, $N_{\rm found}=\langle N_0\rangle+\langle N_H\rangle+\langle N_N\rangle$.

Moreover, let $\{s_i\}$ represent the set of positions of the targets found in a given random search process (i.e., s_i is the location of the i visited target, with $i=1,2,...,N_{\rm found}$). In the present case, these positions must be multiples of λ. For instance, in the asymmetric (non-destructive) search one possible sequence could be

$$\{s_i\}=\{\lambda,2\lambda,\lambda,\lambda,0,-\lambda,-2\lambda,-2\lambda,-3\lambda,-2\lambda\}.$$

According to the definitions, this sequence of $N_{\rm found}=10$ targets has $N_N=6$ (s_1,s_2,s_5,s_6,s_7,s_9), $N_0=2$ (s_4,s_8), and $N_H=2$ (s_3,s_{10}). We now forget for a while the local details of the search walk and focus only on $\{s_i\}_{\rm eff}$, which is a sequence of sites visited by a Brownian random walker with total number of steps $N_{\rm eff}$, step length λ, and equal probabilities to go either to the right or to the left. If we do not consider the events in which the walker repeats the site just visited, the above example could be represented as $\{s_i\}_{\rm eff}=\{\lambda,2\lambda,\lambda,0,-\lambda,-2\lambda,-3\lambda,-2\lambda\}$. Therefore, we observe that this sequence has an effective total of $N_{\rm eff}=N_{\rm found}-N_0=8$ steps. In a classical work, Montroll and Weiss calculated [32] the average number of distinct sites visited in such (i.e. our effective) 1D walk in the large-$N_{\rm eff}$ limit as

$$\langle N_N \rangle = \left(\frac{8N_{\text{eff}}}{\pi} \right)^{1/2}, \tag{13}$$

so that in our case

$$\langle N_N \rangle = \left[\frac{8(N_{\text{found}} - \langle N_0 \rangle)}{\pi} \right]^{1/2}. \tag{14}$$

Since p_0 is the probability to return to the nearby target in a revisit of first order

$$\langle N_0 \rangle = p_0 N_{\text{found}}. \tag{15}$$

Finally, given $\langle N_0 \rangle$, $\langle N_N \rangle$ and N_{found}, we also determine

$$\langle N_H \rangle = N_{\text{found}} - \langle N_N \rangle - \langle N_0 \rangle. \tag{16}$$

From these expressions, the ratio between the average numbers of encounters of the closest (first order revisit) and farthest (new and higher order revisits) targets is

$$Q = \frac{\langle N_0 \rangle}{\langle N_N \rangle + \langle N_H \rangle}. \tag{17}$$

Results

We have applied the general calculations of the theory (Appendix S1 in File S1) to relevant specific reorientation strategies characterized by particular pdfs $p(\ell)$ of move lengths, namely, Lévy power-law, stretched exponential, log-normal and gamma (Appendix S2 in File S1). Our study mostly concerns the case of asymmetric initial condition (see Fig. 1), where nearby and distant targets exist [14,15]. However, results for some other initial conditions are also shown.

Each of the four distributions mentioned above can be characterized by a specific set of parameters (e.g., ϕ and β in the stretched exponential case). In Fig. 2 we show the variability of each pdf for a few choices of parameters. As the parameters vary, changes in the shape of the distribution (i.e. second and higher order moments, see Fig. 2) also determine changes in the search efficiency (Fig. 3). For each pdf there is a unique combination of parameters providing the optimal (highest) search efficiency η. By comparing the plots for the distinct pdfs, we see that the best global maximum for the asymmetric initial condition is the Lévy, followed by the log-normal, stretched exponential and gamma (note in Fig. 3 that all η -axes are in the same scale). Since the family of Lévy power-law pdfs is the one with fatter tails, our results indicate the relevance of long power-law tails (and the associated features, such as superdiffusivity) to achieve large search efficiencies. However, long tails by themselves cannot be regarded as the only relevant ingredient. For instance, the value $\mu = 1.9$ that leads to the globally optimal efficiency in the asymmetric case is less heavy-tailed and less superdiffusive than the one with $\mu \to 1$, for which η is lower. Also, in the case of the stretched exponential, the higher η is not attained in the $\beta \to 0$ limit, equivalent to the Lévy ballistic $\mu \to 1$, but with the intermediate value $\beta = 0.26$.

As we argue below, it is the specific trade-off between visiting nearby and distant regions (while looking for targets) that ultimately sets the most appropriate choice of parameters and, consequently, the shape of the optimal pdf (and the related diffusivity) yielding the best search strategy.

The subtle balance between accessing nearby and distant regions of the search space can be inferred if we factorize the search efficiency into its major components, Eq. (6). First, we observe in Fig. 4 that the qualitative shape of the $\langle L \rangle$ $(= \eta^{-1})$ curve is essentially determined by the term responsible for the distant target encounters $(p_\lambda \langle L_\lambda \rangle)$. This is so because, as we move from ballistic to Brownian motion strategies (see caption of Fig. 3), the mean traveled distance to the distant target $(\langle L_\lambda \rangle)$ increases whereas the probability to find it (p_λ) decreases, so that $\langle L \rangle$ is minimized (i.e., the efficiency is maximized) approximately when the product $p_\lambda \langle L_\lambda \rangle$ is minimized. In contrast, as both $\langle L_0 \rangle$ and p_0 increase going from ballistic to Brownian strategies, the product $p_0 \langle L_0 \rangle$ (responsible for the nearby target encounters) also increases (no counterbalance of opposite factors exists). This latter feature implies a very interesting, but counter-intuitive result: large move lengths and/or an adequate superdiffusive component (heaviness of the pdf) helps to improve the search efficiency in the asymmetric (non-destructive) search condition, where a nearby target is always present. Indeed, as both Figs. 3 and 4 show, a Brownian search strategy actually leads to rather inefficient searches in this case.

Even though the search optimization curve (the shape of η) is strongly influenced by $p_\lambda \langle L_\lambda \rangle$, certainly $p_0 \langle L_0 \rangle$ also plays an important role, as seen in the $\langle L \rangle$ curves in Fig. 4. In fact, we note that the factorized quantity $p_0 \langle L_0 \rangle$ $(p_\lambda \langle L_\lambda \rangle)$ limits the efficiency of the strategies in the Brownian (ballistic) regime. In intermediate regimes, search strategies allow the sum of these two quantities to be minimal, and this is when the maximum efficiency can be achieved. Interestingly, as observed in Fig. 4, the global maximum η for the Lévy pdf with $\mu = 1.9$ occurs after the crossover between the $p_\lambda \langle L_\lambda \rangle$ and $p_0 \langle L_0 \rangle$ curves, in the region where the largest contribution to $\langle L \rangle$ comes from the nearby visits, $p_0 \langle L_0 \rangle$.

In Fig. 5 we show the r.m.s. behavior of the random search process in the asymmetric search condition. A crossover point between two different diffusive regimes can be observed for certain move length pdf parameterizations. Move length pdfs with heavy-enough tails promote superdiffusion over the range of spatiotemporal scales where encounters are negligible (i.e. first-passage time regime). As spatiotemporal scales become larger (long-term regime) move length truncations due to encounters pervade the search process which then becomes Brownian (normal diffusion). Because of this, the crossover takes place around the characteristic density scale of the system, λ (we use $\lambda = 10^3$ in Fig. 5). However, it can occur even before (or simply disappear) if the move length pdf is such that it barely holds, or it cannot hold, superdiffusive properties over some scales range.

Overall, in Fig. 5 a nice agreement is displayed between numerical simulations and the short-term R_{fpt} and the long-term R_{Brownian} regime predictions. We can also check the theoretical estimation of the R_{fpt}–R_{Brownian} crossover. Based on the theory presented in [see Eq. (10)], for the truncated Lévy pdf with $\mu = 1.9$, $\tau = 10^4$, $\lambda = 10^3$, $r_v = 1$, $v = 1$, and $a = 1.2$ (asymmetric case), we find that the crossover time should approach the lower boundary, $t_{\text{crossover}} \approx (\lambda - r_v - a)/v = 997.8$. This prediction compares nicely with the numerical results of Fig. 5, where $t_{\text{crossover}} \approx 1000$.

When we compare in Fig. 6a the r.m.s behavior of the optimal strategy of diverse pdfs $p(\ell)$ in the asymmetric condition $(x_0 = 2r_v = 2)$, we observe a pretty similar scaling exponent in the first-passage time regime $(v = 0.85)$, which is close to the theoretical

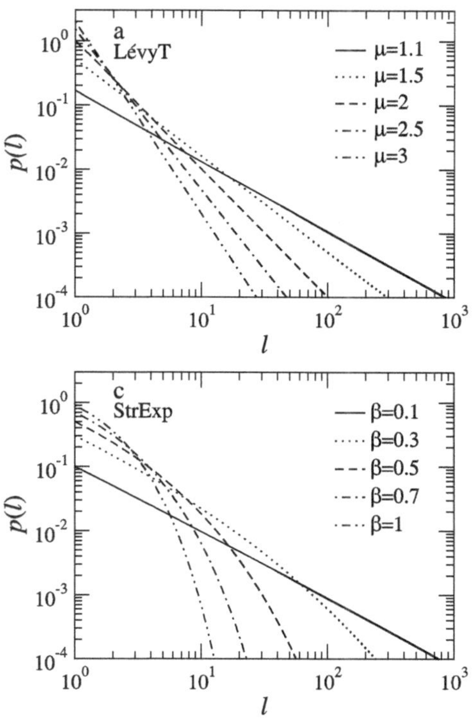

Figure 2. Probability density functions. Different combinations of shape parameters (scale parameters are fixed) for the four different move length probability density functions here considered as reorientation strategies: (a) Lévy truncated ($\tau = 10^4$), (b) log-normal ($\beta = 50$), (c) stretched exponential ($\phi = 1$), and (d) gamma ($\beta = 0.005$). For all the distributions (except for the gamma), the smaller the shape parameter the heavier the tail, hence the larger the probability of large move lengths.

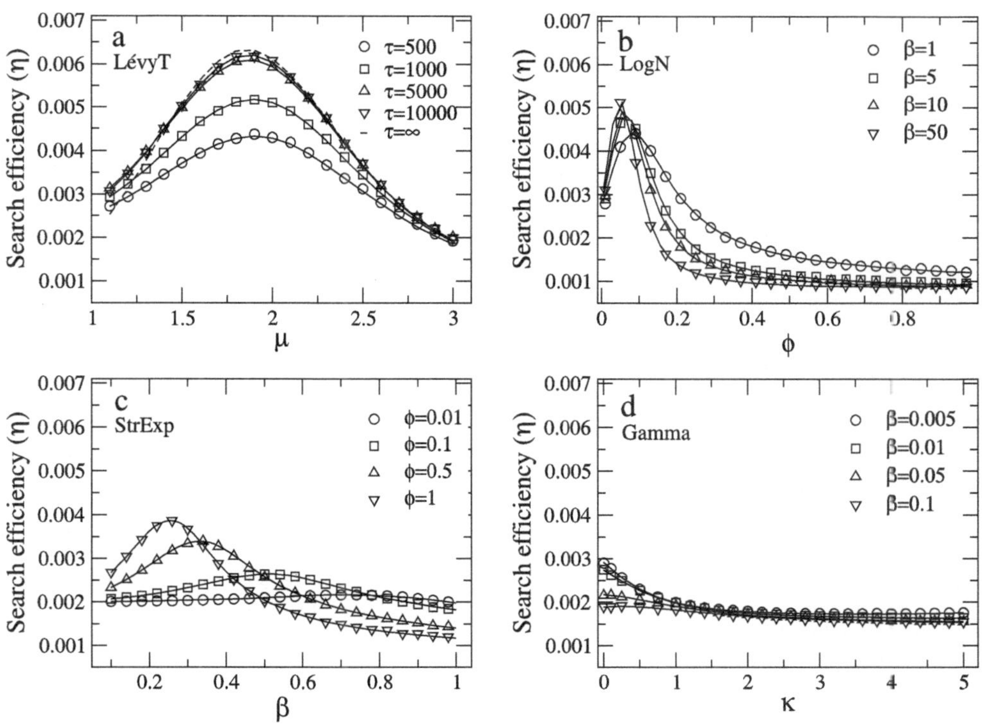

Figure 3. Search efficiency. Search efficiency η for (a) Lévy truncated, (b) log-normal, (c) stretched exponential, and (d) gamma reorientation strategies in the asymmetric search condition. Different combinations of parameters are shown for each strategy On the x-axes, low (large) parameter values represent ballistic (Brownian) regimes, to be compared with the heavy (non-heavy) pdf tails in Fig. 2. Maximum search efficiencies are achieved at some intermediate value of the parameters (x-axes), except for the gamma distribution where the maximum takes place at $k \to 0$. Notice the striking agreement between the analytical (black lines) and the numerical (symbols) results. We used parameters $\Delta x = 0.2$, $r_v = \ell_0 = 1$, $\lambda = 10^3$, and $x_0 = 2r_v$.

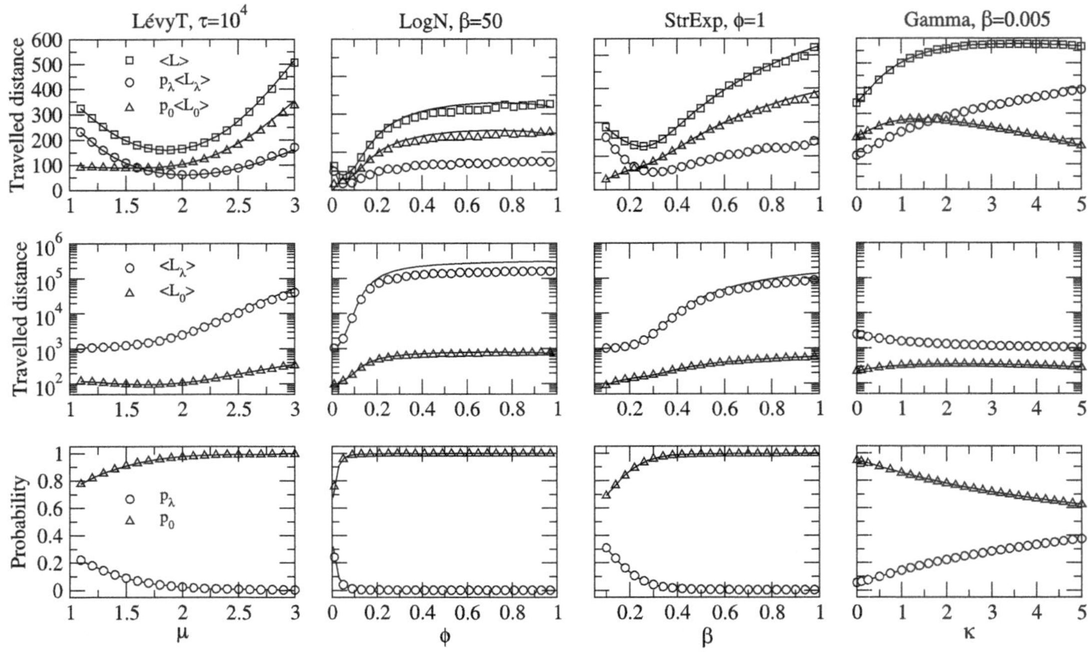

Figure 4. Factorized search efficiency. Factorized search efficiency, $\langle L\rangle = p_0\langle L_0\rangle + p_\lambda\langle L_\lambda\rangle$, analyzed to understand the relative contribution of the encounters of near (subscript 0) and distant (subscript λ) targets to the global search efficiency in the asymmetric search scenario. The behavior of the traveled distances ($\langle L\rangle$, $p_0\langle L_0\rangle$, and $p_\lambda\langle L\rangle$) and the partial quantities ($\langle L_0\rangle$, $\langle L_\lambda\rangle$, p_0, p_λ) is shown for each pdf model. Except for the gamma model, the search dynamics goes from ballistic to Brownian with increasing shape parameters (compare across x-axes with Fig. 2). The scale parameters are fixed at the search optimal. Note that the minimal $\langle L\rangle$ is close to the minimal $p_\lambda\langle L_\lambda\rangle$, suggesting that, in the asymmetric search condition, the encounter efficiency of distant targets is relevant to the global search efficiency. However, the precise optimal strategy in each case results from the subtle balance between exploring nearby areas and accessing faraway regions. Analytical (black lines) and numerical (symbols) results are displayed with nice agreement. We used parameters $\Delta x = 0.2$, $r_v = \ell_0 = 1$, $\lambda = 10^3$, and $x_0 = 2r_v$.

diffusion exponent for a non-truncated Lévy strategy in free space [15,33]. These results suggest that nearly the same unique optimal diffusion *exponent* should exist for the best parametrization choice of different models (Fig. 6b). This optimal exponent clearly depends on the specific initial condition (i.e. x_0). Different models, with different parametrization, should approach the optimal scaling exponent to some extent in order to generate an efficient reorientation strategy.

In contrast, the optimal diffusion *constant* largely varies depending on the reorientation strategy. Figure 6b explicitly compares, across models, the diffusive parameters and the close-to-distant ratio of target encounters Q, Eq. (17). As mentioned above, the diffusion exponents for the four models reach values close to $\nu = 0.85$. On the contrary, the diffusion constants and the Q values show a large range of variation. In general, the smaller the optimal diffusion constant, the larger the Q value and the efficiency of the search strategy (compare Figs. 3 and Fig. 6b). This pattern suggests that small diffusion constants (i.e., Lévy and log-normal models) increase the relative contribution to the search efficiency of close target encounters compared to distant ones, implying larger values of Q. Hence, once an optimal diffusion exponent has been achieved, a small diffusion constant should further improve the search efficiency in the asymmetric condition by providing an optimal balance between visiting nearby and faraway targets through the enhancement of the average encounter rate of close targets. In this case, models incorporating Lévy statistics are the most efficient ones (Fig. 3), precisely because they can generate optimal superdiffusive patterns while keeping the diffusion constant at a minimum rate, which facilitates intensive (localized) search.

We finally comment on the effect of the initial search condition (x_0) on the optimality of random searches. The starting position x_0 sets the initial distance to the closest and farthest targets, determining the degree of asymmetry in the relative distances of targets, bearing connections with landscape degrees of heterogeneity [14]. The highest asymmetric case ($x_0 = \ell_0 + r_v = 2r_v$) is the main focus of this work, but a continuous range of asymmetry can be found as x_0 varies from $x_0 = 2r_v$ to the most symmetrical case $x_0 = \lambda/2$. In Fig. 6c,d we show the search efficiency and the Q ratios for each optimal model at each initial condition x_0 in this range. Across x_0, Lévy reorientation strategies are always the most efficient, but its relative gain in efficiency compared to the other pdfs becomes progressively smaller as $x_0 \to \lambda/2$ (we use $\lambda = 10^3$). In particular, the pure (i.e., non-truncated) Lévy strategy provides the maximum possible search efficiency at $x_0 = \lambda/2$, i.e. $\eta = 1/\langle L\rangle = 2/\lambda$, in which the faraway targets are reached in only one very large (ballistic) move of length $\lambda/2$. However, the same is not always true for truncated Lévy search walks with a maximum cutoff length τ. In this case, other strategies can display higher efficiencies close to the symmetric regime, given that this upper truncation may considerably decrease the ability to reach distant targets (in Fig. 6 we use $\tau = 10\lambda$). Regarding the Q-ratios we observe that in the asymmetric regime optimal search strategies are associated to large values ($Q \gg 1$). As we move from asymmetric to symmetric initial regimes, the Q-ratios tend to become smaller as the probability to encounter the closest target (assigned at $x = 0$) decreases. This trend remains until reaching the limiting value $Q = 1$ in the fully symmetric condition, where it is equiprobable for the searcher to find the targets at $x = 0$ or $x = \lambda$.

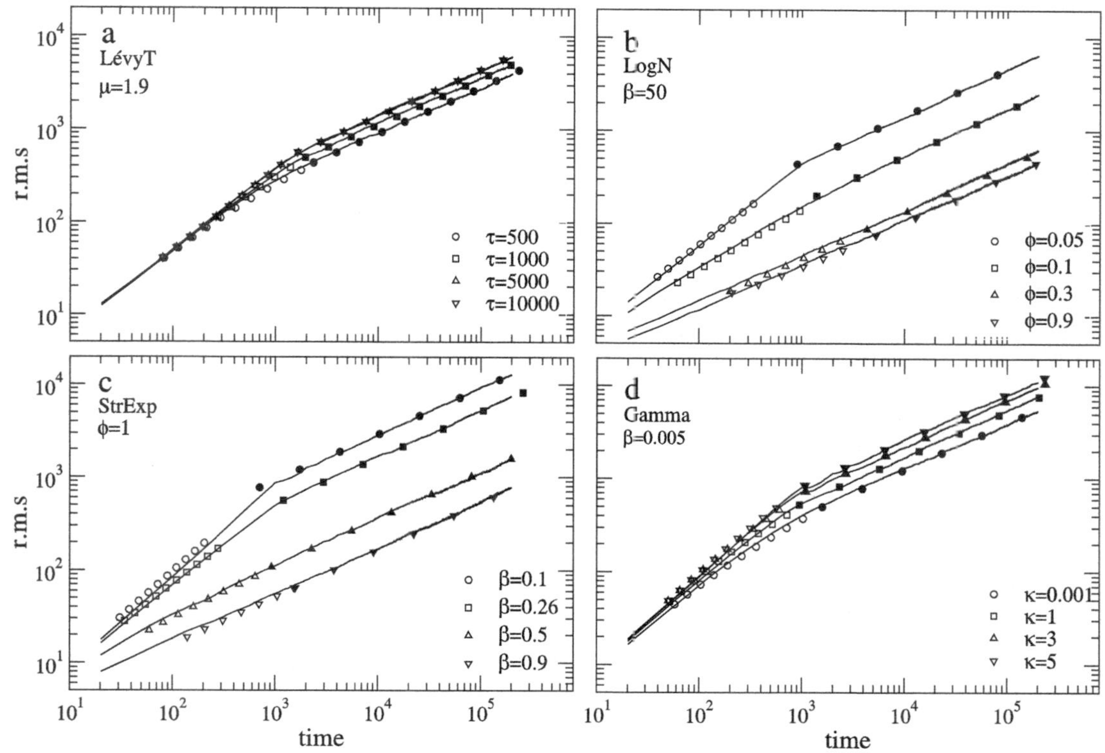

Figure 5. Root mean square behavior. R.m.s. behavior for the different reorientation strategies (different parametrization for each case) in the asymmetric search condition. In each case the fixed parameter is set at the search optimal. Note the switch in the spreading dynamics of the searcher at times $\approx 10^3$ (for some of the parametrization), coinciding with the parameter λ. Solid lines: numerical simulations. Symbols: analytical results for both the short-term first-passage-time regime, R_{fpt}, defined in Eq. (9) (open symbols), and the long-term Brownian regime, defined in Eq. (12) (filled symbols). Notice the nice agreement between numerical and analytical results. We used parameters $\Delta x = 0.2$, $r_v = \ell_0 = 1$, $\lambda = 10^3$, and $x_0 = 2r_v$.

Discussion

A key tension identified in random (stochastic, non-oriented) search strategies is to find the balance between efficiently looking for nearby targets while exploring new areas to find distant targets [14,15,27]. When this trade-off is important, the exact stochastic laws governing reorientation strategies (i.e., move length or interevent reorientation time pdfs) have a clear impact on the search success. The SOFT framework discussed here shows the fundamental principles underlying efficient stochastic searching under the assumption that movement is not governed by cues (but see [18]), and that both close and distant targets are available (asymmetric condition or non-destructive search in [20]). Being simple enough to allow a general and complete study, the 1D case is thus very useful in establishing a coherent rationale to analyze crucial mechanisms of random search processes. Moreover, as discussed in the Introduction, many relevant features of nD random search are akin to the 1D case [14,20].

We have found that the ballistic strategy is very efficient in locating distant targets. However, once the direction towards the distant target has been chosen, the ballistic strategy prevents finding closer targets, and in this sense the overall search efficiency ends up governed or limited by the distant target encounters. On the other hand, the Brownian strategy is assumed to be the most efficient one to locate nearby targets, but entails too much spatial overlap. We explicitly show that the introduction of some rare albeit large move lengths enhance nearby target encounter rates by reducing the spatial overlap and the occurrence of *excessively large and small-stepped walks*. Nearby encounters based on the latter type of walks ultimately limit the search efficiency of Brownian strategies. In the end, neither ballistic nor Brownian strategies alone solve the encounter problem when close and distant targets need to be found.

Based on our comparison across reorientation strategies in the asymmetric initial condition, we found that the essential mechanism leading to an efficient random search consists on both *the combination of an optimal superdiffusive exponent with a minimal diffusion constant*. The search process not only should adequately balance the tension between finding close and distant targets, but it should also be capable of maximizing the chances to fill in nearby spatial voids generated through the search process. In other words, the best strategies are those that not only can promote superdiffusive properties but can also shift the optimal balance towards relatively larger close-target compared to distant-target encounters.

The move length distributions analyzed here have finite moments, and thus all of them fulfill the central limit theorem, generating pure diffusive properties at some scale (generally large). Because of this, it might be preferable the term *enhanced diffusion* or *transient superdiffusion* rather than superdiffusion [14]. Importantly, we show that the distributions with slower convergence rates to the central limit theorem (the ones with fattest tails) have the larger search efficiency. In the long run, all the strategies might show Brownian properties but their search efficiencies are going to be different due to transient superdiffusive properties that can hold over relevant spatiotemporal scales in relation to the search process. In the end, the amount of superdiffusion needed to optimize the search depends also on the specific initial conditions (i.e. landscape properties).

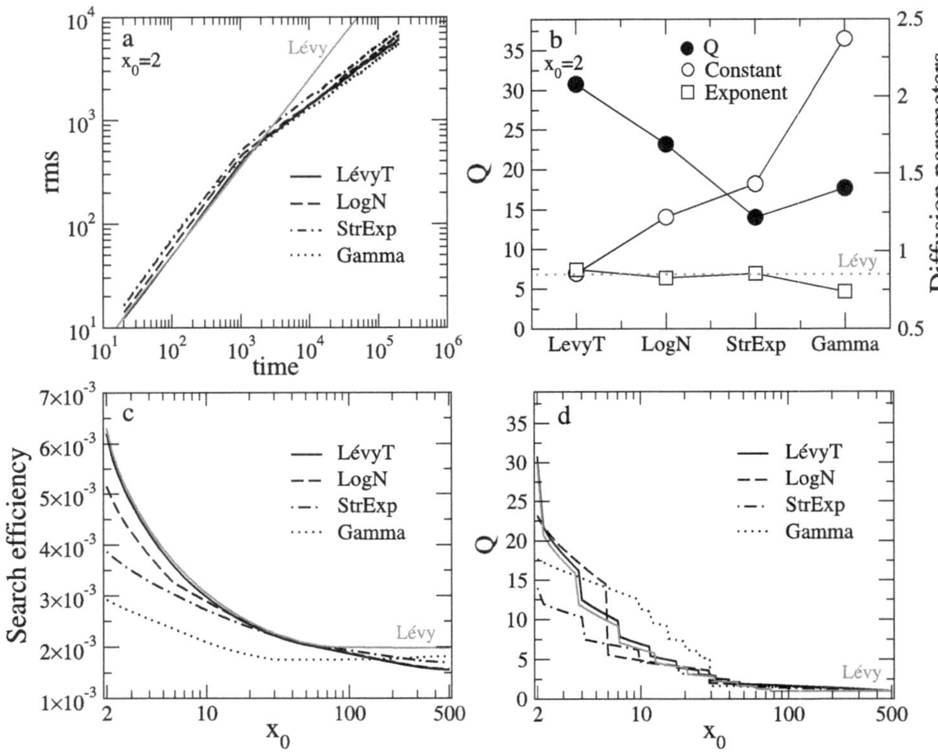

Figure 6. Role of diffusion and the close-to-distant encounter ratio in search efficiency. Comparison of key quantities across *optimal* reorientation strategies at $x_0=2$ (a,b), and for any initial condition x_0 (c,d). Panels a and b: r.m.s. behavior (a) and values for the diffusion constant, diffusion exponent, and close-to-distant encounter ratio Q (b) at the asymmetric search condition ($x_0=2r_v$, $\lambda=10^3$, and $r_v=\ell_0=1$). Parameter values as follows: Lévy truncated ($\mu=1.9$, $\tau=10^4$), log-normal ($\phi=0.05$, $\beta=50$), stretched exponential ($\phi=1$, $\beta=0.26$) and gamma ($\beta=0.005$, $\kappa=0.001$). Solid brown lines show the r.m.s. behavior (a) and the diffusion exponent (b) for a non-truncated Lévy reorientation strategy with Lévy index $\mu=2$. Panels c and d: search efficiency (c) and Q-ratio (d) for the *optimal* reorientation strategies at different initial conditions x_0 in the whole range up to $\lambda/2$. Solid brown lines indicate the results for a pure Lévy walk with $\mu_{opt}(x_0)$. Note that, regardless x_0, Lévy reorientation strategies show the largest search efficiency compared to the other reorientation strategies, though truncation decreases the efficiency when reaching the symmetric limit $x_0=\lambda/2=500$.

The mechanistic rationale exposed above should be valid for any dimensional system. However, we should observe that the relative impact on the search efficiency of Lévy patterns against other strategies decreases when dimensionality increases [17,27,34,35]. More generally, mean field theory predicts that in the limit of large number of dimensions, encounter rates would depend more on the target/searcher relative densities rather than on the search dynamics (type of movement and encounter reactions), initial search position, or target distribution.

Foraging animals are known to switch between intensive and extensive modes of search, producing complex movement patterns [16,36]. The spatiotemporal structure of such *run and tumble* patterns can be adjusted to optimize the compromise of finding both close and distant targets. These elementary results are in general concordance with simulation results exploring the effect of different encounter reactions and system dimensions on optimal search strategies [17,35,37–40]. They are also valid for intermittent search strategies where mixtures of short and fast trips exist but scanning is discontinuous [41–44].

The present work departs from current movement modeling debates by depicting relevant generic properties of the search process, going beyond specific model parametrization (Figure 6). Animals' search behavior may not consist on following specific random walk types, but on accommodating (to the extent they can) key statistical movement properties, ideally reflected in the Lévy model. Commonly used statistical approaches do not always allow distinguishing unambiguously between qualitatively different processes [45] as, for instance, composite random walks from Lévy walks [36,46–50]. Because biological motor-sensory systems are constrained by evolutionary history [51–53] we should not expect Nature to strictly follow mathematical models. Our results suggest that regardless of the specific behavioral processes involved, the incorporation of multiple movement scales in the search (for example by means of complex directional change distributions) should improve the way animals resolve the intensive-extensive search trade-off.

SOFT and current works on random search [20,41,42,44,54,55] show how basic tools of statistical mechanics [13,56–58] can naturally address the concepts of search and uncertainty in behavioral ecology. Routes to integrate classical OFT [2,4,6,59–61], sensory ecology [18,62], and SOFT will be needed in order to satisfactorily answer questions about efficiency and adaptiveness to uncertainty in animal foraging strategies [11,13].

Supporting Information

File S1 Developements in random search theory: mathematical expressions and numerical procedures. Appendix S1 Explicit expressions and numerical procedures to calculate relevant random search quantities in the general case of $p(\ell)$.

Appendix S2 Application of the general theory to specific reorientation strategies, $p(\ell)$'s.

Acknowledgments

All authors thank the hospitality and the fruitful discussions that took place at the Simon Levin Lab, at the Department of Ecology and Evolutionary Biology, Princeton University (USA). The authors also acknowledge the hospitality of CEAB-CSIC community, and Jordi Catalan for his generous and profuse insights on the problem of search.

Author Contributions

Conceived and designed the experiments: FB EPR GV MGEdL. Contributed reagents/materials/analysis tools: FB EPR GV MGEdL. Wrote the paper: FB EPR GV MGEdL. Performed the analytical and numerical analysis: FB EPR.

References

1. Kamil AC, Sargent TD, editors (1981) Foraging Behavior: Ecological, Ethological, and Psychological Approaches. Garland Series in Ethology, Garland STPM Press, New York.
2. Kamil AC, Krebs JR, Pulliam HR, editors (1987) Foraging Behavior. Plenum Press, New York.
3. Shoener TW (1987) A brief history of optimal foraging ecology. In: Kamil AC, Krebs JR, Pulliam H, editors, Foraging Behavior, Plenum Press, New York. pp.5–67.
4. Krebs JR, Davies NB (1993) An Introduction to Behavioural Ecology. Blackwell, Oxford (3rd ed).
5. Stephens DW (2007) Models of inofrmation use. In: Stephens DW, Brown JS, Ydenberg RC, editors, Foraging: Behaviour and Ecology, The University of Chicago Press. Chicago, USA. pp.31–58.
6. Shoener TW (1971) Theory of feeding strategies. Ann Rev Ecol Syst 2: 369–404.
7. Lima SL, Zollner PA (1996) Towards a behavioural ecology of ecological landscapes. Trends Ecol Evol 11: 131–135.
8. Zollner PA, Lima SL (1997) Landscape-level perceptual abilities in white-footed mice: perceptual range and detection of forested habitats. Anim Behav 80: 51–60.
9. Zollner PA, Lima SL (1999) Illumination and the perception of remote habitat patches by white-footed mice. Anim Behav 58: 489–500.
10. Hoffmann G (1983) The random elements in the systematic search behavior of the desert isopod *Hemilepistus reaumuri*. Behav Ecol Sociobiol 13: 81–92.
11. Bartumeus F (2009) Behavioural intermittence, Lévy patterns, and randomness in animal movement. Oikos 118: 488–494.
12. da Luz MGE, Grosberg A, Raposo EP, Viswanathan GM (2009) The random search problem: Trends and perspectives. J Phys A 42: 430301.
13. Viswanathan GM, da Luz MGE, Raposo EP, Stanley HE (2011) The Physics of Foraging: An Introduction to Random Searches and Biological Encounters. Cambridge University Press, New York.
14. Raposo EP, Bartumeus F, da Luz MGE, Ribeiro-Neto PJ, Souza TA, et al. (2011) How landscape heterogeneity frames optimal diffusivity in searching processes. PLoS Comput Biol 7: e1002233.
15. Bartumeus F, Raposo EP, Viswanathan GM, da Luz MGE (2013) Stochastic optimal foraging theory. In: Lewis MA, Maini PK, Petrovskii SV, editors, Dispersal, Individual Movement and Spatial Ecology: A Mathematical Perspective, Springer-Verlag Berlin Heidelberg. pp.3–32.
16. Bell WJ (1991) Searching behaviour: the behavioural ecology of finding resources. Chapman and Hall Animal Behaviour Series. Chapman and Hall. London, UK.
17. James A, Planck MJ, Edwards AM (2011) Assessing Lévy walks as models of animal foraging. J R Soc Interface 8: 1233–1247.
18. Hein AM, McKinley SA (2012) Sensing and decision-making in random search. P Natl Acad Sci USA 109: 12070–12074.
19. Berg HC (1993) Random Walks in Biology. Princeton University Press. New Jersey, USA.
20. Viswanathan GM, Buldyrev S, Havlin S, da Luz MGE, Raposo EP, et al. (1999) Optimizing the success of random searches. Nature 401: 911–914.
21. Anderson DJ (1982) Optimal foraging and the traveling salesman. Theor Popul Biol 24: 145–159.
22. Reynolds AM, Frye MA (2007) Free-flight odor tracking in *Drosophila* is consistent with an optimal intermittent scale-free search. PLoS ONE 4: e345.
23. Dees ND, Bahar S, Moss F (2008) Stochastic resonance and the evolution of *Daphnia* foraging strategy. Phys Biol 5: 044001.
24. Dees ND, Bahar S, Garcia R, Moss F (2008) Patch exploitation in two dimensions: from *Daphnia* to simulated foragers. J Theor Biol 252: 69–76.
25. Liang L, Nrrelykke SF, Cox EC (2008) Persistent cell motion in the absence of external signals: a search strategy for eukaryotic cells. PLoS ONE 3: e2093.
26. Méndez V, Campos D, Bartumeus F (2014) Stochastic Foundations in Movement Ecology. Springer Series in Synergetics. Springer Berlin Heidelberg, 177–205 pp.
27. Bartumeus F, Levin SA (2008) Fractal intermittence in locomotion: linking animal behaviour to statistical patterns of search. P Natl Acad Sci USA 150: 19072–19077.
28. Redner S (2001) A Guide to First-Passage Processes. Cambridge University Press. Cambridge, UK.
29. Buldyrev SV, Havlin S, Kazakov AY, da Luz MGE, Raposo EP, et al. (2001) Average time spent by Lévy flights and walks on an interval with absorbing boundaries. Phys Rev E 64: 041108.
30. Buldyrev SV, Gitterman M, Havlin S, Kazakov AY, da Luz MGE, et al. (2001) Properties of Lévy fights on an interval with absorbing boundaries. Physica A 302: 148–161.
31. Viswanathan GM, Raposo EP, Bartumeus F, Catalan J, da Luz MGE (2005) Necessary criterion for distinguishing true superdiffusion from correlated random walk processes. Phys Rev E 72: 011111(6).
32. Montroll EW, Weiss GH (1965) Random walks on lattices. ii. J Math Phys 6: 167–181.
33. Shlesinger MF, Klafter J (1985) Comment on accelerated diffusion in Josephson junctions and related chaotic systems. Phys Rev Lett 54: 2551–2551.
34. Bartumeus F, Fernández P, da Luz MGE, Catalan J, Solé RV, et al. (2008) Superdiffusion and encounter rates in diluted, low dimensional worlds. Eur Phys J Spec Topics 157: 157–166.
35. Reynolds AM, Bartumeus F (2009) Optimising the success of random destructive searches: Lévy walks can outperform ballistic motions. J Theor Biol 260: 98–103.
36. Benhamou S (2007) How many animals really do the Lévy walk? Ecology 88: 1962–1969.
37. Raposo EP, Buldyrev SV, da Luz MGE, Santos MC, Stanley HE, et al. (2003) Dynamical robustness of Lévy search strategies. Phys Rev Lett 91: 240601.
38. Santos MC, Raposo EP, Viswanathan GM, da Luz MGE (2004) Optimal random searches of revisitable targets: crossover from superdiffusive to ballistic random walks. Europhys Lett 67: 734–740.
39. Plank MJ, James A (2008) Optimal foraging: Lévy pattern or process? J R Soc Interface 5: 1077–1086.
40. James A, Planck MJ, Brown R (2008) Optimizing the encounter rate in biological interactions: ballistic versus Lévy versus brownian strategies. Phys Rev E 78: 051128.
41. Bénichou O, Loverdo C, Moreau M, Voituriez R (2006) Two-dimensional intermittent search processes: an alternative to Lévy flight strategies. Phys Rev E 75: 020102(R).
42. Lomholt MA, Koren T, Metzler R, Klafter J (2008) Lévy strategies in intermittent search processes are advantageous. P Natl Acad Sci USA 105: 1055–1059.
43. Bénichou O, Loverdo C, Moreau M, Voituriez R (2011) Intermittent search strategies. Rev Mod Phys 83: 81–129.
44. Campos D, Méndez V, Bartumeus F (2012) Optimal intermittence in search strategies under speed-selective target detection. Phys Rev Lett 108: 028102.
45. Kawaii R, Petrovskii S (2012) Multi-scale properties of random walk models of animal movement: lessons from statistical inference. P Roy Soc A 468: 1428–1451.
46. Reynolds AM (2008) How many animals really do the Lévy walk? comment. Ecology 89: 2347–2351.
47. Reynolds AM (2010) Bridging the gulf between correlated random walks and Lévy walks: autocorrelation as a source of Lévy walk movement patterns. JR Soc Interface 7: 1753–1758.
48. de Jager M, Weissing FJ, Herman PJ, Nolet BA, van de Koppel J (2011) Lévy walks evolve through interaction between movement and environmental complexity. Science 332: 1551–1553.
49. Jansen VAA, Mashanova A, Petrovskii S (2012) Comment on Lévy walks evolve through interaction between movement and environmental complexity. Science 335: 918.
50. de Jager M, Weissing FJ, Herman PMJ, Nolet BA, van de Koppel J (2012) Response to comment on Lévy walks evolve through interaction between movement and environmental complexity. Science 335: 918.
51. Jacob F (1977) Evolution and tinkering. Science 196: 1161–1166.
52. Kauffman SA, Levin SA (1987) Towards a general theory of adaptive walks on rugged landscapes. J Theor Biol 128: 11–45.
53. Levin SA (1998) Ecosystems and the biosphere as complex adaptive systems. Ecosystems 1: 431–436.
54. Bartumeus F, da Luz MGE, Viswanathan GM, Catalan J (2005) Animal search strategies: a quantitative random-walk analysis. Ecology 86: 3078–3087.
55. Reynolds AM, Smith AD, Menzel R, Greggers U, Reynolds DR, et al. (2007) Displaced honey bees perform optimal scale-free search flights. Ecology 88: 1955–1961.
56. Okubo A, Levin SA (2001) Diffusion and Ecological Problems: Modern Perspectives. Springer Berlin Heidelberg.
57. Shlesinger MF, Klafter J (1985) Lévy walks versus Lévy flights. In: Stanley H, Ostrowsky N, editors, On Growth and Form: Fractal and Non-Fractal Patterns in Physics. Nijhoff, Dordrecht. pp.279–283.

58. Shlesinger MF, Zaslavsky G, Frisch U, editors (1995) Lévy Flights and Related Topics in Physics. Springer, Berlin.
59. Green RF (1987) Stochastic models of optimal foraging. In: Kamil A, Krebs J, Pulliam H, editors, Foraging Behaviour, Plenum Press, New York. pp.273–302.
60. Kacelnick A, Todd IA (1992) Psychological mechanisms and the marginal value theorem: Effect of variability in travel time on patch exploitation. Anim Behav 43: 313–322.
61. Cuthill IC, Haccou P, Kacelnick A (1994) Starlings *sturnus vulgaris* exploiting patches: Response to long-term changes in travel time. Behav Ecol 5: 81–90.
62. Dusenbery DB (1992) Sensory Ecology: How Organisms Acquire and Respond to Information. W.H. Freeman and Company, New York.

Permissions

All chapters in this book were first published in PLOS ONE, by The Public Library of Science; hereby published with permission under the Creative Commons Attribution License or equivalent. Every chapter published in this book has been scrutinized by our experts. Their significance has been extensively debated. The topics covered herein carry significant findings which will fuel the growth of the discipline. They may even be implemented as practical applications or may be referred to as a beginning point for another development.

The contributors of this book come from diverse backgrounds, making this book a truly international effort. This book will bring forth new frontiers with its revolutionizing research information and detailed analysis of the nascent developments around the world.

We would like to thank all the contributing authors for lending their expertise to make the book truly unique. They have played a crucial role in the development of this book. Without their invaluable contributions this book wouldn't have been possible. They have made vital efforts to compile up to date information on the varied aspects of this subject to make this book a valuable addition to the collection of many professionals and students.

This book was conceptualized with the vision of imparting up-to-date information and advanced data in this field. To ensure the same, a matchless editorial board was set up. Every individual on the board went through rigorous rounds of assessment to prove their worth. After which they invested a large part of their time researching and compiling the most relevant data for our readers.

The editorial board has been involved in producing this book since its inception. They have spent rigorous hours researching and exploring the diverse topics which have resulted in the successful publishing of this book. They have passed on their knowledge of decades through this book. To expedite this challenging task, the publisher supported the team at every step. A small team of assistant editors was also appointed to further simplify the editing procedure and attain best results for the readers.

Apart from the editorial board, the designing team has also invested a significant amount of their time in understanding the subject and creating the most relevant covers. They scrutinized every image to scout for the most suitable representation of the subject and create an appropriate cover for the book.

The publishing team has been an ardent support to the editorial, designing and production team. Their endless efforts to recruit the best for this project, has resulted in the accomplishment of this book. They are a veteran in the field of academics and their pool of knowledge is as vast as their experience in printing. Their expertise and guidance has proved useful at every step. Their uncompromising quality standards have made this book an exceptional effort. Their encouragement from time to time has been an inspiration for everyone.

The publisher and the editorial board hope that this book will prove to be a valuable piece of knowledge for researchers, students, practitioners and scholars across the globe.

List of Contributors

Alecia J. Carter
Fenner School of Environment and Society, Australian National University, Canberra, Australian Capital Territory, Australia
Department of Zoology, University of Cambridge, United Kingdom

William E. Feeney
School of Marine and Tropical Biology, James Cook University, Townsville, Queensland, Australia
Research School of Biology, Australian National University, Canberra, ACT, 0200, Australia

Barbora Trubenová and Reinmar Hager
Computational and Evolutionary Biology, Faculty of Life Sciences, University of Manchester, Manchester, United Kingdom

L. Fredrik Sundström, Rasmus Kaspersson, Joacim Näslund and Jörgen I. Johnsson
University of Gothenburg, Department of Biological and Environmental Sciences, Göteborg, Sweden

Jakob Bro-Jørgensen
Mammalian Behaviour & Evolution Group, Department of Evolution, Ecology & Behaviour, Institute of Integrative Biology, Faculty of Health and Life Sciences, University of Liverpool, Neston, United Kingdom

Eckhard W. Heymann
Abt. Verhaltensökologie & Soziobiologie, Deutsches Primatenzentrum, Göttingen, Germany

Dietmar Zinner
Abt. Kognitive Ethologie, Deutsches Primatenzentrum, Göttingen, Germany

Jörg U. Ganzhorn
Abt. Tierökologie, Zoologisches Institut, Universita¨t Hamburg, Hamburg, Germany

Eduardo S. A. Santos
Department of Zoology, University of Otago, Dunedin, New Zealand
Departamento de Ecologia, Universidade de São Paulo, São Paulo, Brazil

Shinichi Nakagawa
Department of Zoology, University of Otago, Dunedin, New Zealand

Benjamin P. Pauli and Patrick A. Zollner
Department of Forestry and Natural Resources, Purdue University, West Lafayette, Indiana, United States of America

Robert Cummings
Rancho Deluxe Consulting, Stevens Point, Wisconsin, United States of America

Nicholas P. McCann
Department of Forestry and Natural Resources, Purdue University, West Lafayette, Indiana, United States of America
Minnesota Zoo, Apple Valley, Minnesota, United States of America

Jonathan H. Gilbert
Great Lakes Indian Fish and Wildlife Commission, Odanah, Wisconsin, United States of America

Eric J. Gustafson
USDA Forest Service, Northern Research Station, Rhinelander, Wisconsin, United States of America

Deborah M. Gordon
Department of Biology, Stanford University, Stanford, California, United States of America

Katherine N. Dektar
Department of Biology, Stanford University, Stanford, California, United States of America
Google, Mountain View, California, United States of America

Noa Pinter-Wollman
Department of Biology, Stanford University, Stanford, California, United States of America
University of California San Diego, BioCircuits Institute, La Jolla, California, United States of America

James R. White and Mark I. McCormick
School of Marine and Tropical Biology, James Cook University, Townsville, Queensland, Australia

Mark G. Meekan
Australian Institute of Marine Science, Botany Building, The University of Western Australia, Crawley, Western Australia, Australia

Cho Nam Ng and Raymond Yu Wang
Department of Geography, The University of Hong Kong, Pokfulam, Hong Kong

Tianjie Zhao
Institute of Remote Sensing and Digital Earth, Chinese Academy of Sciences, Beijing, China

Gunnar Brandt
Systems Ecology, Leibniz Center for Tropical Marine Ecology, Bremen, Germany

Agostino Merico
Systems Ecology, Leibniz Center for Tropical Marine Ecology, Bremen, Germany
School of Engineering and Science, Jacobs University, Bremen, Germany

Björn Vollan
Institutional & Behavioural Economics, Leibniz Center for Tropical Marine Ecology, Bremen, Germany
Institute of Public Finance, University of Innsbruck, Innsbruck, Austria

Achim Schlüter
Institutional & Behavioural Economics, Leibniz Center for Tropical Marine Ecology, Bremen, Germany
School of Humanities and Social Sciences, Jacobs University, Bremen, Germany

Satomi Tsuboko, Atsuko Shimada, Hiroyuki Takeda, Takeo Kubo and Hideaki Takeuchi
Department of Biological Sciences, Graduate School of Science, The University of Tokyo, Bunkyo, Tokyo, Japan

Tetsuaki Kimura and Kiyoshi Naruse
Laboratory of Bioresources, National Institute for Basic Biology, Okazaki, Aichi, Japan
National Institute for Basic Biology Center of the Interuniversity Bio-Backup Project, National Institute for Basic Biology, Okazaki, Aichi, Japan

Minori Shinya
Model Fish Genomics Resource, Genetic Strains Research Center, National Institute of Genetics, Mishima, Shizuoka, Japan
Department of Biology, Keio University, Yokohama, Kanagawa, Japan

Yuji Suehiro
Department of Biological Sciences, Graduate School of Science, The University of Tokyo, Bunkyo, Tokyo, Japan
Department of Physiology, Tokyo Women's Medical University School of Medicine, Shinjuku, Tokyo,

Teruhiro Okuyama
Department of Biological Sciences, Graduate School of Science, The University of Tokyo, Bunkyo, Tokyo, Japan
Laboratory of Bioresources, National Institute for Basic Biology, Okazaki, Aichi, Japan

José M. Rojas
Departamento de Ecología and Center of Applied Ecology & Sustainability (CAPES), Facultad de Ciencias Biológicas, P. Universidad Católica de Chile, Santiago, Chile
Centro de Investigación e Innovación para el Cambio Climático, Universidad Santo Tomás, Santiago, Chile

Simón B. Castillo and Francisco Bozinovic
Departamento de Ecología and Center of Applied Ecology & Sustainability (CAPES), Facultad de Ciencias Biológicas, P. Universidad Católica de Chile, Santiago, Chile

Guillermo Folguera
Filosofía de la Biología Group, Facultad de Ciencias Exactas y Naturales & Facultad de Filosofía y Letras, Universidad de Buenos Aires, Buenos Aires, Argentina

Sebastián Abades
Departamento de Ecología and Center of Applied Ecology & Sustainability (CAPES), Facultad de Ciencias Biológicas, P. Universidad Católica de Chile, Santiago, Chile
Instituto de Ecología y Biodiversidad, Santiago, Chile

Szymon M. Drobniak, Joanna Sudyka and Mariusz Cichoń
Institute of Environmental Sciences, Jagiellonian University, Kraków, Poland

Andrzej Dyrcz
Department of Behavioural Ecology, Wroclaw University, Wroclaw, Poland

Ruixun Zhang
MIT Department of Mathematics, Cambridge, Massachusetts, United States of America

Thomas J. Brennan
Northwestern University School of Law, Chicago, Illinois, United States of America

Andrew W. Lo
MIT Sloan School of Management, CSAIL, and EECS, Cambridge, Massachusetts, United States of America
AlphaSimplex Group, LLC, Cambridge, Massachusetts, United States of America

Nishant M. Srinivasaiah
Postgraduate Program in Wildlife Biology and Conservation, National Centre for Biological Sciences, Tata Institute of Fundamental Research, GKVK Campus, Bangalore, India

Vijay D. Anand
A Rocha India, Bangalore, India

Srinivas Vaidyanathan
Foundation for Ecological Research, Advocacy and Learning, Appavounagar, Vazhakulam, Pondicherry, India

Anindya Sinha
School of Natural Sciences and Engineering, National Institute of Advanced Studies, Indian Institute of Science Campus, Bangalore, India
Nature Conservation Foundation, Mysore, India

Claire M-S Dufour
Université de Rennes I, Unité Mixte de Recherche (UMR) Centre National de la Recherche Scientifique (CNRS) 6553 EcoBio, Rennes, France
Institut des Sciences de l'E´volution de Montpellier, Montpellier, France

Philippe Louâpre and Joan van Baaren
Université de Rennes I, Unité Mixte de Recherche (UMR) Centre National de la Recherche Scientifique (CNRS) 6553 EcoBio, Rennes, France

Véronique Martel
Université de Rennes I, Unité Mixte de Recherche (UMR) Centre National de la Recherche Scientifique (CNRS) 6553 EcoBio, Rennes, France
Natural Resources Canada, Canadian Forest Service, Laurentian Forestry Centre, Quebec City, Canada

Marie-Andrée Giroux, Jean-Pierre Tremblay and Steeve D. Côté
NSERC-Produits Forestiers Anticosti Industrial Research Chair, Département de Biologie and Centre d'E´tudes Nordiques, Université Laval, Québec, Québec, Canada

Christian Dussault
NSERC-Produits Forestiers Anticosti Industrial Research Chair, Département de Biologie and Centre d'E´tudes Nordiques, Université Laval, Québec, Québec, Canada
Direction Générale de l'Expertise sur la Faune et ses Habitats, Ministère des Ressources Naturelles et de la Faune, Québec, Québec, Canada

Nicolas Lecomte
Département de Biologie, Université du Québec à Rimouski, Rimouski, Que´bec, Canada
Department of Environment, Government of Nunavut, Igloolik, Nunavut, Canada

Theresa M. A. Clarin and Björn M. Siemers
Sensory Ecology Group, Max Planck Institute for Ornithology, Seewiesen, Germany

Ireneusz Ruczyński
Sensory Ecology Group, Max Planck Institute for Ornithology, Seewiesen, Germany
Mammal Research Institute PAS, Białowieża, Poland

Rachel A. Page
Smithonian Tropical Research Institute, Panama, Panama

Staŝa Milojević
School of Library and Information Science, Indiana University, Bloomington, Indiana, United States of America

Frederic Vandeperre, Jorge Fontes, Marco Santos, Ricardo Serrão Santos and Pedro Afonso
Centre of IMAR of the University of the Azores; Department of Oceanography and Fisheries, Horta, Portugal
LARSyS Associated Laboratory, Lisboa, Portugal

Alexandre Aires-da-Silva
Inter- American Tropical Tuna Commission, La Jolla, California, United States of America

Jan Zukal
Institute of Vertebrate Biology, Academy of Sciences of the Czech Republic, Brno, Czech Republic
Department of Botany and Zoology, Masaryk University, Brno, Czech Republic

Hana Bandouchova, Jiri Brichta, Veronika Kovacova, Karel Ondracek and Jiri Pikula
Department of Ecology and Diseases of Game, Fish and Bees, University of Veterinary and Pharmaceutical Sciences, Brno, Czech Republic

Tomas Bartonicka, Matej Dolinay and Zdenek Rehak
Department of Botany and Zoology, Masaryk University, Brno, Czech Republic

Hana Berkova
Institute of Vertebrate Biology, Academy of Sciences of the Czech Republic, Brno, Czech Republic

Virgil Brack
Environmental Solutions & Innovations Inc., Cincinnati, Ohio, United States of America

Kamil S. Jaron
Institute of Biostatistics and Analysis, Masaryk University, Brno, Czech Republic

Miroslav Kovarik
Administration of the Moravian Karst Protected Landscape Area, Nature Conservation Agency of the Czech Republic, Blansko, Czech Republic

Natália Martínková
Institute of Vertebrate Biology, Academy of Sciences of the Czech Republic, Brno, Czech Republic
Institute of Biostatistics and Analysis, Masaryk University, Brno, Czech Republic

Gregory G. Turner
Pennsylvania Game Commission, Harrisburg, Pennsylvania, United States of America

James R. White and Mark I. McCormick
School of Marine and Tropical Biology, James Cook University, Townsville, Queensland, Australia

Mark G. Meekan
Australian Institute of Marine Science, Botany Building, The University of Western Australia, Crawley, Western Australia, Australia

Maud C. O. Ferrari
Department of Biomedical Sciences, Western College of Veterinary Medicine, University of Saskatchewan, Saskatoon, Canada

Marco Vasconcelos
School of Psychology, University of Minho, Braga, Portugal

Tiago Monteiro and Alex Kacelnik
Department of Zoology, University of Oxford, Oxford, United Kingdom

Frederic Bartumeus
ICREA Movement Ecology Laboratory, CEAB-CSIC, Blanes, Spain
CREAF, Cerdanyola del Vallè s. Barcelona, Spain

Ernesto P. Raposo
Laboratório de Física Teórica e Computacional, Departamento de Física, Universidade Federal de Pernambuco, Recife-PE, Brazil

Gandhimohan M. Viswanathan
Departamento de Física Teórica e Experimental, Universidade Federal do Rio Grande do Norte, Natal-RN, Brazil

Marcos G. E. da Luz
Departamento de Física, Universidade Federal do Paraná, Curitiba-PR, Brazil

Index